ANNALS OF THE NEW YORK ACADEMY OF SCIENCES

Volume 470

EDITORIAL STAFF
Executive Editor
BILL BOLAND
Managing Editor
JUSTINE CULLINAN
Associate Editor
STEFAN MALMOLI

The New York Academy of Sciences
2 East 63rd Street
New York, New York 10021

THE NEW YORK ACADEMY OF SCIENCES
(Founded in 1817)
BOARD OF GOVERNORS, 1986

WILLIAM S. CAIN, *President*
FLEUR L. STRAND, *President-Elect*

Honorary Life Governors

| SERGE A. KORFF | H. CHRISTINE REILLY | IRVING J. SELIKOFF |

Vice-Presidents

| FLORENCE L. DENMARK | JACQUELINE MESSITE | JAMES G. WETMUR |
| PIERRE C. HOHENBERG | | VICTOR WOUK |

ALAN J. PATRICOF, *Secretary-Treasurer*

Elected Governors-at-Large

| MURIEL FEIGELSON | DENNIS D. KELLY | ROBERT J. PAYTON |
| WILLIAM T. GOLDEN | PETER LAX | ERIC J. SIMON |

Past Presidents (Governors)

| CRAIG D. BURRELL | | KURT SALZINGER |

HEINZ R. PAGELS, *Executive Director*

TWELFTH TEXAS SYMPOSIUM ON RELATIVISTIC ASTROPHYSICS

ANNALS OF THE NEW YORK ACADEMY OF SCIENCES
Volume 470

TWELFTH TEXAS SYMPOSIUM ON RELATIVISTIC ASTROPHYSICS

Edited by Mario Livio and Giora Shaviv

The New York Academy of Sciences
New York, New York
1986

Copyright © 1986 by The New York Academy of Sciences. All rights reserved. Under the provisions of the United States Copyright Act of 1976, individual readers of the Annals are permitted to make fair use of the material in them for teaching or research. Permission is granted to quote from the Annals provided that the customary acknowledgment is made of the source. Material in the Annals may be republished only by permission of The Academy. Address inquiries to the Executive Editor at The New York Academy of Sciences.

Copying fees: For each copy of an article make beyond the free copying permitted under Section 107 or 108 of the 1976 Copyright Act, a fee should be paid through Copyright Clearance Center Inc., 21 Congress St., Salem, MA 01970. For articles of more than 3 pages the copying fee is $1.75.

Cover: The cover shows IPC X-ray contours superimposed on an optical photograph (see page 349).

Library of Congress Cataloging-in-Publication Data

Texas Symposium on Relativistic Astrophysics (12th: 1984: Jerusalem)
  Twelfth Texas Symposium on Relativistic Astrophysics.

  (Annals of the New York Academy of Sciences, ISSN 0077-8923; v. 470)
  Held in Jerusalem from December 17–21, 1984.
  Includes bibliographies and index.
  1. Astrophysics—Congresses. 2. Relativity (Physics)—Congresses. I. Livio, Mario, 1945– . II. Shaviv, Giora, 1937– . III. Title. IV. Series.
  Q11.N5 vol. 470   500 s   86-8779
  [QB460]   [523.01]

ISBN 0-89766-335-7
ISBN 0-89766-336-5 (pbk.)

SP
Printed in the United States of America
ISBN 0-89766-335-7 (cloth)
ISBN 0-89766-336-5 (paper)
ISSN 0077-8923

ANNALS OF THE NEW YORK ACADEMY OF SCIENCES

Volume 470
May 30, 1986

TWELFTH TEXAS SYMPOSIUM ON RELATIVISTIC ASTROPHYSICS[a]

Editors
MARIO LIVIO and GIORA SHAVIV

International Organizing Committee
P. G. BERGMANN, A. G. W. CAMERON, J. EHLERS, D. S. EVANS, E. J. FENYVES, R. GIACCONI, F. C. JONES, S. A. KORFF, M. LIVIO, L. MESTEL, L. MOTZ, Y. NE'EMAN, I. OZSVATH, R. RAMATY, I. ROBINSON, A. SALAM, E. L. SCHÜCKING, M. M. SHAPIRO, G. SHAVIV, L. C. SHEPLEY, H. J. SMITH, J. STACHEL, S. WEINBERG, J. A. WHEELER, J. C. WHEELER, and L. WOLTJER

Local Organizing Committee
Y. AVNI, G. HORWITZ, M. LIVIO, B. Z. KOZLOVSKY, Y. NE'EMAN, T. PIRAN, G. SHAVIV, and G. TAUBER

CONTENTS

Preface. *By* MARIO LIVIO and GIORA SHAVIV ix

Part I. Cosmology, Early Universe, and Particle Physics

Cosmology and Particle Physics: A General Review. *By* KEITH A. OLIVE 1

Cosmic Strings. *By* ALEXANDER VILENKIN 26

The Cosmic Microwave Background at Its Twentieth Anniversary. *By* R. B. PARTRIDGE. 36

Part II. Active Galactic Nuclei and Jets

Toward a Unified Theory of Active Galactic Nuclei. *By* MITCHELL C. BEGELMAN. 51

X-Ray Properties of QSOs and Their Cosmological Implications. *By* Y. AVNI ... 71

From Molecular Clouds to Active Galactic Nuclei—The Universality of the Jet Phenomenon. *By* ARIEH KÖNIGL. 88

Part III. Large-Scale Structure; Clustering

Superclustering of Galaxy Clusters. *By* NETA A. BAHCALL 108

[a]This volume is the result of a conference entitled Twelfth Texas Symposium on Relativistic Astrophysics, held from December 17 to December 21, 1984 in Jerusalem, Israel.

The Galaxy Distribution and the Large-Scale Structure of the Universe. *By* MARGARET J. GELLER, VALÉRIE DE LAPPARENT, and MICHAEL J. KURTZ .. 123

Part IV. General Relativity

Energy and Its Definition in General Relativity. *By* ROGER PENROSE 136

Gravitational Collapse. *By* DEMETRIOS CHRISTODOULOU 147

Part V. Data from Satellites

The IRAS View of the Extragalactic Sky. *By* B. T. SOIFER 156

Recent Results from the Japanese X-Ray Astronomy Satellites. *By* Y. TANAKA 163

Part VI. Cosmic Rays; Gamma Rays

Abundances in Cosmic Rays. *By* MARTIN H. ISRAEL 188

The Origin of Cosmic Rays. *By* DAVID EICHLER 205

Topics in Gamma Ray Astronomy. *By* R. RAMATY and R. E. LINGENFELTER 215

Part VII. Numerical Astrophysics

Simulations of the Formation of Large-Scale Structure. *By* SIMON D. M. WHITE 243

Gravitational Radiation, Gravitational Collapse, and Numerical Relativity. *By* TSVI PIRAN and RICHARD F. STARK 247

Stellar Core Collapse and Supernova. *By* J. R. WILSON, R. MAYLE, S. E. WOOSLEY, and T. WEAVER 267

Part VIII. Late Stages of Stellar Evolution

Evolution of Supernova Progenitors and Supernova Models. *By* KEN'ICHI NOMOTO 294

Accretion onto the White Dwarf and X-Ray Production in Nonmagnetic Cataclysmic Variables. *By* A. R. KING 320

Part IX. Status Reports

The Hubble Space Telescope. *By* NETA A. BAHCALL 331

The Advanced X-Ray Astrophysics Facility. *By* HARVEY TANANBAUM 338

Poster Papers

Strong Pulsar Waves. *By* E. ASSEO, X. LLOBET, and R. PELLAT 358

Mutually Interacting Quantum Fields in an Expanding Universe: Decay of a Massive Particle. *By* J. AUDRETSCH and P. SPANGEHL 359

Thermodynamics and General Relativity Could Determine the Symmetry of the Universe. *By* SELÇUK Ş. BAYIN 360

The Differential Approach to Spinors and Their Symmetries. *By* I. M. BENN and R. W. TUCKER 362

Vacuum Energy in Cosmic Dynamics. *By* H. J. BLOME and W. PRIESTER 363

The Hubble Parameter—An Upper Limit from QSO 0957 + 561 A,B. *By* U. BORGEEST and S. REFSDAL	364
Ultrahigh Energy Gamma Rays and Cosmic Rays from Accreting Degenerate Stars. *By* K. BRECHER and G. CHANMUGAM.	365
Physical Determination and Meaning of the Law of Hubble. *By* ALEXANDRU CEAPA	366
Quantized Magnetic Bremsstrahlung and Gamma-Ray Bursts. *By* TAI L. CHOW.	367
Observations and FRW Models. *By* A. A. COLEY and B. O. J. TUPPER	369
Analysis of Weyl-Affine Theories of Gravity in Terms of the Gravitational Frequency Shift Effect. *By* A. A. COLEY and A. SARMIENTO G.	370
Image Separation Statistics for Multiply Imaged Quasars. *By* C. C. DYER	371
Symbolic Tensor Manipulation on Personal Microcomputers. *By* C. C. DYER and J. F. HARPER	372
Gamma Rays from a Hot Plasma: Application of the Models to 3C 273 and Geminga. *By* F. GIOVANNELLI, S. KARAKUŁA, and W. TKACZYK	373
On Kaluza-Klein Cosmologies. *By* M. GLEISER	374
Gas in Cosmic Voids. *By* P. M. GONDHALEKAR and N. BROSCH	375
Considerations concerning the Definition and Distribution of Gravitational Energy. *By* I. GOTTLIEB and N. IONESCU-PALLAS	376
Gravitational Collapse and Quantum Gravity. *By* P. HAJICEK	377
A Compact Object in the Bimetric Theory. *By* AMOS HARPAZ and NATHAN ROSEN	378
Microwave Measurement of the Galactic Helium-3 Abundance. *By* G. M. HEILIGMAN and D. G. YORK	379
A First Order Phase Transition from Inflationary to Big Bang Universe. *By* GERALD HORWITZ	380
The Virgo Cluster as a High Energy Cosmic Ray Source. *By* S. KARAKUŁA and W. TKACZYK	381
SS 433 Revisited. *By* W. KUNDT	382
Large-Scale Anisotropy of the Cosmic Background Radiation at 3 mm. *By* P. LUBIN and T. VILLELA	383
The Properties of a Generalized Inflation. *By* F. LUCCHIN and S. MATARRESE	384
A Hybrid Model for the Active Source at the Center of Our Galaxy: A Very Massive Star Coupled with a Black Hole. *By* LEONID M. OZERNOY	385
An X-Ray Study of M51 (NGC 5194) and Its Companion (NGC 5195). *By* G. G. C. PALUMBO, G. FABBIANO, C. FRANSSON, and G. TRINCHIERI	386
Relativistic Motion in the Quasar 3C147? *By* E. PREUSS, W. ALEF, N. WHYBORN, P. N. WILKINSON, and K. I. KELLERMANN	387
Analogies between Kruskal Space and de Sitter Space. *By* WOLFGANG RINDLER	388

Connection between Einstein Equations, Nonlinear Sigma Models, and Self-Dual Yang-Mills Theory. *By* NORMA SANCHEZ and BERNARD WHITING 389

Accretion from a Medium Containing a Density Gradient. *By* NOAM SOKER and MARIO LIVIO 390

Statistic of Voids of Galaxies. *By* ANDRZEJ SOŁTAN 391

Unicity of General Relativity in the Field Theoretic Approach. *By* GIANCARLO SPINELLI 392

Recent Developments in the Pulsating Universe. *By* FRANK R. TANGHERLINI 393

The Evidence for Large Gravitational Redshift in Seyfert Galaxy NGC 4151. *By* W. TKACZYK and S. KARAKUŁA 394

On Hydrodynamics of Astrophysical Jets. *By* R. TUROLLA, L. NOBILI, and M. CALVANI 395

Index of Contributors 397

Financial assistance was received from:
- NATIONAL SCIENCE FOUNDATION
- THE ISRAEL MINISTRY OF SCIENCE AND DEVELOPMENT
- THE HEBREW UNIVERSITY
- THE TECHNION
- THE TEL-AVIV UNIVERSITY
- THE WEIZMANN INSTITUTE OF SCIENCE
- PREVIOUS TEXAS SYMPOSIA
- PROF. Y. NE'EMAN, THROUGH THE WOLFSON CHAIR EXTRAORDINARY OF THEORETICAL PHYSICS

The New York Academy of Sciences believes it has a responsibility to provide an open forum for discussion of scientific questions. The positions taken by the participants in the reported conferences are their own and not necessarily those of The Academy. The Academy has no intent to influence legislation by providing such forums.

Preface

MARIO LIVIO AND GIORA SHAVIV

Department of Physics
Technion - Israel Institute of Technology
Haifa 32000, Israel

This symposium, held in Jerusalem, Israel, from December 17–21, 1984, was the twelfth in the series of Texas Symposia on Relativistic Astrophysics and the second to be held outside the United States.

Continuing the tradition of previous Texas Symposia, the present symposium managed to combine a wide spectrum of topics, ranging from particle physics to the physics of collapsed objects. In this sense, the Texas Symposia play an almost unique role in bringing together astronomers and physicists working in the forefront of scientific research.

The program of the symposium consisted of 28 invited presentations and a poster session; 22 of the invited talks are included in this volume, as well as abstracts of the posters.

Organizing a symposium at a time of crisis in Israel's economy has not been an easy task and we are particularly indebted to I. Robinson and E.J. Fenyves for making it possible. We are grateful to the New York Academy of Sciences for publishing this volume.

We cannot end this preface without expressing the feeling that the astounding beauty of Jerusalem has provided the perfect background to a symposium devoted to understanding the universe.

PART I. COSMOLOGY, EARLY UNIVERSE, AND PARTICLE PHYSICS

Cosmology and Particle Physics: A General Review

KEITH A. OLIVE

Astrophysics Theory Group
Fermilab
Batavia, Illinois 60510

INTRODUCTION

At the time of the First Texas Symposium in 1963, the field that I am now about to discuss did not yet exist. Major topics at that symposium revolved around quasi-stellar sources, massive stars, and gravitational collapse. The state of cosmology at that time was also in sharp contrast with our present views. After all, the First Texas Symposium was held before the discovery of the 3-K microwave background radiation by Penzias and Wilson.[1] Discussions on cosmology still centered on trying to determine a basic cosmological model, either the steady state theory or the big bang. It is only since the discovery of the microwave background that efforts have been concentrated on the big bang model; i.e., our universe has evolved from a once very hot and very dense epoch.

Traced backwards, the 3-K photon background gives us information about the universe when those photons were last in thermal contact with each other; when the age of the universe was about 10^5 years and its temperature about 10^4 K. Our next single most important piece of evidence regarding our hot past comes from the abundances of the light elements. As early as the 1940s, nuclear physics began to play a role in cosmology when it was realized[2] that if our universe began with temperatures in excess of 1 MeV or 10^{10} K, nucleosynthesis would have occurred and in particular the abundance of ^{4}He could be calculated. The fact that big bang nucleosynthesis agrees with the observed abundances of the light elements gives us added proof of our hot beginnings. Big bang nucleosynthesis began when the age of the universe was about one second.

Cosmology today does not stop at $t = 1$ s. Indeed, "reasonable" statements begin at the Planck epoch or when $t \simeq 10^{-44}$ s. In this review, I hope to highlight our current understanding of the various stages in the evolution of the universe from $t \sim 10^{-4}$ s to the period of galaxy formation at $t \gtrsim 10^5$ years. I will try to follow a chronological order for the discussion. Therefore, I will begin briefly (as I do not believe that too much can actually be said realistically) with the Planck epoch. In the section entitled INFLATION, I will discuss the inflationary epoch. In the section, GUTs AND COSMOLOGY, I will review the present status of big bang baryosynthesis, i.e., the origin of the apparent slight excess of baryons over antibaryons. This is perhaps our third most reliable piece of evidence indicating a hot big bang. I will then review the present status of big bang nucleosynthesis and discuss why I feel it is one of the greatest successes of the standard big bang model. Finally, in the last section, I will review the present role of particles in the universe; that is, their effects on galaxy formation and constraints from present observations that can be placed on particle properties.

Throughout this paper, I will be using units such that $\hbar = c = k_B = 1$ and all masses will be given in GeV (unless specifically noted otherwise).

THE PLANCK ERA

The Planck era is defined as the epoch when energy scales become comparable to the Planck mass, $M_p = 1.22 \times 10^{19}$ GeV. At these energies, one expects gravitational interactions to become comparable to other particle interactions, $G_N = M_p^{-2}$. The age of the universe at this point is $t \approx M_p^{-1} \simeq 5 \times 10^{-44}$ s. Cosmology at or above M_p must rely on a deep understanding of gravitational interactions. At the present, such theories are not yet available and cosmological models begin to get very fuzzy (or foamy?[3]). There are, however, several approaches to attack this problem, which I will very briefly describe. These include quantum gravity, Kaluza-Klein theories, and supergravity.

Quantum gravity[4] is an attempt to describe gravitational interactions at the same level as is possible for the strong, weak, and electromagnetic interactions. In the context of the early universe, quantum gravitational effects on particle production have been discussed[5] for isotropic as well as anisotropic models. Initial conditions such as the primordial wave function of the universe[6] have also been put forward in this context.

Kaluza-Klein theories,[7] which began in the 1920s, have awoken interest again recently.[8] The basic idea is that one associates gauge interactions with extra dimensions or rather one begins with a $4 + d$ dimensional theory and tries to reduce it to a four-dimensional space-time with the compactified dimensions acting like gauge interactions. For example, to account for the $U(1)$ gauge group for electromagnetic interactions, one must simply go to five dimensions. For a $SU(5)$ grand unified theory (GUT), one needs ≥ 11 dimensions. A major problem concerning these theories is that they contain only real-field representations for particles.[9] The standard low energy theory of electroweak interactions, however, contains chiral fields (i.e., there are left-handed and right-handed helicity representations). Hence, Kaluza-Klein theories could not provide a true unified theory that could include the standard low energy world. These theories have been useful, however, in that they have begun to establish a formalism for treating (and compactifying) extra dimensions that may be present in the more hopeful supergravity or superstring theories. For a review of the cosmological applications of Kaluza-Klein theories, see reference 10.

Supersymmetry[11] is a symmetry between bosons and fermions. Besides its beauty as a symmetry of nature, it has gained most of its popularity through its resolution of what is known as the gauge hierarchy problem in standard GUTs. Very simply, the gauge hierarchy problem is the problem concerning mass scales that arise in a theory. For example, the weak interaction scale is $M_w \sim 10^2$ GeV, while the GUT scale is $M_x \sim 10^{15}$ GeV. The problem is why are these scales so different. Furthermore, a technical problem arises when one considers radiative corrections to these scales. Radiative corrections to the weak mass scale will tend to be as large as the GUT scale and must therefore be cancelled with enormous precision and to many orders in perturbations theory.

Although such a cancellation is possible, it is not at all natural. Supersymmetry resolves[12] this difficulty in the sense that one can show that these radiative corrections vanish exactly.[13] Thus, the weak scale of 10^2 GeV is said to be stable with respect to radiative corrections.

The most striking effect of making a model supersymmetric is that one essentially doubles the number of known particles. To all spin 1 particles such as the photon or gluons, one adds spin ½ partners called photinos and gluinos; to all spin ½ leptons and quarks, one adds spin 0 partners, sleptons (such as the selectron) and squarks. Spin 0 Higgs bosons are paired up with spin ½ Higgsinos. If supersymmetry is made local (supergravity), then the theory incorporates gravity as well and hence the spin 2 graviton is joined to a spin 3/2 gravitino.

If supersymmetry were an exact symmetry of nature, the supersymmetric partners would have identical properties (except for spin) and hence the selectron mass would be degenerate with the electron mass. However, charged spin 0 particles with a mass of 511 keV have not been observed. Hence, supersymmetry must be broken. In order to preserve the gauge hierarchy, the corrections to scalar masses must be kept as small as 10^2 GeV. These corrections turn out to be proportional to the gravitino mass,[14] which is related to the supersymmetry breaking scale by

$$m_{3/2} = M_s^2/M_p, \qquad (2.1)$$

where $M_p = 1.2 \times 10^{19}$ GeV is the Planck mass. Thus, $M_s \lesssim 10^{10}$ GeV.

Low energy local N = 1 supergravity has been studied extensively.[15] At the Planck scale, there are many questions still to be answered. For example, where did this effective low energy N = 1 theory come from. Several possibilities have been put forward. The true theory might be an N = 8 supergravity[16] (the largest possible in four dimensions) that breaks down to an N = 1. N > 1 supergravity theories, however, do not contain chiral fields and must be written in terms of constituent particles (preons), which confine at the Planck scale to ordinary particles.[17] More recently, N = 1 supergravity theories expressed in terms of ten-dimensional strings (superstrings)[18] have drawn much interest in that they seem to be free of problems concerning infinities that are present in all other theories.

If the reader has not yet determined this, my personal bias lies with supergravity. In the next section, I discuss how supersymmetry is helpful to inflation and in the last section, I discuss how supersymmetry may provide an answer to the dark matter problems that plague cosmology.

INFLATION

As there are already several reviews[19] about inflation,[20] I will try to be brief here. However, any review of the very early universe would be incomplete if it did not at least touch upon inflation. In short, what is meant by inflation is the effect of exponential expansion due to a supercooled phase transition in order to resolve several fine-tunings regarding the initial conditions in the standard big bang model.

As examples of these problems, I will briefly describe what is known as the horizon problem and the curvature problem. The horizon volume (or causally connected volume today) is just related to the age of the universe, $V_0 \propto t_0^3$. The microwave background radiation with temperature, $T_0 \sim 3$ K, has been decoupled from itself since the epoch of recombination at $T_d \sim 10^4$ K. The horizon volume at that time was $V_d \propto t_d^3$. Now, the present horizon volume scaled back to the period of decoupling will be $V_0' = V_0(T_0/T_d)^3$ and the ratio of this volume to the horizon volume at decoupling is

$$V_0'/V_d \sim (V_0/V_d)(T_0/T_d)^3$$
$$\sim (t_0/t_d)^3 (T_0/T_d)^3 \sim 10^5, \quad (3.1)$$

where I have used $t_d \sim 3 \times 10^{12}$ sec and $t_0 \sim 5 \times 10^{17}$ sec. The ratio in equation (3.1) corresponds to the number of regions that were causally disconnected at recombination and grew into our present visible universe.

The microwave background radiation appears to be highly isotropic. In fact, the limits on the anisotropy put[21]

$$\Delta T/T \lesssim (2-5) \times 10^{-5}. \quad (3.2)$$

This means that on large scales, the universe must be very isotropic and homogeneous (any inhomogeneities would also produce fluctuations in the microwave background). The horizon problem, therefore, is the lack of an explanation as to why 10^5 causally disconnected regions at t_d all had the same temperature to within one part in 10^4.

The curvature problem (also known as the flatness or oldness problem) stems from the fact that although the universe is very old, we still do not know whether it is open or closed. If we look at the Freidmann equation for the expansion of the universe,

$$H^2 = \left(\frac{\dot{R}}{R}\right)^2 = \frac{8\pi G\rho}{3} - \frac{k}{R^2} + \frac{\Lambda}{3}, \quad (3.3)$$

where H is the Hubble parameter, R is the Robertson-Walker scale factor, ρ is the total mass energy density, k is the curvature constant ($k = 0, \pm 1$ for a flat, closed, or open universe), and Λ is the cosmological constant. Neglecting Λ, the curvature term can be expressed in terms of the density parameter,

$$\Omega = \rho/\rho_c, \quad (3.4)$$

$$\rho_c = 3H_0^2/8\pi G_N, \quad (3.5)$$

and the present value of the Hubble parameter, H_0, as

$$k/R^2 = (\Omega - 1) H_0^2. \quad (3.6)$$

If we now use the limits of $\Omega < 4$ and $H_0 < 100$ km s^{-1} Mpc^{-1}, we can form a dimensionless constant,

$$\hat{k} = k/R^2T^2 = (\Omega - 1) H_0^2/T^2 \lesssim 3H_0^2/T_0^2 < 2 \times 10^{-58}, \quad (3.7)$$

where I have used $T_0 > 2.7$ K. In an adiabatically expanding universe, $\hat{k}$ is absolutely constant ($R \sim T^{-1}$) and thus the limit (6.4) represents an initial condition that must be imposed so that the universe would have been able to live this long, while still looking so flat.

A more natural initial condition might have been $\hat{k} \sim O(1)$. In this case, the universe would have become curvature dominated at $T \sim 10^{-1} M_p$. For $k = +1$, this would signify the onset of recollapse. Even for k as small as $O(10^{-40})$, the universe would have become curvature dominated when $T \sim 10$ MeV or when the age of the universe was only $O(10^{-2})$ sec. Thus, not only is equation (3.7) a very tight constraint, it must also be strictly obeyed. Of course, it is also possible that $k = 0$ and the universe is actually spatially flat.

These are the two main problems that led Guth[20] to consider inflation. In the problems that were just discussed, it was assumed that the universe has always been expanding adiabatically. During a phase transition, however, this is not necessarily the case. If we look at a potential describing a phase transition from a symmetric false vacuum state, $\langle \Sigma \rangle = 0$, to the broken true vacuum at $\langle \Sigma \rangle = v$ as in FIGURE 1, we see that because of the barrier separating the two minima, the phase transition is a first order transition. If in addition, the transition takes place at T_c such that $T_c^4 < V_0$, the energy stored in the form of vacuum energy will be released. If released fast enough, it

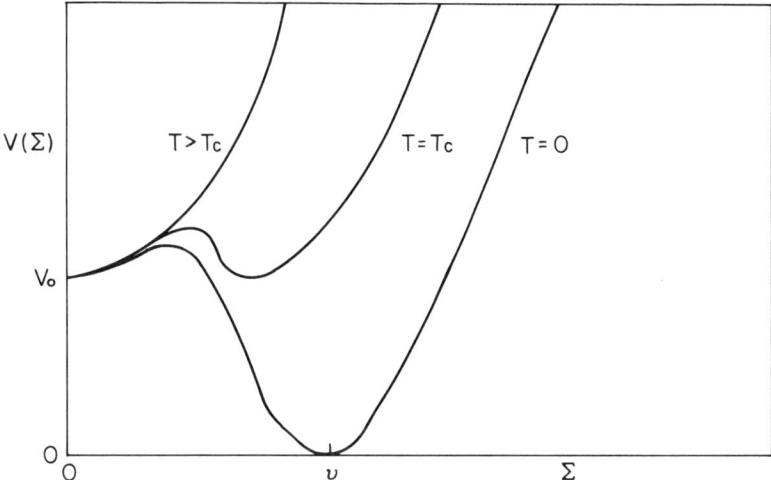

FIGURE 1. The scalar potential for a first order phase transition.

will produce radiation at a temperature, $T_R^4 \sim V_0$. In this reheating process, entropy has been created and

$$(RT)_f \sim (T_R/T_c)(RT)_i, \qquad (3.8)$$

provided that T_c is not too low. Therefore, we see that during a phase transition, the relation, $RT \sim$ constant, need not hold true and thus our dimensionless constant, $\hat{k}$, may not have actually been constant.

The inflationary universe scenario[20] is based on just such a situation. If during some phase transition, the value of RT changed by a factor of $O(10^{29})$, these two cosmological problems would be solved. The isotropy would in a sense be generated by the immense expansion; one small causal region could get blown up and hence our entire visible universe would have been at one time in thermal contact. In addition, the parameter, $\hat{k}$, could have started out at $O(1)$ and could have been driven small by the expansion.

If, in an extreme case, a barrier as in FIGURE 1 caused a lot of supercooling, such that $T_c^4 \ll V_0$, then the dynamics of the expansion would have greatly changed. In the example of FIGURE 1, the energy density of the symmetric vacuum, V_0, acts as a cosmological constant with

$$\Lambda = 8\pi V_0/M_p^2. \tag{3.9}$$

If the universe is trapped inside the false vacuum with $\Sigma = 0$, eventually the energy density due, to say, radiation will fall below the vacuum energy density, $\rho \ll V_0$. When this happens, the expansion rate will be dominated by the constant V_0 and we will get the de Sitter type of expansion,

$$R \sim \exp[Ht], \tag{3.10}$$

where

$$H^2 = \Lambda/3 = 8\pi V_0/3M_p^2. \tag{3.11}$$

The cosmological problems could be solved if

$$H\tau \gtrsim 65, \tag{3.12}$$

where τ is the duration of the phase transition, and if the vacuum energy density was converted to radiation so that the reheated temperature is found by

$$\frac{\pi^2}{30} N(T_R) T_R^4 = V_0, \tag{3.13}$$

where $N(T_R)$ is the number of degrees of freedom at T_R.

If such a barrier persists down to low temperatures, the phase transition must proceed via the formation of bubbles of the broken phase. The bubble formation rate per unit volume is given by[22]

$$p \sim Ae^{-B}, \tag{3.14}$$

where $A^{1/4}$ is generally taken to be the overall mass scale in the problem ($A \sim T^4$ or $A \sim M^4$) and B is tunneling action. The transition will take place in such a way so as to minimize the action. The phase transition will be completed when $p > H^4$.

The scenario just described is the original idea of Guth[20] for cosmological inflation. In this scenario, the universe would undergo a phase transition, say $SU(5) \rightarrow SU(3) \times SU(2) \times U(1)$, in which the potential resembled that in FIGURE 1. The universe would then get hung up in the $SU(5)$ phase down to a very low temperature. After completion of the phase transition, the universe would reheat to

$$T_R \sim M_x/[N(T_R)]^{1/4}. \tag{3.15}$$

Baryon generation would then follow so long as T_R was not too low (see next section).

It is now known that there is a problem with Guth's original idea for inflation.[23] It turns out that the requirement that the universe supercool for a long time ($H\tau > 65$) is not compatible with $p > H^4$ (i.e., the phase transition does not finish). In order to have a long inflationary time scale, a large barrier was necessary so as to be sure that the action for tunneling was also large. It is necessary in this scheme that the initial probability for tunneling be very small. The problem is that under these conditions the tunneling probability never catches up with the expansion rate. As a whole, the universe remains in the de Sitter state trapped in the symmetric $SU(5)$ vacuum, with only a few isolated bubbles containing the true $SU(3) \times SU(2) \times U(1)$ vacuum. Not only is the resulting universe very nonhomogeneous, but each bubble remains empty as all of the energy is stored in the bubble walls and is only released through collisions, which in this case do not occur.

The solution to this problem is called the new inflationary universe[24] and its basic and simple idea is this: tunnel first and inflate later. To realize this type of inflation, one must have a long flat scalar potential. If one can argue (e.g., by thermal effects) that at early times or at high temperatures the universe was in the symmetric phase, $\Sigma = 0$, and then at some lower temperature, $T \ll T_c$, a bubble is formed. The

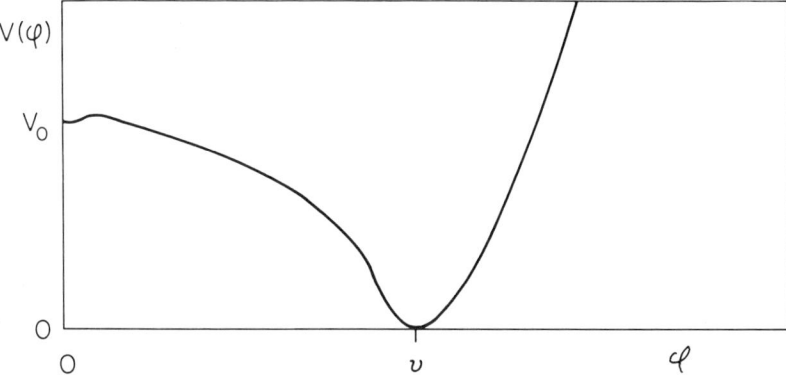

FIGURE 2. A schematic view of the type of scalar potential needed for new inflation.

supercooling may be due to either a barrier as in the previous case or a suppression of thermal fluctuations so that the field Σ rests near the origin. In the case of a barrier, once a bubble is formed, if the potential is very long and flat at values of Σ past the barrier, then the potential energy density (approximately constant) will again act like a cosmological constant. If a single bubble were to expand by 29 orders of magnitude, the phase transition need not be completed as in the previous case. The entire visible universe would be contained within one bubble. The bubble would be filled in this case not by bubble collisions, but by dissipation of the kinetic energy of the scalar field as it finally reaches its global minimum. A generic example of such a potential is shown in FIGURE 2.

Popular examples of flat potentials considered for inflation have been the Coleman-Weinberg[25] potentials, which are derived by taking first order radiative corrections to the tree potential. If scalar self-couplings are small enough, the tree potential can be neglected and we can concentrate on the corrections. Using an $SU(5)$

Coleman-Weinberg potential, though, has been shown to present several insurmountable difficulties.[26-29] These range from unnatural fine-tunings of the self-couplings[27] to transitions occurring to the wrong vacuum state.[29]

The most serious blow to Coleman-Weinberg type inflation comes from the density perturbations that are produced during the rollover.[28] The isotropy of the microwave background radiation tells us that any perturbations produced on large scales must have $\delta\rho/\rho \lesssim O(10^{-4})$. Ideally, what one would want from inflation is what is known as the Harrison-Zeldovich[30] spectrum of density fluctuations. They are also known as scale independent perturbations, which are the type most desired for the purposes of galaxy formation. Their magnitude, however, must be $O(10^{-4})$. Any perturbations stronger than this would produce visible anisotropies in the microwave background radiation, while weaker perturbations would not have had enough time to grow during the present period of matter domination (since decoupling).

As it turns out, phase transitions, such as the $SU(5)$ transition described above, produce[31] very nearly the Harrison-Zeldovich spectrum that is desired. The perturbations are formed because the field ϕ does not roll down to its global minimum homogeneously. There will, in general, be a time spread over which certain regions roll down faster or slower than others. The density perturbations have been calculated[28] for $SU(5)$ and turn out to be $\delta\rho/\rho \sim 50$, i.e., nearly five orders of magnitude too large.

Supersymmetry has led to the resolution of some of these problems.[27,32] In ways similar to those in which supersymmetry resolves the gauge hierarchy problem discussed in the previous section, it relieves the problems of fine-tuning mass scales in inflation and hence allows for flatter potentials. Another way to make flat potentials is to increase the value of the vacuum expectation value, v, from $\sim 10^{15}$ GeV in GUT models to $v \sim M_p \sim 10^{19}$ GeV. This class of models is called primordial inflation.[32] Considering primordial inflation and supersymmetry naturally leads one to inflation in supergravity.[33]

The actual phase transition responsible for inflation is no longer the GUT phase transition and may happen before, during, or after the breaking of GUTs. The scale of supersymmetry breaking, M_s, must be much lower than the GUT or inflation scale ($M_s \sim 10^{10}$ GeV) in order to preserve the gauge hierarchy. Therefore, inflation should not be associated with the supersymmetry breaking transition. Indeed, there have been efforts to associate the two, but these models have required large amounts of fine-tunings.[34]

The simplest SUSY-preserving inflationary models, however, run into certain difficulties regarding initial conditions. In other words, if the initial conditions for inflation are determined by high temperature effects,[35] it has been shown that these are inconsistent with the requirements for a long inflationary period.[36] Nonminimal supergravity models offer a resolution to this problem.[37] In fact, a very simple inflationary model can be written[38] in the context of so-called no-scale nonminimal supergravity.[39] These models are attractive in the sense that only one fundamental scale is put in by hand, namely the Planck scale, and the others are determined through radiative corrections. Such nonminimal models are also thought to stem from extended supergravity models of the type discussed in the previous section and perhaps from recent superstring theories.[40]

Finally, further recent work regarding the role of initial conditions and the possibility of inflation seems to heavily favor models of primordial inflation.[41] It has been claimed[42] that at very high temperatures, the field responsible for inflation will not be found near $\Sigma = 0$, but rather will be spread very far from the high temperature minimum. This, however, is true only for certain circumstances and it has been shown that this effect does not occur in models of primordial inflation.

The actual origin of inflation and the exact identity of the phase transition producing the desired supercooling are not known. Inflation exists as a possibility within the context of supergravity models. Its apparent beauty still reaches out further than our apparent ignorance.

GUTs AND COSMOLOGY

The origins of the modern connection between particle physics and cosmology really began with the generation[43] of a small, but finite, baryon-to-entropy ratio using grand unified theories (GUTs).[44] The problem in cosmology is basically that there is apparently very little antimatter in the universe and the number of photons greatly exceeds the number of baryons. If we define

$$\eta = (n_B - n_{\bar{B}})/n_\gamma, \quad (4.1)$$

where $n_{B,\bar{B},\gamma}$ is the number density of baryons, antibaryons and photons, we find that

$$\eta \approx n_B/n_\gamma \sim 10^{-10}\text{–}10^{-9} \quad (4.2)$$

(see the section entitled BIG BANG NUCLEOSYNTHESIS). In a standard model, the entropy density today is related to n_γ by

$$s \simeq 7 n_\gamma, \quad (4.3)$$

so equation (4.2) implies $n_B/s \sim 10^{-11}\text{–}10^{-10}$. This ratio is conserved however and hence represents another undesirable initial condition with its origin unknown.

Let us for the moment assume that in fact $\eta = 0$. We can compute the final number density of nucleons left over after annihilations have frozen out. At very high temperatures (neglecting a quark-hadron transition) such as $T > 1$ GeV, nucleons were in thermal equilibrium with the photon background and $n_N = n_{\bar{N}} = \frac{3}{2}n_\gamma$ (a factor of two accounts for neutrons and protons and the factor of $\frac{3}{4}$ for the difference between fermi and bose statistics). As the temperature fell below m_N, annihilations kept the nucleon density at its equilibrium value, $(n_N/n_\gamma) = (m_N/T)^{3/2} \exp(-m_N/T)$, until the annihilation rate, $\Gamma_A \approx n_N m_\pi^{-2}$, fell below the expansion rate. This occurred at $T \approx 20$ MeV. However, at this time the nucleon number density has already dropped to

$$n_N/n_\gamma - n_{\bar{N}}/n_\gamma \sim 10^{-18}, \quad (4.4)$$

which is eight orders of magnitude too small,[45] aside from the problem of having to separate the baryons from the antibaryons. If any separation did occur at higher temperatures (so that annihilations were as yet incomplete), then the maximum distance scale on which separation could occur is the causal scale related to the age of the universe at that time. At $T = 20$ MeV, the age of the universe was only $t \approx 2 \times 10^{-3}$

sec. At that time, a causal region (with distance scale defined by $2ct$) could only have contained $10^{-5} M_\odot$, which is very far from the galactic mass scales that we are asking for separations to occur, $10^{12} M_\odot$.

A final possibility might be statistical fluctuations, but in a region containing $10^{12} M_\odot$, there are $\sim 10^{80}$ photons and so one would only expect statistical fluctuations to produce an asymmetry of $\eta \sim 10^{-40}$. Thus, we are left with the problem as to the origin of a small nonzero value for η. We can assume that it was an initial condition to start off with and in a baryon number conserving theory it would remain nearly constant. [The production of entropy (photons) could cause it to fall.] In this case, however, we must still ask ourselves, why is it so small? A more attractive possibility, however, is to suppose that the baryon asymmetry was in some way generated by the microphysics. Indeed, if one can show that a small nonzero value for η developed from $\eta = 0$ (or any other value) as an initial condition, we could consider the question solved. In the rest of this section, we will look at this second possibility for generating a nonzero value of η using GUTs.[44]

There are three basic ingredients necessary[43] to generate a nonzero η. They are:

(1) baryon number violating interactions;
(2) C and CP violation;
(3) a departure from thermal equilibrium.

The first condition is rather obvious; unless there is some mechanism for violating baryon number conservation, baryon number will be conserved and an initial condition such as $\eta = 0$ will remain fixed. C and CP violation indicate a direction for the asymmetry; that is, should the baryon number violating interactions produce more baryons than antibaryons? If C or CP were conserved, no such direction would exist and the net baryon number would remain at zero. The final ingredient is necessary in order to insure that not all processes are actually occurring at the same rate. For example, in equilibrium if every process that produced a positive baryon number was accompanied by an equivalent process that destroyed it, again no net baryon number would be produced.

The first two of these ingredients are contained in GUTs, with the third in an expanding universe where it is not uncommon that interactions come in and out of equilibrium. In $SU(5)$, the fact that quarks and leptons are in the same multiplets allows for baryon nonconserving interactions such as $e^- + d \leftrightarrow \bar{u} + \bar{u}$, etc., or decays of the supermassive gauge bosons, X and Y, such as $X \to e^- + d, \bar{u} + \bar{u}$. Although today these interactions are very ineffective because of the masses of the X and Y bosons, in the early universe, when $T > M_X \sim 10^{15}$ GeV, these types of interactions should have been very important. C and CP violation is very model dependent. In the minimal $SU(5)$ model, the magnitude of C and CP violation is too small to yield a useful value of η. The C and CP violation in general comes from the interference between tree level and first loop corrections.

The departure from equilibrium is very common in the early universe when interaction rates cannot keep up with the expansion rate. In fact, the simplest (and most useful) scenario for baryon production makes use of the fact that a single decay rate goes out of equilibrium. It is commonly referred to as the out of equilibrium decay scenario.[46] The basic idea is that the gauge bosons, X and Y, (or Higgs bosons) may have a lifetime that is long enough to insure that the inverse decays have already ceased so that the baryon number is produced by their free decays.

More specifically, let us call X either the gauge boson or Higgs boson that produces the baryon asymmetry through decays. Let α be its coupling to fermions. For X, a gauge boson, α will be the GUT fine structure constant, while for X, a Higgs boson, $(4\pi\alpha)^{1/2}$ will be the Yukawa coupling to fermions. The decay rate for X will be

$$\Gamma_D \sim \alpha M_X. \tag{4.5}$$

However, decays can only begin occurring when the age of the universe is longer than the X lifetime, Γ_D^{-1}, i.e., when $\Gamma_D > H$,

$$\alpha M_X \gtrsim N(T)^{1/2} T^2 / M_p, \tag{4.6}$$

or at a temperature,

$$T^2 < \alpha M_X M_p N(T)^{-1/2}. \tag{4.7}$$

Scatterings on the other hand proceed at a rate,

$$\Gamma_s \sim \alpha^2 T^3 / M_X^2, \tag{4.8}$$

and hence are not effective at lower temperatures. In equilibrium, therefore, decays must have been effective as T fell below M_X in order to track the equilibrium density of X's (and $\bar{X}$'s). Thus, the condition for equilibrium is that at $T = M_X$, $\Gamma_D > H$ or

$$M_X \lesssim \alpha M_p [N(M_x)]^{-1/2} \sim 10^{18} \alpha \text{ GeV}. \tag{4.9}$$

In this case, we would expect no net baryon asymmetry to be produced.

For masses, $M_X \gtrsim 10^{18} \alpha$ GeV, the lifetime of the X bosons is longer than the age of the universe when $T \sim M_X$. When the decays finally begin to occur at $T < M_X$, however, the density of X's is still comparable to photons, $n_X/n_\gamma \sim 1$, whereas the equilibrium density at $T < M_X$ is $n_X/n_\gamma \sim (M_X/T)^{3/2} \times \exp[-M_X/T] \ll 1$. Hence, the decays are occurring out of equilibrium (inverse decays are not occurring) and we have the possibility for producing a net asymmetry.

Let us now look at what happens during the decay of an X, $\bar{X}$ pair. If we consider the example of the X gauge boson and its decays to $\bar{u}$, $\bar{u}$ (with branching ratio, r, and net baryon number change, $\Delta b_1 = -\frac{2}{3}$) and to e^-, d (with branching ratio, $1 - r$, and net baryon number change, $\Delta b_2 = +\frac{1}{3}$), then

$$X \xrightarrow[r]{} \bar{u} + \bar{u}, \quad \Delta b_1 = -\tfrac{2}{3}, \tag{4.10a}$$

$$X \xrightarrow[1-r]{} e^- + d, \quad \Delta b_2 = +\tfrac{1}{3}. \tag{4.10b}$$

A similar set of decays will occur for $\bar{X}$:

$$\bar{X} \xrightarrow[\bar{r}]{} u + u, \quad \Delta b_{\bar{1}} = +\tfrac{2}{3}, \tag{4.11a}$$

$$\overline{X} \xrightarrow[1-\bar{r}]{} e^+ + \bar{d}, \quad \Delta b_{\bar{2}} = -\tfrac{1}{3}. \tag{4.11b}$$

If C and CP are violated, then $r \neq \bar{r}$ and we can define the total net baryon number produced per decay of X and $\overline{X}$ as

$$\Delta B = (\Delta b_1)r + (\Delta b_2)(1-r) + (\Delta b_{\bar{1}})\bar{r} + (\Delta b_{\bar{2}})(1-\bar{r}) = \bar{r} - r. \tag{4.12}$$

The value of $\bar{r} - r$ will, of course, depend on the specific model for C and CP violation.

The total baryon density that will have been produced by the X, $\overline{X}$ pair [provided equation (**4.9**) is not satisfied] is

$$n_B = (\Delta B)n_X, \tag{4.13}$$

and since we also have $n_X = n_{\overline{X}} = n_\gamma$,

$$n_B = (\Delta B)n_\gamma. \tag{4.14}$$

Although the net baryon number is conserved during the subsequent evolution of the universe, the photon number density is not. A more useful quantity just after baryon generation is the baryon-to-specific entropy ratio, n_B/s. The entropy density is

$$s = \frac{2\pi^2}{45} N(T) T^3. \tag{4.15}$$

At $T \lesssim M_X \sim 10^{15}$ GeV, we expect $N(T) \gtrsim O(100)$ so that $s \sim O(100) n_\gamma$. Thus, the baryon-to-entropy ratio we would expect to produce in the out-of-equilibrium decay scenario would be

$$n_B/s \sim 10^{-2}(\Delta B). \tag{4.16}$$

The value of n_B/s that we are looking for must be related to the limits on η, which will be discussed in the next section. η in the range of $(3-10) \times 10^{-10}$ corresponds to a value of n_B/s in the range of $(4.3-14) \times 10^{-11}$. Comparing this with the expected production, equation (**4.16**) gives us a lot of hope that GUTs may provide us with a viable mechanism for generating a small (but not too small) value for η.

Although we can be encouraged by the above scenario, we must still show that given a GUT, after the full set of Boltzmann equations have been integrated, an acceptable and definite value of η emerges. In particular, most GUTs do satisfy equation (**4.9**) for $\alpha = \tfrac{1}{41}$ and $M_X \sim 10^{15}$ GeV, and decays will be occurring at $T \sim M_X$, but in, at best, partial equilibrium. Thus, the estimate of equation (**4.16**) is not a good one.

In FIGURE 3, we look at the typical results that one finds after a complete numerical integration[47] of the Boltzmann equations. These particular results are for an $SU(5)$ model, but their behavior is generic for most any GUT. What is plotted is the time development of the baryon-to-entropy ratio, n_B/s, normalized to the net baryon number produced by pair decay, ΔB. The horizontal scale, M_x/T, is proportional to $t^{1/2}$. The three curves correspond to different choices for the mass of the boson X. In curve 1, we have chosen a mass that we expect to satisfy the out-of-equilibrium condition, $M_x \simeq 3 \times 10^{18}\alpha$, and we indeed find that the maximum asymmetry has been generated,

$n_B/s \approx 10^{-2}$ (ΔB), as we expected in equation (**4.16**). This in itself confirms the original idea.

The good news that we find from FIGURE 3 is that even for lower masses, an asymmetry is still produced. In curve 2, we have chosen $M_x = 3 \times 10^{17}\alpha$ and we still find a substantial asymmetry, $n_B/s \sim 10^{-4}$ (ΔB). What is happening is that at $T \sim M_x$, inverse decays are still effective in trying to restore equilibrium. Eventually, they too freeze out and any X's and $\overline{X}$'s still present decay freely to produce a net baryon number. If we continue to lower the mass as in curve 3, with $M_x = 3 \times 10^{16}\alpha$, then scatterings begin to play a role in driving things further towards equilibrium. Again,

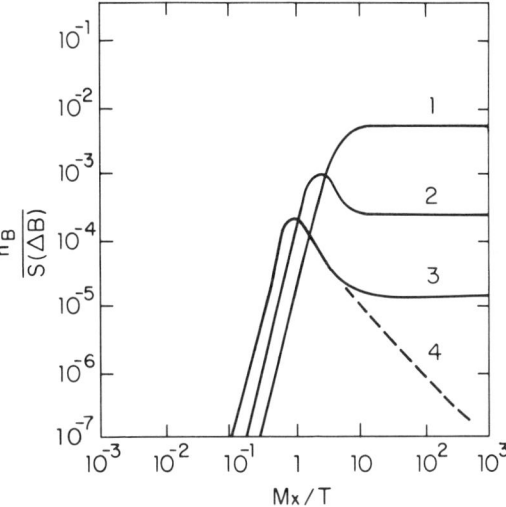

FIGURE 3. The time evolution of the baryon asymmetry in units of (ΔB) for (1) $M_x \simeq 3 \times 10^{18}\alpha$; (2) $3 \times 10^{17}\alpha$; (3) $3 \times 10^{16}\alpha$; and (4) if scatterings remain very effective.

when they freeze out, the remaining X, $\overline{X}$ pairs decay, thus leaving an asymmetry. If scatterings become dominant, however, the resulting asymmetry in the standard model will become exponentially small with decreasing M_x as shown in the dashed curve. In FIGURE 4, we have plotted the final asymmetry that is produced as a function of $K = 3 \times 10^{17} \alpha/M_x$, where K is defined by

$$K \equiv \Gamma_D/H|_{T=M_x}. \qquad (4.17)$$

Depending on whether or not X is a gauge or Higgs boson, the resulting final asymmetry can be approximated by

$$n_B/s \approx 2 \times 10^{-3}(\Delta B)/[1 + (3K)^{1.2}] \qquad (4.18)$$

for Higgs bosons and

$$n_B/s \approx 8 \times 10^{-3}(\Delta B)/[1 + (16K)^{1.3}] \qquad (4.19)$$

for gauge bosons.

Thus, we see that GUTs do indeed offer an explanation to the small, but finite baryon-to-entropy ratio. In supersymmetric theories, the ideas are generally the same, although the details may be somewhat different.[48,49]

BIG BANG NUCLEOSYNTHESIS

As was noted in the introduction, the two most important pieces of evidence in support of the standard big bang model are the observation[1] of the 3-K microwave background radiation and the explanation[2] of the origin of the light elements and their

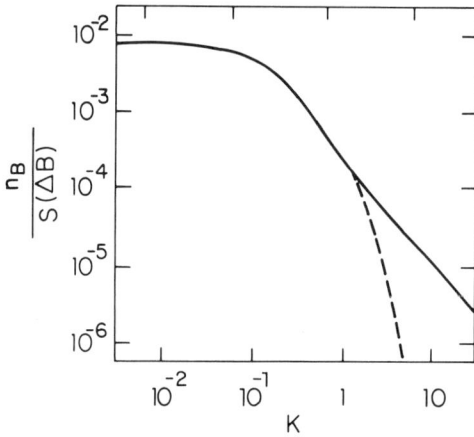

FIGURE 4. The final baryon asymmetry as a function of $K = 3 \times 10^{17} \alpha / M_X$ in units of (ΔB). The dashed curve assumes effective scatterings.

abundances. Because of the initially high temperatures and densities and the large abundance of neutrons relative to protons, the chains of nuclear reactions similar to those occurring in stars might have occurred. Indeed, in the simplest model of nucleosynthesis, one can compute the produced abundances of deuterium, ^{3}He, ^{4}He, and ^{7}Li, and one finds an amazing degree of agreement with the observed abundances. (The observations that must be compared with the big bang abundances must be from sources where little or no subsequent nucleosynthesis has taken place.) In this section, I will review the predictions of big bang nucleosynthesis and its cosmological consequences in terms of limits on particle physics.

The temperature region of interest is one typical of nuclear energies, i.e., $T \sim 1$ MeV. The initial conditions for the problem will therefore be set at $T \gg 1$ MeV. Once again, because the asymmetry between baryons and antibaryons is so small and since we do not expect very different asymmetries among the leptons (standard GUT models even predict their similarity), we will take all chemical potentials to be zero. One of the chief quantities of interest will be the neutron-to-proton ratio (n/p). At very high temperatures ($T \gg 1$ MeV), the weak interaction rates for the processes,

$$n + \nu_e \leftrightarrow p + e^-,$$
$$n + e^+ \leftrightarrow p + \bar{\nu}_e,$$
$$n \leftrightarrow p + e^- + \bar{\nu}_e, \tag{5.1}$$

were all in equilibrium, i.e., $\Gamma_w > H$. Thus, we would expect that initially $(n/p) \approx 1$. Actually, in equilibrium, the ratio is essentially controlled by the Boltzmann factor so that

$$(n/p) \simeq \exp(-\Delta m/T), \tag{5.2}$$

where $\Delta m = m_n - m_p$ is the neutron-proton mass difference. For $T \gg \Delta m$, $(n/p) \simeq 1$.

At temperatures $T \gg 1$ MeV, nucleosynthesis cannot begin to occur even though the rate for forming the first isotope, deuterium, is sufficiently rapid. To begin with, at $T \gtrsim 1$ MeV, deuterium is photodissociated because $E_\gamma > 2.2$ MeV (the binding energy of deuterium; $E_\gamma \approx 2.7\,T$ for a blackbody). Furthermore, the density of photons is very high ($n_\gamma/n_B \sim 10^{10}$). Thus, the onset of nucleosynthesis will depend on the quantity,

$$\eta^{-1} \exp[-2.2 \text{ MeV}/T], \tag{5.3}$$

where η is defined as before. When this quantity in equation (5.3) becomes $\lesssim O(1)$, the rate for $p + n \rightarrow D + \gamma$ finally becomes greater than the rate for dissociation, $D + \gamma \rightarrow p + n$. This occurs when $T \sim 0.1$ MeV or when the universe is a little over two minutes old.

Because nucleosynthesis begins when $T < 1$ MeV, the rates for the processes that control (n/p) in equations (5.1), as well as those that keep neutrinos in equilibrium, are frozen out. Furthermore, because the rates for the processes in equations (5.1) also freeze out (at $T \lesssim 1$ MeV), the neutron-to-proton ratio must be adjusted from its equilibrium value. When the freeze-out occurs, the ratio, (n/p), is relatively fixed at

$$(n/p) \sim 1/6. \tag{5.4}$$

This equilibrium value is adjusted by taking into account the free neutron decays up until the time at which nucleosynthesis begins. This reduces the ratio to

$$(n/p) \sim 1/7. \tag{5.5}$$

Since virtually all the neutrons available end up in deuterium, which gets quickly converted to ^{4}He, we can estimate the ratio of the ^{4}He nuclei formed as compared with the number of protons left over,

$$X_4 \equiv (N_{^4\text{He}}/N_\text{H}) = \tfrac{1}{2}(n/p)(1 - [1 - (n/p)]), \tag{5.6}$$

or more importantly, the ^{4}He mass fraction,

$$Y_4 = 4X_4/(1 + 4X_4) = 2(n/p)/[1 + (n/p)]. \tag{5.7}$$

For $(n/p) \simeq 1/7$, we estimate that $Y_4 \simeq 0.25$, which is very close to the observed value.

The actual calculated value of Y_4 will depend on a numerical calculation that runs through the complete sequence of nuclear reactions.[50] The nuclear chain is temporarily halted because there are gaps at the masses of $A = 5$ and $A = 8$ (i.e., there are no stable nuclei with those masses). There is some further production, however, that accounts for the abundances of ^{6}Li and ^{7}Li. Once again, because of the gap at $A = 8$, there is very little subsequent nucleosynthesis in the big bang. A second chief factor in the ending of nucleosynthesis is that during this whole process the universe continues to expand and cool. At lower temperatures, it becomes exponentially difficult to overcome the coulomb barriers in nuclear collisions. In spite of these effects, numerical calculations of the elemental abundances continue the chain up until Al.

Before reviewing the results of the big bang nucleosynthesis[50-53] calculations, it is important to realize that there are three additional parameters that have a very strong effect on the results.[a] They are (1) the baryon-to-photon ratio, η; (2) the neutron half-life, $\tau_{1/2}$; (3) the number of light particles or, in particular, the number of neutrino flavors, N_ν.

As we have seen above, the value of η controls the onset of nucleosynthesis [see equation (5.3)]. Basically what happens is that for a larger baryon-to-photon ratio, η, the quantity in equation (5.3) becomes smaller, thus allowing nucleosynthesis to begin earlier at a higher temperature. Remember also that a key ingredient in determining the final mass fraction of ^{4}He, Y_4, was (n/p) [see equation (5.7)] and that the final value of (n/p) was determined by the time at which nucleosynthesis begins, thus controlling the time available for free decays after the freeze-out. If nucleosynthesis begins earlier, this leaves less time for neutrons to decay and the values of (n/p) and hence Y_4 are increased.

The value of η cannot be determined directly from observations. If we break η up, we find that

$$n_B = \rho_B/m_B = \Omega_B \rho_c/m_B$$
$$= 1.13 \times 10^{-5} \Omega_B h_0^2 \text{ cm}^{-3}, \quad (5.8)$$

where ρ_B is the energy density in baryons, m_B is the nucleon mass, Ω_B is that part of Ω which is in the form of baryons, and ρ_c is the critical energy density. The number density of photons is just

$$n_\gamma = 400 \, (T_0/2.7)^3 \text{ cm}^{-3}, \quad (5.9)$$

where T_0 is the present temperature of the microwave background radiation. Putting η back together, we find

$$\eta = 2.81 \times 10^{-8} \Omega_B h_0^2 \, (2.7/T_0)^3. \quad (5.10)$$

Thus, we could determine η if we knew Ω_B, h_0, and T_0.

The second parameter, $\tau_{1/2}$, is important in that it also plays a role in determining the value of Y_4. Although we don't usually consider $\tau_{1/2}$ a parameter, the uncertainties in its measured value are significant from the point of view of nucleosynthesis. After all, it is this quantity that will control the weak interaction rates and hence determine the freeze-out temperature. The common value of $\tau_{1/2} \simeq 10.6$ minutes is actually uncertain by about two percent and this is enough to affect the production of ^{4}He. The range we will consider is

$$10.4 \text{ min} < \tau_{1/2} \leq 10.8 \text{ min}. \quad (5.11)$$

As in the case of η, increasing $\tau_{1/2}$ leads to a larger value of Y_4. We can see this by looking again at a comparison between the weak interaction rates and the expansion

---

[a] I am not considering the effects of a chemical potential, which can also greatly vary the results.[54]

rate. If we parameterize the weak interaction rate by $\Gamma_{wk} = AT^5$ and the expansion rate by $H = BT^2$, then the freeze-out temperature is determined by

$$H(T_D) \simeq \Gamma_{wk}(T_D) \tag{5.12}$$

or

$$T_d^3 = B/A. \tag{5.13}$$

If we now increase $\tau_{1/2}$, this corresponds to decreasing $\Gamma_{wk} \sim \tau_{1/2}^{-1}$ or decreasing the value of A. This in turn gives a higher value for T_d. Now if T_d is larger, this will give a larger value of (n/p) at freeze-out via equation (5.2) and hence more ^{4}He via equation (5.7).

The final input parameter that we stated was the number of light particles. Specifically, what we mean is the number of degrees of freedom corresponding to particles that are still relativistic ($m \ll T$) when $T < O(1)$ MeV. In addition, we must require that these particles be relatively stable so that they will be present when freeze-out occurs; thus, $\tau >$ few seconds. As we hinted to above, likely candidates for these particles are neutrinos and thus the number of neutrino flavors, N_ν, becomes important. Of course, any other types of light particles such as photinos or axions, etc., may also be important.

The number of neutrino flavors, N_ν, will also affect the primordial abundance of ^{4}He, and like η and $\tau_{1/2}$, increasing N_ν, increases Y_4. The expansion rate is proportional to $[N(T)]^{1/2}$. At $T \gtrsim 1$ MeV, $N(T)$ is given by

$$N(T) = 2 + \frac{7}{2} + \frac{7}{4} N_\nu, \tag{5.14}$$

which takes into account the contribution of γ's, $e^{\pm}$'s, and N_ν flavors of neutrinos. Thus increasing N_ν, increases B in the notation of equation (5.13) and again leads to a higher value of T_d, with the same effect of producing more ^{4}He.

Let us now look at the observations[55] that tell us the abundances of the light elements. In particular, we will be interested in the abundances of D, ^{3}He, ^{4}He, and ^{7}Li. Deuterium is the most easily destroyed of the light elements. It is also very difficult to produce in astrophysical systems where it is not further processed to form ^{3}He. Therefore, any of the observed D is generally assumed to be primordial. Furthermore, because deuterium is so easily destroyed (or burned), we must assume that the abundance of D produced in the big bang is greater than the observed value or

$$(D/H)_{BB} \gtrsim (D/H)_{OBS}, \tag{5.15}$$

where (D/H) is the ratio (by number) of deuterium to hydrogen.

Unlike deuterium, ^{3}He is very difficult to destroy in its entirety in stellar systems. Pre-main-sequence stars are very efficient in burning deuterium to ^{3}He via D + p → ^{3}He + γ. ^{3}He is only destroyed at high temperatures ($T > 7 \times 10^6$ K) through ^{3}He + ^{3}He → ^{4}He + 2p and ^{3}He + ^{4}He → ^{7}Be + γ. At higher temperatures ($T > 10^8$ K), ^{4}He is burned to carbon and oxygen. The point is that, in general, some fraction, g, of the initial ^{3}He abundance will survive stellar processing. If one takes into account the fact that some of this ^{3}He is redeposited in the interstellar medium (pre-solar), then in terms of g, we have

$$(D + {}^3He/H)|_{BB} \leq (D + {}^3He)/H|_{pre\odot} + (1/g - 1) \, {}^3He/H|_{pre\odot}. \tag{5.16}$$

The value of g, however, can only be determined[56] by models of stellar evolution and in fact may differ depending on the mass of the star. In low mass stars ($M < 8M_\odot$), $g > 0.7$ is not unreasonable, while for high mass stars ($8M_\odot < M < 100 \, M_\odot$), g may be as low as $\frac{1}{4}$. Since an initial spectrum of stellar masses would cover all ranges, perhaps a lower limit on g of $\frac{1}{2}$–$\frac{1}{4}$ would be safe.

Using the observational limits on D/H and ^{3}He/H (see references 55 and 57–60),

$$(D/H) \geq (1–2) \times 10^{-5}, \tag{5.17a}$$

$$(D + {}^3He)/H|_{pre\odot} \leq 4 \times 10^{-5}, \tag{5.17b}$$

$$^3He/H|_{pre\odot} \leq 2 \times 10^{-5}, \tag{5.17c}$$

we find from the results of the big bang nucleosynthesis calculations[53] (shown in FIGURE 5) that

$$(3–4) \times 10^{-10} \leq \eta \leq (7–10) \times 10^{-10}, \tag{5.18}$$

which is consistent with both ^{3}He and D.

^{7}Li is another isotope that is in principle difficult to draw solid conclusions from. The main difficulty is that ^{7}Li is both easily produced as well as destroyed. Recently, however, there have been some measurements[61] of the ^{7}Li abundance in some very old Population II stars. Since some ^{7}Li might have been destroyed before the formation of these stars, we might expect $(^7Li/H)_{Pop\,II} \leq (^7Li/H)_{BB}$. (The present ^{7}Li abundance would still be larger, representing the contribution from stellar processing.) The observed limit of the ^{7}Li abundance is

$$(^7Li/H)_{Pop\,II} \leq 1.5 \times 10^{-10} \tag{5.19}$$

and is consistent with big bang nucleosynthesis for

$$10^{-10} < \eta < 7 \times 10^{-10}, \tag{5.20}$$

which agrees well with equation (**5.18**).

This brings us to ^{4}He, which is probably the most important of the isotopes studied. The main reason ^{4}He is so important is that there is so much of it. Next to hydrogen, it is the most abundant element around and its abundance is quite well known. Unlike the other light elements that have observational uncertainties of $\gtrsim 100\%$, the ^{4}He abundances are measured to within a few percent. The main problem is that it is also produced in stars and care must be taken in trying to derive the "observed" primordial abundance.

To be sure, one can place an upper limit on the primordial abundance by making $Y_{4BB} < Y_{4OBS}$ (Y_4, remember is the total ^{4}He mass fraction). However, in order to use big bang nucleosynthesis to set limits on particle physics (e.g., N_ν), a much more accurate determination of Y_{4BB} is needed. Spectral measurements[55] of galactic HII regions give very accurate values of Y_4, but there they have been contaminated with by-products of stellar processing. The observations of galaxies with low metal abundances could in principle yield an accurate value of Y_{4BB}, but these measurements are difficult because these galaxies are typically very far away. It is not possible within the scope of these lectures to cover completely the discussion of Y_4. The best estimates consistent with observations place Y_4 in the range of

$$0.22 \leq Y_4 \leq 0.25. \tag{5.21}$$

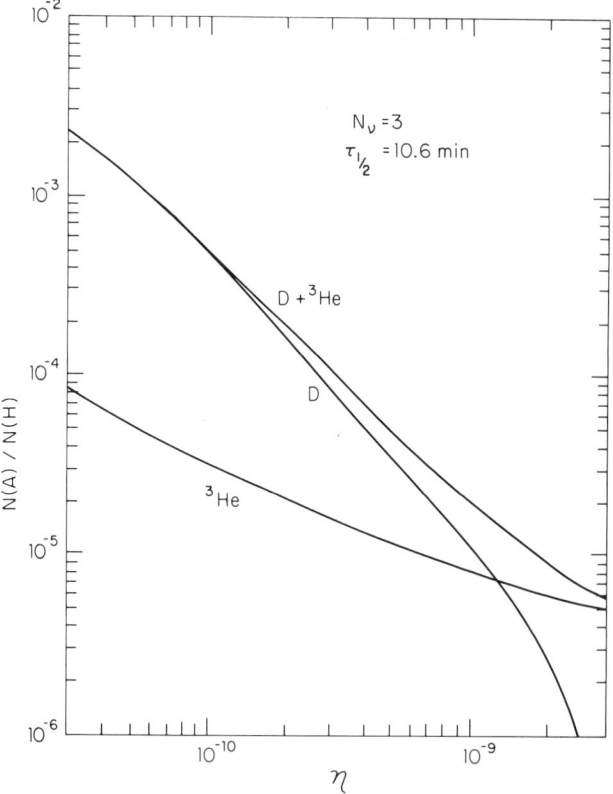

FIGURE 5. The abundances (by number relative to hydrogen) of D, ^{3}He, and their sum as a function of η for $N_\nu = 3$ and $\tau_{1/2} = 10.6$ min.

If we restrict ourselves as before to $N_\nu = 3$ and $\tau_{1/2} = 10.6$ min, then the upper limit on Y_4 implies an upper limit on η from FIGURE 6,

$$\eta \leq 5 \times 10^{-5}, \tag{5.22}$$

which is once again consistent with the previous limits in equation (5.18). (The lower limit on Y_4 does not give an interesting bound on η.)

FIGURE 6 actually contains significantly more information than just a limit on η. In FIGURE 6, we see clearly the behavior of Y_4 with respect to all three parameters: η, $\tau_{1/2}$, and N_r. It is clear how Y_4 increases with increasing values of any of the three parameters. It is also immediately clear that we can set a limit[51-53] on N_ν provided that we have a lower limit on η. Using $\eta > 3 \times 10^{-10}$ and $Y_4 < 0.25$, we find that $N_\nu \leq 4$ with the equality being at best marginal. This implies that, at most, one more generation is allowed, assuming that the neutrinos associated with each generation are light and stable.

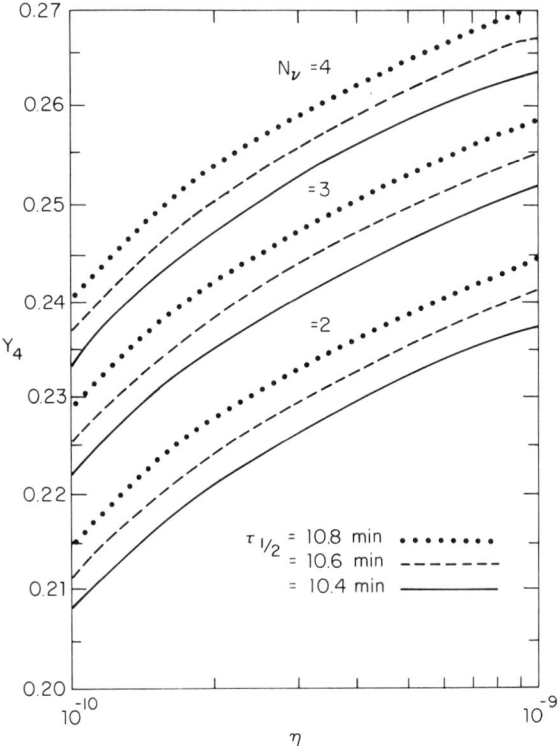

FIGURE 6. The abundance (by mass) of ^{4}He as a function of η for N_ν = 2, 3, and 4 and for $\tau_{1/2}$ = 10.4 min (solid), 10.6 min (dashed), and 10.8 min (dotted).

The strong dependence of Y_4 on the three parameters requires great precision to strengthen the limits due to nucleosynthesis. Strictly speaking, $\eta > 3 \times 10^{-10}$ and $\tau_{1/2} >$ 10.4 min allows $N_\nu = 4$ only if $Y_4 \gtrsim 0.253$; however, we are not yet in a position to believe the third decimal place. For $\tau_{1/2} \gtrsim 10.4$ min, the limit in equation (**5.22**) on η can be relaxed so that $N_\nu \leq 3$ and $Y_4 < 0.25$, which implies $\eta < 7 \times 10^{-10}$. We can also turn the limits around and set a lower limit on the helium abundance by assuming $\eta > 3 \times 10^{-10}$ and $N_\mu \geq 3$, thus giving $Y_4 > 0.24$. If future observations actually yield $Y_4 < 0.24$, one would have to argue that perhaps ν_τ is heavy and/or unstable (the present limit is only $m_{\nu_\tau} < 250$ MeV). If we only assume $N_\nu \geq 2$, then the lower limit on Y_4 becomes $Y_4 \geq 0.22$. Any observation of the primordial helium abundance less than 0.22 would indicate an inconsistency with the standard model.

There is still one important consequence of the above limits and this is that the limit on η can be converted to a limit on the baryon density and Ω_B. If we turn around equation (5.10), we have

$$\Omega_B = 3.56 \times 10^7 \eta h_0^{-2}(T_0/2.7)^3, \qquad (5.23)$$

and using the limits on η in equation (5.18), h_0, and T_0 from (2.7–3)K, we find a range for Ω_B of

$$0.01 \leq \Omega_B \leq 0.19. \qquad (5.24)$$

Recall that for a closed universe, $\Omega > 1$, thus [from equation (5.24)] we can conclude that the universe is not closed by baryons. This does not exclude the possibility that other forms of matter (e.g., massive neutrinos, etc.) exist in large quantities to provide for a large Ω. In fact, if large clusters of galaxies were representative of Ω, the limit from nucleosynthesis would indicate that some form of dark matter must exist.

DARK MATTER AND GALAXY FORMATION

Inflation predicts that $\Omega = 1$. Big bang nucleosynthesis requires $\Omega_B \lesssim 0.2$. The obvious resolution to this conflict is to suppose that there exists nonbaryonic dark matter, $\Omega_D \gtrsim 0.8$. Observational determinations of Ω, however, always prove to be less than a few tenths,[62] thus indicating that perhaps a large fraction of the dark matter is unclustered.

The above represents the first part of a growing chain of dark matter problems. Other dark matter problems become evident when one considers smaller scales. On the scale of galaxies, the observation of flat rotation curves strongly implies the existence of a dark halo component for spiral galaxies; that is, it appears that a substantial fraction of the galactic mass is nonvisible. Because on galactic[63] scales of $\Omega \sim 0.1$, one might think that baryons are the logical candidate. However, unless the baryons are carefully hidden in clouds distributed over the halo or in low mass stars with a special mass distribution, baryons are unlikely[64] to provide the missing mass in galactic halos. (Very massive black holes, $M \geq 100 M_\odot$, that leave little or no ejecta, though, remain a possibility.)

On smaller scales, those of dwarf spheroidal galaxies, there appears to be a dark matter problem as well.[65] Even locally, in the solar neighborhood, there appears to be as much dark material as there is in the form of stars and gas.[66] The dark matter in the galactic disk, however, is presumably in the form of baryons, as they must have undergone dissipational processes to get them into the disk.

Once we accept that nonbaryonic dark matter is needed, we can distinguish[67] three forms of dark matter depending on their impact on the growth of the density perturbations needed to account for galaxy formation. Perturbations grow primarily during the matter dominated phase of expansion. Particles that are relativistic just before matter dominance are called hot particles. Neutrinos or light Higgsinos are examples. Because these particles are relativistic at relatively late times, their free streaming wipes out perturbations out to scales,[68]

$$M_J = 3 \times 10^{18} \, M_\odot/m_\nu^2(\text{ev}), \qquad (6.1)$$

where M_J is the Jeans mass and is the minimum scale on which clustering occurs. In the hot scenario of galaxy formation, large-scale structures form first and must fragment down to galactic scales.[69] A second class of dark matter candidates is referred to as warm matter and includes more massive particles up to O(1) keV. In this case, initial mass scales are somewhat smaller.[70,71] If they exist, right-handed neutrinos might be warm particle candidates.[70]

Particles that have been nonrelativistic long before perturbation growth began are called cold particles. Particles more massive than 1 keV, such as heavy neutrinos, photinos, Higgsinos, sneutrinos, etc., are cold particle candidates. In the cold dark matter scenario,[72] small scale structures ($M > 10^6 M_\odot$) form first and larger scales are built in a hierarchical manner. For a summary of the pros and cons of these three possibilities in relation to models of galaxy formation, see reference 73 in these proceedings.

The existence of the dark matter candidate depends, of course, on the particular particle physics model that one employs. Supersymmetric models are interesting in this context because they guarantee one stable and probably massive particle. If the particle is neutral, there are basically four possibilities: (1) the spin 0 partner of the neutrino or sneutrino; (2) the spin ½ partner of an axion-like particle or axino; (3) the spin 3/2 gravitino; and (4) the spin ½ partner of the photon or photino.

Similar to limits on neutrino masses, there are limits on most of the supersymmetric dark matter candidates. The exception is the sneutrino, whose mass is not constrained by cosmology.[74] The axino mass limits depend on the temperature at which axino interactions dropped out of equilibrium, but roughly, $m_{\tilde{a}} \lesssim 1$ keV. Similarly, the gravitino, if stable, must have[75] $m_{3/2} \lesssim 1$ keV unless the number of gravitinos was reduced by inflation,[76] in which case $m_{3/2}$ is unconstrained. Photinos are most like neutrinos in the sense that their mass must be[77,78] $m_{\tilde{\gamma}} \gtrsim \frac{1}{2}$ GeV. This last possibility has been suggested[79] to account for the low-energy antiprotons observed in cosmic rays through photino annihilations. In addition, photinos in the sun may provide[80] a direct observational test of dark matter through annihilations to energetic neutrinos.

The gravitino is also quite interesting cosmologically in that if it is not stable it must be very long-lived. Its decay must be purely gravitational and hence its decay rate is given by[81]

$$\Gamma \sim m_{3/2}^3/M_p^2, \qquad (6.2)$$

which shows that decays may have been occurring fairly recently. These late decays may be of interest[82] to galaxy formation models using a decaying particle scenario.[82,83] Decays into photons and photinos can highly constrain the abundance of gravitinos,[84] again necessitating an inflationary solution,[76] but with a low reheat temperature so as not to reproduce gravitinos after inflation. Such low abundances of gravitinos decaying into $\gamma + \tilde{\gamma}$, however, may be able to account for observed features in the γ-ray spectrum.[85]

In this last section, we have seen some of the recent results and activities taking place between particle physics and cosmology and those that will probably remain the most active area in the interface in the near future. Planck era cosmology must still rely heavily on breakthroughs in particle physics and quantum gravity. Inflation seems to have reached a plateau. Inflation is possible, but its origin is unclear. Perhaps some new idea from superstring theory will replace it. In the last sections of this paper, it was seen that big bang nucleosynthesis and baryosynthesis have laid the cornerstone of modern cosmology. Therefore, this leaves us with the dark matter problems, their solutions, and the tests of the existence of dark matter. The detection[79,80,86] of dark matter offers cosmologists as well as particle physicists with a promising challenge.

REFERENCES

1. PENZIAS, A. A. & R. W. WILSON. 1965. Astrophys. J. **142**: 419.
2. GAMOW, G. 1946. Phys. Rev. **70**: 572; ALPHER, R. A., H. BETHE & G. GAMOW. 1948. Phys. Rev. **73**: 803.
3. HAWKING, S. W. 1978. Nucl. Phys. **B144**: 349.
4. For reviews, see articles by DEWITT, B. S. & S. W. HAWKING. 1979. *In* General Relativity: An Einstein Centenary Survey. S. W. Hawking & W. Israel, Eds. Cambridge University Press. London/New York.
5. FISCHETTI, M. V., J. B. HARTLE & B-L. HU. 1979. Phys. Rev. **D20**: 1757; HARTLE, J. B. & B-L. HU. 1979. Phys. Rev. **D20**: 1772; 1980. Phys. Rev. **D21**: 2756; HARTLE, J. B. 1980. Phys. Rev. **D22**: 2091; 1981. Phys. Rev. **D23**: 2121.
6. HARTLE, J. B. & S. W. HAWKING. 1983. Phys. Rev. **D28**: 2960.
7. KALUZA, T. 1921. Sitzungsber. Preuss. Akad. Wiss. Phys. Math. **K1**: 966; KLEIN, O. 1926. Z. Phys. **37**: 895.
8. WITTEN, E. 1981. Nucl. Phys. **B186**: 412; SALAM, A. & J. STRATHDEE. 1982. Ann. Phys. **141**: 316.
9. WITTEN, E. 1984. *In* Proc. June 1983 Shelter Island II Conference. N. Khuri *et al.*, Eds. MIT Press. Cambridge, Massachusetts.
10. KOLB, E. 1985. Fermilab preprint 85-17.
11. GOL'FAND, Y. A. & E. P. LIKHTMAN. 1971. Pis'ma Zh. Eksp. Teor. Fiz. **13**: 323; VOLKOV, D. & V. P. AKULOV. 1973. Phys. Lett. **46B**: 109; WESS, J. & B. ZUMINO. 1974. Nucl. Phys. **B70**: 39; for a review, see: FAYET, P. & S. FERRARA. 1977. Phys. Rep. **32C**: 249.
12. MAIANI, L. 1979. *In* Proc. of the Summer School of Gif-sur-Yvette, p. 3; WITTEN, E. 1981. Nucl. Phys. **B188**: 513; DIMOPOULOS, S. & H. GEORGI. 1981. Nucl. Phys. **B193**: 150; SAKAI, N. 1982. Z. Phys. **C11**: 153.
13. WESS, J. & B. ZUMINO. 1974. Phys. Lett. **49B**: 52; ILIOPOULOS, J. & B. ZUMINO. 1974. Nucl. Phys. **B76**: 310; FERRARA, S., J. ILIOPOULOS & B. ZUMINO. 1974. Nucl. Phys. **B77**: 413; GRISARU, M. T., W. SIEGEL & M. ROCEK. 1979. Nucl. Phys. **B159**: 420.
14. ELLIS, J. & D. V. NANOPOULOS. 1982. Phys. Lett. **116B**: 133.
15. See reviews by: ARNOWITT, R., A. H. CHAMSEDDINE & P. NATH. 1983. Northeastern University preprints 2597, 2600, and 2613; ELLIS, J. 1983. CERN preprint TH-3718; NANOPOULOS, D. V. 1983. CERN preprint TH-3699; NILLES, H-P. 1984. Phys. Rep. **110C**: 1; POLCHINSKI, J. 1983. Harvard University preprint HUTP-83/A036.
16. CREMMER, E. & J. JULIA. 1978. Phys. Lett. **80B**: 48; 1979. Nucl. Phys. **B159**: 141; DEWIT, B. & H. NICOLAI. 1981. Phys. Lett. **108B**: 285; 1982. Nucl. Phys. **B208**: 323.
17. ELLIS, J., M. K. GAILLARD & B. ZUMINO. 1980. Phys. Lett. **94B**: 343; ELLIS, J., M. K. GAILLARD, L. MAIANI & B. ZUMINO. 1980. *In* Unification of the Fundamental Particle Interactions. S. Ferrara, J. Ellis & P. Van Nieuwenhuizen, Eds.: 69. Plenum Press. New York.
18. SCHWARZ, J. H. 1982. Phys. Rep. **89**: 223; GREEN, M. B. & J. H. SCHWARZ. 1984. Caltech preprints CALT-68-1182 and CALT-68-1194.
19. LINDE, A. D. 1984. Rep. Prog. Phys. **47**: 925; BRANDENBERGER, R. 1985. Rev. Mod. Phys. **57**: 1.

20. GUTH, A. H. 1981. Phys. Rev. **D23**: 347.
21. USON, J. & D. WILKENSON. 1985. *In* Proc. of the Inner Space/Outer Space Conference. E. Kolb, M. S. Turner, D. Lindley, K. A. Olive & D. Seckel, Eds. University of Chicago Press. Chicago.
22. COLEMAN, S. 1977. Phys. Rev. **D15**: 2929; CALLAN, C. & S. COLEMAN. 1977. Phys. Rev. **D16**: 1762.
23. GUTH, A. H. & E. WEINBERG. 1981. Phys. Rev. **D23**: 826; 1983. Nucl. Phys. **B212**: 321.
24. LINDE, A. D. 1982. Phys. Lett. **108B**: 389; ALBRECHT, A. & P. J. STEINHARDT. 1982. Phys. Rev. Lett. **48**: 1220.
25. COLEMAN, S. & S. WEINBERG. 1973. Phys. Rev. **D7**: 1888.
26. LINDE, A. D. 1982. Phys. Lett. **116B**: 335.
27. ELLIS, J., D. V. NANOPOULOS, K. A. OLIVE & K. TAMVAKIS. 1982. Phys. Lett. **118B**: 335.
28. HAWKING, S. W. 1982. Phys. Lett. **115B**: 295; GUTH, A. H. & S-Y. PI. 1982. Phys. Rev. Lett. **49**: 1110; SAROBINSKI, A. A. 1982. Phys. Lett. **117B**: 175; BARDEEN, J. M., P. J. STEINHARDT & M. S. TURNER. 1983. Phys. Rev. **D28**: 679.
29. BREIT, J., S. GUPTA & A. ZAKS. 1983. Phys. Rev. Lett. **51**: 1007.
30. HARRISON, E. R. 1970. Phys. Rev. **D1**: 2726; ZEL'DOVICH, YA. B. 1972. Mon. Not. R. Astron. Soc. **160**: 1P.
31. PRESS, W. H. 1980. Phys. Scr. **21**: 702.
32. ELLIS, J., D. V. NANOPOULOS, K. A. OLIVE & K. TAMVAKIS. 1983. Nucl. Phys. **B221**: 224.
33. NANOPOULOS, D. V., K. A. OLIVE, M. SREDNICKI & K. TAMVAKIS. 1983. Phys. Lett. **123B**: 41.
34. OVRUT, B. & P. J. STEINHARDT. 1984. Phys. Rev. Lett. **53**: 732; BINETRUY, P. & S. MAHAJAN. 1985. LBL preprint.
35. GELMINI, G., D. V. NANOPOULOS & K. A. OLIVE. 1983. Phys. Lett. **131B**: 53.
36. OVRUT, B. A. & P. J. STEINHARDT. 1983. Phys. Lett. **133B**: 161.
37. JENSEN, L. & K. A. OLIVE. 1985. Fermilab preprint 85-53.
38. ELLIS, J., E. ENQVIST, D. V. NANOPOULOS, K. A. OLIVE & M. SREDNICKI. 1985. Phys. Lett. **152B**: 175.
39. CREMMER, E., S. FERRARA, C. KOUNNAS & D. V. NANOPOULOS. 1983. Phys. Lett. **133B**: 287; ELLIS, J., A. B. LAHANAS, D. V. NANOPOULOS & K. TAMVAKIS. 1984. Phys. Lett. **134B**: 29; ELLIS, J., C. KOUNNAS & D. V. NANOPOULOS. 1984. Nucl. Phys. **B241**: 406 and **B247**: 373.
40. WITTEN, E. February 1985. Princeton University preprint.
41. ALBRECHT, A. & R. BRANDENBERGER. 1985. Phys. Rev **D31**: 1225; COUGHLAN, G. D. & G. G. ROSS. 1985. Phys. Lett. **157B**: 151; JENSEN, L. & K. A. OLIVE. 1985. Phys. Lett. **159B**: 99.
42. MAZENKO, G., W. UNRUH & R. WALD. 1985. Phys. Rev. **D31**: 273.
43. SAKHAROV, A. D. 1967. Zh. Eksp. Teor. Fiz. Pis'ma. Red. **5**: 32.
44. For a review, see: KOLB, E. & M. S. TURNER. 1983. Annu. Rev. Nucl. Part. Sci. **33**: 645.
45. STEIGMAN, G. 1976. Annu. Rev. Astron. Astrophys. **14**: 339.
46. WEINBERG, S. 1979. Phys. Rev. Lett. **45**: 850; TOUSSAINT, D., S. B. TREIMAN, F. WILCZEK & A. ZEE. 1979. Phys. Rev. **D19**: 1036.
47. KOLB, E. W. & S. WOLFRAM. 1978. Phys. Lett. **B91**: 217; KOLB, E. W. & S. WOLFRAM. 1980. Nucl. Phys. **B172**: 224; FRY, J. N., K. A. OLIVE & M. S. TURNER. 1980. Phys. Rev. **D22**: 2953, 2977.
48. NANOPOULOS, J. V. & K. TAMVAKIS. 1982. Phys. Lett. **114B**: 235.
49. MASIERO, A., D. V. NANOPOULOS, K. TAMVAKIS & T. YANAGIDA. 1983. Z. Phys. **C17**: 33.
50. WAGONER, R. V., W. A. FOWLER & F. HOYLE. 1967. Astrophys. J. **148**: 3; WAGONER, R. V. 1969. Astrophys. J. Suppl. Ser. **18**: 247; WAGONER, R. V. 1972. Astrophys. J. **179**: 343; SCHRAMM, D. N. & R. V. WAGONER. 1977. Annu. Rev. Nucl. Part. Sci. **27**: 37.
51. STEIGMAN, G., D. N. SCHRAMM & J. E. GUNN. 1977. Phys. Lett. **B66**: 202; YANG, J., D. N. SCHRAMM, G. STEIGMAN & R. T. ROOD. 1979. Astrophys. J. **227**: 697.
52. OLIVE, K. A., D. N. SCHRAMM, G. STEIGMAN, M. S. TURNER & J. YANG. 1981. Astrophys. J. **246**: 547.
53. YANG, J., M. S. TURNER, G. STEIGMAN, D. N. SCHRAMM & K. A. OLIVE. 1984. Astrophys. J. **281**: 493.

54. DAVID, Y. & H. REEVES. 1980. *In* Physical Cosmology. R. Balian, J. Audouze & D. N. Schramm, Eds.: 443. North-Holland. Amsterdam.
55. For a recent compilation of observations, see: THE PROCEEDINGS OF THE ESO WORKSHOP ON PRIMORDIAL HELIUM. 1983. P. A. Shaver, D. Kunth & K. Kajär, Eds. Garching, Germany.
56. IBEN, I. 1967. Astrophys. J. **147**: 624; ROOD, R. T. 1972. Astrophys. J. **177**: 681; IBEN, I. & J. W. TRURAN. 1978, Astrophys. J. **220**: 980; DEARBORN, D. S. P., J. B. BLAKE, K. L. HAINEBACH & D. N. SCHRAMM. 1978. Astrophys. J. **223**: 552; BRUNISH, W. & J. W. TRURAN. 1983. In preparation.
57. BLACK, D. C. 1971. Nat. Phys. Sci. **234**: 148; BLACK, D. C. 1972. Geochim. Cosmochim. Acta **36**: 347; GEISS, J. & H. REEVES. 1972. Astron. Astrophys. **18**: 126.
58. TRAUGER, J. T. *et al.* 1973. Astrophys. J. Lett. **184**: L137; KUNDE, V. *et al.* 1982. Astrophys. J. **263**: 443.
59. YORK, D. G. & J. B. ROGERSON, JR. 1976. Astrophys. J. **204**: 378; VIDAL-MADJAR, A. *et al.* 1977. Astrophys. J. **211**: 91; VIDAL-MADJAR, A. *et al.* 1983. Astron. Astrophys. **120**: 58.
60. ANDERS, E., D. HEYMANN & E. MAZOR. 1970. Geochim. Cosmochim. Acta **34**: 127.
61. SPITE, F. & M. SPITE. 1982. Astron. Astrophys. **115**: 357; SPITE, M. & F. SPITE. 1982. Nature **297**: 483.
62. DAVIS, M. & P. J. E. PEEBLES. 1983. Annu. Rev. Astron. Astrophys. **21**: 109.
63. FABER, S. M. & J. J. GALLAGHER. 1979. Annu. Rev. Astron. Astrophys. **17**: 135.
64. HEGYI, D. & K. A. OLIVE. 1983. Phys. Lett. **126B**: 28; 1985. Fermilab preprint 85/86.
65. AARONSON, M. 1983. Astrophys. J. **266**: L11; FABER, S. M. & D. N. C. LIN. 1983. Astrophys. J. **266**: L17.
66. BAHCALL, J. 1984. Astrophys. J. **276**: 169.
67. BOND, J. R. & A. S. SZALAY. 1983. Astrophys. J. **274**: 443.
68. BOND, J. R., G. EFSTATHIOU & J. SILK. 1980. Phys. Rev. Lett. **45**: 1980; ZELDOVICH, YA. B. & R. A. SUNYAEV. 1980. Pis'ma Astron. Zh. **6**: 451.
69. MELOTT, A. 1983. Mon. Not. R. Astron. Soc. **202**: 595; BOND, J. R., A. S. SZALAY & S. D. M. WHITE. 1983. Nature **301**: 584.
70. OLIVE, K. A. & M. S. TURNER. 1982. Phys. Rev. **D25**: 213.
71. BOND, J. R., A. S. SZALAY & M. S. TURNER. 1982. Phys. Rev. Lett. **48**: 1636.
72. PEBBLES, P. J. E. 1982. Astrophys. J. **263**: L1; 1984. Astrophys. J. **277**: 440; BLUMENTHAL, G., S. M. FABER, J. R. PRIMACK & M. REES. 1984. Nature **311**: 517.
73. WHITE, S. D. M. 1986. Ann. N.Y. Acad. Sci. **470**: 243–246.
74. IBANEZ, L. F. 1984. Phys. Lett. **137B**: 160; HAGELIN, J., G. L. KANE & S. RABY. 1984. Nucl. Phys. **B241**: 638.
75. PAGELS, H. R. & J. R. PRIMACK. 1982. Phys. Rev. Lett. **48**: 223.
76. ELLIS, J., A. D. LINDE & D. V. NANOPOULOS. 1982. Phys. Lett. **118B**: 39.
77. WEINBERG, S. 1983. Phys. Rev. Lett. **50**: 387.
78. GOLDBERG, H. 1983. Phys. Rev. Lett. **50**: 1419; KRAUSS, L. 1983. Nucl. Phys. **B227**: 556; ELLIS, J., J. HAGELIN, D. V. NANOPOULOS, K. A. OLIVE & M. SREDNICKI. 1984. Nucl. Phys. **B238**: 453.
79. SILK, J. & M. SREDNICKI. 1984. Phys. Rev. Lett. **53**: 624.
80. SILK, J., K. A. OLIVE & M. SREDNICKI. 1985. Phys. Rev. Lett. **55**: 257.
81. WEINBERG. S. 1982. Phys. Rev. Lett. **48**: 1303.
82. OLIVE, K. A., D. SECKEL & E. VISHNIAC. 1985. Astrophys. J. **292**: 1; OLIVE, K. A., D. N. SCHRAMM & M. SREDNICKI. 1985. Nucl. Phys. **B255**: 495.
83. DAVIS, M., M. LECAR, C. PRYOR & E. WITTEN. 1981. Astrophys. J. **250**: 423; HUT, P. & S. D. M. WHITE. 1984. Nature **301**: 637; TURNER, M. S., G. STEIGMAN & L. KRAUSS. 1984. Phys. Rev. Lett. **52**: 2090; GELMINI, G., D. N. SCHRAMM & J. W. E. VALLE. 1984. Phys. Lett. **146B**: 311.
84. NANOPOULOS, D. V., K. A. OLIVE & M. SREDNICKI. 1983. Phys. Lett. **127B**: 30; KHLOPOV, M. YU. & A. D. LINDE. 1984. Phys. Lett. **138B**: 265; ELLIS, J., J. KIM & D. V. NANOPOULOS. 1984. Phys. Lett. **145B**: 181; ELLIS, J., D. V. NANOPOULOS & S. SARKER. 1985. Nucl. Phys. **B259**: 175; JUSZKIEWICZ, R., J. SILK & A. STEBBINS. 1985. Phys. Lett. B. Submitted.
85. OLIVE, K. A. & J. SILK. 1985. Phys. Rev. Lett. **55**: 2362.
86. DRUKIER, A. & L. STODOLSKY. 1984. Phys. Rev. **D30**: 2295; GOODMAN, M. & E. WITTEN. 1984. Princeton University preprint.

Cosmic Strings[a]

ALEXANDER VILENKIN[b]

Lyman Laboratory of Physics
Harvard University
Cambridge, Massachusetts 02138

INTRODUCTION

Just like phase transitions in condensed matter systems, cosmological phase transitions can give rise to various kinds of defects.[1,2] Depending on the symmetries of the states before and after the phase transition, these defects can be in the form of domain walls, strings, or monopoles (or various combinations of the above). In this talk, I shall concentrate on strings, which appear to be the most interesting type of defects for cosmology. Since this is a meeting on astrophysics, I shall emphasize the astrophysical and observational aspects of strings (at the expense of field theory aspects). For a detailed discussion of strings and other defects, see, for example, reference 3.

PROPERTIES OF STRINGS

Strings arise in gauge theories with spontaneously broken symmetries. Such theories have a vacuum state of higher symmetry that is unstable and is, therefore, called false vacuum, and a stable vacuum of lower symmetry that is called true vacuum. If the symmetry groups satisfy certain topological conditions, the field equations of the theory will have stable stringlike solutions with true vacuum everywhere except in a thin tube. Thus, we can say that a string is a thin tube of false vacuum. The strings do not have ends; they are either infinite or closed. It can be shown also that there is a unit flux of a massive gauge field running along each string. (I am describing here only the most common type of strings. For more details, see reference 3.) Unlike monopoles, strings are not mandatory for grand unified theories. However, they are not at all uncommon and a number of "realistic" models with strings have been suggested.[4-6]

A close condensed matter analogue of strings is a quantized tube of magnetic flux in superconductors. The metal is superconducting outside the tube and is normal inside. Normal and superconducting states correspond to false and true vacua, respectively.

Some properties of strings can be derived just from dimensionality and Lorentz invariance arguments. Strings are typically characterized by one dimensional parame-

---

[a]This research was supported in part by the National Science Foundation under grants nos. PHY83-51860, PHY82-06202, and PHY82-15249, and by contributions from General Electric Company, Dennison Manufacturing Company, and Massachusetts Electric Company.
[b]Permanent address: Physics Department, Tufts University, Medford, Massachusetts 02155.

ter, η, which is the energy scale of the symmetry breaking. (For the grand unification phase transition, η is $\sim 10^{16}$ GeV.) Then, from dimensionality, the thickness of the string is $\delta \sim \eta^{-1}$. The false vacuum energy density is $\rho_v \sim \eta^4$ and the mass per unit length of string is

$$\mu \sim \rho_v \delta^2 \sim \eta^2 . \tag{1}$$

(I use the system of units in which $\hbar = c = 1$ everywhere, except in the section entitled SUPERCONDUCTING STRINGS.) With $\eta \sim 10^{16}$ GeV, we get $\delta \sim 10^{-30}$ cm and $\mu \sim 10^{22}$ g/cm.

Consider a static, straight string lying along the z-axis. It is described by a solution of Lorentz invariant field equations that is independent of z and t and should therefore be invariant under Lorentz boosts along the z-axis. Hence, it is meaningless to talk about the motion of the string along its length; only the transverse motion has physical meaning. The energy-momentum tensor of the string should have the same invariance. Neglecting the width of the string, we can write

$$T^\nu_\mu = \mu \, \delta(x) \, \delta(y) \, \text{diag}(1, 0, 0, 1) . \tag{2}$$

We see that the tension (or negative pressure) along the string is equal to the mass density.

Under the action of tension, curved strings develop an acceleration inversely proportional to the local curvature radius. Closed loops oscillate and it can be shown that their motion is periodic with a period equal to half of the loop's length.[7] More exactly, $T = M/2\mu$, where M is the mass of the loop.

FORMATION AND EVOLUTION OF STRINGS

Strings are formed at a phase transition because the directions of symmetry breaking are different in different regions of space. When these regions "try" to match together, they sometimes run into topological problems; a string is a topological defect resulting from just that. One can introduce a correlation length, ξ, such that the directions of symmetry breaking are uncorrelated over distances greater than ξ. Causality requires that ξ cannot exceed the causal horizon: $\xi \lesssim t$. The actual value of ξ, then, depends on the dynamics of the phase transition.[1,2]

The typical scale of the system of strings at formation is $\sim \xi$. There is about one string segment of length, $\sim \xi$, per volume, $\sim \xi^3$, and the mass density of the system is $\rho_s \sim \mu \xi^{-2}$. Strings have the shape of Brownian trajectories of step $\sim \xi$, so the length along the string between two points separated by a distance, R, is $\ell \sim R^2/\xi$. Numerical simulations of the phase transition show[8] that infinite strings contribute about 80% of the total string density. The rest is in the form of closed loops with a scale-invariant distribution,

$$n_R \sim R^{-3} . \tag{3}$$

Here, R is the size of the loop, which can be defined as the diameter of the smallest sphere enclosing the loop, and n_R is the number of loops of size, $\sim R$, per unit volume. In a closed universe, all strings are closed and the role of infinite strings is played by the loops of size comparable to the size of the universe.

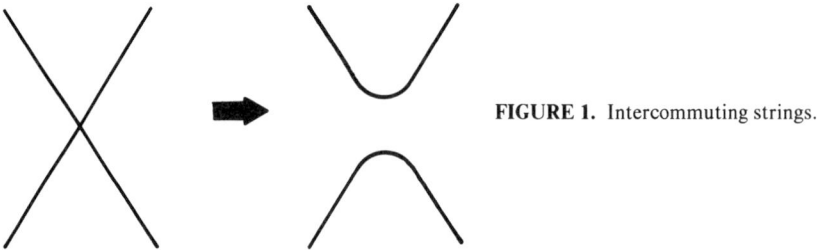

FIGURE 1. Intercommuting strings.

We now turn to the cosmological evolution of strings.[1-3,9-13] Expansion of the universe straightens out long strings on scales smaller than the horizon (so that the persistence length of Brownian strings at time, t, is $\sim t$) and has very little effect on small loops of size, $R \ll t$. Under the action of tension, strings move with relativistic speeds and frequently intersect. At the point of intersection, the strings can intercommute (or "change partners;" see FIGURE 1). This process is very important for the energy dissipation in the system of strings since it results in closed loop formation and the loops eventually radiate away their energy. Shellard[14] has solved numerically the field equations for two strings colliding with various velocities. His results indicate that the intercommuting probability for intersecting strings is close to one.

The loop formation by intercommuting strings is illustrated in FIGURE 2. The typical curvature radius of strings at time, t, is $\sim t$ and the size of the loops formed by their intercommuting is also $\sim t$. The number of loops formed per unit volume per Hubble time is $n \sim t^{-3}$.

The dominant energy loss mechanism for the loops is the gravitational radiation. A loop of size, R, oscillates with a typical frequency of $\omega \sim R^{-1}$ and the quadrupole radiation formula gives

$$\dot{M} \sim -\gamma \, GM^2 \, R^4 \, \omega^6 \sim -\gamma \, G\mu^2 \,, \qquad (4)$$

where $M \sim \mu R$ is the mass of the loop and γ is a numerical coefficient. The value found from computer calculations is[15] $\gamma \sim 100$. The lifetime of the loop is

$$\tau \sim M/|\dot{M}| \sim R/\gamma G\mu \,. \qquad (5)$$

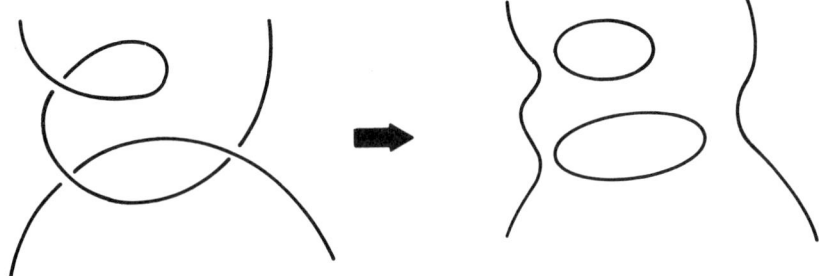

FIGURE 2. Closed loops are formed by pairs of strings intercommuting at two points and by self-intercommuting of individual strings.

$G\mu$ is an important dimensionless parameter characterizing the gravitational interactions of strings. From equation (1),

$$G\mu = (\eta/m_p)^2, \tag{6}$$

where $m_p \sim 10^{19}$ GeV is the Planck mass. For $\eta \sim 10^{16}$ GeV, this gives $G\mu \sim 10^{-6}$.

We can now describe the system of strings at an arbitrary cosmic time, t. Consider a horizon volume of size, $\sim t$. It will typically contain one or a few long segments of string stretching across the volume and a large number of closed loops with sizes from t down to $\gamma G\mu t$. (Smaller loops have already radiated away their energy.) To find the size distribution of loops, we note that loops of size, R, are formed at $t \sim R$ with a number density of $n_R \sim R^{-3}$. At later times, the density is diluted as

$$n_R(t) \sim [a(R)/a(t)]^3 R^{-3}. \tag{7}$$

In a radiation-dominated universe, this gives

$$n_R(t) \sim (tR)^{-3/2}. \tag{8}$$

The dominant contribution to the mass density of strings is given by the smallest loops with $R \sim \gamma G\mu t$:

$$\rho_s/\rho_\gamma \sim (G\mu)^{1/2} \ll 1, \tag{9}$$

where

$$\rho_\gamma \sim (30Gt^2)^{-1} \tag{10}$$

is the radiation density.

Albrecht and Turok[13] have recently confirmed this picture of string evolution by a direct numerical simulation. They used the method of Vachaspati and Vilenkin[8] to simulate the phase transition and evolved the resulting system of strings by solving numerically the dynamical equations of motion. The intercommuting probability for intersecting strings was assumed equal to one. A snapshot of the system of strings is shown in FIGURE 3.

GALAXY FORMATION

Oscillating loops of string can serve as seeds for galaxies and clusters of galaxies.[11] On scales greater than the horizon, the density fluctuations due to strings are balanced by the corresponding variations in matter and radiation density. The density fluctuation on each scale is produced at the time of horizon crossing. One loop of size, $\sim t$, formed per horizon volume ($\sim t^3$), introduces a density fluctuation, $\delta\rho \sim \mu t/t^3 \sim \mu t^{-2}$. Using equation (10), we obtain the magnitude of density fluctuations at the horizon crossing:[16]

$$(\delta\rho/\rho)_{hor} \sim 30\, G\mu. \tag{11}$$

Reasonable galaxy formation scenarios are obtained for $(\delta\rho/\rho)_{hor} \sim 10^{-4}$–$10^{-5}$. This corresponds to $G\mu \sim 10^{-6}$ and $\eta \sim 10^{16}$ GeV. It is encouraging that the required value of $G\mu$ falls in the grand unification range.

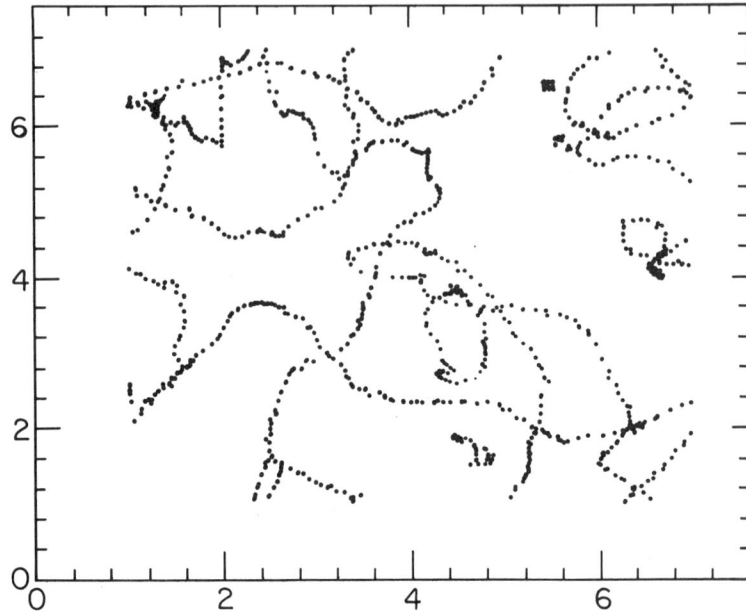

FIGURE 3. A snapshot of the system of strings (from Albrecht and Turok[13]). The number of closed loops in this picture is much smaller than it would be in actual evolution. The reason is that the simulation covered only a small interval of cosmic time (the scale factor changed by a factor of ten).

The following evolution of the density fluctuations in this scenario is similar to that of isothermal fluctuations. During the radiation era, $\delta\rho/\rho$ grows like $t^{1/2}$ because the radiation density decreases faster than that of the loops. This growth terminates when either the loops on the corresponding scale decay or when the universe becomes matter-dominated, whichever comes first. In the matter-dominated era, the density fluctuations grow like $t^{2/3}$ due to the gravitational instability.

Loops of size, R, introduce a density fluctuation on a co-moving scale, $\ell \sim n_R^{-1/3}$. Using equation (7) for the density of loops, one can find the spectrum of density fluctuations with various assumptions about the dark matter.[3,6] The shapes of resulting spectra for baryon-, neutrino-, and axion-dominated scenarios are shown in FIGURE 4. Now I would like to mention some interesting new features introduced by the string scenario of galaxy formation:

(1) In the standard model of galaxy formation with a primordial spectrum of adiabatic fluctuations, all density perturbations are damped below the Silk mass ($M \sim 10^{14} M_\odot$) or the neutrino free-streaming mass ($M \sim 4 \times 10^{14} M_\odot$) in the baryon- and neutrino-dominated cases, respectively. The same is true if the density fluctuations are due to inflation. In the string scenario, the fluctuations are preserved on smaller scales. Baryons and neutrinos erase their own fluctuations, but when the Jeans mass drops down to the corresponding scale,

they pick up the perturbations produced by surviving loops.[6] It is interesting that the galactic mass of $\sim 10^{12}\, M_\odot$ naturally arises as a lower cutoff of the spectra.[3]

(2) Closed loops have sizes much smaller than those of the galaxies condensing around them. A loop representing a small density fluctuation on the galactic scale produces a large density contrast in its immediate vicinity. This results in accretion of matter onto the loops and formation of massive compact objects, which can be identified with quasars and active galactic nuclei.[17,18] Early formation of quasars can reionize the universe, smoothing out small-scale temperature fluctuations, and thus resolve one of the difficulties of the baryon-dominated scenario. It can also explain the existence of quasars at $z \sim 1$ in the neutrino-dominated model, which requires the pancake collapse at $z < 0.5$. (See Simon White's talk in these proceedings.[19])

(3) Wakes formed behind long, rapidly moving strings can help to explain the observed large-scale structure in the axion-dominated case.[17] On scales $\gtrsim 10^{14}\, M_\odot$, the density fluctuations due to loops and due to wakes are of comparable magnitude.

(4) Density fluctuations produced by strings are not in the form of waves with random phases; this can explain the observed deviations from the Gaussian

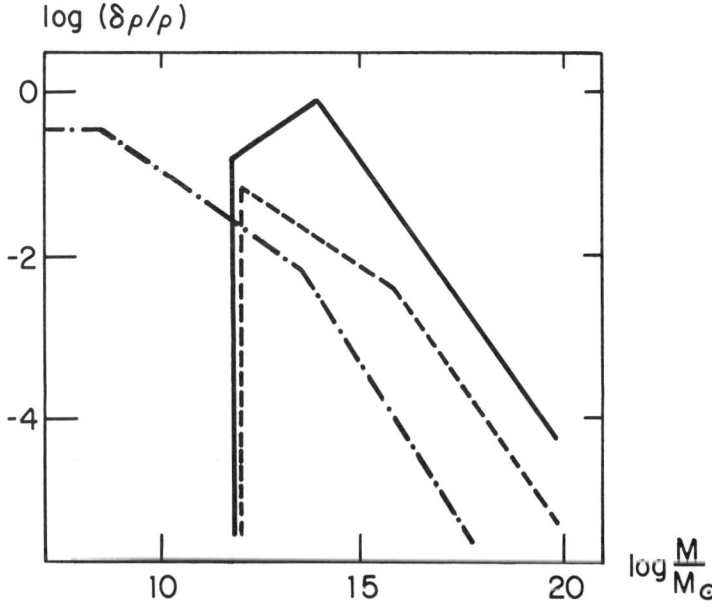

FIGURE 4. The spectrum of density fluctuations in the (solid curve) neutrino-, (dashed curve) baryon-, and (dashed-dotted curve) axion-dominated universe. The scale of $\delta\rho/\rho$ is arbitrary. All three spectra have $\delta\rho/\rho \propto M^{-2/3}$ on the largest scales. On intermediate scales, $\delta\rho/\rho$ scales like $M^{-1/3}, M^{1/3}$, and $M^{-1/3}$ in the baryon-, neutrino-, and axion-dominated cases, respectively. In the axion-dominated case, there is also a flat part of the spectrum on small mass scales ($M < 10^9\, M_\odot$).

behavior. For example, rare supergiant loops can produce localized regions of density contrast much greater than one would expect from Gaussian fluctuations. Occasional splitting of loops can be responsible for the observed cluster–cluster correlations.[20]

It should be noted that the string scenario of galaxy formation assumes that the universe is initially homogeneous and isotropic on scales much greater than the horizon. Such initial conditions can be explained if we assume that there was a period of inflation before the string formation. An example of a grand unified model that gives both a satisfactory inflationary scenario and strings of required energy scale has been given in reference 21.

GRAVITATIONAL FIELD OF STRINGS

The gravitational field of long strings is rather peculiar.[22–24] The solution of Einstein's equations for a straight string with the energy-momentum tensor [see

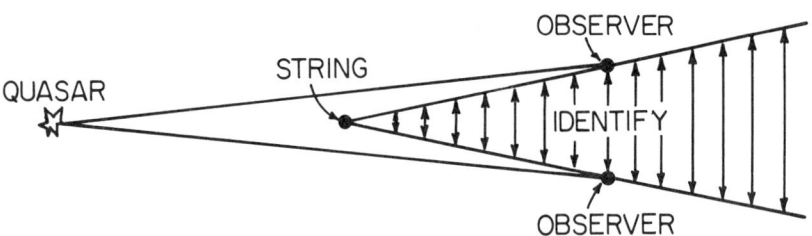

FIGURE 5. The conical space around a straight string can be obtained from a Euclidean space by cutting out a wedge of an angular size of $8\pi G\mu$ and identifying the exposed surfaces. Light rays emitted by a quasar intersect behind the string and the observer sees two images of the same quasar. (The string is perpendicular to the page.)

equation (2)] is

$$ds^2 = dt^2 - dz^2 - dr^2 - (1 - 4G\mu)^2 r^2 d\phi^2. \qquad (12)$$

A coordinate transformation of $\phi' = (1 - 4G\mu)\phi$ brings the metric to a locally Minkowskian form, but now the angle, ϕ', changes from 0 to $(1 - 4G\mu)2\pi$. This is a "conical space," that is, a flat space with a wedge of angular size of $8\pi G\mu$ taken out and the two faces of the wedge identified. In the coordinates (t, z, r, ϕ'), the geodesics are just straight lines and we see immediately that a particle initially at rest relative to the string will remain at rest and will not experience any gravitational attraction.

Although the metric in equation (12) is locally flat, its global structure is different from that of Minkowsky space. The most striking effect of this difference is the formation of double images of objects located behind the string. This is illustrated in FIGURE 5. The light rays from a quasar intersect behind the string and the observer sees two images of the quasar. The angular separation between the images is

$$\delta\phi = 8\pi\, G\mu\ell(d + \ell)^{-1} \sin\theta , \qquad (13)$$

where ℓ and d are the distances from the string to the quasar and to the observer, respectively, and θ is the angle between the string and the line of sight.

When a straight string moves through a medium, it produces a wake of an angular width of $8\pi G\mu$. In the frame of the string, the trajectories of particles on the two sides of the string are deflected by an angle of $8\pi G\mu$ relative to one another. This deflection is purely kinematic in nature and does not result in any frictional force on a string.

To avoid confusion, I would like to emphasize that equation (**12**) applies only to relatively straight segments of string at distances small compared to their curvature radius. A closed loop of size, R, produces a regular Schwarzschild field at distances $\gg R$.

OBSERVATIONAL EFFECTS OF STRINGS

The string scenario of galaxy formation predicts a number of distinctive observational effects. Some of these effects will soon be within the experimental capabilities and will allow one to rule out or to confirm the string scenario.

The loops that served as seeds for galaxies were formed at $t \sim 10^9$ s when the co-moving scale of a galaxy crossed the horizon. They had sizes of $R \sim 20$ pc and masses of $M \sim 10^9 M_\odot$. Unfortunately, these loops cannot be seen at the galactic centers; they all decayed at $t \sim 10^{13}$ s. The smallest presently surviving loops have sizes of ~ 100 kpc and are separated by distances of ~ 100 Mpc.

Furthermore, the gravitational waves emitted by oscillating loops add up to a stochastic gravitational wave background with a scale-invariant spectrum:[15,25,26]

$$\Omega_g(\omega) \sim 10^{-4}(G\mu)^{1/2} h^{-2}. \qquad (14)$$

Here, $\Omega_g(\omega) = (\omega/\rho_c)(d\rho_g/d\omega)$, ρ_g is the energy density of gravitational waves, and $\rho_c = 2 \times 10^{-29}\, h^2$ g cm^{-3} in the critical density. Equation (**14**) applies in a wide range of frequencies:

$$10^{-2}\, \text{yrs}^{-1} \lesssim \omega \lesssim 10^5\, \text{s}^{-1}. \qquad (15)$$

The present upper bound on the density of gravitational waves from the observations of the millisecond pulsar is $\Omega_g \lesssim 10^{-5}$ for waves with periods of ~ 1 year. The accuracy grows rapidly with the time of observation and $\Omega_g \sim 10^{-7}$ will probably become detectable within several years.[26,27]

Another observational prediction of the string scenario is the formation of double images of objects located behind the strings. With $G\mu \sim 10^{-6}$, the typical angular separation of the images is ~ 10 arc sec, which is comparable to that in the known double quasars. The expected number of double quasars due to strings is about one.[23,28,29] Galaxies are much more numerous than quasars and one can search for lines of double galaxies along open strings or large closed loops. A natural place to start the search is near the known double quasars.

Kaiser and Stebbins[30] have pointed out that strings should leave a characteristic signature on the microwave background; the background temperature should have steplike discontinuities on curves on the sky. This effect is similar to the wake formation behind a moving string: photons are blue-shifted in the wake. The resulting temperature fluctuation is $\delta T/T \sim 8\pi G\mu v$, where $v \sim 1$ is the velocity of the string. Present observational limits are consistent with $G\mu \lesssim 10^{-5}$.

You may be wondering what happens if a string passes through a person.[31] In the rest frame of the string, the person moves towards the string with velocity, $v \sim 1$. The head and feet of the person are deflected just like particles or light and they start moving towards one another with velocity, $8\pi G\mu v \sim 5$ km/s. This is unhealthy. However, don't panic; the probability for a string to pass through the earth is only $\sim 10^{-23}$ per year.

SUPERCONDUCTING STRINGS

As I said at the beginning, strings are thin tubes of false vacuum. The symmetries of the false vacuum are different from those of the true vacuum outside the string. In particular, electromagnetic gauge invariance can be spontaneously broken inside the string. Solid-state physicists know very well what a spontaneously broken gauge invariance means: it means superconductivity. Hence, strings can behave as superconducting wires and Witten[32] has demonstrated that this is the case in a wide class of grand unified models.

If an electric field, E, is applied along a superconducting string, the current builds up at the rate of

$$di/dt = \beta(ce^2/\hbar)E,$$

where e is the electron charge and β is a model-dependent numerical coefficient. When the current reaches a critical value, i_{max}, its growth terminates and the string starts producing particles at the rate of

$$d^2N/dt\, d\ell \sim (e/\hbar)E.$$

The magnitude of i_{max} is typically

$$i_{max} \sim e\eta/\hbar,$$

where η is the energy scale of symmetry breaking, but it can be much smaller in some models.[32]

Before Witten's paper, it appeared that strings could manifest themselves only through their gravitational interactions. However, superconducting strings can have very interesting interactions with cosmic magnetic fields. These interactions and their observational consequences are not well understood, but I shall mention some preliminary thoughts (some of them due to Witten[32]). Consider a string moving through a magnetized plasma with a velocity, $\vec{v}$. In the frame of the string, there is an electric field, $\vec{E} = c^{-1}\vec{v} \times \vec{B}$, and the current builds up at the rate of $di/dt \sim (e^2/\hbar)v B$. This current creates its own magnetic field, $B_s \sim i/cr$, which in turn acts on the plasma. It appears that the plasma will develop a shock at a distance, r_0, from the string where the magnetic pressure is equal to the dynamical pressure of the plasma: $B_s^2 \sim \rho v^2$, where ρ is the mass density of the plasma. This gives $r_0 \sim i/c\, v\, \rho^{1/2}$. The magnetic field of the string at this distance is $B_s \sim \rho^{1/2} v$. The radius of the shock grows at the rate of

$dr_0/dt \sim (e^2/\hbar c) B \rho^{-1/2}$. In the galactic disc, $\rho \sim 10^{-24}$ g cm^{-3} and $B \sim 10^{-6}$ gauss, and we obtain $B_s \sim 10^{-2}$ gauss (at the shock) and $dr_0/dt \sim 10^4$ cm/s.

The interaction of closed loops with cosmic plasmas is much more complicated. It is possible that periodic oscillations of the loops can give rise to a dynamo effect. When superconducting strings are formed, weak stochastic magnetic fields are generated at the same time and it is possible that the loop dynamos can amplify these initial fields to cosmologically interesting magnitudes. As I said, all these thoughts are very preliminary and much work will be needed to understand the complex astrophysical processes involved.

[**Note added in proof:** Astrophysics of superconducting strings has been recently discussed in reference 33, where it is shown that such strings can be observed as sources of synchrotron radiation.]

ACKNOWLEDGMENTS

It is a pleasure to thank the organizers of this meeting for their hospitality.

REFERENCES

1. KIBBLE, T. W. B. 1976. J. Phys. **A9:** 1387.
2. KIBBLE, T. W. B. 1980. Phys. Rep. **67:** 183.
3. VILENKIN, A. 1985. Phys. Rep. **121:** 263.
4. KIBBLE, T. W. B., G. LAZARIDES & Q. SHAFI. 1982. Phys. Rev. **D26:** 435.
5. OLIVE, D. & N. TUROK. 1982. Phys. Lett. **117B:** 193.
6. VILENKIN, A. & Q. SHAFI. 1983. Phys. Rev. Lett. **51:** 1716.
7. KIBBLE, T. W. B. & N. TUROK. 1982. Phys. Lett. **116B:** 141.
8. VACHASPATI, T. & A. VILENKIN. 1984. Phys. Rev. **D30:** 2036.
9. KIBBLE, T. W. B. 1985. Nucl. Phys. **B252:** 227.
10. VILENKIN, A. 1981. Phys. Rev. Lett. **46:** 1169, 1469(E).
11. VILENKIN, A. 1981. Phys. Rev. **D24:** 2082.
12. TUROK, N. 1984. Nucl. Phys. **B242:** 520.
13. ALBRECHT, A. & N. TUROK. 1984. Phys. Rev. Lett. **54:** 1868.
14. SHELLARD, P. 1984. Unpublished.
15. VACHASPATI, T. & A. VILENKIN. 1985. Phys. Rev. **D31:** 3052.
16. ZEL'DOVICH, Y. B. 1980. Mon. Not. R. Astron. Soc. **192:** 663.
17. SILK, J. & A. VILENKIN. 1984. Phys. Rev. **D29:** 1700.
18. HOGAN, C. J. 1984. Phys. Lett. **143B:** 87.
19. WHITE, S. 1986. Ann. N.Y. Acad. Sci. **470:** 243–246.
20. TUROK, N. & D. N. SCHRAMM. 1984. Nature **312:** 598.
21. SHAFI, Q. & A. VILENKIN. 1984. Phys. Rev. **D29:** 1870.
22. VILENKIN, A. 1981. Phys. Rev. **D23:** 852.
23. GOTT, J. R. 1985. Astrophys. J. **288:** 422.
24. LINET, B. 1984. Institute Henri Poincaré preprint.
25. VILENKIN, A. 1981. Phys. Lett. **107B:** 47.
26. HOGAN, C. J. & M. J. REES. 1984. Nature **311:** 109.
27. WITTEN, E. 1984. Phys. Rev. **D30:** 272.
28. VILENKIN, A. 1984. Astrophys. J. Lett. **282:** L51.
29. HOGAN, C. J. & R. NARAYAN. 1984. Mon. Not. R. Astron. Soc. **211:** 575.
30. KAISER, N. & A. STEBBINS. 1984. Nature **310:** 391.
31. PEEBLES, P. J. E. 1984. Unpublished.
32. WITTEN, E. 1985. Nucl. Phys. **B249:** 557.
33. CHUDNOVSKY, E. M., G. FIELD, D. SPERGEL & A. VILENKIN. 1985. Center for Astrophysics preprint.

The Cosmic Microwave Background at Its Twentieth Anniversary

R. B. PARTRIDGE

Department of Astronomy
Haverford College
Haverford, Pennsylvania 19041

I am here to describe a set of astronomical observations that have had a major impact on cosmology over the past two decades. Aside from a few speculative remarks about the X-ray and infrared backgrounds, I will restrict my attention to the so-called microwave background radiation, whose spectrum extends from tens of centimeters down to about 1 mm wavelength. As we all know, this radiation was discovered 20 years ago by Arno Penzias and Robert Wilson at the Bell Telephone Laboratories.[1] It was rapidly interpreted as residual heat from an initially hot big bang of the universe.[2] The radiation reaches us from very high redshifts, probably of the order of 1000–1500.

I should like to begin my discussion by expressing some disappointment, as an observer, in the remarkable blandness of or lack of peculiar features in this radiation. First, all current observations of the intensity of the radiation are consistent with a purely thermal or Planck spectrum with a current temperature of 2.7–2.8 K. In addition, despite 15 years of observational effort, there appears to be no *cosmological* anisotropy in the radiation. (I shall return later, though, to the question of one *astrophysical* source of anisotropy.) Nevertheless, even in the absence of surprises, this radiation has proven remarkably useful in determining the large-scale properties of the universe. I note, for instance, that many other contributions here mention one or another aspect of the cosmic microwave background. Four general properties seem particularly important in refining and constraining our theories about the properties of the universe: the thermal spectrum; the large entropy (or ratio of the number density of photons to that of baryons) in the universe; the large-scale isotropy of the radiation; and the small-scale isotropy or homogeneity of the radiation. To be accepted, any cosmological theory must agree with these four general observational results. Indeed for the purposes of this discussion, let me represent theory and theorists as a sort of expansionist power that is continually trying to extend its borders. This "expansionist power" is hemmed in by the four observational results I have referred to above (see FIGURE 1). As observers, it is our hope, of course, to restrict the allowable territory of theorists and hence to move closer to the appropriate model for the large-scale properties of the universe.

In the twenty years since its discovery, the microwave background has taken on a major role in the renaissance of cosmology that, in part, this series of symposia celebrates. Where do we stand twenty years after the discovery of the microwave background? Furthermore, what may we hope for in the next few years? Let me turn first to an analysis of the present observational situation and then proceed on to examine some of the consequences that follow from the observational results we now have in hand. Other consequences have been, and will be, pursued by other speakers at this symposium.

THE PRESENT OBSERVATIONAL SITUATION

The Spectrum

Three observational programs have recently been completed to provide us with accurate ($\sim 5\%$) measurements of the intensity of the radiation over a wavelength range of more than 100, from 1 mm to 12 cm.

The first of these programs I wish to discuss is a large-scale collaborative venture between the University of California at Berkeley, Haverford College, TE.S.R.E. in Bologna, LFCTR in Milano, and the University of Padua (Smoot et al., 1983;[3] Partridge et al., 1985;[4] Smoot et al., 1985[5]). We sought to remeasure the intensity of the microwave background at five wavelengths: 12, 6.3, 3.0, 0.91, and 0.33 cm. We took particular care to reduce known and suspected sources of systematic error. For

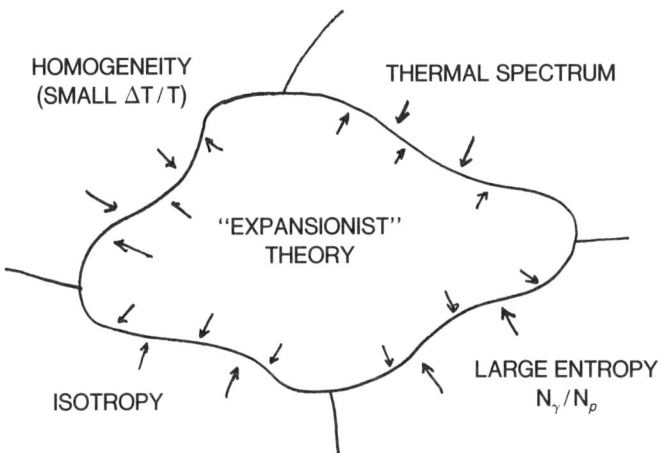

FIGURE 1. The (admittedly biased) metaphor of theory as an "expansionist power" hemmed in by observations.

instance, we used corrugated-horn antennas to reduce side- and back-lobe pickup from the ground and we used a carefully calibrated, helium-cooled cold load as a temperature reference for all the radiometers. In addition, we attempted to measure and subtract the two major sources of competing emission: emission from the galaxy by making multifrequency scans of the galactic plane[6,7] and emission from the Earth's atmosphere by making zenith scans throughout the measurements.[8] Emission from the Earth's atmosphere was also reduced by making the observations from a high, dry mountain site, the White Mountain Research Station of the University of California, at 3800 meters. The observations were made in the summers of 1982 and 1983 since the summer is the only time of the year the site is accessible. The combined results of two years' observations are displayed in TABLE 1. Since the temperature contributed by the Earth's atmosphere varied with water vapor content, especially at the two shortest wavelengths, approximate figures are given in the second column of the table. If we take the weighted average of all five measurements, we find $T = 2.73 \pm 0.05$ K for the

TABLE 1. Results[a] of Measurements of the Spectrum of the Cosmic Background Radiation (CBR)[b]

Wavelength (cm)	Typical Temp. of Atmosphere	Thermodynamic Temp. of CBR	Combined Results
12.0	0.95 ± 0.05	2.54 ± 0.24 2.87 ± 0.16	2.77 ± 0.13
6.3	1.00 ± 0.07	2.73 ± 0.22 2.70 ± 0.08	2.70 ± 0.08
3.0	1.20 ± 0.13	2.91 ± 0.17 2.64 ± 0.14	2.75 ± 0.08
0.91	~4.5	2.82 ± 0.21 2.81 ± 0.14	2.81 ± 0.12
0.33	~10	2.42 ± 0.00 2.57 ± 0.14	2.57 ± 0.14

[a]See Smoot et al., 1985.[5]
[b]In the third column, the top row is for our 1982 measurements and the bottom row is for the 1983 measurements. The fourth column gives the weighted averages of the two values for each wavelength.

temperature of the microwave background in the Rayleigh-Jeans portion of its spectrum.[5] The higher frequency measurements at λ = 3.0, 0.91, and 0.33 cm were repeated in the summer of 1984, with results agreeing with those of TABLE 1.[9]

The second new observational result is a more accurate measurement of the excitation temperature of interstellar cyanogen (CN) molecules.[10,11] The fact that excited rotational states of this molecule are populated in interstellar space was first noticed in the 1940s by McKellar.[12] These rotational states lie at energies corresponding to wavelengths of 2.64 and 1.32 mm above the ground state of cyanogen. If one assumes that they are kept populated by being bathed by the microwave background, one can then determine the temperature of the radiation at those two specific wavelengths. After applying some small corrections for effects due to the interstellar environment of these molecules, Meyer (1985)[11] finds at λ = 2.64 mm, T = 2.70 ± 0.04 K, and at λ = 1.32 mm, T = 2.76 ± 0.20 K. These results are very nicely consistent with the longer wavelength measurements. When they first appeared, however, they appeared to lie somewhat below other short wavelength measurements available in the literature, particularly those of Woody and Richards.[13,14] These latter measurements had been made with a bolometer flown to high altitudes by a balloon. Woody and Richards' data suggested a temperature about 0.2 K higher than the Meyer and Jura results[10] at about 2 mm wavelength. Richards and his collaborators[15] have now repeated their bolometric observations. Their new experiment once again employed a balloon-borne detector. For the more recent flight, they made measurements in five spectral bands defined by filters and they included in-flight calibration. At the time of this conference, the results have not yet appeared in the literature, but I can summarize them by saying that at wavelengths from 1–3.5 mm, the average temperature derived from these bolometric measurements is about 2.8 K. Again, the measurements seem to reveal a slightly higher temperature at about 2 mm than at the shortest wavelength measured. However, the bolometric results are now consistent with the cyanogen observations reported above and thus also with the longer wavelength measurements.

Therefore, the present observational situation is that the spectrum of the microwave background appears to be thermal (with perhaps only the most marginal deviation at about 2 mm wavelength) over a wide range of its spectrum. These results are summarized in FIGURE 2 in which the observed thermodynamic temperature is presented as a function of wavelength. For future reference, it is worth noting that while the majority of the energy of the radiation is concentrated in the region near the peak of the Planck spectrum at 1–3 mm wavelength, the predicted spectral distortions of the radiation are most likely to emerge at the longer wavelengths, $\lambda \gtrsim 3$ cm.

Large-Scale Anisotropy

We have known for roughly a decade that on the largest possible angular scale, the microwave background does display an anisotropy. This is the dipole (or 24-hour) anisotropy in which there is a small sinusoidal variation of temperature in the radiation around the great circle in the sky, with a maximum intensity observed to lie at $\sim 11^h$ right ascension and $-5°$ declination. We now know[16,17] the amplitude of the dipole

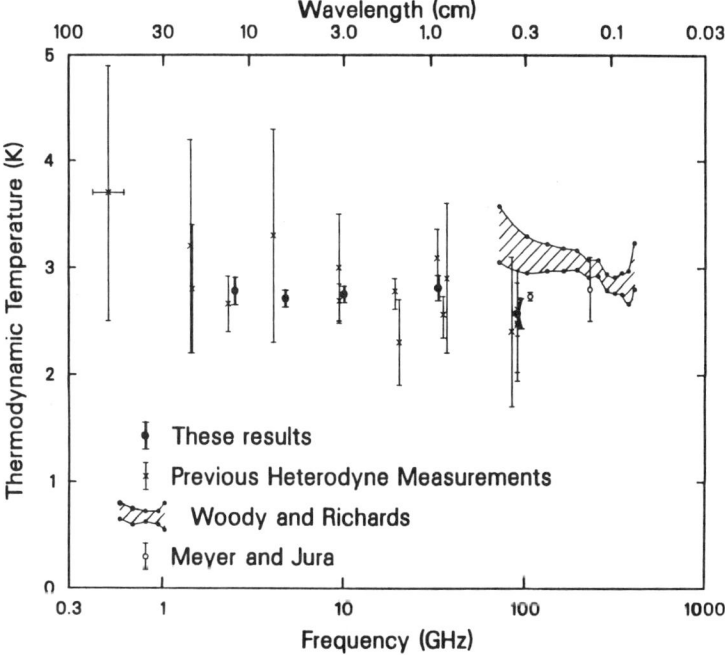

FIGURE 2. Currently available measurements of the spectrum of the cosmic microwave background. The filled circles ("these results") refer to our recent measurements (Smoot *et al.*, 1985).[5] As noted in the text, new measurements by Richards and his colleagues[15] give values ~ 0.2 K below the Woody and Richards results.[13,14]

anisotropy to an accuracy of ~ 5%; it is $\Delta T/T = 1.3 \times 10^{-3}$. This anisotropy is ascribed to the motion of the Earth. Although there were some published reports, about five years ago, of a detection of a quadrupole anisotropy, it now appears that there is no significant detectable quadrupole anisotropy in the radiation. The most recent results on this question will be presented in a Poster Session by Philip Lubin; here I can state the essential result that the quadrupole moment appears to be small, $\Delta T/T \leq 6 \times 10^{-5}$.

These results on the large-scale distribution of the cosmic microwave background have recently been confirmed by the first satellite devoted to the cosmic microwave background, the Soviet Prognoz 9 satellite.[18] It made observations at a wavelength of 8 mm, covering much of the celestial sphere, and the measurement of the dipole anisotropy and the upper limits on the quadrupole anisotropy are consistent with the results I have just cited.

It is worth noting that this instrument and the balloon-borne instruments of Lubin et al.[16] and Fixsen et al.[17] also provide measurements of the isotropy or homogeneity of the radiation on angular scales corresponding to the size of the antenna patterns employed, approximately 5° and 7°. This remark leads me to my next topic.

Small-Scale Anisotropy

The crucial point here is that there is as yet no detected anisotropy on any angular scale ranging from 6 arcseconds to 90° (the quadrupole scale). Measurements have been made at wavelengths from below 1 mm to above 10 cm. While disappointing, this lack of results has had an important impact on theories of the origin of large-scale structure in the universe, as we shall see. Now let me turn my attention to the newest results (or rather lack thereof).

First, let me call your attention to the gap in the observational results on angular scales between roughly 1° and 5° (see FIGURE 3). Why is this particular angular scale of interest? It is because 1°–3° is the angular scale corresponding to the largest regions of the universe that were causally connected at the epoch from which microwave background reaches us. Observations on scales of $\theta > 1°–3°$ span regions larger than the causal horizon at the time the radiation began its journey to us. Classical big bang models offer no explanation for the homogeneity of the universe on these large scales; the inflationary model does (see Steinhardt's paper here). Hence, observations on scales of $\theta > 1°–3°$ are of particular interest. While observation at any scale greater than 1°–3° would be of interest, most models predict a decrease in the amplitude of fluctuations as the angular scale increases. Hence, the most interesting observations and the most useful limits would be just at the angular scale corresponding to the size of the causal horizon at the epoch from which the radiation reaches us, that is a few degrees. Unfortunately, it is just at this angular scale that observations are lacking. Their absence may be explained by two problems facing observers. The first is emission from the Earth's atmosphere, particularly from water vapor. Clouds, which have a typical angular scale of a few degrees, can produce fluctuating signals as large as several degrees in the microwave region of the spectrum. Hence, observations beneath the Earth's atmosphere are difficult. Also, a scale of 1°–3° is an awkward one for conventional radio astronomy—smaller than can easily be achieved using corrugated-horn antennas of the type used for the absolute measurements of the spectrum, yet, on the other hand, larger than the angular scale available at most conventional radio telescopes. Diffraction determines the resolving power of an instrument and to reach an angular scale of a few degrees at wavelengths of a few centimeters requires an aperture of roughly a meter or two. Instruments of this size have not yet been carried above the Earth's atmosphere. The best current limits we have on anisotropies at scales of a few degrees or more come as a by-product of searches for dipole and quadrupole anisotropy. Examples are the results of Melchiorri and his colleagues[19] at an angular scale of 6°, the limit set by the Soviet Prognoz 9 satellite at an angular scale of 5°,[18] and the observations of Lubin et al.[16] referred to above (see also Lasenby and Davies).[20]

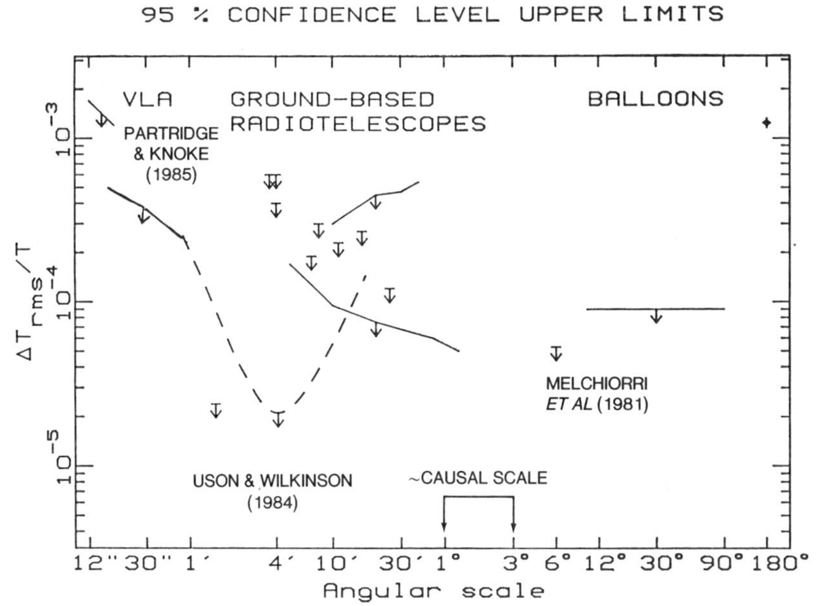

FIGURE 3. Upper limits on $\Delta T/T$ fluctuations in the background, adopted from Uson and Wilkinson.[24,25] Only the dipole moment (180°) is a measurement, as opposed to an upper limit.

A group of us from Italy, Norway, and the United States have made observations on an angular scale of 3° and these results should appear shortly in *Nature*. In addition, I expect substantially better limits over the interesting range of angular scale of $1/2°-3°$ from both ground-based and balloon-borne experiments in the near future.

Now let us turn to the observational results for the smallest angular scales so far observed, 6"–60". To make observations on such small angular scales at microwave wavelengths requires the use not of a single radio antenna, but of an array of antennas. By using aperture synthesis or radio interferometry, one achieves an angular resolution determined by the size of the array, not the individual elements of the array. On the other hand, there is a substantial price to be paid: the effective collecting area of the array is the sum of the collecting areas of the individual elements, not the area spanned

TABLE 2. Upper Limits (2σ) on $\Delta T/T$ at λ = 6 cm Using the Very Large Array[a]

Angular Scale		$\Delta T/T \times 10^3$	
	Fomalont et al.	Knoke et al.	Partridge et al.
6"	—	3.2	—
12"	—	1.7	—
18"	1.0	1.2	0.5
30"	0.8	—	0.4
60"	0.5	—	0.25

[a]At this wavelength, the equivalent value of $\Delta T/T$ on an angular scale of 60" from discrete radio sources alone is $\sim 0.12 \times 10^{-3}$.

by the array as a whole. As a consequence, the efficiency of the arrays is small and our limits on fluctuations in the microwave background are not as sensitive as may be achieved by using single, filled-aperture telescopes. Nevertheless, two groups (Fomalont et al., 1984;[21] and Knoke et al., 1984[22] and Partridge and Knoke, 1985[23]) have used the Very Large Array in New Mexico to set limits on the amplitude of fluctuations in the microwave background on angular scales of 6"–60". Those results are summarized in TABLE 2; the wavelength employed was 6 cm.[a]

As may be seen, these limits are roughly an order of magnitude less sensitive than the limits established at larger angular scales. On the other hand, aperture-synthesis observations do offer one substantial advantage: they provide us with a real two-dimensional map of the microwave background. A second side benefit is that observational programs of this sort also provide information about very weak discrete radio sources (that is, they extend the radio source counts to smaller fluxes).

Just to show how featureless the cosmic background truly is, I am including here a map of the CBR made at a wavelength of 6 cm at the VLA with all the discrete radio sources removed (see FIGURE 4). It is in effect a snapshot of a $\sim (43 \text{ arcmin})^2$ region of the background. With the exception of a set of faint rings concentric with the map's center, there is little to see. The rings are an instrumental artifact and in fact limit the precision of such searches. Other than the rings, one sees essentially noise and it is from this noise that one extracts the upper limits displayed in TABLE 2 (see Knoke et al.[22]).

By far the most interesting observational limit on fluctuations in the microwave background is the new result of Uson and Wilkinson,[24,25] which has already received considerable attention at this conference because of the difficulty it presents to those making models of galaxy formation. The observations were carried out at a 40-m telescope at a wavelength of 1.5 cm. The angular resolution achieved was 1.'5, but upper limits were also obtained at somewhat larger angular scales, as shown in FIGURE 3, which is borrowed from the 1985 paper of Uson and Wilkinson.[25] At 1.'5, their work limits fluctuations in the microwave background at $\Delta T/T \leq 2.5 \times 10^{-5}$ at the 95% confidence level. This remarkable degree of isotropy on the small scale is one of the most important features of the cosmic microwave background.

[a]**Note added in proof:** Both groups have made further observations, resulting in upper limits lower than those given here.

WHAT DO THE OBSERVATIONS TELL US?

I should now like to turn to the second part of my discussion, in which I will try to summarize what we learn from these observations of the microwave background. My comments will necessarily be brief and somewhat selective; other participants at this conference have, and no doubt also will, use the observations I outlined to draw additional conclusions about the large-scale properties and evolution of the universe.

The Spectrum

The mere fact that the radiation is present at all, at least at a temperature exceeding 0.003 K, poses the so-called entropy problem: the large ratio of the photon number to baryon number in the universe. It is regarded as a considerable success of the new Grand Unified Theories that they offer a natural solution to the entropy problem (see Steinhardt's paper here and Gibbons *et al.*, 1983,[26] for reviews).

FIGURE 4. A "photograph" of the cosmic microwave background at a wavelength of 6 cm, made at the Very Large Array. The r.m.s. noise level is 12–13 microJansky on angular scales of $\sim 18''$ and it can be shown to be mostly instrumental noise. The faint concentric rings are instrumental artifacts. The upper limit (95% confidence) on possible background fluctuations at $\theta = 18''$ is $\Delta T/T < 5 \times 10^{-4}$ (Partridge and Knoke[23]). Discrete radio sources were subtracted from this image.

As is well known, the synthesis of light elements in the first few minutes of the expansion of the universe depends in part on the temperature.[27,28] A precise measurement of the thermodynamic temperature of the radiation pins down primordial nucleosynthesis. Indeed, at present, the major uncertainties in the predictions of light element production early in the universe lie with particle physics. The fraction of ^{4}He emerging from the big bang, for instance, depends on the number of neutrino flavors[29] and also on the half-life of the neutron. While astronomical observations of the helium in primordial matter or at least very old objects are quite difficult (see Greenstein[30] and Shaver et al.[31] for reviews; also Pagel[32] and Kunth and Sargent[33]), there is now good agreement with the observations of several light elements (e.g., ^{7}Li; see Spite and Spite[34]) and from these observations we can draw several important conclusions (reviewed by Yang et al.[35]). The first is that the baryon density in the universe must be low, of the order of a few percent of the critical density. Another intriguing result is that the observed temperature of the cosmic microwave background and the observed helium abundance, taken together, appear to limit the number of neutrino flavors to three (the number already known) or just possibly four.[29,36] Finally, let me be slightly more radical in my interpretation of the measurements. Let us accept that there are three neutrino flavors and let us combine that assumption with the new measurements of the microwave background temperature and of the helium abundance of primordial matter. Putting these data together, we may derive a value of the neutron half-life more accurate than any currently in the literature (my estimate is 10.3 ± 0.1 min). I recognize that this is a contentious argument and that the laboratory results are crucial, but I make it to emphasize the potential benefit to other branches of physics of these astronomical measurements.

As I have noted, the spectrum of the background radiation appears to be thermal over a wide range of wavelengths up to and beyond the wavelength of peak intensity of a 2.73-K blackbody. The uniformity of this spectrum presents problems for "cold" big bang models that invoke astrophysical processes at redshifts of 1000 or below to generate the observed background radiation. In general, these processes involve absorption and reemission of higher temperature radiation from an earlier generation of stars or supermassive stars (see, for instance, Rees,[37] Negroponte et al.,[38] and Carr et al.[39]). In order to match the observed spectrum, some essentially wavelength-independent absorber and emitter is needed; such a mechanism is hard to arrange. In fairness to those working on this problem (e.g., Negroponte et al.[38]), it should be noted that one impetus behind the development of such models was the apparent departure from a pure thermal spectrum reported five years ago by Woody and Richards.[13,14] As noted above, the newer measurements of the spectrum do not reveal such large departures from a pure thermal spectrum.

Of most current interest to me is the apparent lack of distortions in the spectrum at wavelengths of a few centimeters or longer. The absence of such distortions sets constraints on energy-releasing processes over a large and interesting scale of redshifts, namely $\sim 10-10^6$ (see, for instance, Illarionov and Sunyaev;[40] Danese and de Zotti[41,42]). An energy release of any of a number of forms whose total amplitude exceeded $\sim 5\%$ of the energy in the radiation field itself would have distorted the spectrum to a significant degree.[5] Whatever mechanism one has in mind for the origin of galaxies, stars, and other bound systems, it must not violate this constraint. As one quick example, let me mention the possibility that massive stars were formed at a redshift of

~ 100. Could such stars have generated the observed helium abundance? No, because the release of thermonuclear energy would exceed the limits set by the microwave background observations.

Large-Scale Isotropy

The conventional explanation of the observed dipole anisotropy in the microwave background is the peculiar motion of the Earth. If so, we know with good precision the velocity of the sun with respect to the frame established by matter at large distances. From other astronomical data, we can calculate the component of the sun's velocity due to the rotation of our Galaxy; a subtraction then permits us to find the velocity of the center of the Galaxy with respect to matter at large distances. This number is being actively exploited to tell us about the large-scale distribution of matter in the universe since the velocity of the Galaxy seems to be at least in part produced by gravitational acceleration of a nearby clump of matter. Such arguments (reviewed by Davis and Peebles[43]) can even be used to constrain values of the mean mass density in the universe. While these results are somewhat model dependent, they again illustrate the utility of the basic measurements of the properties of the cosmic background radiation.

In addition, both the lack of an observed quadrupole anisotropy and the fact that the dipole anisotropy is relatively small and has a natural explanation tell us that the expansion of the universe is isotropic and shear-free to a remarkable degree.[44,45] Although the isotropic Friedman-Robertson-Walker models are a subset of measure zero of possible cosmological solutions to Einstein's equations, it appears that they provide a remarkably good model for the universe as we observe it.

Small-Scale Isotropy

As already noted, the high degree of isotropy of the radiation on angular scales larger than a few degrees directly raises the causality problem in classical cosmological models, a problem neatly solved by the new inflationary models.

Limits on fluctuations in the microwave background on arcminute scales play a crucial role in constraining or guiding theories of galaxy formation, a phrase I use to refer to the formation of large, gravitationally-bound structures in the universe, whether they be pancakes, galaxies, supermassive stars, or what have you. Any inhomogeneous distribution of matter at the surface of last scattering will produce fluctuations in the microwave background; consequently, measurements of $\Delta T/T$ constrain the scales and amplitudes of density inhomogeneities. The actual connection between the amplitude of density inhomogeneities and $\Delta T/T$ is quite model dependent and has been the subject of a great deal of theoretical work. The field is even richer when we include the possibility that the matter density of the universe is dominated by dark matter, since different assumptions about the nature of the dark matter produce quite different predictions for $\Delta T/T$ in the microwave background.[46,47] FIGURE 5, taken from the work of Bond and Efstathiou,[46] shows the predicted $\Delta T/T$ fluctuations for a number of dark matter models. A single observational point is that of Uson and Wilkinson.[24,25] Note that it appears to exclude some, but not all of the models. For an observer, such a situation is of particular interest—the observations are already useful in constraining the theories and relatively modest improvements in their sensitivity could make them much more valuable still.

To return for a moment to the metaphor I introduced earlier of an "expansionist power," one should note that the theorists have recently established a new beachhead by introducing two new ideas that in effect produce lower values for $\Delta T/T$ for a given degree of inhomogeneity. The first of these is the use of "strings" (Vilenkin, this conference; Schramm, 1985[48]) and the second is "biased" galaxy formation (Bar-

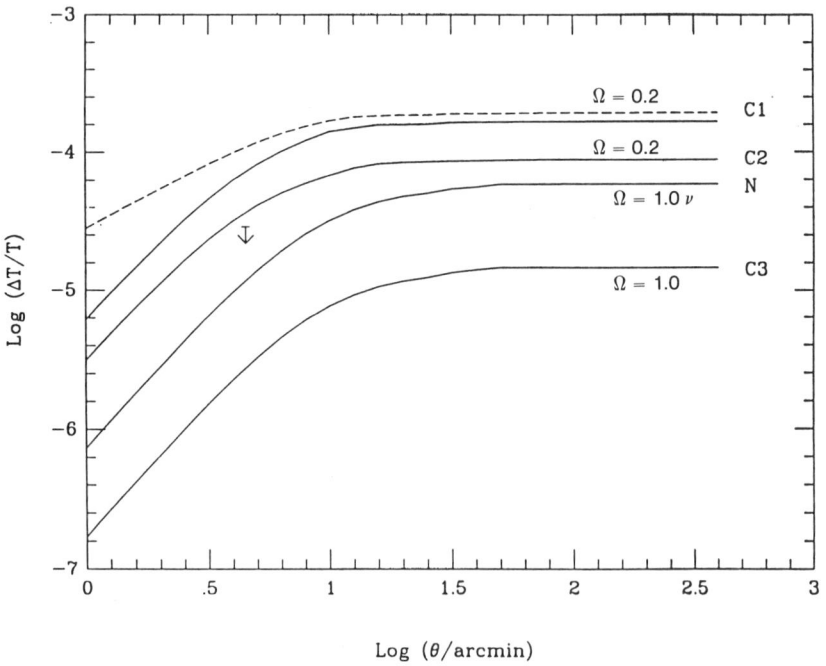

FIGURE 5. An example of the predictions of $\Delta T/T$ for different assumptions about the nature of the matter content of the universe, particularly the "dark matter." ("C" is for cold particles; "N" is for massive neutrinos.) Note that the single data point shown[24] excludes some models and that measurements at comparable sensitivity, but $\theta \sim 30'$, would exclude most. Also, $H_0 = 75$ km/sec per Mpc, except for C1 ($H_0 = 50$). (From Bond and Efstathiou, 1984.)[46]

deen;[49] White[50]). It remains to be seen what the future of these new models will be; I believe it is not too outrageous to say that their introduction owes at least something to the difficulties presented to earlier models by the very stringent limits on $\Delta T/T$.

I would like next to turn to a point raised earlier by Paul Steinhart at this symposium. Again it is a question of the constraints set by the observed limits on $\Delta T/T$ on theory, but this time I have particle theory in mind. One relatively well-known example is that the minimal $SU(5)$ Grand Unified Theory predicts density perturbations of the order of $\Delta\rho/\rho = 50$, which is a situation not consistent with the observations. Put more baldly, $SU(5)$ will not work as the basis for a picture of the earliest universe. Paul Steinhart went on to note that there is a connection between the super-symmetry breaking scale and both the observed electroweak energy scale and (indirectly) $\Delta T/T$. Will it eventually turn out that the limits on $\Delta T/T$ will permit us to calculate both the SUSY scale and the electroweak scale? We are not yet at that point, but who would have thought five years ago that radio astronomers would have had anything at all to say about fundamental particle physics?

Next, as noted by several authors including Hogan (1984),[51] observations of the small-scale isotropy of the background radiation constrain nonconventional models of Galaxy formation, such as the "cold" big bang models or the hydrodynamic models of Ostriker and Cowie (1981)[52] or Ikeuchi (1981).[53] A general feature of such models is to move the surface of last scattering of the microwave background photons to lower redshifts, in the interval of $10 < z < 1000$. New features in the predicted angular spectrum of $\Delta T/T$ fluctuations result, including the possibility of higher amplitudes at arcsecond scales and the intriguing possibility of wavelength-dependent values of $\Delta T/T$.[54,55] Here I want only to mention that the observed small-scale isotropy of the radiation, like the closely thermal spectrum of the radiation, pose difficult, but probably not insuperable problems for such nonconventional models.[22,51]

Finally, searches for fluctuations in the microwave background, particularly those carried out at $\lambda = 6$ cm at the VLA, are coming close to detecting the statistical fluctuations in intensity expected from distant unresolved radio sources alone.[56] As noted long ago by Longair and Sunyaev (1969),[57] the fluctuating background due to faint radio sources may prove to be a fundamental limit to searches for true cosmological fluctuations, especially at wavelengths of $\lambda \gtrsim 3$ cm. On the other hand, the searches for microwave fluctuations provide important radio astronomical information in their own right since such observations are probing the radio source counts at levels somewhat fainter than direct source counts can yet reach. From both the upper limits on fluctuations at $\lambda = 6$ cm and from direct source counts,[58] we can draw the conclusion that the counts of faint radio sources continue to fall below the expectation for a static Euclidean universe and that there is as yet no convincing evidence for any new species of source to appear at very low fluxes. A new population of flat spectrum sources at high redshifts has been suggested earlier by Danese *et al.*[59] and the possibility that ordinary "radio quiet" galaxies at low redshifts would begin to dominate the source counts has been suggested by Sunyaev and Partridge. At the levels we have currently reached, we do not yet see strong evidence for either, but aperture-synthesis observations of the background are still in their infancy and improvements in sensitivity of a factor of 2–5 can be expected.

The Future

Let me close by speculating a bit on the future of the observational programs I have sketched out today. Given the progress (often in unexpected directions) in this field in the past twenty years, I cannot imagine being able to say much of value about the full span of the next twenty years. However, there are a few predictions I feel confident in making.

The Cosmic Background Explorer (COBE) will refine measurements of the spectrum of the radiation in the important spectral range below 3 mm wavelength. This satellite, due for launch in 1987, will also improve the precision of our limits on $\Delta T/T$ on angular scales $\gtrsim 7°$.

Ground-based measurements will (slowly, I fear) improve limits on $\Delta T/T$ on arcminute scales. It will be easier, I believe, to increase the angular scale of the observations (say to $10'-20'$) at the current levels of sensitivity than to increase the sensitivity by more than a factor of three. Even a factor of three, however, to $\Delta T/T \leq 10^{-5}$, might finally provide an all-important detection as opposed to yet another upper limit.

I would like to end with the thought that the power of these observational results may be enhanced when they are used in combination with other astronomical observations. An instance is the use of the value of the temperature of the background radiation in tandem with observations of ^{4}He abundance to set constraints on important parameters of both cosmology and particle physics. To return to my analogy of an "expansionist power," what I have in mind is "binding alliances" of several observational results to hem in the theorists more tightly. Let me mention two possibilities, neither of them very fully explored. The first concerns galaxy formation, in the loose sense in which I am using the phrase. Can the upper limits we have on fluctuations in the background, on spectral distortions, and on the optical[60,61] and IR backgrounds be combined fruitfully to tell us more about galaxy formation? The second concerns the extragalactic X-ray background. The searches for fluctuations in the background and the faint source counts, combined with estimates of the X-ray luminosity of sources as a function of their radio luminosity (see Avni, this volume), can tell us something about the contribution of discrete sources to the X-ray background (see Danese et al.[59]). On the other hand, the absence of spectral distortion in the background constrains the temperature and density of any intergalactic plasma and hence the possible contribution that bremsstrahlung from hot plasma might make to the X-ray background.[62]

These are only illustrative (and also speculative) examples. The calculations I have in mind may turn out to be uninteresting. However, I am confident others will still be making fruitful use of the observations I have described for years to come.

ACKNOWLEDGMENTS

Especially on behalf of those enjoying their first trip to Jerusalem, I would like to thank all the organizers of the Texas Symposium here for their invitation to come to Jerusalem, a city that is a spiritual "home" for so many of us.

REFERENCES

1. PENZIAS, A. A. & R. W. WILSON. 1965. Astrophys. J. **142:** 419.
2. DICKE, R. H., P. J. E. PEEBLES, P. G. ROLL & D. T. WILKINSON. 1965. Astrophys. J. **142:** 414.
3. SMOOT, G., G. DE AMICI, S. FRIEDMAN, C. WITEBSKY, N. MANDOLESI, R. B. PARTRIDGE, G. SIRONI, L. DANESE & G. DE ZOTTI. 1983. Phys. Rev. Lett. **51:** 1099.
4. PARTRIDGE, R. B., G. SIRONI, G. BONELLI, L. DANESE, G. DE ZOTTI, G. DE AMICI, S. D. FRIEDMAN, G. F. SMOOT, N. MANDOLESI, S. CORTIGLIONI & G. MORIGI. 1985. The Cosmic Background Radiation and Fundamental Physics. F. Melchiorri, Ed. Italian Physical Society. Bologna.

5. SMOOT, G. F., G. DE AMICI, S. FRIEDMAN, C. WITEBSKY, G. SIRONI, G. BONELLI, N. MANDOLESI, S. CORTIGLIONI, G. MORIGI, R. B. PARTRIDGE, L. DANESE & G. DE ZOTTI. 1985. Astrophys. J. Lett. **291**: L23.
6. MANDOLESI, N., P. CALZOLARI, S. CORTIGLIONI & G. MORIGI. 1984. Phys. Rev. **D29**: 2680.
7. SIRONI, G., P. INZANI & A. FERRARI. 1984. Phys. Rev. **D29**: 2686.
8. PARTRIDGE, R. B., J. CANNON, R. FOSTER, C. JOHNSON, E. RUBINSTEIN & A. RUDOLPH. 1984. Phys. Rev. **D29**: 2683.
9. SMOOT, G. F., G. DE AMICI, S. LEVIN & C. WITEBSKY. 1985. The Cosmic Background Radiation and Fundamental Physics. F. Melchiorri, Ed. Italian Physical Society. Bologna.
10. MEYER, D. M. & M. JURA. 1984. Astrophys. J. Lett. **276**: L1.
11. MEYER, D. M. 1985. See reference 4.
12. MCKELLAR, A. 1941. Publ. Dom. Astrophys. Obs. Victoria, B.C. **7**: 251.
13. WOODY, D. P. & P. L. RICHARDS. 1979. Phys. Rev. Lett. **42**: 925.
14. WOODY, D. P. & P. L. RICHARDS. 1981. Astrophys. J. **248**: 18.
15. PETERSON, J. B., P. L. RICHARDS & T. TIMUSK. 1985. Phys. Rev. Lett. **50**: 332.
16. LUBIN, P. M., G. L. EPSTEIN & G. F. SMOOT. 1983. Phys. Rev. Lett. **50**: 616.
17. FIXSEN, D. F., E. S. CHENG & D. T. WILKINSON. 1983. Phys. Rev. Lett. **50**: 620.
18. STRUKOV, I. A. & D. P. SKULACHEV. 1984. Sov. Astron. Lett. **10**: 1.
19. MELCHIORRI, F., B. MELCHIORRI, C. CECCARELLI & L. PIETRANERA. 1981. Astrophys. J. Lett. **250**: L1.
20. LASENBY, A. N. & R. D. DAVIES. 1983. Mon. Not. R. Astron. Soc. **203**: 1137.
21. FOMALONT, E. B., K. I. KELLERMANN & J. V. WALL. 1984. Astrophys. J. Lett. **277**: L23.
22. KNOKE, J. E., R. B. PARTRIDGE, M. I. RATNER & I. I. SHAPIRO. 1984. Astrophys. J. **284**: 479.
23. PARTRIDGE, R. B. & J. E. KNOKE. 1985. In Inner Space/Outer Space, Proceedings of a Fermilab Conference.
24. USON, J. M. & D. T. WILKINSON. 1984. Astrophys. J. Lett. **277**: L1 (see also Astrophys. J. **283**: 471).
25. USON, J. M. & D. T. WILKINSON. 1985. Nature **312**: 427.
26. GIBBONS, G. W., S. W. HAWKING & S. T. C. SIKLOS, Eds. 1983. The Very Early Universe. Cambridge University Press. London/New York.
27. WAGONER, R. V. 1973. Astrophys. J. **179**: 343 (see also WAGONER, R. V., W. A. FOWLER & F. HOYLE. 1967. Astrophys. J. **148**: 3).
28. PEEBLES, P. J. E. 1971. Physical Cosmology. Princeton University Press. Princeton, New Jersey.
29. YANG, J., D. N. SCHRAMM, G. STEIGMAN & R. T. ROOD. 1979. Astrophys. J. **227**: 697.
30. GREENSTEIN, J. L. 1980. Phys. Scr. **21**: 759 (see also GRY, C., G. MALINIE, J. AUDOUZE & A. VIDAL-MADJAR. 1984. Formation and Evolution of Galaxies. J. Audouze & J. Tran Thanh Van, Eds.: 279. Reidel. Dordrecht).
31. SHAVER, P. A., D. KUNTH & K. KJÄR, Eds. 1983. ESO Workshop on Primordial Helium.
32. PAGEL, B. J. E. 1982. Phil. Trans. R. Soc. London **A307**: 19.
33. KUNTH, D. & W. L. W. SARGENT. 1983. Astrophys. J. **273**: 81.
34. SPITE, F. & M. SPITE. 1982. Astron. Astrophys. **115**: 357.
35. YANG, J., M. S. TURNER, G. STEIGMAN, D. N. SCHRAMM & K. OLIVE. 1984. Astrophys. J. **281**: 493.
36. BARROW, J. D. & J. MORGAN. 1983. Mon. Not. R. Astron. Soc. **203**: 393.
37. REES, M. J. 1978. Nature **275**: 35.
38. NEGROPONTE, J., M ROWAN-ROBINSON & J. SILK. 1981. Astrophys. J. **248**: 38.
39. CARR, B. J., J. R. BOND & W. D. ARNETT. 1984. Astrophys. J. **277**: 445.
40. ILLARIONOV, A. F. & R. A. SUNYAEV. 1975. Sov. Astron. AJ. **18**: 691.
41. DANESE, L. & G. DE ZOTTI. 1977. Riv. Nuovo Cimento **7**: 277.
42. DANESE, L. & G. DE ZOTTI. 1982. Astron. Astrophys. **107**: 39.
43. DAVIS, M. & P. J. E. PEEBLES. 1983. Annu. Rev. Astron. Astrophys. **21**: 109.
44. COLLINS, C. B. & S. W. HAWKING. 1973. Mon. Not. R. Astron. Soc. **162**: 307.
45. BARROW, J. D., R. JUSZKIEWICZ & D. H. SONODA. 1984. Universal rotation: How large can it be? (preprint).

46. BOND, J. R. & G. EFSTATHIOU. 1984. Astrophys. J. Lett. **285:** L45.
47. VITTORIO, N. & J. SILK. 1984. Astrophys. J. Lett. **285:** L41.
48. SCHRAMM, D. N. 1985. Nuovo Cimento. To be published.
49. BARDEEN, J. M. 1985. *In* Inner Space/Outer Space, Fermilab Conference Proceedings.
50. WHITE, S. D. M. 1985. *In* Inner Space/Outer Space, Fermilab Conference Proceedings.
51. HOGAN, C. J. 1984. Astrophys. J. Lett. **284:** L1.
52. OSTRIKER, J. P. & L. L. COWIE. 1981. Astrophys. J. Lett. **243:** L127.
53. IKEUCHI, S. 1981. Publ. Astron. Soc. Japan **33:** 211.
54. HOGAN, C. J. 1980. Mon. Not. R. Astron. Soc. **192:** 891.
55. HOGAN, C. J. 1982. Astrophys. J. **252:** 418; see also Astrophys. J. Lett. **256:** L33.
56. DANESE, L., G. DE ZOTTI & N. MANDOLESI. 1983. Astron. Astrophys. **121:** 114.
57. LONGAIR, M. S. & R. A. SUNYAEV. 1969. Nature **223:** 719.
58. FOMALONT, E. B., K. I. KELLERMANN, J. V. WALL & D. E. WEISTROP. 1984. Science **225:** 23.
59. DANESE, L., G. DE ZOTTI & N. MANDOLESI. 1983. *In* The Birth of the Universe, 17th Rencontre de Moriond. J. Audouze & J. Tran Thanh Van, Eds.
60. DUBE, R. R., W. C. WICKES & D. T. WILKINSON. 1979. Astrophys. J. **232:** 333.
61. TOLLER, G. N. 1983. Astrophys. J. Lett. **266:** L79.
62. FIELD, G. B. & S. C. PERRENOD. 1977. Astrophys. J. **215:** 717.

PART II. ACTIVE GALACTIC NUCLEI AND JETS

Toward a Unified Theory of Active Galactic Nuclei[a]

MITCHELL C. BEGELMAN[b,c]

Joint Institute for Laboratory Astrophysics
University of Colorado
and
National Bureau of Standards
Boulder, Colorado 80309

INTRODUCTION

Most astrophysicists believe that the energetic phenomena that characterize active galactic nuclei (AGNs) result from the interaction of matter with a massive black hole. This conclusion rests on several indirect lines of argument, the most persuasive of which couples the compactness of the primary energy source (inferred from variability) with the large mass-energy processed through the source during its lifetime (estimated from source statistics or the energy content of radio lobes). One is forced to conclude that the "central engine" contains a large mass—typically 10^6–10^9 $M_\odot$—in a region that may be comparable in size to the solar system. Of the hypothetical systems that could satisfy these mass and size constraints (e.g., "spinars" or relativistic star clusters), black holes are conceptually and physically the most robust; hence they are regarded as the most likely candidates for "prime mover" in all AGNs.

However, the mass and size constraints that motivate the black hole hypothesis are determined for only a small subset of the objects that are now classified as AGNs, with the classification of the majority being based on "secondary" characteristics such as nonthermal spectra, the presence of jets, and evidence for high-velocity line-emitting gas. None of these features point unambiguously to the workings of the central engine—indeed, all of them may result from reprocessing of the primary energy supply on scales many orders of magnitude larger than the event horizon of the black hole. How then are we to understand such diverse phenomena as manifestations of a single generic process?

In this article, I will discuss recent progress toward a unified theory of active galactic nuclei, based on the physics of accretion flows around massive black holes. The qualitative "mode" of accretion depends on the radiative properties of the accreting gas through a single parameter, essentially the ratio of the accretion rate to the mass of the hole. Each of the modes of accretion has unique spectral properties that may be

---

[a]This work was supported in part by the National Science Foundation under grant no. AST83-51997, and by grants from Ball Aerospace Systems Division, Rockwell International Corporation, and the Exxon Education Foundation.
[b]Presidential Young Investigator.
[c]Also at Department of Astrophysical, Planetary, and Atmospheric Sciences, University of Colorado, Boulder, Colorado 80309.

compared with the observed properties of various types of AGN. The results of this comparison suggest a scheme in which the different modes of accretion account for the widely differing manifestations of activity in the nuclei of galaxies.

MODES OF ACCRETION

The energy that powers an AGN may consist of gravitational binding energy released during the accretion of gas with angular momentum[1] and spin energy extracted electromagnetically (hydromagnetically) from the hole by magnetic stresses.[2-4] In order for electromagnetic extraction to occur, a substantial amount of plasma must surround the hole to provide inertial confinement for the magnetic fields. Unless the currents that generate the fields are confined to a thin skin on the surface of the plasma, it is likely that at least as much energy will be released by accretion as by the Blandford-Znajek process.[5] If the specific angular momentum exceeds $\sqrt{12}\, r_g c$ with $r_g = GM/c^2$ (for a Schwarzschild hole of mass, M), then the gas must give up some of its angular momentum via viscous (or other) torques before it can cross the event horizon. The binding energy dissipated in this process may be either radiated away or retained by the gas in the form of heat. An accretion flow that is able to radiate away most of its binding energy forms a thin accretion disk. Thin disks have been studied extensively in connection with cataclysmic variables and X-ray binaries.[6]

Radiation pressure and electron scattering play important roles in disks around stellar-mass objects and are equally important in determining the structure of disks around massive black holes.[7] The structures of such disks are largely governed by the ratio ($\dot{m}$) of the actual accretion rate to the "Eddington accretion rate," $\dot{M}_E \equiv L_E/c^2$, where $L_E = 1.3 \times 10^{38}\,(M/M_\odot)$ ergs s^{-1} = $1.3 \times 10^{46}\,m_8$ ergs s^{-1} is the Eddington limit and $m_8 \equiv M/10^8\,M_\odot$. If the ratio of inflow speed to Keplerian speed is fixed, then flows with similar values of $\dot{m}$ should be qualitatively similar. In particular, accretion rates, timescales, t, and length scales, ℓ, should all scale with M for similar disks. Thus, the characteristic density in a disk scales as $\rho \propto \dot{M}t/\ell^3 \propto \dot{m}/M$. The cooling timescale is inversely proportional to density, so the ratio of cooling time to dynamical time is inversely proportional to $\dot{m}$ and is independent of M. In other words, an accretion flow may not be able to cool and form a thin disk if $\dot{m}$ is too small. Such a flow may instead puff up to form an ion torus supported by gas pressure.[5] Alternatively, consider a flow with $\dot{m}$ very large. The electron scattering optical depth scales as $\tau \propto \ell\rho \propto \dot{m}$. This is also proportional to the ratio of photon diffusion time (Kelvin-Helmholtz time) to dynamical time, so radiation is unable to leak out of a flow with $\dot{m}$ too large. Such a flow is also unable to cool and it puffs up to form a radiation torus supported by radiation pressure. Therefore, because the luminosity required to inflate a radiation torus is $\gtrsim L_E$, such structures may form whenever $\dot{m}$ is larger than a few.

The generic properties of the three modes of accretion are summarized in TABLE 1 and are discussed in greater detail below. One obvious difference between them is in the quality of spectra they produce. While ion tori probably radiate nonthermally, thin disks and radiation tori are both primarily thermal emitters, with spectra peaking in the ultraviolet. However, either may possess a corona that can add a substantial power-law continuum to the underlying thermal spectrum. The decomposition of quasar and Seyfert spectra into thermal and nonthermal components[8,9] does not by

TABLE 1. Modes of Accretion

	(Two-Temperature) Ion Torus	Thin Disk	Radiation Torus
DETERMINANTS			
Inflow time	$\alpha^{-1} t_{kep} \propto \alpha^{-1} M$	$\sim \alpha^{-1} (h/R)^{-2} t_{kep}$	$\sim \alpha^{-1} t_{kep} \propto \alpha^{-1} M$
Local cooling time	$\propto \dfrac{1}{\rho} \simeq \dfrac{\alpha M^2}{\dot{M}}$	Short	Short
Radiative transfer time	Short	Short	$\sim \dfrac{r_T}{c} \propto \rho r^2 \propto \dfrac{\alpha M^2}{\dot{M}}$
$\dot{M}$	$<50 \alpha^2 \dot{M}_E$	$\leftarrow$ overlap $\rightarrow$ $\dot{M}_E$	$> \dot{M}_E$
RADIATIVE EFFICIENCY	$0.1 \left(\dfrac{\dot{M}}{\dot{M}_E}\right) \left[1 + 3 \times 10^{-3} \dfrac{\dot{M}_E}{\dot{M}}\right]$ ($\ll 1$)	~ 0.1	$\gtrsim (\dot{M}/\dot{M}_E)^{-1}$ ($\ll 1$) [$\lesssim (\dot{M}/\dot{M}_E)^{-2/3}$ if trapped radiation released from wind]
RADIATION			
Basic input radiation	Mostly nonthermal (synchrotron and Compton)	Thermal (Comptonized bremsstrahlung)	Thermal (Comptonized bremsstrahlung)
Basic radiation output	IR (self-absorbed synchrotron); γ-rays (Comptonized synchrotron and bremsstrahlung)	Opt $\rightarrow$ UV (soft $\rightarrow$ X)	UV $\rightarrow$ X (soft)
Coronae	Basic flow resembles corona	Ion corona (inner radii) Electron corona, wind (outer radii) (X-ray heated wind)	Electron corona, wind
Nonthermal/thermal ratio	High	Variable $\gtrsim 1$ (Nonthermal synchrotron, Compton, SSC in ion corona, thermal Compton in electron corona)	Variable, $\lesssim 1$
OPTIONAL ACCESSORIES	Funnels, exotic MHD and plasma processes, jets		Funnels, radiation pressure driven wind, jets

itself discriminate between thin disks and radiation tori, although the details of the decomposition (when they can be agreed upon) may. Specifically, thin disks are broadband emitters, with a gently sloping spectrum built up from emission at a range of temperatures. The integrated spectrum of a radiation torus cannot be predicted with such theoretical certainty, but Blandford[10] has suggested that radiation tori may have starlike photospheres, with spectra resembling single temperature blackbodies. Malkan[9] has fitted his data for six luminous quasars to both broadband and blackbody spectra, and prefers the broadband model because of the flatness in the thermal component just shortward of the Balmer continuum. Narayan, however, has replotted some of Malkan's published data in terms of νF_ν versus ν,[11] revealing a well-defined hump that looks very much like single-temperature blackbody at 27,500 K. Both results are sensitive to the assumed spectrum of the nonthermal continuum and other observational uncertainties, but it is curious that all of the Malkan's quasars can be fitted by single-temperature blackbodies in the range of 25,000–30,000 K. Polarization may be a further discriminant between radiation tori and thin disks: we expect disks to exhibit a higher degree of polarization (at least a few percent due to scattering) than nearly spherical tori, which is a feature that is not seen in quasars[12] and may or may not be present in Type I Seyferts.[13]

A second difference between the modes of accretion is in the radiative efficiency of the flow. For example, the weakness of the radiation from the nuclei of many radio galaxies, compared with the inferred kinetic energy fluxes into the jets and lobes, supports the ion torus as the dominant mode of accretion in these objects.[5]

It would clearly be advantageous if we had some independent means of determining $\dot{m}$ in a given object. The best quantitative diagnostics for quasars and Seyferts are provided by the broad emission lines, which appear to have remarkably similar characteristics (density, ionization parameter, velocity dispersion, covering factor) over a huge range of luminosities. At present, we do not know whether the velocity dispersions, σ_{BLR}, of the broad emission lines are local Keplerian velocities or whether they substantially exceed v_{Kep}. However, writing the Keplerian speed in the form of

$$v_{Kep}(BLR) \sim 900 \left(\frac{L_E}{L}\right)^{1/2} \left(\frac{L}{10^{46} \text{ ergs s}^{-1}}\right)^{1/4} \left(\frac{n}{10^{10} \text{ cm}^{-3}}\right)^{1/4} \Xi^{1/4} \text{ km s}^{-1}, \quad (1)$$

where Ξ is the ionization parameter (~ 0.1–1)[14] and n is the density of the line-emitting gas, reveals that at least we have a clear choice: If $\sigma_{BLR} \sim v_{Kep}$, then quasars and Seyferts must be substantially sub-Eddington, with $\dot{m} \ll 1$. The appearance of the thermal component in the spectrum then argues for a thin disk/corona model. If, on the other hand, it can be shown that the clouds substantially exceed the local escape speed (e.g., they are swept out in a wind or injected by supernova explosions), then L is much closer to L_E and could even exceed L_E by a small factor. The conclusion that $L \sim L_E$ for all quasars and Seyfert I's would have important implications for quasar evolution since it would rule out downward luminosity evolution in a single source. Alternatively, if quasars are all very sub-Eddington, then there must be some $10^{10} M_\odot$ black holes lurking in contemporary galaxies; perhaps Space Telescope will tell us where they are or are not. A third possibility is that quasars are radiation tori, while Seyfert I's are sub-Eddington; we should then expect to find systematic differences between the spectra and polarization properties of quasars and Seyferts.

At least we do not expect σ_{BLR} to be smaller than v_{Kep}; this places a lower limit on $\dot{m}$ in the range of 10^{-2}–10^{-3}.

THIN DISKS AND CORONAE

A thin disk is by definition an efficient radiator (efficiency, $\epsilon \equiv L/\dot{M}c^2 \sim 0.1$). Hence, most of the radiation comes from deep within the potential well of the accreting black hole, at $r \sim 20\, r_g$ (for a Kerr hole with $a/M \ll 1$). I will concentrate on the quality of the radiation that may be emitted in this region rather than on the integrated emission of the entire disk or secondary emission resulting from processes at $r \gg r_g$. The radiation energy density above and below the disk is insensitive to the value of the viscosity and is given approximately by

$$U_{disk} \sim \frac{m_p c^2 \dot{m}}{100\, \sigma_T r_g} \sim 1.5 \times 10^6 \frac{\dot{m}}{m_8}\, \text{ergs cm}^{-3}, \tag{2}$$

where σ_T is the Thomson cross section and I have assumed an efficiency of 0.1. This quantity provides a fiducial value for energy densities in a disk corona, such as the energy density in magnetic fields, relativistic electrons, or thermal plasma. A second fiducial value is associated with the mean energy density inside the disk, but this, unfortunately, depends on the still unknown viscosity law.

I will follow the standard practice of using the α-parametrization for the viscosity,[6,15] in which the viscous stress is assumed to be αp, where $\alpha \lesssim 1$ and p is the pressure inside the disk. The radial behavior of this parametrization is not crucial since I am only considering conditions over a limited range in r. Since electron scattering provides most of the opacity inside the disk, the characteristic internal radiation density is $\tau_{disk} U_{disk}$, where τ_{disk} is the electron scattering optical depth integrated vertically from the central plane of the disk. Characteristic densities and opacities inside the disk depend further on whether the internal energy density is dominated by gas pressure or radiation pressure. If radiation pressure dominates the central pressure, then the half-thickness of the disk is $h \sim \dot{m} r_g$ and parameters characterizing the internal disk structure are given by

$$\text{inflow time:}\ t_{in} \sim \alpha^{-1} \left(\frac{h}{r}\right)^{-2} \frac{r}{v_{Kep}} \sim 0.6 \frac{m_8}{\alpha \dot{m}^2}\, \text{yr}, \tag{3a}$$

$$\text{electron scattering opacity:}\ \tau_{disk} \sim \frac{100}{\alpha \dot{m}}, \tag{3b}$$

$$\text{electron density:}\ n \sim \frac{100}{\alpha \dot{m}^2 \sigma_T r_g} = 10^{13} (\alpha \dot{m}^2 m_8)^{-1}\, \text{cm}^{-3}. \tag{3c}$$

Gas pressure dominates at $20\, r_g$ if the central temperature, T_c, exceeds $T_g = 1.4 \times 10^9 \dot{m}^2$ K, in which case the values of t_{in}, τ_{disk}, and n given in equations (3) must be multiplied by $(T/T_g)^{-1}$, $(T/T_g)^{-1}$, and $(T/T_g)^{-3/2}$, respectively.

One can determine the quality of the emitted radiation in principle by comparing the various emission and absorption processes. The temperature is determined self-consistently by balancing heating and cooling. In practice, this procedure is complicated by the uncertain role of nonthermal processes. However, by using the fiducial energy densities judiciously, we can suggest plausible scaling laws for these mechanisms.

The dominant radiation mechanisms in a disk consisting primarily of electron-ion plasma are thermal bremsstrahlung and nonthermal synchro-Compton radiation produced by a small population of highly relativistic electrons. Much of the nonthermal radiation may arise in a disk corona that is heated mechanically.[16] If the disk or corona is heated to temperatures in excess of a few $\times$ 10^9 K, an abundance of electron-positron pairs will be created. The observable consequences of this are being explored by several groups; they will not be discussed further here.

Thermal Radiation

Bremsstrahlung is not only able to cool a thin disk, but for $\dot{m}$ or $\alpha \lesssim O(1)$, the radiation is thermalized with a central temperature,

$$T_c \sim \begin{cases} 4.5 \times 10^5 \, (\alpha m_8)^{-1/4} \text{ K} & \text{radiation pressure dominated} \\ 2.2 \times 10^6 \, (\alpha m_8)^{-1/5} \dot{m}^{2/5} \text{ K} & \text{gas pressure dominated.} \end{cases} \quad (4)$$

Gas pressure dominates in the disk if $\dot{m} \lesssim 0.016 \, (\alpha m_8)^{-1/8}$. Electron scattering opacity exceeds the free-free opacity at all heights above the disk plane, so the escaping thermal radiation will be a diluted blackbody, with a color temperature, T_{disk}, in the range of $T_c > T_{disk} > T_c/\tau_{disk}^{1/4} \sim 1.4 \times 10^5 \, (\dot{m}/m_8)^{1/4}$ K. Temperatures at the lower end of this range are consistent with the broadband spectral fits to the "UV bump" preferred by Malkan,[9] but they are much higher than the 25,000–30,000 K indicated by the single-temperature blackbody fits,[8,9,11] unless $\dot{m}/m_8 < 10^{-3}$. In any case, the insensitivity of T_{disk} to $\dot{m}$, m_8, or α makes it very risky to use the observed spectrum to infer anything about disk parameters or the mass of the black hole.

A thermal corona at $T_{cor} \lesssim 10^9$ K may produce a quasi-power-law component of the spectrum by Comptonizing soft radiation from the disk.[17,18] The resulting spectral index depends sensitively on the "Compton y-parameter," $y = 4k T_{cor}/m_e c^2$ max $[\tau, \tau^2]$, which must be $\sim O(1)$ to form a power law over an appreciable range of frequencies shortward of the injection frequency. Production of a standardized spectrum would require some kind of sensitive feedback that regulates the optical depth and/or temperature of the corona.[19] For a special index, $s \equiv -[d \log F_\nu/d \log \nu]$ between 0 and 1, the ratio of Comptonized (power law) to thermal flux from the disk is

$$\frac{F_{comp}}{F_{disk}} \sim \min[1, \tau_{cor}] e^{\tau_{cor}} \left(\frac{T_{cor}}{T_{disk}}\right)^{1-s}, \quad (5)$$

where τ_{cor} is the electron scattering optical depth of the corona. The catastrophic cooling that the corona must suffer if $s < 1$ (for $y \gtrsim 1$) suggests a natural mechanism for producing a standard ν^{-1} power law. If T_{cor} is maintained at $\sim 10^9$ K (e.g., by pair processes), then the value of τ_{cor} required to give $y = 1$ will hover around unity. The sensitivity of F_{comp}/F_{disk} to both τ_{cor} and s would account for the wide range of "nonthermal to thermal" flux ratios inferred by Malkan and Sargent.[8] Presumably, the heating of the corona is accomplished through shocks or magnetic phenomena in the underlying disk. Progressive quenching of this energy supply at increasing $\dot{M}$ would show up observationally as a positive correlation between the total luminosity and the ratio of quasi-blackbody to power-law flux. The existence of such a correlation is claimed by Malkan and Sargent.[8]

Nonthermal Radiation

It is likely that any accretion disk corona heated by shocks or magnetic flares would contain a healthy admixture of relativistic electrons. Indeed, there is no reason why a substantial fraction of the energy density within the disk itself could not reside in relativistic electrons. We must therefore examine the contributions of such particles to the overall spectrum.

Relativistic particles are generally thought to have nonthermal distribution functions, with energies distributed according to a power law, $n(\gamma) \propto \gamma^{-p}$, where γ is the Lorentz factor. A power-law distribution of particles produces a power-law spectrum, with a spectral index, $(p - 1)/2$, regardless of whether the emission is associated with synchrotron radiation, inverse Compton scattering, or the so-called "synchrotron self-Compton" process.

The emissivity of a group of relativistic electrons in a logarithmic bin of Lorentz factors around γ may be written as

$$\epsilon(\gamma) \sim \gamma U_{rel}(\gamma) U_{inj} \frac{c\sigma_T}{m_e c^2}, \tag{6}$$

where $U_{rel}(\gamma) \sim n(\gamma) \gamma^2 m_e c^2$ and U_{inj} is the "injected" energy density. For synchrotron radiation, $U_{inj} = U_{mag} = B^2/8\pi$, while for inverse Compton scattering of the thermal disk spectrum, $U_{inj} = U_{disk}$ outside the disk and $\sim \tau U_{disk}$ at optical depth, τ, inside the disk. First, consider nonthermal emission from particles mixed in with a disk corona. Such a corona is unlikely to be very optically thick in thermal particles, so it is safe to assume that the scattering optical depth in highly relativistic electrons is quite small. We can scale the magnetic field strength in terms of the radiation energy density, $U_{mag} = \eta_{mag} U_{disk}$; η_{mag} may be as large as τ_{disk} if the field is anchored in a radiation pressure dominated disk and larger if the disk is gas pressure dominated. The ratio of synchrotron to Compton emissivity for a given group of electrons is simply η_{mag}. The field strength is given by

$$B \sim 6 \times 10^3 \left(\frac{\dot{m}\eta_{mag}}{m_8}\right)^{1/2} G \tag{7}$$

and electrons with Lorentz factors, $\sim \gamma$, produce synchrotron radiation at a characteristic frequency of

$$\nu(\gamma) \sim 7.4 \gamma^2 \left(\frac{\dot{m}\eta_{mag}}{m_8}\right)^{1/2} GHz. \tag{8}$$

For $s \sim 1$, the integrated emissivity is given by

$$\epsilon_{synch} \sim \gamma_{min}\eta_{mag}\left(\frac{U_{rel}}{U_{disk}}\right)\frac{U_{disk}^2 c\sigma_T}{m_e c^2}, \qquad (9)$$

where γ_{min} is the smallest Lorentz factor that contributes to the power law and U_{rel} is the total energy density in relativistic electrons with $\gamma > \gamma_{min}$.

In practice, γ_{min} will be set by synchrotron self-absorption, which limits the brightness temperature to

$$kT_b \sim \epsilon_{synch} h_{cor} \frac{c^2}{2\nu^2(\gamma)} < \frac{\gamma m_e c^2}{3}, \qquad (10)$$

where h_{cor} is the scale height of the corona. Without explicit knowledge of h_{cor}, we can express γ_{min} (and ν_{min}) in terms of the total energy density of synchrotron emission in the corona, $U_{synch} \sim \epsilon_{synch} h_{cor}/c$:

$$\frac{U_{synch}}{U_{disk}} \sim \frac{\nu_{min}^3 (\gamma_{min})\gamma_{min} m_e c^2}{c^3 U_{disk}}. \qquad (11)$$

Using equations (7), (8), and (9) with equation (2), we obtain

$$\nu_{min} \sim 7.8 \times 10^{13} \eta_{mag}^{1/14} \left(\frac{\dot{m}}{m_8}\right)^{5/14} \left(\frac{U_{synch}}{U_{disk}}\right)^{2/7} \text{ Hz} \qquad (12a)$$

and

$$\frac{U_{synch}}{U_{disk}} = \frac{F_{synch}}{F_{disk}} \sim 2.2 \times 10^5 \eta_{mag}^{11/12} \dot{m}^{13/12} m_8^{1/12} \left(\frac{U_{rel}}{U_{disk}}\right)^{7/6} \left(\frac{h_{cor}}{20 r_g}\right)^{7/6}. \qquad (12b)$$

If the corona is dominated by thermal particles at $T \sim 10^9$ K, then the factor of 2.2×10^5 $(h_{cor}/20\, r_g)^{7/6}$ in equation (12b) can be replaced by 5.6×10^3. Thus, a very modest energy density of $U_{rel}/U_{disk} \sim 10^{-4} \dot{m}^{-1}$ in relativistic electrons is all that is required to produce a synchrotron flux comparable with the thermal disk flux, shortward of infrared wavelengths.[d]

Now consider inverse Compton scattering of the thermal disk radiation by the same relativistic particles. The emissivity is again given by equation (9), with $\eta_{rad} = 1$ replacing η_{mag}, and this time there is no lower limit on γ imposed by self-absorption. For an $s \sim 1$ spectrum, the Compton emissivity is essentially η_{mag}^{-1} times the synchrotron emissivity, but $U_{rel}(\gamma)$ must continue to Lorentz factors below γ_{min} to produce photons with energies in the range of $kT_{disk} < h\nu < \gamma_{min}^2 kT_{disk}$. Photons in this energy range could also be produced by the synchrotron process if $n(\gamma)$ extends to high enough energies.

The synchrotron self-Compton process, in which the same relativistic electrons produce the synchrotron photons and subsequently Comptonize them, will also operate

---

[d] It is well known that observable radio and submillimeter flux cannot come from stationary gas this near the black hole, but rather must come from gas at larger radii or from an ultrarelativistic jet.

in the corona and will be particularly important if $U_{synch} > U_{disk}$. The power laws produced by the first order (synchrotron) and second order (Compton) processes will be the same, but the two continua will join to form a single power law only if $U_{synch} \sim U_{mag}(=\eta_{mag}U_{disk})$. Thus, necessary conditions for the synchrotron self-Compton process to form a single power law from the infrared to X rays include (1) that $\eta_{mag} > 1$ and (2) that the nonthermal flux at least be comparable with the thermal flux from the underlying disk.

Nonthermal energy production inside the disk is much more complicated than analogous processes in the corona since photons must run the gamut of free-free absorption and multiple electron scattering before they can escape. We do not expect U_{mag} to exceed the radiation energy density in this region unless the disk is gas pressure dominated (and some authors[20,21] have proposed even more stringent limits on U_{mag}), and in any case, most synchrotron-produced photons with $h\nu < kT_{disk}$ will be absorbed. On the other hand, if a thermal photon suffers one relativistic Compton scattering at a modest optical depth, τ, within the disk, it is unlikely to be absorbed subsequently since the free-free absorption cross section declines rapidly with frequency ($\propto \nu^{-3}$). The real enemy of a Comptonized photon trying to escape the disk is Compton cooling by the thermal gas, which is effective if

$$\frac{\tau^2 h\nu}{m_e c^2} > 1. \tag{13}$$

If we define a γ-dependent optical depth to Compton scattering,

$$\tau(\gamma) \sim \frac{\gamma n_{rel}(\gamma)}{n_{thermal}}\tau \sim \frac{U_{rel}(\gamma)}{U_{thermal}}\frac{kT_{disk}}{\gamma m_e c^2}\tau, \tag{14}$$

and set $h\nu \sim \gamma^2 kT_{disk}$ in equation (13), then we obtain an upper limit on the fraction, $f(\gamma)$, of photons that may escape with an energy of $\sim \gamma^2 kT_{disk}$:

$$f(\gamma) \lesssim [\tau(\gamma)\tau]_{max} \sim \left(\frac{U_{rel}(\gamma)}{U_{thermal}}\right)\gamma^{-3}. \tag{15}$$

If $\tau(\gamma)\tau < [\tau(\gamma)\tau]_{max}$, then the Comptonized part of the spectrum will be a power law with $s = (p-1)/2$; where $\tau(\gamma)\tau > [\tau(\gamma)\tau]_{max}$, the Comptonized flux will saturate at

$$F_{Comp}(\gamma) \sim \left(\frac{U_{rel}(\gamma)}{U_{thermal}}\right)\gamma^{-1}F_{disk} \propto \gamma^{1-p}F_{disk}, \tag{16}$$

giving a spectrum,

$$(F_\nu)_{Comp} \propto \nu^{-[(p+1)/2]} \propto \nu^{-(s+1)}, \tag{17}$$

where s is the "normal" spectral index, $(p-1)/2$.

RADIATION TORI

The fate of a rotating accretion flow with $\dot{m} \gg 1$ is still hotly debated. Shakura and Sunyaev[15] conjectured that any influx in excess of $\dot{m}$ of about a few would be blown away in a radiation pressure driven wind as the flux of binding energy released by the accreted material exceeded L_E. Meier[22–24] computed models for the radiative properties of such winds, but did not delve into the details of their creation. An alternative viewpoint was developed by Abramowicz, Paczyński, Wiita, and collaborators, who suggested that inflowing gas with arbitrarily large $\dot{m}$ could settle into a quasi-stationary "radiation torus."

The qualitative features of radiation tori are sketched in FIGURE 1. A transition from thin disk to torus occurs where pressure forces become comparable with gravity. Pressure forces point outward in the outer parts of a torus, but they must point inward in the innermost region because the black hole is a perfect drain for bound material with angular momentum smaller than a threshold value. Thus, a torus must contain a pressure maximum (FIGURE 1c) where material orbits the hole at the local (relativistic) Keplerian speed. This pressure maximum is usually assumed to lie at a few gravitational radii, but it must lie outside the radius of the marginally stable orbit ($r_{ms} = 6\,r_g$ for a Schwarzschild black hole) because there are no pressure forces acting on the gas at the pressure maximum. Inside the pressure maximum, pressure forces augment gravity and material may be able to orbit the black hole at radii as small as the marginally bound orbit ($r_{mb} = 4r_g$ for a Schwarzschild black hole).

In practice, a quasi-stationary torus is expected to have an inner surface with an equatorial cusp[25–27] located somewhere between r_{mb} and r_{ms}. The exact location of the cusp, r_{cusp}, determines (or is determined by) the amount of binding energy that must be

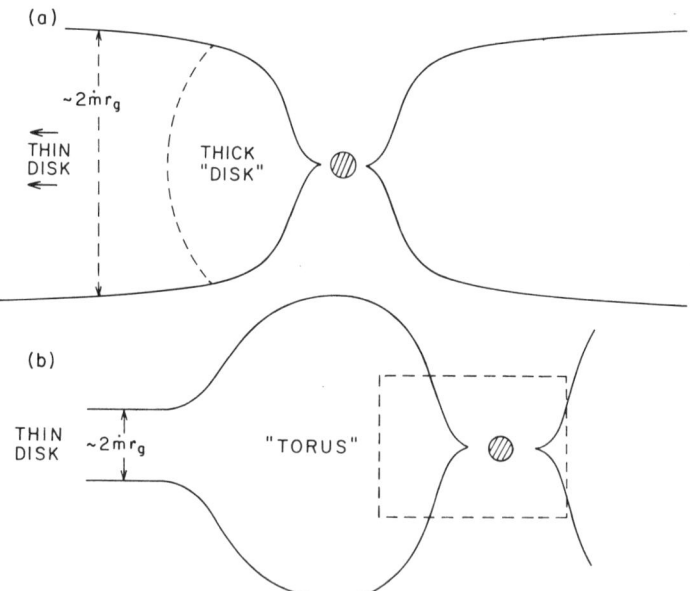

FIGURE 1a, b. It is unclear how a radiation "torus" would join onto an outer, thin accretion disk. Outside the torus, the disk will be supported by radiation pressure and have a uniform thickness, $\sim 2\,\dot{m}r_g$. The torus may begin at a radius comparable to that thickness (a) or at a larger radius (b). Dashed line in (b) encloses region shown in FIGURE 1c. Reprinted with permission from University Science Books.

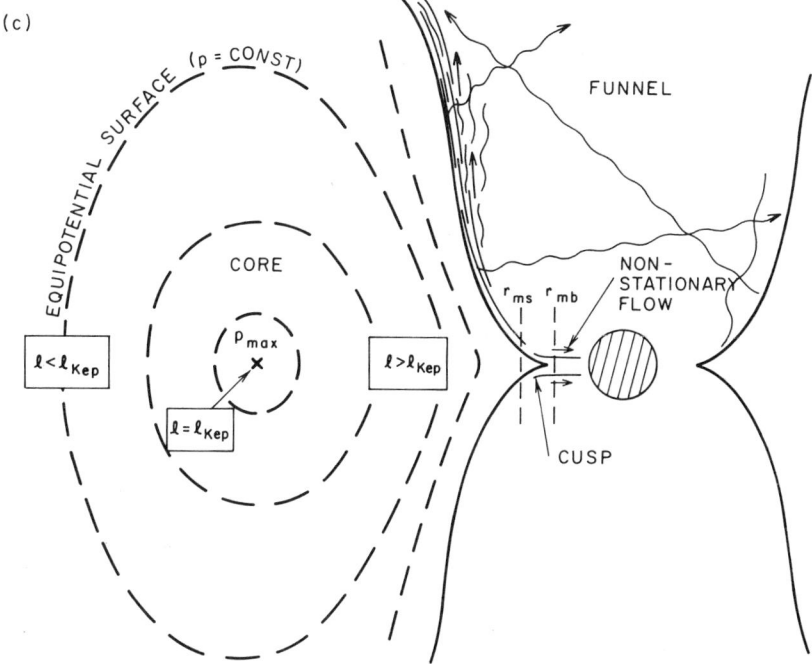

FIGURE 1c. The inner part ("core") of a stationary torus has a pressure maximum where the specific angular momentum, ℓ, equals the Keplerian value. Inside this radius, $\ell > \ell_{Kep}$ and inward-directed pressure forces allow material to remain in stable orbits down to a radius that is somewhat smaller than the normal marginally stable orbit (r_{ms}), but larger than the marginally bound orbit (r_{mb}). The cusp is like a Lagrangian point in a mass-transferring binary and marks the transition from a slow inward drift to nearly radial free-fall. The centrifugally dominated funnel is filled with radiation and is a possible site of jet formation. Outside the pressure maximum, the pressure forces point outward and $\ell < \ell_{Kep}$. At large r, the structure of a radiation torus may resemble that of a radiation pressure supported star. Reprinted with permission from University Science Books.

radiated away by the torus, ranging from $>0.057 \dot{M}c^2$ for a cusp at r_{ms} (thin disk limit) to an arbitrarily small fraction of $\dot{M}c^2$ as the cusp nears r_{mb}.[28] The total luminosity of a radiation torus can exceed the Eddington limit at most by a factor of $\sim \ln(h_t/r_{cusp})$, where h_t is the total height of the torus;[29] consequently, tori with $\dot{m} \gg 1$ must have cusps lying close to r_{mb}.[e] The radiative efficiency of a radiation torus is therefore $\sim \dot{m}^{-1}$ times the logarithmic factor.

What is the quality of the radiation produced by a radiation torus? We can make

---

[e] Most of the logarithmic enhancement of L over L_E occurs in the "funnel" that surrounds the rotation axis.[30] For a torus with a small binding energy, this funnel may be quite narrow and may contain a high radiation flux density, thus giving rise to the hope that it would engender and collimate a radiation pressure driven jet.[31] However, recent investigations[32,33] have revealed severe limitations on jet powers, speeds, and collimations obtainable in this way. It now appears that more complicated and less well-understood processes must be operating if powerful jets do emerge from radiation tori.[34,35]

some crude guesses by thinking of the torus as a toroidal star supported by radiation pressure. The "core" will be a toroidal region surrounding the pressure maximum, with a total thickness comparable with the radius of the pressure maximum (say, $6\,r_g$). The condition of hydrostatic equilibrium then implies a relation between the maximum

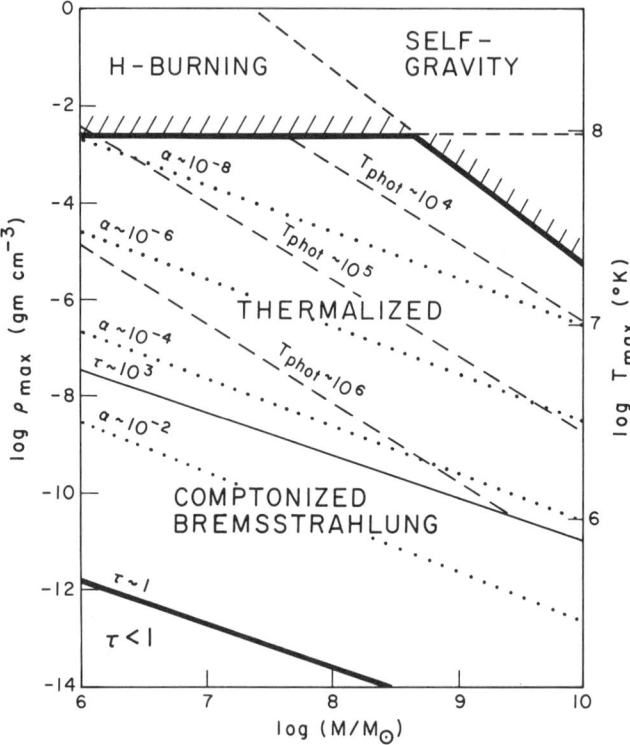

FIGURE 2. Conditions in the "core" of a radiation torus, in terms of the maximum density (ρ_{max}) and the mass of the black hole. Tori are excluded from the upper region of the parameter plane by the onset of nuclear energy generation at a rate exceeding L_E and from the upper right-hand corner by the onset of gravitational instabilities. Radiation in the interior of the torus is thermalized when the electron scattering optical depth is $\gtrsim 10^3$, above the solid line marked "$\tau \sim 10^3$." Central core temperatures in the thermalized regime are marked on the right-hand ordinate. Photospheric temperatures for an isentropic envelope are indicated by dashed sloping lines; if the torus is not isentropic (but is convectively stable), then these lines will be shifted upwards. Lines of constant α (consistent with dissipative energy generation at a rate of $\sim L_E$) are dotted. Reprinted with permission from University Science Books.

pressure and maximum density, $p_{max} \sim 0.1\,\rho_{max}c^2$. FIGURE 2 shows the physical state of the core as a function of ρ_{max} and $M/M_\odot$. In the lower left-hand corner, where the optical depth to electron scattering is <1, the viscous α parameter exceeds one, so radiation tori do not really occur in this region of the parameter plane. Below and to the left of the line labeled "$\tau \sim 10^3$," the radiation field is that of Comptonized bremsstrahlung; only above the line is the opacity great enough to fully thermalize the radiation. In this region, the right-hand ordinate is labeled to show the maximum temperature in the torus. When T_{max} exceeds $\sim 10^8$ K (for $10^6 < M/M_\odot < 10^9$), the energy output due to thermonuclear fusion exceeds L_E. Since the efficiency of fusion is nearly 0.01, it is clear that inefficient tori cannot exist above this line in the parameter plane. Finally, in the upper right-hand corner of the graph, the mass of the torus becomes comparable with the mass of the black hole and dynamical instabilities will prevent the establishment of a quasi-stationary flow.

The dashed lines in FIGURE 2 are lines of constant photospheric temperature for an isentropic torus. Note that temperatures in the range associated with the "UV bump"[8,9] require rather high central densities. For AGNs with $L \sim L_E < 10^{46}$ ergs s^{-1}, the central mass must be less than $10^8 \, M_\odot$, which implies that these tori fall not too far below the hydrogen-burning limit. Begelman[36] suggested that the onset of hydrogen burning regulates the central density in radiation tori. If the entropy in the torus is not uniform, it must increase outward (entropy decreasing outward would lead to convection and/or large-scale circulation under most circumstances). The density would then drop off more rapidly than in the isentropic torus and the photospheric temperature would be higher than for an isentropic torus with the same central density. Thus, lack of isentropy exacerbates the difficulty of finding a niche in parameter space that will accommodate Malkan's results, particularly if one adopts the single-temperature blackbody fits.[10]

Now consider radiation tori as accretion flows. The viscous dissipation rate per unit volume equals the viscous stress times the rate of strain and we expect the total emissivity of the core to be of the order of L_E, i.e.,

$$10\alpha\rho_{max}c^3r_g^2 \sim L_E \tag{18}$$

for an α-model viscosity. Thus, α is uniquely determined by ρ_{max} and $M/M_\odot$; loci of constant α are plotted as dotted lines in FIGURE 2. Apparently, a radiation torus capable of producing a "UV bump" must have an extremely small viscosity, $\alpha \lesssim 10^{-8}$. Such small values of α may seem unpalatable since turbulent or magnetic stresses are generally thought to yield $\alpha \gtrsim 10^{-3}$.[20,37,38] However, they do not seem unreasonably small if an analogy is made with the molecular viscosity in a differentially rotating star, such as the sun.

The relation between α, ρ_{max}, and $M/M_\odot$ implies that the inflow speed scales as $\alpha\dot{m}v_{Kep}$. This contrasts with the extrapolation of thin disk theory, according to which $v_{in} \sim \alpha(h/r)^2 v_{Kep} \sim \alpha v_{Kep}$ for a torus with $h \sim r$. The difference arises from the fact that the inflow speed is proportional to the viscous stress divided by the angular momentum gradient. For a given angular momentum, the gradient in the core of a radiation torus must be smaller than that in a thin disk by a factor of $\sim \dot{m}^{-1}$. Since only a small amount of angular momentum need be transferred to drive the inflow, a large $\dot{M}$ can be sustained without the dissipation rate exceeding a few times L_E.

Like thin disks, radiation tori probably possess coronae containing a mixture of thermal and nonthermal particles. The radiative properties of these coronae can be analyzed in much the same way as those that surround thin disks, so the details will not be repeated here.

ION TORI

The thin disk flow described above can persist to very low values of $\dot{m}$. As $\dot{m}$ decreases, so does the flux of energy that needs to be radiated. As a result, the disk is cooler, (h/r) is smaller, and the density inside the disk remains sufficiently large that bremsstrahlung can continue to cool the disk. However, if a disk with $\dot{m} < 50 \, \alpha^2$ min$[1,(2000 \, r_g/r)^{1/2}]$ is ever heated to its local virial temperature, $T_{vir} \sim (GMm_p/3kr)$ at r, then the regions inside this radius may never cool back down. In other words, accretion flows with sufficiently low $\dot{m}$ may have two stable thermal states.

Wherever cooling is ineffective, the disk must swell into a torus supported by gas pressure, with a temperature comparable to the virial temperature of the protons. Inside 2000 r_g, however, this temperature approaches and exceeds the temperature associated with the rest mass energy of the electrons, and a range of very efficient cooling processes (especially relativistic synchrotron emission and relativistic Compton scattering) "turn on." Thus, if the electrons and protons are thermally coupled ($T_e \sim T_p \sim T_{vir}$), the electrons will rapidly drain energy from the protons at $r < 2000 \, r_g$ and radiate it away, and the torus will deflate. If, however, the electrons and protons are thermally decoupled ($T_e \ll T_p$), then the torus can survive, supported almost entirely by the thermal pressure of the ions. "Two-temperature" bistable disks were first proposed[39] to account for the "high" and "low" states of the galactic X-ray source, Cygnus X-1.

Electrons and protons are coupled at least by Coulomb collisions and the minimal condition for thermal coupling is roughly the same as for the failure of bremsstrahlung cooling at 2000 r_g,[5] namely

$$\dot{m} < 50 \, \alpha^2. \tag{19}$$

It is not known whether plasma collective effects, perhaps resulting from shear-induced anisotropies, will couple the electrons and protons on a shorter timescale.

Structurally, ion tori share many features in common with radiation tori, including a core (surrounding the pressure maximum), a cusp, and (if the binding energy is small) a funnel. Like radiation tori, ion tori are inefficient radiators of the energy released through accretion. If the electron-ion coupling is entirely due to Coulomb collisions, then the radiative efficiency declines linearly with $\dot{m}$ from 0.1 at $\dot{m} \sim 50 \, \alpha^2$, but levels off when it reaches $\epsilon \sim 3 \times 10^{-4}$ because energy input into the electrons is then dominated by adiabatic heating rather than Coulomb energy transfer. The quality of the radiation emitted by an ion torus is very different from that of a radiation torus or thin disk. The torus will be optically thin to electron scattering and opaque to incoherent synchrotron or cyclotron emission (in fact, to any photons longward of the far infrared). Relativistic thermal bremsstrahlung may be a source of γ-ray photons and Compton upscattering of synchrotron and bremsstrahlung photons may produce photons at all wavelengths shortward of the far IR. Much of the Compton power may be contributed by a nonthermal distribution of ultrarelativistic electrons; hence, spectral details will depend sensitively on particle acceleration mechanisms. However, it is fair to say that the spectrum of an ion torus will have a nonthermal character.

The important feature of an ion torus is not the spectral quality per se, but the low radiative efficiency. This property makes ion tori prime candidates for powering the jets in powerful radio galaxies,[5] where the radiative output at all accessible frequencies is puny compared with the inferred kinetic energy fluxes in the jets. In addition to the direct conversion of energy released through accretion into jet energy, perhaps through a hydromagnetic wind,[40] ion tori (as well as radiation tori and thin disks) can catalyze the extraction of spin energy from the black hole through the Blandford-Znajek effect.[2] Here, organized magnetic fields that pierce the ergosphere of the hole are supported by currents in the disk or torus, while the dragging of inertial frames by the spinning hole causes a Maxwell stress that carries energy (and angular momentum) away from the hole. In effect, the disk or torus merely provides inertia to hold the field in place, and in principle no accretion is necessary. In practice, however, it is likely that sufficient field will penetrate the disk or torus so that the resulting accretion (due to the

a)
```
─────────── ION TORI ──?──→ THIN
                       ←─?─ DISKS ←─ RADIATION TORI
──────────────────────────────•──────────────────────→ ṁ
     RADIO GALAXIES           ṁ~1 ←───── QUASARS ─
                      ←───?── SEYFERT I's ─────────
```

b)
```
─────────── ION TORI ──?─→
        ?─────THIN DISKS ────────→ ←─ RADIATION TORI
──────────────────────────────────•─────────────────→ ṁ
     RADIO GALAXIES              ṁ~1    ?
                          ←───── QUASARS ─────→
                          ─────── SEYFERT I's ─→
```

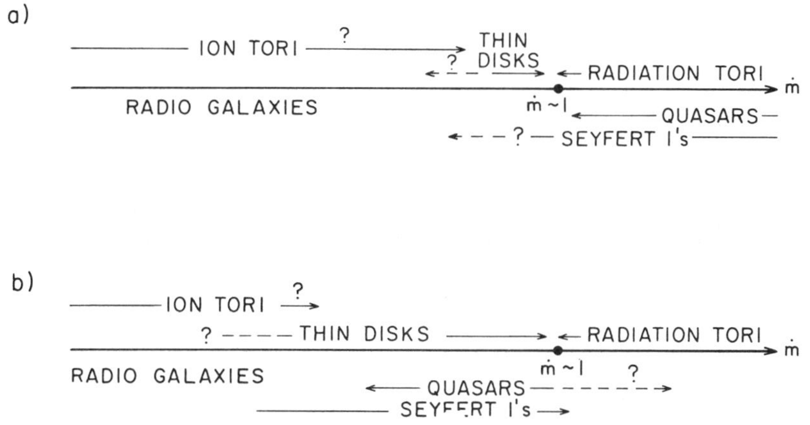

FIGURE 3. One-dimensional "unified" schemes for quasars, Seyfert I's, and radio galaxies, in terms of $\dot{m}$. In (a), quasars and most Seyfert I's are assumed to contain radiation tori, which account for their spectral similarities. If this scheme is adopted, broad-line clouds must be in unbound orbits and Seyferts could not have evolved from quasars. Assumed scarcity of thin disks could be accounted for if $\alpha \gtrsim 0.1$ and sources with $\dot{m} < 1$ contained ion tori. In (b), some quasars and most Seyferts are assumed to contain thin disks, which occur over a wider range in $\dot{m}$. Seyferts could have evolved from quasars in this picture and broad-line clouds may or may not be unbound. Reprinted with permission from University Science Books.

transfer of angular momentum to the field lines) will release energy at a rate comparable with the rate of energy extraction from the hole.

UNIFIED AGN SCHEMES

If dissipative accretion flows and their characteristic spectra can be classified largely on the basis of one parameter, $\dot{m}$, it is tempting to construct a one-dimensional classification scheme for all forms of AGN, from optically quiet radio galaxies (e.g., M87 or Cygnus A) to the most luminous quasars. Such a scheme (FIGURE 3) would associate increasing ratio of nonthermal to thermal emission with decreasing $\dot{m}$. Bright quasars and possibly some Seyfert I's might be associated with radiation tori (FIGURE

3a). This would imply that the broad-line clouds are moving at speeds in excess of the local Keplerian speed. Other implications of radiation tori include:

(1) The low radiative efficiency of the accretion flow ($\sim \dot{m}^{-1}$) exacerbates the already tough problem of finding enough fuel to "feed the monster."
(2) The luminosity of $\sim L_E$ essentially depends only on the mass of the black hole. If Seyfert I's contain radiation tori, they could not have evolved from more luminous quasars, as Weedman's analysis[41] suggests. If one skirts this problem by positing that the Seyferts are thin disks while quasars are radiation tori, then one might expect to find a higher ratio of nonthermal to thermal flux and a higher degree of polarization in Seyferts. Malkan and Sargent[8] report such a trend in their thermal versus nonthermal decomposition; the polarization trend is uncertain because Seyferts are more heavily contaminated by starlight than are quasars.[13] It will be interesting to see whether the thermal components in Seyfert spectra are better characterized as broadband or single-temperature blackbodies (provided that this issue can be resolved for the quasars).
(3) Radiation tori can produce a thermal spectrum with a color temperature compatible with the "UV bump"[8,9] only if the viscosity is very small. Such small values of α are at odds with the current wisdom on magnetic and turbulent viscosity in disks. More worrisome are the global nonaxisymmetric instabilities demonstrated recently by Papaloizou and Pringle[42,43] for tori with idealized boundary conditions. Several groups are now trying to evaluate the importance of this dynamical instability in more realistic systems. If the instability occurs in all tori, it may place a lower bound on α, which would effectively rule out the production of the "UV bump" by a radiation torus.

There is presently no compelling reason to prefer radiation tori to thin disks plus coronae in any individual object. One might therefore array the sources as shown in FIGURE 3b. In this picture, the broad-line clouds may or may not move with super-Keplerian velocities. If Seyfert I's evolved from quasars, then they would have systematically smaller values of $\dot{m}$, but it is not clear that this would show up in the spectral data since the factors governing energy injection into the corona are unknown. One difficulty with the thin disk picture is that it leads one to expect relatively high percentage polarizations of $\lesssim 10\%$, whereas most quasars have polarizations of $\lesssim 1\%$.[12] Standard thin disks would also be in trouble if it is confirmed that quasar (and Seyfert I) spectra contain a single-temperature blackbody component at 25,000–30,000 K.

The overlap between Seyferts and quasars in FIGURE 3 points out the fundamental limitation in this type of diagram: different classes of AGN are distinguished not only by radiative efficiency and spectral qualities, but also by absolute energy output. What is really needed is a two-parameter diagram, say, in the $\dot{m}$-M or $\dot{M}$-M plane. FIGURE 4 is an illustrative example of the latter, with diagonal lines of constant $\dot{m}$ labeled. Seyferts that have evolved from quasars lie to the left of their progenitors, while Seyferts that have never been quasars would tend to lie below and to the left of spectrally similar quasars, along lines of constant $\dot{m}$. Presumably, LINERS are also found at similar values of $\dot{m}$, but at much lower values of M.

Predominantly nonthermal sources, such as radio galaxies and the nucleus of the Milky Way, would contain ion tori and occupy the region with $\dot{m} \ll 1$. Even if most of the energy powering these objects were extracted electromagnetically from a "black hole flywheel,"[4,44] one would still expect a strong correlation between $\dot{M}$ and the total energy output, $\dot{E}$, because inflowing material is still needed to "anchor" the magnetic field in the ergosphere of the hole.[5] Naive scaling arguments suggest that the total efficiency, $\dot{E}/\dot{M}c^2$, is of the order of unity. If this is true then the most powerful radio

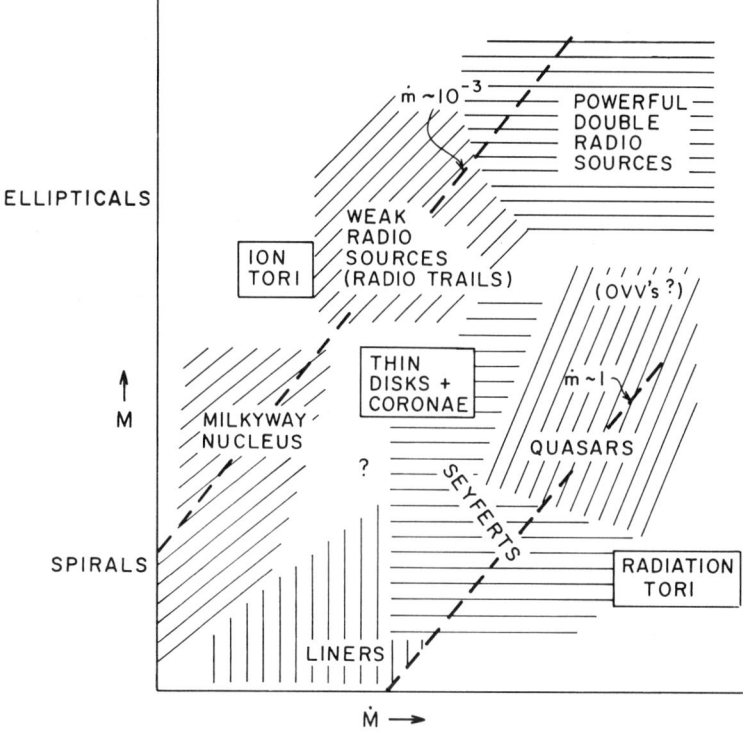

FIGURE 4. A two-dimensional "unified" AGN scheme is displayed schematically in the $\dot{M}$-M plane. Possible parameter space for optically active AGNs (quasars, Seyferts, LINERS) lies on either side of the diagonal (dashed) line of constant $\dot{m} \sim 1$. Seyferts that evolved from quasars would lie to the left of the quasar region (cf., FIGURE 3). Parameter space for predominantly nonthermal sources, including giant elliptical radio galaxies and the nucleus of the Milky Way, straddles the dashed line of constant $\dot{m} \sim 10^{-3}$. Note that powerful double radio sources must have very massive black holes if they are powered by accretion at $\dot{m} \ll 1$. Optically violently variable quasars (OVVs) are shown tentatively at an intermediate point between quasars and radio sources. Ordinate labels indicating galaxy type suggest that correlations between galaxy type and types of activity may result from a correlation of galaxy type with hole mass. See text for a critical discussion of this type of unified scheme. Reprinted with permission from University Science Books.

galaxies (e.g., Cygnus A), which appear to spew forth in jet kinetic energy what bright quasars emit in radiation, should lie vertically above these quasars in FIGURE 4. These radio galaxies would have to contain black holes that are more massive than quasar black holes by a factor of $\dot{m}_{quasar}/\dot{m}_{rad.\,gal.} \gtrsim 10^3$. Even if quasars were tucked up against the Eddington limit, the most powerful radio galaxies would have to have black hole masses in excess of $10^{11}\ M_\odot$. This conclusion may be tested by taking spectra of the weak emission lines often found in the nuclei of active ellipticals and/or measurements of stellar velocity dispersions. There are some conjectural ways out of this conclusion,[5] but their plausibility remains to be seen. Weaker radio galaxies do not present these difficulties and indeed there is evidence for mass concentrations in excess of $10^9\ M_\odot$ in the nuclei of M87[45] and NGC 6251.[46]

If we believe that these sources are ion tori, then from the fact that strong radio galaxies are ellipticals while Seyferts and quasars seem to be in spirals, we could draw the conclusion that ellipticals contain more massive central black holes than do spirals, while the supply of accreting matter is comparable or smaller. The properties may also be correlated with the ability of the galaxy to produce a large-scale collimated jet.[47]

Nevertheless, there remains the unsolved problem of whether ion tori can exist at all. In addition to the fundamental question of ion-electron coupling, there is also the question of why most of the material would go into an ion torus at low $\dot{m}$, when an ordinary thin accretion disk appears to be a viable alternative. This question was addressed by Shapiro, Lightman, and Eardley[39] in their two-temperature disk model for Cygnus X-1. Their motivation for studying an apparently bistable system was the observed switching of Cyg X-1 between "high" and "low" states, and their model assumed that the system was capable of switching spontaneously between the two modes of accretion. However, there is no evidence that giant elliptical radio galaxies ever become quasars and it is virtually certain that Seyferts and radio galaxies are very different beasts.

In view of these uncertainties, it is worth considering the possible existence of "nondissipative" accretion disks. Blandford and Payne[40] showed that open magnetic field lines piercing a disk could efficiently remove angular momentum and binding energy, transferring it to a hydromagnetic wind. The magnetic field required to drive a mass flux, $\dot{M}$, through a radius, R (actually, the geometric mean of the poloidal and toroidal components, or, equivalently, the square root of the Reynolds stress), scales as

$$B \sim \frac{\dot{M}^{1/2}}{R}\left(\frac{GM}{R}\right)^{1/4} \propto R^{-5/4} \qquad (20)$$

and attains a value comparable with that given by equation (7) at $R \sim 20\ r_g$, $B(20\ r_g) \sim 10^4\ (\dot{m}/m_8)^{1/2}$ G.

Fields of this strength or higher (since the pressure associated with this field is still smaller than the pressure inside the disk) could have been created by shear and reconnection within the disk,[20,48,49] but it is not clear that these processes would lead to the desired forest of open field lines. However, if the accreted gas starts out with a substantial net magnetic flux, then it might retain this flux all the way into the nucleus. The typical magnetic flux density found in the interstellar medium of the Milky Way would be sufficient to drive a respectable accretion rate. The ratio of dissipative to dissipationless energy loss might be related to the ratio of magnetic flux density to column density in the accreted gas and differences in the latter ratio could determine whether a given flow produces an extended radio source or a quasar. Admittedly, there is a great deal of doubt about whether the sheared magnetic field could carry away so much angular momentum and energy without producing a comparable amount of dissipative heating within the disk.

To complete a classification scheme for AGNs, one would like to insert all of the other beasts in the AGN zoo—Lacertids, Type II Seyferts, etc.—into their niches in the $\dot{M}$-M plane. Additional parameters, such as viewing angle, will very likely prove to be important, leading to more complicated diagrams with three or more axes.

ACKNOWLEDGMENTS

Many of the views discussed above have been developed collectively during a long-term collaboration with Roger Blandford, Sterl Phinney, and Martin Rees. Portions of the text are excerpted from an article by the author entitled, "Accretion Disks in Active Galactic Nuclei," which appeared in *Astrophysics of Active Galaxies and Quasi-Stellar Objects,* edited by J. S. Miller (1985). Copyrighted materials are reprinted with permission from University Science Books.

REFERENCES

1. LYNDEN-BELL, D. 1969. Nature **223:** 690.
2. BLANDFORD, R. D. & R. L. ZNAJEK. 1977. Mon. Not. R. Astron. Soc. **179:** 433.
3. MACDONALD, D. & K. S. THORNE. 1982. Mon. Not. R. Astron. Soc. **198:** 345.
4. PHINNEY, E. S. 1983. Ph.D. Thesis. University of Cambridge.
5. REES, M. J., M. C. BEGELMAN, R. D. BLANDFORD & E. S. PHINNEY. 1982. Nature **295:** 17.
6. PRINGLE, J. E. 1981. Annu. Rev. Astron. Astrophys. **19:** 137.
7. PRINGLE, J. E., M. J. REES & A. G. PACHOLCZYK. 1973. Astron. Astrophys. **29:** 179.
8. MALKAN, M. A. & W. L. W. SARGENT. 1982. Astrophys. J. **254:** 22.
9. MALKAN, M. A. 1983. Astrophys. J. **268:** 582.
10. BLANDFORD, R. D. 1985. *In* Numerical Astrophysics. J. Centrella, J. LeBlane & R. Bowers, Eds.: 6. Jones & Bartlett. Portola Valley, California.
11. BLANDFORD, R. D. 1985. *In* Active Galactic Nuclei. J. Dyson, Ed. Manchester University Press. Manchester, England.
12. ANGEL, J. R. P. & H. S. STOCKMAN. 1980. Annu. Rev. Astron. Astrophys. **18:** 321.
13. ANTONUCCI, R. R. J. 1983. Nature **303:** 158.
14. KROLIK, J. H., C. F. MCKEE & C. B. TARTER. 1981. Astrophys. J. **249:** 422.
15. SHAKURA, N. I. & R. A. SUNYAEV. 1973. Astron. Astrophys. **24:** 337.
16. LIANG, E. T. P. & R. H. PRICE. 1977. Astrophys. J. **218:** 247.
17. KATZ, J. I. 1976. Astrophys. J. **206:** 910.
18. TAKAHARA, F., S. TSURUTA & S. ICHIMARU. 1981. Astrophys. **251:** 26.
19. IONSON, J. A. & M. KUPERUS. 1984. Astrophys. J. **284:** 389.
20. CORONITI, F. V. 1981. Astrophys. J. **244:** 587.
21. STELLA, L. & R. ROSNER. 1984. Astrophys. J. **277:** 312.
22. MEIER, D. L. 1982. Astrophys. J. **256:** 386.
23. MEIER, D. L. 1982. Astrophys. J. **256:** 681.
24. MEIER, D. L. 1982. Astrophys. J. **256:** 693.
25. KOZLOWSKI, M., M. JAROSZYŃSKI & M. A. ABRAMOWICZ. 1978. Astron. Astrophys. **63:** 209.
26. ABRAMOWICZ, M. A., M. JAROSZYŃSKI & M. SIKORA. 1978. Astron. Astrophys. **63:** 221.
27. JAROSZYŃSKI, M., M. A. ABRAMOWICZ & B. PACZYŃSKI. 1980. Acta Astron. **30:** 1.
28. PACZYŃSKI, B. & P. WIITA. 1980. Astron. Astrophys. **88:** 23.
29. ABRAMOWICZ, M. A., M. CALVANI & L. NOBILI. 1980. Astrophys. J. **242:** 772.
30. SIKORA, M. 1981. Mon. Not. R. Astron. Soc. **196:** 257.

31. ABRAMOWICZ, M. A. & T. PIRAN. 1980. Astrophys. J. Lett. **241:** L7.
32. SIKORA, M. & D. B. WILSON. 1981. Mon. Not. R. Astron. Soc. **197:** 529.
33. NARAYAN, R., R. NITYANDA & P. J. WIITA. 1983. Mon. Not. R. Astron. Soc. **205:** 1103.
34. BEGELMAN, M. C. & M. J. REES. 1983. *In* Astrophysical Jets. A. Ferrari & A. G. Pacholczyk, Eds.: 215. Reidel. Dordrecht.
35. BEGELMAN, M. C. & M. J. REES. 1984. Mon. Not. R. Astron. Soc. **206:** 209.
36. BEGELMAN, M. C. 1984. *In* Proceedings of IAU Symposium 110, VLBI and Compact Radio Sources. R. Fanti, K. Kellermann & G. Setti, Eds.: 227. Reidel. Dordrecht.
37. SAKIMOTO, P. J. & F. V. CORONITI. 1981. Astrophys. J. **247:** 19.
38. LYNDEN-BELL, D. & J. E. PRINGLE. 1974. Mon. Not. R. Astron. Soc. **108:** 603.
39. SHAPIRO, S. L., A. P. LIGHTMAN & D. M. EARDLEY. 1976. Astrophys. J. **204:** 187.
40. BLANDFORD, R. D. & D. G. PAYNE. 1982. Mon. Not. R. Astron. Soc. **199:** 883.
41. WEEDMAN, D. W. 1985. *In* Astrophysics of Active Galaxies and Quasi-Stellar Objects. J. S. Miller, Ed.: 497. University Science Books. Mill Valley, California.
42. PAPALOIZOU, J. C. B. & J. E. PRINGLE. 1984. Mon. Not. R. Astron. Soc. **208:** 721.
43. PAPALOIZOU, J. C. B. & J. E. PRINGLE. 1985. Mon. Not. R. Astron. Soc. **213:** 799.
44. PHINNEY, E. S. 1983. *In* Astrophysical Jets. A. Ferrari & A. G. Pacholczyk, Eds.: 201. Reidel. Dordrecht.
45. YOUNG, P. J., J. A. WESTPHAL, J. KRISTIAN, C. J. WILSON & F. P. LANDAUER. 1978. Astrophys. J. **221:** 721.
46. YOUNG, P. J., W. L. W. SARGENT, J. KRISTIAN & J. A. WESTPHAL. 1979. Astrophys. J. **234:** 76.
47. SPARKE, L. S. & F. H. SHU. 1980. Astrophys. J. Lett. **241:** L65.
48. EARDLEY, D. M. & A. P. LIGHTMAN. 1975. Astrophys. J. **200:** 187.
49. PUDRITZ, R. E. 1981. Mon. Not. R. Astron. Soc. **195:** 881.

X-Ray Properties of QSOs and Their Cosmological Implications[a]

Y. AVNI

Harvard-Smithsonian Center for Astrophysics
Cambridge, Massachusetts 02138
and
Weizmann Institute of Science
Rehovot 76100, Israel

INTRODUCTION

The properties of QSOs as X-ray emitters in the ~2 keV region have been the subject of intensive study in recent years. The flux sensitivity and angular resolution of the Einstein Observatory[1] have made it possible to detect X-ray fluxes from many previously known QSOs or to derive meaningful upper limits for the X-ray flux from such QSOs. Thus, the population properties of QSOs at X-ray energies and their cosmological implications could be investigated.

The first study of a sample of previously known QSOs with the Einstein Observatory[2] yielded two fundamentally important conclusions—that X-ray emission from QSOs is frequent and strong:

(1) A detection rate of ~50% led to the conclusion that a large fraction of QSOs emit X-rays.
(2) The derived integrated X-ray luminosities for the detected QSOs were found to be comparable, by order of magnitude, to the integrated optical luminosities.

These findings implied that QSOs may contribute an appreciable fraction of the extragalactic X-ray background at 2 keV and that QSOs can be efficiently discovered through their X-ray emission. Intensive research efforts were undertaken in these areas and more generally in problems related to the luminosity function of QSOs, to its cosmological evolution, and to the correlations of QSO X-ray luminosity with redshift, optical luminosity, and radio luminosity. It has also become evident that accounting for the X-ray emission and its systematic properties must be included in any model for the QSO phenomena.

X-ray luminosities of QSOs are commonly described by the parameter, $\alpha_{o,x}$,[2] defined by

$$\frac{L_x}{L_o} = \left[\frac{\nu_x}{\nu_o}\right]^{-\alpha_{o,x}} = 10^{-(2.605\alpha_{o,x})}, \qquad (1)$$

---

[a]This research was supported by NASA contract no. NAS8-30751 and by the MINERVA Foundation, Munich, Germany.

where L_x is the monochromatic X-ray luminosity at 2 keV at the source rest frame (erg sec^{-1} Hz^{-1}), L_o is the monochromatic optical luminosity at 2500Å, and ν_x and ν_o are the corresponding frequencies. We consider only Friedmann cosmologies and assume $H_o =$ 50 km sec^{-1} Mpc^{-1}. The dynamic range of the detected values of $\alpha_{o,x}$ is 1.0–2.0, which corresponds to about three orders of magnitude in L_x/L_o. The choice of the parameter, $\alpha_{o,x}$, was motivated by the finding that L_x is positively correlated with L_o.

A larger sample of previously known QSOs has been further studied with the Einstein Observatory by Zamorani et al.[3] (see also Ku, Helfand, and Lucy[4]). The correlations of $\alpha_{o,x}$ with redshift, z, L_o, and radio luminosity, L_R (monochromatic at 5000 MHz), have been specifically addressed and two main conclusions have been derived:

(1) $\alpha_{o,x}$ is different for "radio-loud" QSOs and "radio-quiet" QSOs: QSOs that are relatively brighter in the radio (higher L_R/L_o) are relatively brighter in X rays (higher L_x/L_o, i.e., lower $\alpha_{o,x}$).
(2) $\alpha_{o,x}$ depends on redshift and/or optical luminosity, such that for higher z and/or higher L_o, L_x/L_o is lower ($\alpha_{o,x}$ is higher).

Since most QSOs are presently not detected in the radio, we consider from now on only optically selected QSOs and thereby only the dependence of $\alpha_{o,x}$ on z and L_o. The dependence of $\alpha_{o,x}$ on L_R is not "neglected": optically selected QSOs have a distribution of radio luminosities (some are "radio-loud") and any dependence on L_R is automatically taken care of statistically.

The interest in and importance of studying the dependence of $\alpha_{o,x}$ on z and L_o and the distribution function of $\alpha_{o,x}$ were historically motivated by calculations of the contribution of QSOs to the extragalactic X-ray background. Any such calculation that starts from the optical luminosity function and its cosmological evolution requires the detailed properties of $\alpha_{o,x}$. However, the importance of the $\alpha_{o,x}$ correlations and distribution is much more general. The differential number density of QSOs can be written in the form,

$$dN = dV(z)\, \Psi_o(z, L_o)\, dL_o\, \Phi_x(L_x|z, L_o)\, dL_x, \qquad (2)$$

where $dV(z)$ is an element of co-moving volume, $\Psi_o(z, L_o)$ is the optical luminosity function (including cosmological evolution), and $\Phi_x(L_x|z, L_o)$ is the conditional X-ray luminosity function given z and L_o. For a given L_o, $\alpha_{o,x}$ can replace L_x as an independent variable and therefore dN can also be written as

$$dN = dV(z)\, \Psi_o(z, L_o)\, dL_o\, \phi_x(\alpha_{o,x}|z, L_o)\, d\alpha_{o,x}, \qquad (3)$$

where $\phi_x(\alpha_{o,x}|z, L_o)$ is the conditional distribution function of $\alpha_{o,x}$ given z and L_o. It follows that a study of the distribution of $\alpha_{o,x}$ and its dependence on z and L_o is equivalent to studying the conditional X-ray luminosity function, and this will yield information on such questions as the correlations between X-ray and optical luminosities and the relations between the X-ray and optical evolution rates. These are fundamental ingredients for physical models of QSOs and for cosmological implications of QSOs.

This review summarizes recent results obtained at the Harvard-Smithsonian Center for Astrophysics on the X-ray properties of optically selected QSOs as reflected in the $\alpha_{o,x}$ systematics, lists open problems, and indicates directions for further study.

OPTICALLY SELECTED HETEROGENEOUS SAMPLE

The explicit dependence of $\alpha_{o,x}$ on z and L_o for optically selected QSOs has been studied by Avni and Tananbaum (1982)[5] using a heterogeneous sample of previously known QSOs. This sample consists of a "mixed bag" of QSOs from the CFA survey, Seyfert 1 nuclei from Kriss, Canizares, and Ricker,[6] and several QSOs that form a complete ($B_{lim} = 19.2$) sample from Marshall et al.[7] As a whole, the heterogeneous sample does not have any well defined completeness properties in the optical. Such a completeness is not required for the type of analysis carried out, but it should be kept in mind that complete samples have the advantage of being manifestly free from any biases that may be introduced by "human" selection.

This heterogeneous sample has a total of 73 QSOs, with 41 of them positively detected in X rays by the Einstein Observatory. The distribution of those QSOs in the (z, L_o) plane has a rather strong correlation between z and L_o, which implies that it is difficult to separate a dependence of $\alpha_{o,x}$ on L_o from a dependence on z. It is interesting to note that in this respect a heterogeneous sample has an advantage over a complete optically selected sample with a single magnitude limit since such a complete sample would have an even stronger correlation between z and L_o.

A simple linear dependence of $\alpha_{o,x}$ on cosmological look-back time, $\tau(z)$ (in units of the present age of the universe, with $q_o = 0$ used), and on log L_o was assumed:

$$\alpha_{o,x} = \{A_z[\tau(z) - 0.5] + A_o[\log L_o - 30.5] + A\} + \{\text{residual}\}. \tag{4}$$

A_z and A_o are the coefficients of the linear dependence and the subtraction terms of 0.5 and 30.5 are chosen to make A be the typical value of $\alpha_{o,x}$ for the central values of $\tau(z)$ and log L_o in the sample. The terms in the first curly brackets represent the average dependence of $\alpha_{o,x}$ as a function of z and L_o, $\bar{\alpha}_{o,x}(z, L_o) = \langle \alpha_{o,x}|z, L_o\rangle$. The term in the second curly brackets represents the distribution of the residuals of $\alpha_{o,x}$ relative to the average dependence, namely the distribution of $\alpha_{o,x} - \bar{\alpha}_{o,x}(z, L_o)$. This distribution must have an average of zero by construction and was assumed to be Gaussian (with a mean of zero) characterized by a dispersion, σ. The regression analysis was carried out using a parametric version of the Detections and Bounds (DB) method.[8] The choice of the Gaussian form for the distribution of the residuals was motivated, in part, by the fact that it defines the most straightforward generalization of the standard regression analysis to the domain of the Detections and Bounds problem. The validity of the Gaussian distribution was confirmed *a posteriori*.

The results of the DB regression-analysis for the two coefficients, A_z and A_o, are described in FIGURE 1 (adapted from Avni and Tananbaum, 1982),[5] where the best-estimate parameters and the $\Delta S = 4$ error-contour are displayed. The best estimate corresponds to $A_z \simeq 0$, which indicates that $\alpha_{o,x}$ depends predominantly on L_o rather than on z; however, a substantial dependence on z cannot be ruled out. The narrow and elongated shape of the error region is due to the intrinsic z-L_o correlation in the sample and means that, in fact, it is difficult to separate a dependence of $\alpha_{o,x}$ on L_o from a dependence on z.

If the explicit dependence of $\alpha_{o,x}$ is only on L_o, not on z, as given by the best-estimate parameters, then L_x depends on L_o as

$$L_x \propto L_o^{1-2.605A_o} \propto L_o^{0.7}. \tag{5}$$

Such a dependence was shown by Tucker[9] to be consistent with a class of models for QSOs that involves accretion onto massive black holes. We note, however, that the functional dependence that we have derived refers specifically to the conditional properties of L_x at a given L_o, a point that was not explicitly addressed by Tucker.

The range of allowed parameter values (A_z, A_o) implies that, for a wide class of models for the cosmological evolution of QSOs, the X-ray evolution rate is expected to be slower than the optical evolution rate. For example, in the case of pure luminosity

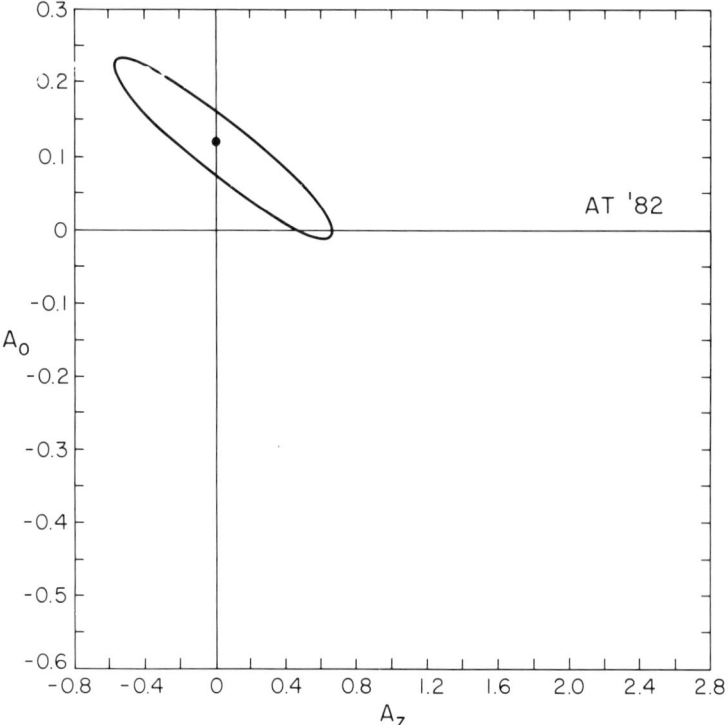

FIGURE 1. Best estimate and $\Delta S = 4$ error-contour for the correlation parameters, A_z and A_o, that describe the average dependence of $\alpha_{o,x}$ on $\tau(z)$ and log L_o for the old heterogeneous sample. (Adapted from Avni and Tananbaum, 1982.)[5]

evolution with an exponential dependence on look-back time,

$$L_o \propto e^{\gamma_o \tau(z)}, L_x \propto e^{\gamma_x \tau(z)}, \tag{6}$$

the evolution parameters, γ_o and γ_x, are related through

$$\gamma_x = \gamma_o - 2.605 \, (\gamma_o A_o + 2.3 \, A_z). \tag{7}$$

Within the error region given in FIGURE 1 for (A_z, A_o), one finds $\gamma_x < \gamma_o$ for typical observed values of γ_o. This has been supported by a comparison of the optical evolution rate[10] with the X-ray evolution rate[11] in the framework of pure luminosity evolution. Such a difference in evolution rates would have far-reaching implications. For example, calculations of the QSO contribution to the X-ray background could not make use of the evolution rate determined in the optical in conjunction with the local X-ray volume emissivity; also, the bolometric luminosity could evolve still differently and so care must be exerted in comparing theoretical evolution models with observations.

The analysis of Avni and Tananbaum (1982)[5] raised several issues that required further study:

(1) The heterogeneity of the sample used, which was "humanly" selected: While it could be formally argued that the sample was adequate for the type of analysis carried out, was it still possible that a dependence of $\alpha_{o,x}$ on some yet unidentified QSO property introduced a "hidden" bias into the sample?

(2) The assumed Gaussian distribution of the $\alpha_{o,x}$ residuals: The actual distribution of the residuals relative to the best-estimate average dependence was compared with the expected best-estimate Gaussian distribution and shown to be acceptable. However, this comparison indicated (but did not require) a skew distribution of residuals. Would a larger data set require a non-Gaussian distribution, and if so, what would be the implications for the derived average dependence? In particular, would the undetected QSOs indicate the existence of a separate population of "X-ray quiet" QSOs that are not properly described by a localized distribution?

(3) The numerical values used for some of the parameters: The analysis assumed traditional values for the optical and the X-ray spectral indices, $\alpha_o = 0.5$ and $\alpha_x = 0.5$, as well as $q_o = 0.0$. To what extent do the results depend substantially on the choice of these values?

In addition, when the $\alpha_{o,x}(z, L_o)$ correlation was combined with a specific model for the optical evolution and luminosity functions (assuming, in particular, a pure power-law luminosity function), the calculated slope for the X-ray luminosity function did not agree with the slope derived from the Einstein Observatory Medium Sensitivity Survey.[11] What is the source of this discrepancy and how can it be resolved?

These questions have now been addressed by a more recent study (Avni and Tananbaum, 1985),[12] which is based on new observational data and which is described in the following sections.

COMPLETE OPTICALLY SELECTED SAMPLES

The recent study of Avni and Tananbaum (1985)[12] is based on two samples with well-defined completeness properties in the optical. The first sample is the Bright Quasar Sample (BQS) of Schmidt and Green,[13] which is a complete optically selected sample to an average B_{lim} of 16.2 magnitudes. Sixty-six of the 82 QSOs in this sample outside of the declination zone of $30° < \delta < 60°$ have been observed by the Einstein Observatory in a collaborative effort of the CfA group and Schmidt and Green (Tananbaum et al.[14]). The choice of the observed QSOs has been dictated by considerations of scheduling of the Einstein Observatory. Thus, the 66 observed QSOs can be considered as a fair, unbiased, representative group of the full BQS sample vis-à-vis their redshifts, optical properties, and X-ray properties, and they can therefore be viewed as a complete optically selected sample. Of the 66 QSOs, 57 have been positively detected in X-rays and the other 9 have yielded flux upper bounds.

The second sample is the faint Braccesi sample (BF) of Marshall et al.[15] This is a complete optically selected sample to a B_{lim} of 19.8 magnitudes, observed with the Einstein Observatory. This sample contains 35 QSOs, with 13 X-ray detections and 22 flux upper bounds.

The distribution of QSOs from the BQS sample and from the BF sample in the (z, L_o) plane is described in FIGURE 2, adapted from Avni and Tananbaum (1985),[12] where QSOs from different samples are represented by different symbols and where positive X-ray detections are distinguished from X-ray upper bounds. The two samples are based upon UV excess selection and are therefore confined to $z \leq 2.2$. We note the very strong correlation between z and L_o for the QSOs of the BQS sample, which was expected for a single complete sample. Similarly, we note the very strong correlation between z and L_o for the QSOs of the BF sample. When the two samples are taken together, however, they span a rather wide dynamic range in the two-dimensional (z, L_o) plane and the intrinsic $z - L_o$ correlation is much weaker.

FIGURE 2 also describes the distribution in the (z, L_o) plane of QSOs from a third sample, which has been incorporated into the analysis at a later stage. This sample is an updated version of the original heterogeneous CfA survey and will be called henceforth

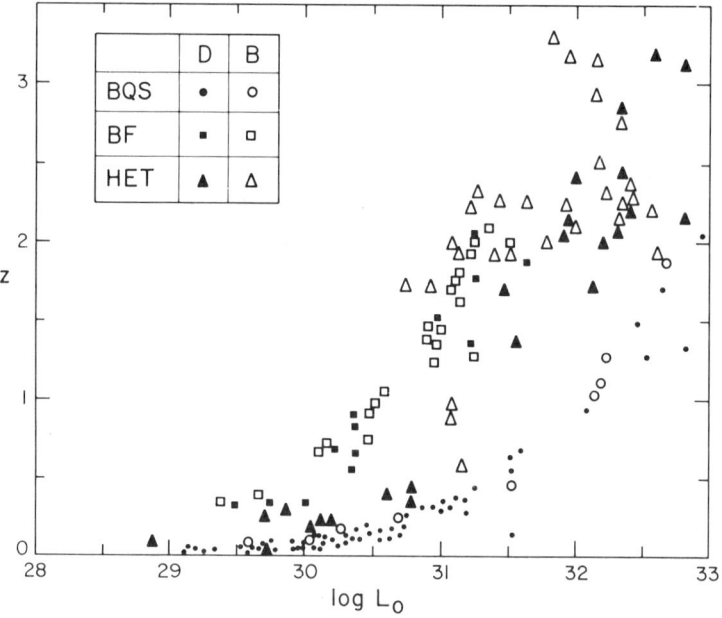

FIGURE 2. The distribution in the (z, L_o) plane of the QSOs for the new samples of optically selected QSOs. Circles: Bright Quasar Sample. Squares: Braccesi Faint sample. Triangles: new heterogeneous sample. Filled symbols denote X-ray detections; empty symbols denote X-ray nondetections. (Adapted from Avni and Tananbaum, 1985.)[12]

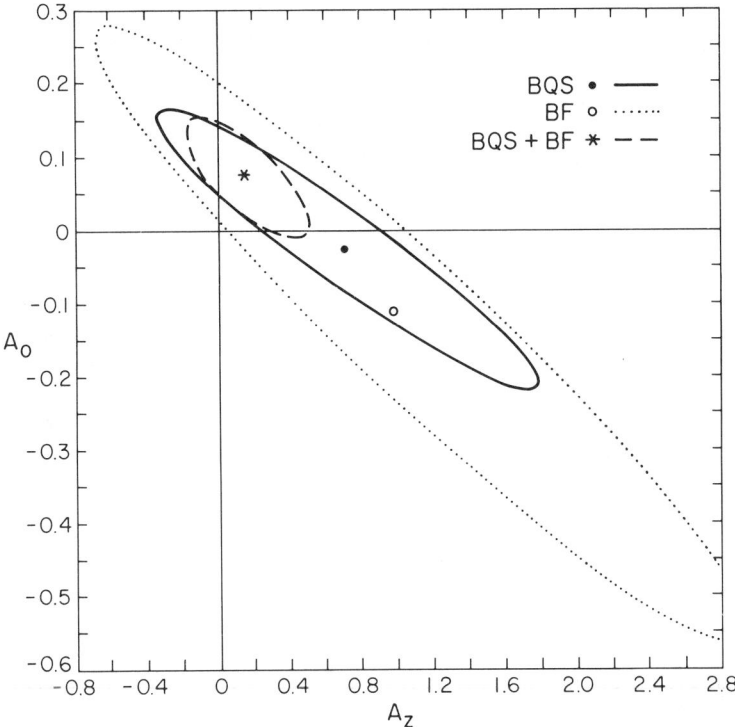

FIGURE 3. Best estimate and $\Delta S = 4$ error-contour for the correlation parameters, A_z and A_o, that describe the average dependence of $\alpha_{o,x}$ on $\tau(z)$ and $\log L_o$. Filled circle and solid line: Bright Quasar Sample. Open circle and dotted line: Braccesi Faint sample. Star and broken line: joint BQS and BF sample. (Adapted from Avni and Tananbaum, 1985.)[12]

the "new heterogeneous" sample.[12] It consists of 53 QSOs, with 24 X-ray detections and 29 upper bounds. In the (z, L_o) plane, this sample partially fills in the gap between the BQS and BF samples and also provides QSOs at $z > 2.2$. In total, the three samples contain 154 QSOs with 94 X-ray detections and 60 upper bounds.

The explicit dependence of $\alpha_{o,x}$ on z and L_o, using the functional form given by equation (**4**) and assuming a Gaussian distribution of residuals, has been studied separately for each of the two complete samples, BQS and BF. The best estimate coefficients, A_z and A_o, and the $\Delta S = 4$ error-contours in the (A_z, A_o) plane are described in FIGURE 3, which is adapted from Avni and Tananbaum (1985).[12] The results for the two samples are consistent with each other and the samples can therefore be joined together. The best-estimate parameters and error region for the joint BQS + BF sample are also described in FIGURE 3.

By comparing FIGURE 3 with FIGURE 1, we find that the recent results for each of the two complete samples as well as for their joint sample are consistent with the previous results of Avni and Tananbaum (1982)[5] for the "old" heterogeneous sample. In particular, the results for the BQS + BF joint sample are very close to the results for the old heterogeneous sample. This establishes that in fact a heterogeneous sample is adequate for the type of analysis carried out and that there are no strong dependencies of $\alpha_{o,x}$ on any yet unidentified QSO properties that would have introduced a hidden bias into the heterogeneous sample.

Several additional comments can be made on the basis of the comparison of these results. The error region for (A_z, A_o) derived from the BQS sample is substantially longer and wider than the region derived from the old heterogeneous sample, in spite of the facts that the two samples are comparable in size and that the number of detections in the BQS sample is larger than the number of detections in the old heterogeneous sample. This results, in part, from the tighter intrinsic $z - L_o$ correlation in the complete BQS sample relative to the old heterogeneous sample. It highlights the specific disadvantage of analyzing any single complete sample by itself. If one considers, in addition, the enormous reduction of the size of the error region when BQS and BF are analyzed jointly, it follows that it would be better to increase the size of the sample by studying a larger number of intermediate size complete samples with different magnitude limits rather than a smaller number of larger complete samples.

We also note the apparently puzzling fact that the best estimate for (A_z, A_o) from the BQS + BF sample does not lie between the corresponding best estimates from BQS and BF separately. This follows from the existence of two additional parameters in the regression analysis, A and σ. In particular, an offset in the best estimate for A from BQS relative to that from BF causes the "sideways" shift in the best estimate for (A_z, A_o). We have explicitly checked that all the results from the separate samples are consistent with each other in the full four-dimensional parameter space.

It is also interesting to note that the best estimate for (A_z, A_o) from the BQS sample (as well as that from the BF sample) lies outside the error region derived from the joint BQS + BF sample. This highlights the importance of using all available information and the advantage of combining all available samples. In addition, it cautions against using best-estimate results from any single sample.

FIGURE 4, adapted from Avni and Tananbaum (1985),[12] describes the results for (A_z, A_o) derived from the new heterogeneous sample (HET). These results are consistent with those obtained from the joint BQS + BF sample, also plotted in FIGURE 4. Since we have established that heterogeneous samples are adequate for this type of analysis, we form the joint BQS + BF + HET sample, which we call henceforth the "total" sample. The results for (A_z, A_o) from the total sample are also described in FIGURE 4. These results are now fully consistent with and very similar to the results obtained from the old heterogeneous sample by Avni and Tananbaum (1982),[5] described in FIGURE 1. The new results, however, are largely based on complete optically selected samples and the error region is now smaller by better than a factor of two. Again, the best estimate indicates that $\alpha_{o,x}$ depends predominantly on L_o rather than on z, but, again, a substantial dependence on z cannot be ruled out.

The principal consequences from the study of the new samples presented so far are: that heterogeneous samples have been established as a viable source of data for this type of analysis; that a wide dynamic range in the two-dimensional (z, L_o) plane is very important and calls for observations of several complete samples at different limiting magnitudes rather than one large sample at one limiting magnitude; and that the previous results of Avni and Tananbaum (1982)[5] have been confirmed by new independent data from complete samples. We have furthermore considered how the results are affected when q_o is changed from 0.0 to 0.5 and found that there is no appreciable qualitative effect. We have also studied the effect on the results of changing the assumed optical spectral index from its traditional value of $\alpha_o = 0.5$. We found that varying α_o in the range from 0.0 to 1.0 does not have an appreciable effect for the present sample; however, for samples containing about four times the number of QSOs, a more precise knowledge of the optical spectrum will be required. At this point,

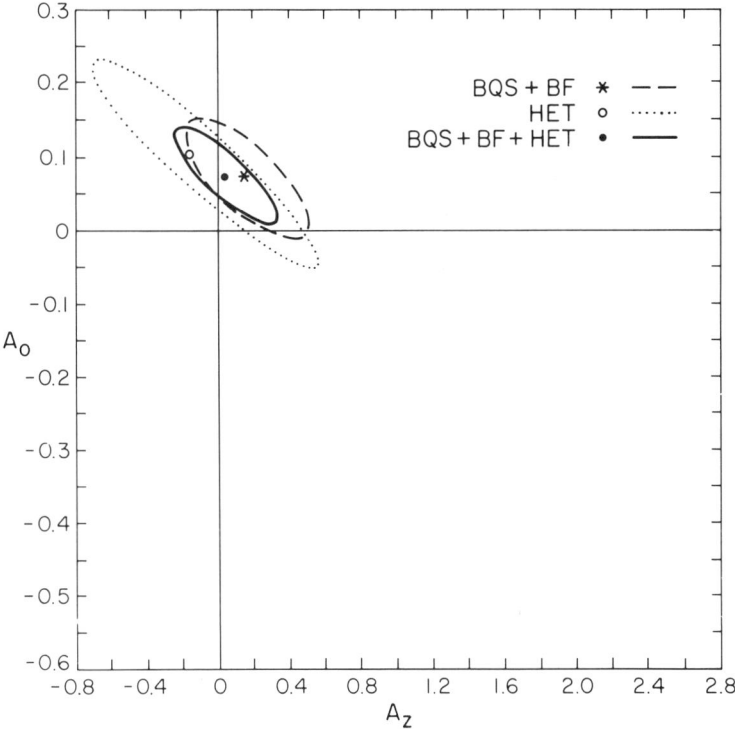

FIGURE 4. Best estimate and $\Delta S = 4$ error-contour for the correlation parameters, A_z and A_o, that describe the average dependence of $\alpha_{o,x}$ on $\tau(z)$ and log L_o. Star and broken line: BQS and BF sample. Open circle and dotted line: HET (new heterogeneous sample). Filled circle and solid line: joint BQS and BF and HET sample. (Adapted from Avni and Tananbaum, 1985.)[12]

we have not studied the effect of changing the X-ray spectral index from its traditional value of $\alpha_x = 0.5$; the observed diversity of QSO X-ray spectra[16] and the lack of systematic results for the spectral indices do not enable a meaningful quantitative study at present. The results for (A_z, A_o) from the present total sample imply the same general conclusions for the X-ray properties of optically selected QSOs as those of Avni and Tananbaum (1982),[5] namely, that if $\alpha_{o,x}$ depends only on L_o as indicated by the best estimate, then

$$L_x \propto L_o^{1-2.605A_o} \propto L_o^{0.8} \qquad (8)$$

(with a slight change in the value of the power-law index) and that from the full range of permissible values for (A_z, A_o), the X-ray evolution rate is expected to be slower than the optical evolution rate within the framework of a wide class of QSO evolution models, including pure luminosity evolution.

NO X-RAY QUIET QSOs

The availability of the new joint sample, BQS + BF + HET, which spans a wider two-dimensional dynamic range in the (z, L_o) plane and contains a larger number of QSOs than the old heterogeneous sample, makes it possible to address more detailed questions regarding the $\alpha_{o,x}$ (z, L_o) dependence and the distribution of residuals. We consider here specifically two related questions. The first question is whether a population of optically selected, X-ray quiet QSOs exists. The second question is whether the distribution of $\alpha_{o,x}$ residuals departs from a simple Gaussian in the sense of requiring significant skewness.

It may be recalled that the study of Avni and Tananbaum (1982)[5] has indicated (but did not require) a skew, non-Gaussian distribution of residuals, in the sense of a shorter tail at low $\alpha_{o,x}$ (high L_x) and a longer tail at high $\alpha_{o,x}$ (low L_x). This could mean either of two extreme possibilities or some combination of them. It is possible that the dynamic range of the $\alpha_{o,x}$ values is rather narrow, approximately 1.0 to 2.0, which corresponds to the positively detected values of $\alpha_{o,x}$, and that the distribution of the $\alpha_{o,x}$ residuals is correspondingly "localized" (approximately -0.5 to $+0.5$) and skew. However, it is also possible that a certain fraction of the population of optically selected QSOs has much lower X-ray luminosities and correspondingly much higher values of $\alpha_{o,x}$ than implied by the above dynamic range. We refer to such a possible part of the QSO population as "X-ray quiet." It is not easy to distinguish a priori between these two possibilities given the prevalence of X-ray nondetections, which amount to $\sim 40\%$ of the QSOs in the sample, since the actual values of $\alpha_{o,x}$ for the nondetections could either correspond to a localized distribution of $\alpha_{o,x}$ or be well above the values of 1.0 to 2.0.

To answer those questions, we have carried out a regression analysis of $\alpha_{o,x}(z, L_o)$ assuming the same average dependence as before, but allowing for a more general distribution of residuals. Formally, we write [as in equation (4)]

$$\alpha_{o,x} = \{A_z[\tau(z) - 0.5] + A_o[\log L_o - 30.5] + A\} + \{\text{residual}\}. \qquad (9)$$

The distribution of the residuals is now assumed to include two components and is described by FIGURE 5, adapted from Avni and Tananbaum (1985).[12] A fraction, $1 - P$, of the population of optically selected QSOs is "X-ray loud" and corresponds to a localized distribution of residuals with a mean of zero by construction. A fraction, P, of the population is "X-ray quiet" and is described formally by a δ-function contribution to the distribution of residuals at very high values of $\alpha_{o,x}$. The precise formal description of the X-ray quiet population is not important as long as the corresponding values of the $\alpha_{o,x}$ residuals are well above the dynamic range of the localized distribution describing the X-ray loud population. With this representation, the first curly brackets of equation (9) define the average $\overline{\alpha}_{o,x}(z, L_o|\text{X-ray loud})$ for the "loud" component of the population.

We describe the distribution of residuals for the X-ray loud component by a functional form that is a simple modification of the Gaussian form, that admits a transparent measure of skewness, and that is convenient for analytic and numerical manipulations. This functional form is made out of two "half-Gaussians" with a common height that are "glued together" at their maxima (see FIGURE 5). Since this distribution must have a mean of zero by definition, it is characterized by two independent parameters, which can be chosen to be σ_L and σ_R (the usual σ's for the left and right half-Gaussians, respectively.) It is also possible to characterize the distribution by another set of two independent parameters: σ, defined as the formal r.m.s. width of the distribution, and a skewness parameter, $R = \sigma_R/\sigma_L$. Thus, the full representation of $\alpha_{o,x}$ is defined by six parameters: A_z, A_o, A, P, R, and σ.

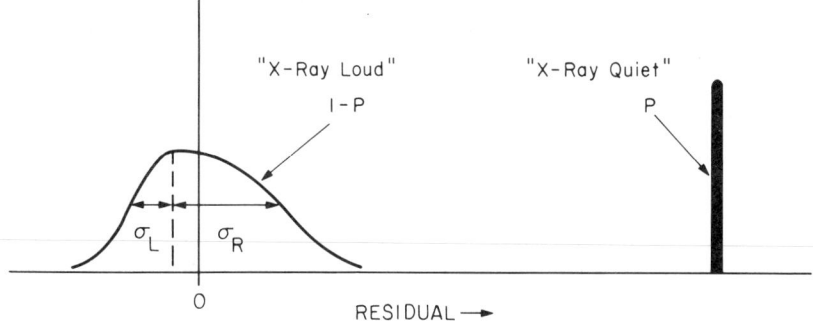

FIGURE 5. A schematic representation of the distribution of $\alpha_{o,x}$ residuals that allows for the existence of X-ray quiet QSOs and allows for a skew distribution for X-ray loud QSOs. (Adapted from Avni and Tananbaum, 1985.)[12]

An analysis of the total sample, BQS + BF + HET, using the DB method,[8] assuming the traditional values of $q_o = 0.0$, $\alpha_o = 0.5$, and $\alpha_x = 0.5$, and considering all the six regression parameters as free parameters, has yielded the following best-estimate values: $P = 0$, $R = 3.3$, $\sigma = 0.21$, $A_z = 0.02$, $A_o = 0.09$, and $A = 1.54$. The best-estimate value of $P = 0$ means that the data indicate that there are no X-ray quiet QSOs. Furthermore, treating P as a single interesting parameter,[17] the DB regression analysis yields an upper limit of $P \leq 8\%$ at the 95% confidence level. Thus, no more than a few percent of optically selected QSOs can be X-ray quiet. While these specific quantitative results depend on the functional representation chosen and on the numerical values of the assumed parameters, the qualitative conclusion is not very sensitive to these assumptions. Thus, the large majority (probably all) of the optically selected QSOs are X-ray loud, in the sense of having $\alpha_{o,x}$ values in the approximate range of 1.0 to 2.0. In this sense, X-ray emission seems to be a universal property of QSOs. This affirms the importance of X-ray observations for studying QSOs and confirms the importance of X-ray emission as a tool for discovering or selecting QSOs.

The best-estimate value of $R = 3.3$ means that the data indicate a large amount of skewness in the distribution of $\alpha_{o,x}$ residuals, with a longer tail at high $\alpha_{o,x}$ (low L_x) and a shorter tail at low $\alpha_{o,x}$ (high L_x). Treating R as a single interesting parameter, the DB regression analysis yields a very significant skewness ($R > 1$) at the 99.999% confidence level. This result is also supported by an *a posteriori* comparison of the nonparametric distribution of residuals relative to the best-estimate average dependence (determined by A_z, A_o, and A) with the expected best-estimate parametric distribution of residuals (determined by σ and R). From such a comparison, a KS test marginally rejects the pure Gaussian ($R = 1$) best fit at $\geq 94\%$ confidence, while the skew ($R = 3.3$) distribution is perfectly acceptable.

We note that while the new samples require a non-Gaussian distribution of residuals, such a skewness has an insignificant effect on the average dependence, $\bar{\alpha}_{o,x}(z, L_o)$. When the values of $P = 0$ and $R = 1$ are enforced in the regression analysis, the best-estimate parameters that result are $A_z = 0.05$, $A_o = 0.07$, $A = 1.54$, and $\sigma = 0.20$. These are not significantly different from the values derived in the full, six-dimensional analysis. Thus, the previous conclusions regarding the correlation of X-ray and optical luminosities and regarding the relative X-ray and optical cosmological evolution rates remain unaffected by the finding of the significant skewness. This skew, non-Gaussian behavior is a new qualitative property of the conditional X-ray luminosity function, $\Phi_x(L_x|z, L_o)$. It has important implications for constructing or constraining the full bivariate X-ray–optical luminosity and evolution function.

COMPARISON WITH QSO X-RAY NUMBER COUNTS

The average dependence, $\bar{\alpha}_{o,x}(z, L_o)$, together with the distribution of $\alpha_{o,x}$ residuals, or, equivalently, the conditional X-ray luminosity function, $\Phi_x(L_x|z, L_o)$, provide a "bridge" that connects the optical luminosity and evolution function, $\Psi_o(z, L_o)$, with the X-ray luminosity and evolution function, $\Psi_x(z, L_x)$. Thus if one assumes a specific model for $\Psi_o(z, L_o)$, one can calculate $\Psi_x(z, L_x)$, which can then be compared with observations of samples of X-ray selected QSOs.

The choice of model for $\Psi_o(z, L_o)$ for such a comparison is not clear-cut. In fact, several simple functional representations of $\Psi_o(z, L_o)$ suggested in the literature do not provide an adequate description of all presently available samples of optically selected QSOs. Several models that assume pure density evolution are in conflict with optical number counts at faint magnitudes and violate constraints imposed by the X-ray background. The specific luminosity dependent density evolution models suggested by Schmidt and Green[13] are not consistent with the Braccesi faint (BF) sample.[18,19] This inconsistency could be interpreted as due to a fluctuation in the number density of QSOs in the BF field, but if this point of view is taken, the validity of other samples, upon which the derivation of $\Psi_o(z, L_o)$ is based, can also be questioned. Several simple models that assume pure luminosity evolution are not fully consistent with the detailed properties of the available complete samples.[20] Thus, a great deal of care must be exerted when drawing conclusions that are based on any specific choice of representation for $\Psi_o(z, L_o)$. Still, such comparisons with X-ray selected samples are very valuable since they are important for obtaining a better understanding of all the issues involved and for eventually leading to a consistent representation of the full bivariate X-ray–optical luminosity and evolution function, $\Psi(z, L_o, L_x)$.

As mentioned earlier, the slope of the X-ray luminosity function calculated from a pure luminosity evolution model for $\Psi_o(z, L_o)$ with an exponential dependence on look-back time and with a pure power-law luminosity function, combined with the $\alpha_{o,x}(z, L_o)$ dependence derived with Gaussian residuals, does not agree with the slope determined from the Einstein Observatory Medium Sensitivity Survey.[11] Furthermore, it has been known for some time to a number of workers in the field that the QSO X-ray number counts calculated from the above representations do not agree with the counts determined from the Medium Survey (This inconsistency is mentioned and briefly discussed by Maccacaro, 1984.[21]) However, no specific quantitative details have yet been published.

We have calculated the expected X-ray number counts for a few cases so as to determine the sensitivity of the counts to some of the input ingredients. (We assume $q_o = 0$). We have used two pure luminosity evolution models from Marshall[20] for $\Psi_o(z, L_o)$. The first model invokes an exponential dependence of the optical luminosity on look-back time, $\exp[5.9\tau(z)]$, a pure power-law present epoch luminosity function, $\tilde{L}_o^{-3.6}$ ($\tilde{L}_o$ is the present epoch optical luminosity), and a normalization of 8.7 Gpc^{-3}. The range of validity of this model in the (z, L_o) plane is characterized by a luminosity cutoff of $L_o \geq 0.6 \times 10^{30}$ erg sec^{-1} Hz^{-1}, by a present epoch luminosity cutoff of $\tilde{L}_o \geq$

TABLE 1. Comparison of Calculated and Observed X-Ray Number Counts

	S_x (0.3–3.5 keV) erg sec^{-1} cm^{-2}	
Model or Data	1.56×10^{-13}	1.0×10^{-12}
$e^{\tau\tau(z)}$, $R = 1$	1.5×10^4 str^{-1}	7.0×10^2 str^{-1}
$e^{\tau\tau(z)}$, $R = 3.3$	1.2×10^4	2.3×10^2
$(1 + z)^\delta$, $R = 3.3$	0.78×10^4	1.7×10^2
MSS	0.38×10^4	1.3×10^2

10^{29} erg sec^{-1} Hz^{-1}, and by a redshift range of $0 \leq z \leq 2.2$. The second model invokes a power-law dependence of the optical luminosity on redshift, $(1 + z)^{3.5}$, a pure power-law present epoch luminosity function, $\tilde{L}_o^{-3.6}$, and a normalization of 22 Gpc^{-3}. The range of validity of this model is characterized by $L_o \geq 0.6 \times 10^{30}$ erg sec^{-1} Hz^{-1}, $\tilde{L}_o \geq 2 \times 10^{29}$ erg sec^{-1} Hz^{-1}, and $0 \leq z \leq 2.2$.

The range of validity of any given model for $\Psi_o(z, L_o)$ is determined by the properties of the optical samples on which the derivation of $\Psi_o(z, L_o)$ is based and by the functional form of $\Psi_o(z, L_o)$. This range must be taken into account when a comparison is made with the observed QSO X-ray number counts. Only those X-ray selected QSOs whose optical luminosities and redshifts are within the range of validity should be included in the X-ray number count for such a comparison. This is an important effect since about half of the QSOs of the Medium Sensitivity Survey from Maccacaro et al. (1984)[22] are excluded by the requirement of $L_o \geq 0.6 \times 10^{30}$ erg sec^{-1} Hz^{-1}. For the comparisons to made here, we have therefore rederived the X-ray number counts from the Medium Survey data, taking into account the constraints imposed by the optical evolution models.

TABLE 1, adapted from Avni and Tananbaum (1985),[12] presents a comparison between calculated and observed QSO X-ray number counts, $N(>S_x)$ (per steradian), for two values of S_x, where S_x is the integrated 0.3–3.5 keV flux in erg sec^{-1} cm^{-2}. The notation of "$e^{\gamma\tau(z)}$" denotes the first model for $\Psi_o(z, L_o)$ described above, with an exponential dependence of optical luminosity on look-back time. Similarly, "$(1 + z)^\delta$" denotes the second model for $\Psi_o(z, L_o)$ described above, with a power-law dependence of optical luminosity on redshift. "$R = 1$" denotes the best-estimate $\alpha_{o,x}(z, L_o)$ average dependence and distribution of residuals, obtained when $P = 0$ and $R = 1$ are enforced, i.e., with a Gaussian distribution of residuals. "$R = 3.3$" denotes the overall best-estimate $\alpha_{o,x}(z, L_o)$ average dependence and distribution of residuals, allowing for the more general skew distribution of residuals. "MSS" denotes the observed number counts that we have derived from the Medium Survey data.

The calculated number counts for "$e^{\gamma\tau(z)}$, $R = 1$" are the values corresponding to the "traditional" exponential model for $\Psi_o(z, L_o)$ and Gaussian residuals. Those values are higher than the observed number counts by factors of 4–5. The calculated number counts for "$e^{\gamma\tau(z)}$, $R = 3.3$" correspond to the same optical luminosity and evolution function, combined with the skew distribution of residuals. The effect of replacing the "$R = 1$" Gaussian distribution by the "$R = 3.3$" skew distribution is very large (reduction by a factor of three) for $S_x = 10^{-12}$ erg sec^{-1} cm^{-2}, but is moderate (reduction by 20%) for $S_x = 1.56 \times 10^{-13}$ erg sec^{-1} cm^{-2}. We therefore find a strong numerical sensitivity to the shape of the distribution of residuals for a large part of the dynamic range of S_x covered by the Medium Survey. The calculated number counts for "$(1 + z)^\delta$, $R = 3.3$" correspond to the power-law model for $\Psi_o(z, L_o)$, combined with the same skew distribution of residuals. The effect of replacing the exponential evolution model by the power-law evolution model is substantial—the calculated X-ray number counts are reduced by 30%.

Two conclusions emerge from the few cases we have studied. First, the calculated number counts are sensitive both to the shape of the distribution of $\alpha_{o,x}$ residuals and to the functional form of the optical luminosity and evolution function. Second, the specific combination of $\Psi_o(z, L_o)$ and $\alpha_{o,x}(z, L_o)$ considered is not consistent with the observed X-ray number counts (by 25% at 10^{-12} erg sec^{-1} cm^{-2} and by a factor of two at 1.56×10^{-13} erg sec^{-1} cm^{-2}). Thus, while the new results go a significant way towards resolving the number-count discrepancy, much detailed work still has to be done to fully understand the bivariate optical–X-ray luminosity and evolution function.

DISCUSSION

The study of Avni and Tananbaum (1985)[12] has yielded a number of consequences regarding the X-ray properties of optically selected QSOs. Heterogeneous samples of optically selected QSOs were established as adequate for the study of the explicit dependence of $\alpha_{o,x}$ on z and L_o. It was shown that for such studies it is better to increase the data base by observing a larger number of intermediate size complete samples with different magnitude limits rather than a smaller number of large complete samples. The previous results of Avni and Tananbaum (1982)[5] regarding the correlations between X-ray and optical luminosities and regarding the relative rates of X-ray and optical cosmological evolution have been confirmed. It was shown that changing the value of q_o from 0.0 to 0.5 does not have an appreciable qualitative effect on the results.

It was found that varying the assumed slope of the optical spectrum, α_o, between 0.0 and 1.0 does not have an important effect for the present size sample. It was shown that the large majority (probably all) of the optically selected QSOs are X-ray loud: no more than a few percent of optically selected QSOs can be X-ray quiet. The distribution of the $\alpha_{o,x}$ residuals relative to the $\alpha_{o,x}(z, L_o)$ average dependence was shown to be significantly skew, which is a property that is important for constructing or constraining the bivariate luminosity and evolution function, $\Psi(z, L_o, L_x)$.

However, there remains a residual discrepancy of about a factor of two when the calculated QSO X-ray number counts for the presently available simple pure luminosity optical evolution model, $\Psi_o(z, L_o)$, combined with the best-estimate $\alpha_{o,x}(z, L_o)$ average dependence and distribution of residuals, are compared with the observed counts. This lack of consistency needs to be resolved. We now discuss briefly some directions for further research that are important for fully understanding the optical-X-ray luminosity and evolution function.

Regarding the optical $\Psi_o(z, L_o)$, perhaps the most pressing issue is whether the BF sample is a fluctuation in number density. This calls for observations of randomly chosen fields of about the same area to about the same limiting magnitude with the same selection criteria. One needs first to understand the data base before detailed representations for $\Psi_o(z, L_o)$ are constructed. Another problem that needs to be addressed is the shape of the luminosity function at low L_o and the connection of the "QSO" luminosity function and evolution rate to the "Seyfert" (and related low luminosity AGNs) luminosity function and evolution rate. There is a large numerical sensitivity to the details of the luminosity function at low L_o. In addition, effects of measurement errors of the observed magnitudes have not yet been fully and conclusively evaluated.

With respect to the study of the explicit dependence $\alpha_{o,x}(z, L_o)$, there are three factors that affect directly the shape and width of the distribution of $\alpha_{o,x}$ residuals, that affect indirectly the average $\bar{\alpha}_{o,x}(z, L_o)$ dependence, and that require detailed and rigorous study: measurement errors, time variability, and anisotropic emission. Time variability has an effect because the optical and X-ray luminosities have not been observed at the same time. When the explicit dependence of L_x on L_o is studied, time variability of L_o needs to be dealt with. Anisotropic emission has an effect if the angular distribution of L_x is different from the angular distribution of L_o since this will cause the observed "directed" luminosities to be different from the intrinsic direction-averaged luminosities. The potential importance of these effects stems mostly from the large numerical sensitivity to the shape of the distribution of $\alpha_{o,x}$ residuals that we have shown above. As a further demonstration of this point, if we recalculate the QSO X-ray number counts for the representation denoted by "$(1 + z)^\delta$, $R = 3.3$" in TABLE 1, but change (in an ad hoc fashion) the value of σ from 0.21 to 0.18, we find 0.55×10^4 sources str^{-1} for S_x of 1.56×10^{-13} erg sec^{-1} cm^{-2} and 1.1×10^2 str^{-1} for 10^{-12} erg sec^{-1} cm^{-2}. This discrepancy at 1.56×10^{-13} erg sec^{-1} cm^{-2} reduces to 50% and at 10^{-12} erg sec^{-1} cm^{-2} to 20%, but with the opposite sign.

Further progress in the study of $\alpha_{o,x}(z, L_o)$ can also be obtained by using X-ray observations of previously known QSOs from the Einstein Observatory data bank to increase substantially the size of the sample. Observations by other X-ray satellites will also be beneficial for that purpose. Progress can also be made by improving the treatment of the nondetections by using a detailed probability distribution for the observed counts in the detector rather than the discrete 3σ upper limits. The effects of the departure of the X-ray spectral index from the traditional value of 0.5 need also be studied and the systematics of QSO X-ray spectra are required for that.

Regarding the derivation of the X-ray $\Psi_x(z, L_x)$, much progress will be made by increasing the size of the Einstein Observatory Medium Sensitivity Survey and by completing the optical identifications for the Einstein Deep Surveys. For the Medium Survey QSOs, observations of the color indices, which are used to select QSOs in UV excess samples of optically selected QSOs, are important. Comparisons of optically selected samples with X-ray selected samples should also take into account the range of validity in the color domain. Effects of measurement errors, time variability, and anisotropic emission should be considered when studying the $\alpha_{o,x}$ properties of X-ray selected QSOs.

In conclusion, significant progress has been made in the study of the X-ray properties of QSOs, in particular in showing that most, probably all, optically selected QSOs are X-ray loud (no more than a few percent can be X-ray quiet). Much has yet to be learned and some fairly well-defined avenues for obtaining additional observational data and for further analyses are ahead of us.

ACKNOWLEDGMENTS

It is a pleasure to thank H. Tananbaum for the collaboration that has yielded the results described here. I thank M. Schmidt and G. Zamorani for helpful discussions and comments. T. Maccacaro and I. Gioia kindly provided an unpublished, expanded version of the sky area coverage table for the Einstein Observatory Medium Sensitivity Survey.

REFERENCES

1. GIACCONI, R. *et al.* 1979. Astrophys. J. **230:** 540.
2. TANANBAUM, H., Y. AVNI, G. BRANDUARDI, M. ELVIS, G. FABBIANO, E. FEIGELSON, R. GIACCONI, J. P. HENRY, J. P. PYE, A. SOLTAN & G. ZAMORANI. 1979. Astrophys. J. Lett. **234:** L9.
3. ZAMORANI, G., J. P. HENRY, T. MACCACARO, H. TANANBAUM, A. SOLTAN, Y. AVNI, J. LIEBERT, J. STOCKE, P. A. STRITTMATTER, R. J. WEYMANN, M. G. SMITH & J. J. CONDON. 1981. Astrophys. J. **245:** 357.
4. KU, W. H. M., D. HELFAND & L. B. LUCY. 1980. Nature **288:** 323.
5. AVNI, Y. & H. TANANBAUM. 1982. Astrophys. J. Lett. **262:** L17.
6. KRISS, G. A., C. R. CANIZARES & G. R. RICKER. 1980. Astrophys. J. **242:** 492.
7. MARSHALL, H. L., H. TANANBAUM, G. ZAMORANI, J. P. HUCHRA, A. BRACCESI & V. ZITELLI. 1983. Astrophys. J. **269:** 42.
8. AVNI, Y., A. SOLTAN, H. TANANBAUM & G. ZAMORANI. 1980. Astrophys. J. **238:** 800.
9. TUCKER, W. H. 1983. Astrophys. J. **271:** 531.
10. MARSHALL, H. L., Y. AVNI, H. TANANBAUM & G. ZAMORANI. 1983. Astrophys. J. **269:** 35.
11. MACCACARO, T., Y. AVNI, I. M. GIOIA, P. GIOMMI, R. GRIFFITHS, J. LIEBERT, J. STOCKE & J. DANZIGER. 1983. Astrophys. J. Lett. **266:** L73.
12. AVNI, Y. & H. TANANBAUM. 1985. Astrophys. J. Submitted.
13. SCHMIDT, M. & R. F. GREEN. 1983. Astrophys. J. **269:** 357.
14. TANANBAUM, H., Y. AVNI, R. F. GREEN, M. SCHMIDT & G. ZAMORANI. 1985. Astrophys. J. Submitted.

15. MARSHALL, H. L., Y. AVNI, A. BRACCESI, J. P. HUCHRA, H. TANANBAUM, G. ZAMORANI & V. ZITELLI. 1984. Astrophys. J. **283:** 50.
16. ELVIS, M., B. J. WILKES & H. TANANBAUM. 1985. Astrophys. J. **292:** 357.
17. AVNI, Y. 1976. Astrophys. J. **210:** 642.
18. MARSHALL, H. L. 1983. Ph.D. Thesis. Harvard University.
19. LANDMAN, U. 1984. M.Sc. Thesis. Weizmann Institute of Science.
20. MARSHALL, H. L. 1985. Astrophys. J. **299:** 109.
21. MACCACARO, T. 1984. In Proceedings of X-ray and UV Emission from Active Galactic Nuclei (Garching, Germany, July 1984). W. Brinkman & J. Trumper, Eds.: 63. Max Planck Institut für Physik und Astrophysik. Munich.
22. MACCACARO, T., I. M. GIOIA & J. T. STOCKE. 1984. Astrophys. J. **283:** 486.

From Molecular Clouds to Active Galactic Nuclei—The Universality of the Jet Phenomenon

ARIEH KÖNIGL

Department of Astronomy and Astrophysics
University of Chicago
Chicago, Illinois 60637

INTRODUCTION

Jets are among the most remarkable astrophysical phenomena explored in recent years. The term "jets" was originally coined to describe the narrow, elongated features that had been discovered in radio maps (and, in some cases, also by X-ray and optical observations) of extragalactic sources. Similar features have subsequently been found, however, also in our own galaxy, with the relativistic beams of SS433 being probably the most celebrated example. While the SS433 beams are still unique, there is now mounting evidence that oppositely directed jets are very frequently associated with nascent stars embedded in dense molecular clouds. The purpose of this article is, in essence, to "bridge the gap" between these smallest-scale jets and their enormously larger extragalactic counterparts. By concentrating on the similarities between molecular-cloud and extragalactic jets, I shall try to extract some of the basic dynamical principles that could account for the apparent universality of this phenomenon. Following an observational overview, I shall consider, in turn, the general hydrodynamic and magnetohydrodynamic (MHD) aspects of the production, the collimation, and the propagation of jets in protostellar and in active-galactic-nuclei (AGN) environments. Although the distinction between hydrodynamic and hydromagnetic effects is not always clear-cut (in fact, a fluid dynamics approach is valid only if the jet is magnetized), I make this differentiation here deliberately in order to emphasize the likely global role of magnetic fields in jets. I conclude with a brief summary.

OBSERVATIONAL OVERVIEW

Extragalactic Radio Jets

The current status of the observations of extragalactic radio jets (both extended and compact) and of the physical parameters deduced from these observations has recently been surveyed in a number of excellent review articles and conference proceedings.[1-8] Here I merely list some of the salient facts that are relevant to the comparison with molecular-cloud jets. Jets (defined as distinct radio features that appear to emanate from a galactic nucleus and are at least four times longer than they are wide) are detected in 65–80% of weak radio galaxies and in 40–70% of extended radio quasars. They are less common in the more powerful radio galaxies, which tend to be edge-brightened double-lobe sources (possessing high-surface-brightness "hot spots" near their outer edges). By contrast, the lobes in the weaker extended sources are generally edge-darkened. The lobes are typically several hundred kiloparsecs in size (although they reach a few megaparsecs in certain weak doubles) and in most cases have a detectable central component associated with the parent galaxy. The radio luminosity of the central core appears to be correlated with the total radio power and both are correlated with the jet structure: jets in weak sources are mainly two-sided, while most jets associated with powerful sources are one-sided. This observation is consistent with the notion that all double sources are supplied with mass and energy by bipolar outflows emanating from the central galaxy, but that only the weaker, more lossy flows become visible as radio jets. One-sided jets are also typically associated with the compact nuclear sources (projected scales of a few parsecs) that are resolved by VLBI techniques. The one-sidedness of these jets has been attributed to Doppler brightening arising from relativistic motions near the origin, an interpretation supported by the frequent occurrence of apparent superluminal expansions in the brightest compact sources.

The physical conditions in extragalactic radio jets are still rather uncertain. While the superluminal compact sources are most likely relativistic (Lorentz factors of 5–10), there are arguments both in favor and against relativistic velocities in the extended jets. Various indirect estimates typically yield values in the range of 10^3–10^4 km sec^{-1}. In several cases, direct measurements of optical emission lines give values of several hundred km sec^{-1}, but these probably arise in regions where the jet interacts with the ambient interstellar medium. Therefore, they give only lower limits on the outflow velocity.[9] In the case of Cen A, the nearest (5 Mpc) radio galaxy, there are radial-velocity measurements for several optical emission-line knots that, after allowing for projection effects, imply velocities of 1000–2000 km sec^{-1} away from the nucleus.[10,11] In one of these knots (located at a projected distance of 30 kpc from the nucleus), there are differential velocities of at least 800 km sec^{-1}, which correspond to a cross-knot travel time significantly shorter than the inferred mean travel time from the galactic center.[12] These optical emission knots are aligned in the general direction of the inner ($\lesssim 2$ kpc) jet, which has been detected by X-ray,[13] radio,[14] and optical[15] observations.

The values of the pressure and the density in jets are also quite uncertain. The minimum pressures inferred from synchrotron-radiation equipartition arguments in extended jets are typically in the range of 10^{-11}–10^{-13} dyne cm^{-2} and are comparable to the inferred values for the surrounding interstellar medium. However, in a number of cases the deduced minimum pressures may well exceed the ambient values, implying that the jets are either free or magnetically self-confined. The inferred equipartition values are typically also rather high in the hot spots (10^{-8}–10^{-10} dyne cm^{-2}), which probably represent regions where the jets are shocked as they impact on the ambient medium. For the densities, on the other hand, there usually exist only upper limits, which are obtained from Faraday-rotation arguments. These typically give values in the range of 10^{-2}–10^{-4} cm^{-3} for thermal electrons in extended jets. In some highly polarized compact sources, similar arguments imply an almost total lack of low-energy ($\lesssim 10$ MeV) electrons (or else the presence of an electron-positron plasma). Direct measurements of the density exist in a number of optical emission-line regions, but in most cases the derived values are too high (10^2–10^4 cm^{-3}) to correspond to the outflowing jet material and thus probably represent the density of compressed interstellar matter.

Two recent developments have already begun to provide qualitatively new observational input on extragalactic radio jets. One is the initiation of multifrequency, high-dynamic-range, and high-resolution observations of extended jets and double radio lobes that utilize the full observing power of aperture-synthesis telescopes such as the Very Large Array. These observations have led to the discovery of morphologically intricate structures in the lobes of sources like Her A[16] and Cyg A,[17] the detailed mapping and classification of hot spots,[18,19] the detection of emission from hitherto invisible "counterjets" or from low-surface-brightness regions of jets like 3C449[20] (which have previously been classified as "gaps"), and the high-resolution mapping of the projected intrinsic magnetic field vectors and the Faraday rotation-measure distribution in such prototypical jets as NGC 6251[21] and NGC 1265.[22] The other new development is the acquired capability to produce VLBI polarization maps.[23,24] This rather unexpected achievement has been facilitated by the fact that, in contrast to the bright, optically thick cores of compact sources which exhibit relatively little polarization, the fainter, optically thin parts of VLBI jets are often highly polarized. The significance of this development lies in the fact that it may enable the monitoring of magnetic field effects from the milliarcsecond to the 100-arcsecond scale, perhaps even continuously in the same source (as has already been done with the total intensity data in a source like 3C120[25]).

Molecular-Cloud Jets

Jets were first discovered in regions of active star formation only about three or four years ago, so naturally these outflows have not yet been studied as extensively as extragalactic jets. Nevertheless, a number of good reviews have already appeared in the literature,[26-28] so again I confine myself here to a brief summary of the main facts. On the smallest scale, the jets are detected in high-resolution optical CCD images[29] and by radio continuum measurements.[30] These observations indicate that the outflows are typically collimated within $\lesssim 10^{15}$ cm from the origin to an opening angle of 3°–10° and that their velocities are 100–400 km sec^{-1}. On the large scale ($\sim 10^{18}$ cm), these outflows are manifested as bipolar CO emission-line lobes (with masses in the range of 0.1–100 $M_\odot$), which straddle the central star and which separate with velocities of $\sim$10–50 km sec^{-1}.[31] These molecular emission lobes are the direct analogs of extragalactic double radio lobes, but, unlike the latter (for which there exist only upper limits on the velocity, typically $\sim 0.1c$), their velocities are directly measurable. The measured values are consistent with the interpretation of the lobes as "working surfaces,"[32] where the ram pressure of the flow is balanced by the pressure of the shocked, swept-up ambient medium. On the basis of the number of detected high-velocity molecular emission lobes and of their dynamical lifetimes ($\sim 10^4$ yr), it has been estimated[33] that an energetic outflow phase leading to the formation of such lobes could occur during the birth of every star above a solar mass or so. This conclusion is supported by the relatively low infrared luminosities (0.1–100 $L_\odot$) measured for most of the molecular flow sources.

Within the CO lobes, one often detects high-velocity optical emission-line condensations, the so-called Herbig-Haro (HH) objects.[27] These objects tend to be aligned in the general direction of the lobe separation axis, and in those cases for which proper motion measurements are available, one finds that the condensations move away from the central star with typical speeds of a few times 10^2 km sec^{-1}. One natural interpretation of these objects is, therefore, that they represent dense clumps that have been accelerated by the jet.[34] Among the best-studied examples of Herbig-Haro objects are HH 1 and 2 in the Orion molecular cloud. Each one of these has been found to consist of several subcondensations that exhibit large proper motions away from a central radio-continuum source, but which also show large differential velocities inconsistent with coherent motion over the inferred travel time from that source.[35] In that respect, these condensations resemble some of the optical emission knots in Cen A.[10–12] A possible resolution of the apparent inconsistency implied by the large velocity dispersions is suggested by recent high-resolution Balmer emission-line maps of HH1, which can be modeled in terms of a bow shock that forms downstream from a dense clump.[36] HH1 could thus represent an "interstellar bullet,"[37] originally accelerated by a jet, which now undergoes fragmentation and differential deceleration after running into a dense parcel of cloud material. The analogy between HH objects and the optical emission knots in Cen A is strengthened by the fact that they also have similar spectra that are characteristic of excitation by shocks[10,11,15,27] and by the fact that in a number of cases they also have comparable inferred values of density and shock velocity. This analogy may conceivably extend to many of the bright emission knots detected in extragalactic jets. In both cases, the knots are often found to be strung out quasi-periodically along the jet; particularly good examples are provided by the jet associated with the Herbig-Haro objects HH 7–11 in NGC 1333 (for which a high proper motion away from an embedded infrared source has been measured[38]) and by the extragalactic jet associated with M87 (in which radio observations[39] indicate the presence of a large-scale shock upstream of the brightest knot). Possible interpretations of this quasi-periodic pattern are discussed below.

The first object identified as a bipolar outflow source, IRS-5 in the molecular cloud L1551, still serves as the prototype of the class. This source has associated CO lobes,[40] high-velocity HH objects,[41] and an inner radio-continuum[30] and optical-emission[42] jet. The jet appears to bend continuously by $\sim 20°$ over the first few arcseconds until it is approximately aligned with the outer lobes, a behavior reminiscent of the bends exhibited by several compact extragalactic jets that appear to curve by as much as 40° over the first few milliarcseconds from the core.[4] If the dynamical origin of the bends is the same in both types of jet, then the IRS-5 observations indicate that the large position-angle variations observed in compact extragalactic sources may not be simply the product of line-of-sight projection effects. Another morphological similarity between the IRS-5 jet and extragalactic jets can be found in the transverse oscillations of the intensity ridge line; a particularly striking example of this behavior in molecular clouds is provided by the HH 12 jet,[43] whereas NGC 6251[21] is a representative extragalactic source. Recent high-resolution spectroscopic observations[44,45] have provided direct measurements of the jet radial velocity (~ 200 km sec^{-1}) and of the mean electron density ($\gtrsim 10^3$ cm^{-3}), and they have indicated that the observed flow is mostly neutral. The optical observations have also revealed that the jet is visible only on one side of IRS-5 (unlike the radio jet, which is two-sided) and that it is surrounded by a wide, biconical cavity. The absence of an optical counterjet is probably due to an obscuration by a molecular disk surrounding the infrared object, whereas the cavity is perhaps best interpreted as representing the expanding "cocoon" of shocked jet material[2] that has evacuated the surrounding cloud material.

Direct support for the presence of a dense disk around IRS-5 has been provided by CS line observations (probing densities $\gtrsim 10^4$ cm^{-3}), which revealed a ~ 0.1 pc torus rotating with an apparent velocity of 0.35 km sec^{-1}.[46] Evidence is now accumulating that other bipolar sources are also surrounded by slowly rotating molecular disks on similar scales.[47-49] Further evidence for a circumstellar ($\lesssim 10^3$ AU) disk around IRS-5 has been obtained by submillimeter observations (probing densities in excess of 10^6 cm^{-3}).[50] In the case of the Kleinmann-Low nebula in Orion, an inner disk extending from $>10^3$ AU down to a distance of 35 AU from the infrared source IRc2 has been detected through millimeter-wavelength interferometry.[51] The significance of these discoveries lies in the fact that they provide the first direct information on the immediate environment where cosmic jets are produced. These observations are being supplemented by optical and infrared spectrophotometric measurements of the dense envelopes surrounding the central young stellar objects.[52,53] Interestingly enough, the physical conditions indicated by these measurements are similar to those that have been inferred to exist in the broad emission-line regions of Seyfert 1 galaxies and QSOs.

Molecular-cloud jets are distinguished also by the fact that the physical properties of the larger-scale ambient medium are directly accessible to observations. For example, the molecular hydrogen density in the outer ($\gtrsim 0.1$ pc) envelopes of dark clouds is generally found to scale with distance, r, from the center as r^{-2},[54] which is in accordance with the predictions of the singular isothermal sphere model.[55] In addition, optical and infrared polarization measurements provide information on the structure of the embedded magnetic field in such clouds. These measurements have shown that in many bipolar sources the ambient magnetic field is aligned (to within 20°) with the outflow axis.[56-58] Together with direct (using Zeeman-splitting measurements)[59,60] and indirect (using MHD shock models)[61,62] estimates of the magnetic field strength (which imply values of the order of a milligauss in regions with densities $\gtrsim 10^6$ cm^{-3}), these observations provide significant clues to the dynamical role of magnetic fields in these sources.

HYDRODYNAMIC EFFECTS

Nozzles and Funnels

The production of oppositely directed, high-velocity outflows clearly involves the establishment of two narrow channels through which the gas can escape from the vicinity of the compact source. Two basic mechanisms for the formation of such channels have been proposed: the transonic nozzle, which arises from the dynamical interaction between a high-entropy gas and a slightly flattened, confining mass distribution, and the centrifugally supported funnel, which forms a preexisting conduit for any hot gas injected near the origin. The transonic flow idea is based on the well-known de Laval nozzle principle: a subsonic gas expanding into a convergent-divergent nozzle will undergo a continuous reduction in its pressure and will become supersonic after passing through the narrowest portion of the channel. In the astrophysical application,[32] the decreasing external pressure distribution (rather than the nozzle geometry) is assumed to be given and it is hypothesized that the flow still undergoes a transonic transition as the channel walls adjust to maintain pressure equilibrium with the surrounding medium. This scenario has been verified by numerical calculations that indicated that the nozzle should be stable for at least some limited range of jet kinetic powers.[63,64] If the jet material first expands isotropically in the form of a supersonic wind, then, in order to collimate it, the ambient pressure must decrease with the distance r from the center slower than r^{-2}.[65] Furthermore, once a directed supersonic flow has been established (along the z-direction, say), it can remain confined only so long as the ambient pressure decreases along the jet axis more slowly than z^{-2}.[2] These arguments suggest that a stationary nozzle configuration can be maintained only if the confining pressure distribution is not too steep. In the case of extragalactic jets that form in the r^{-1} potential well of a massive black hole, the ambient pressure falls off too rapidly for this requirement to be satisfied unless the confining cloud also has significant rotational support.

Although the transonic nozzle mechanism has originally been proposed in the context of AGN jets, it is just as likely to operate in other astrophysical settings since it is independent of the source of the hot gas and of the particular distribution of the confining pressure. In particular, de Laval nozzles may be expected to form when an isotropic wind from an embedded young star expands into a surrounding "torus" of molecular matter. The interstellar bubble driven by such a wind is susceptible to the formation of nozzles when the ambient density distribution decreases as r^{-m}, with m close to 2.[66] Power-law distributions of this form are likely to occur in protostellar environments; in fact, the value of m in a star-forming cloud should range from 2 in the outer envelope (as often observed) to 1.5 in the vicinity of an accreting protostar[55] (although m could decrease to 0.5 in the immediate vicinity of the star if rotational effects are important[67]). The hot gas that undergoes the transonic transition is identified in this scenario with the shocked stellar wind. In order for this mechanism to operate effectively, the cooling time of the shocked gas must be long compared to the flow transit time between the shock and the nozzle; this condition could be satisfied even if the cooling time were short in comparison with the dynamical time of the original bubble. The required flattening of the confining molecular cloud could be due to rotation. This has been proposed also in the original model for AGN jets[32] and it is consistent with the current evidence from molecular-line observations. However, the inferred alignment of the bipolar flow axes with the projected magnetic field vectors in the associated molecular clouds suggests an alternative possibility,[66] namely, that the flattening is due to the preferential contraction of a self-gravitating cloud along the original direction of the "frozen-in" magnetic field.[68] The presence of either rotation or lateral magnetic stresses could also "flatten" the protostellar pressure distribution, which, as in the case of a black hole environment, would otherwise be too steep to support a stationary nozzle configuration.

Funnels near an accreting, compact object can be formed in one of two ways: either by the action of centrifugal forces, which exclude angular-momentum-carrying, inflowing matter from the vicinity of the rotation axis, or, in the context of thick accretion disks around black holes, by a purely relativistic effect that gives rise to an evacuated, roughly paraboloidal "zone of nonstationarity" in the innermost regions of the disk. In the latter case, the disks develop sharp cusps at the inner edges through which low-angular-momentum matter can flow into the hole;[69] this is in contrast to the situation in Newtonian disks, where even a small amount of angular momentum prevents material from falling in. The "centrifugal barrier" mechanism for funnels is likely to operate on both galactic and extragalactic scales, and it could manifest itself in quasi-spherical inflows[70] as well as in flattened accretion disks supported by either radiation[71] or gas[72] pressure. This mechanism thus provides a universal means for producing two oppositely directed jets that, just as in the case of transonic nozzles, is independent of the source of the hot gas introduced near the center. Although centrifugally supported funnels are generally divergent (and so cannot give rise to a de Laval transonic transition), the jets produced by this mechanism could still become supersonic after passing through the effective nozzle induced by the central gravitational field.[73]

In both nozzles and funnels, a substantial fraction of the outflowing mass could be composed of material that was torn off the channel walls and mixed into the jet.[64,74] Another possibility, indicated by recent numerical experiments,[75] is that the very process of funnel formation through centrifugal deceleration may cause some fraction of the ingoing flow to be redirected into an outgoing jet. For example, in the case of a nearly isotropic initial inflow endowed with a fixed specific angular momentum, it was found that a standoff shock forms in advance of the funnel, intersecting the equatorial plane at the radius where the effective (gravitational plus centrifugal) potential attains a minimum. (When the accreting object is a black hole, such a minimum exists only for specific angular momenta in excess of the "marginally stable" value.) Material crossing the standoff shock was found to flow along the shock surface towards the equatorial plane and to then pass through another shock that redirected it into the space between the standoff shock and the funnel surface, away from the central object. Although in these calculations the shocked material did not attain the escape velocity, it is conceivable that additional acceleration (associated, for example, with radiation pressure[76]) could result in some of the redirected matter leaving as a high-energy jet. Such a jet would initially be hollow and would possess a dynamically significant distribution of internal angular momentum.[77] Other schemes for extracting a portion of the inflowing mass before it reaches the central source have also been proposed. These include disk outflows that are induced by the deposition of momentum (e.g., by a stellar wind that ablates material from the disk[78]) or energy (e.g., by high-energy photons that evaporate matter off the disk surface[79]) transported from the central object, as well as locally powered outflows in the form of radiatively driven winds[80] and hydromagnetic flows (see discussion below).

Accretion from a nearly isotropic cloud characterized by a constant initial angular velocity, Ω_0, is probably a more realistic representation of the conditions in protostellar environments than the constant-specific-angular-momentum scenario.[81] In this case, the specific angular momentum at any fixed radius from the central object scales with the initial angle, θ_0, between the radius vector and the rotation axis as $\sin^2\theta_0$. Material with this angular momentum distribution would fall towards the equatorial plane from all directions and hence would not (so long as the supply of mass is not exhausted, i.e., on time scales $\ll \Omega_0^{-1}$) give rise to an evacuated central funnel.[67] However, the ram pressure distribution of the supersonic inflow would be anisotropic, with the lowest values attained along the rotation axis. Such an inflow could therefore provide the necessary conditions for the formation of nozzles.

Confinement and Collimation

Jets produced by the transonic nozzle mechanism may at first be rather poorly collimated. This is because the transonic transition generally takes place at a distance of the order of a pressure scale height from the central source, where the nozzle is still comparatively wide.[64] A jet emerging from a funnel could likewise have a large opening angle, which might even exceed the original opening angle of the funnel if the jet pressure is sufficiently high. However, such jets could still be collimated at large distances, r, from the origin by the pressure of the ambient medium. For this to take place, it is necessary that the external pressure decrease sufficiently slowly with r so that the jet can remain in pressure equilibrium with its surroundings. In the case of a supersonic, narrow jet and a pressure distribution that scales as r^{-n}, this condition becomes simply $n \leq 2$.[2] This condition should be satisfied in the outer envelopes of molecular clouds, which are characterized by $n \approx 2$.[54,55] Interestingly, this is also the value that has been inferred for the mean slope of the pressure distribution in extragalactic radio jets between the origin and ~ 100 kpc.[2] The pressure distribution is probably steeper closer in to the central source (for example, $n = 2.5$ in thermal-pressure-supported thick disks and in supersonic accretion flows) and in some of the transition zones encountered along the way (e.g., when the jet emerges from a galaxy or from a molecular-cloud core). A supersonic jet that could not achieve pressure equilibrium would expand freely with a constant opening angle until the pressure distribution became flatter again, at which stage the jet would be shocked back into confinement and recollimated.[82] This behavior has, in fact, been inferred in both galactic[45] and extragalactic[20,21] jets. In certain weak extragalactic sources, the reconfinement radius, r_*, has been identified with the distance to the bright emission knot that marks the end of the surface-brightness "gap" at the base of the jet. For those jets in this class that also show curvature due to motion through a dense intergalactic medium, one can express r_* in terms of the radius of curvature, R_c, and the jet radius, R_j, by $r_* \approx (R_c R_j)^{1/2}$,[83] a relation that is consistent with the observations in several sources.

An alternative interpretation of the recollimation "shoulders" observed in some radio jets is that they represent regions where the flow is subsonic and where therefore the channel radius decreases with decreasing external pressure (as in the convergent portion of a de Laval nozzle).[84] The transition of a supersonic jet back to the subsonic regime could be triggered, for example, by the relatively rapid decrease in the confining pressure that may accompany its exit from the galactic core. This would cause the jet to become "underexpanded" and could lead to its subsequent deceleration by internal shocks and through the entrainment of ambient material.[85] Entrainment into a supersonic jet may be initiated by surface instabilities, which should be most effective in a low-Mach-number flow propagating through a relatively tenuous medium (see further discussion below). The entrainment process can be described in terms of a viscous coupling between the jet and its surroundings.[86] The flow pattern could, in some cases, remain laminar,[87] but for sufficiently strong jets (large momentum fluxes), it is expected to become turbulent.[85] When the entrainment is efficient (in the sense that the mass flux in the jet increases with distance from the origin) and the flow is well mixed, then the jet may approach a self-similar evolution. In that case, the jet opening angle could remain approximately constant even if the ambient pressure continued to decrease,[83,88] which provides an alternative to the "free expansion" interpretation of conical jets. In addition to entrainment along the length of the jet, there is also the possibility in high-Mach-number, low-density jets for efficient entrainment to be initiated near the terminus of the flow. This process is induced by periodic vortex shedding into the cocoon of shocked jet material and should be most efficient in jets where the gas flow in the cocoon remains subsonic relative to the ambient medium.[89]

Knots and Wiggles

Surface instabilities, mentioned above in connection with their likely role in initiating entrainment into the flow, are also likely to be responsible for the common occurrence of emission knots and ridge-line wiggles in astrophysical jets. Specifically, these two morphological features could be associated with, respectively, the pinch ($m = 0$) mode and the kink ($m = 1$) mode of the Kelvin-Helmholtz instability (where m denotes the azimuthal wave number). Recent analytical results on the linear development of these modes,[90-93] together with numerical results on their nonlinear evolution,[94-96] have provided a better understanding of the possible effects of the instability on cylindrical, supersonic jets. In the linear regime, both of these modes can be analyzed in terms of the number, n, of internal nodes in the radial component of the perturbation functions. A natural distinction then arises between the fundamental or "ordinary" mode ($n = 0$) and the higher-order ($n \geq 1$) "reflection" modes. The latter modes are due to the resonant interaction between acoustic waves that propagate from one side of the jet to the other; at the boundary of the jet, these waves are partially transmitted out and in part reflected back with increased amplitudes, with the enhanced energy deriving from the relative motion between the jet and the ambient medium. In the case of the pinch mode, the linear analysis indicates that the pressure perturbations associated with the reflection modes should grow more rapidly than the ordinary-mode perturbations when the flow velocity exceeds the sum of the internal and external sound speeds.[93] This condition can be expressed in terms of the Mach number, M, of the jet and the ratio, η, of the internal to the external densities as $M > 1 + \eta^{1/2}$. The intriguing result of various numerical experiments[94,95] is that this relation also distinguishes between two very different regimes of the nonlinear evolution of pinched jets. These experiments have indicated that jets that originally lie in the "ordinary mode" regime tend to develop strong (planar) internal shocks and a progressively broadening mixing layer (governed by long-wavelength perturbations), which eventually lead to their disruption, but that jets in the "reflection mode" regime produce only weak (oblique) shocks and relatively little entrainment, so that they remain stable over large distances. The different behavior in these two regimes is consistent with the results of the linear analysis,[93] which showed that the pressure perturbations for the ordinary mode are essentially longitudinal oscillations with wavelengths in excess of the jet diameter, but that the corresponding perturbations for the reflection modes are primarily transverse and of shorter wavelengths.

On the basis of these results, one could interpret the aligned, quasi-periodic emission knots in cosmic jets as representing the high-pressure regions that have formed along the axis of the flow as a result of the nonlinear evolution of Kelvin-Helmholtz reflection pinch modes. In this picture, the knots arise downstream of the intersection points of biconical shocks[94] and hence should have a characteristic morphology that could presumably be detected by high-resolution observations. Depending on the excitation mechanism and the properties of the flow, the velocity of the knots could range from zero to a substantial fraction of the jet speed. For example, if the jet were perturbed by a spatially localized transition into a region of lower pressure, then the resulting knot pattern would be stationary with respect to the ambient medium,[66] corresponding to the point of marginal stability for the first reflection pinch mode.[93] Subsequent evolution within the jet might, however, transform the pattern into a faster growing, convected perturbation of shorter wavelength.[90] Environmental factors could also play an important role in the "parameter evolution" of the flow. For example, if a stable jet propagated into a medium where the sound speed increased with distance from the source, then the above-mentioned condition on the Mach number M might eventually be violated and the jet would enter the disruptive "ordinary mode" regime. This could perhaps account for the relative shortness of radio jets in certain clusters of galaxies that have been inferred to possess cooling accretion flows.[97]

Linear stability analyses of helical kink perturbations in supersonic jets indicate that they should dominate over pinch perturbations in the "ordinary mode" regime (wavelengths greater than the jet diameter).[90,92] Numerical simulations of the non-linear development of this instability[95,96] (performed, however, for a slab geometry rather than for a cylindrical beam) have confirmed that the fastest growing kink perturbation is indeed the fundamental ($n = 0$) mode. It was found that this mode evolves into an ever-steepening pattern of zigzagging internal shocks, which reinforce the initial perturbation by transmitting it from one side of the jet to the other and which ultimately lead to the disruption of the jet. This instability could thus be responsible for the appearance of noncollinear emission knots and of the associated transverse oscillations of the ridge line in certain jets. Indeed, recent high-resolution radio maps of the Cen A jet,[98] which revealed side-to-side limb brightening in the inner emission knots, could be interpreted in terms of this mechanism. Another consequence of the nonlinear steepening observed in the numerical experiments is that the supersonic motion of the kinks relative to the ambient gas may eventually lead to the formation of a series of fairly strong oblique shocks in the external medium. A similar pattern has been found also during the nonlinear evolution of the pinch mode; in that case, the shocks can be attributed to the steepening of acoustic waves transmitted through the boundary. These results suggest that external shocks should be a common feature in astrophysical jets and imply that jets could have a strong dynamical coupling to the surrounding medium (which would be particularly effective in jets that do not possess extensive cocoons, i.e., in relatively high-density or low-Mach-number flows).[95] Such shocks could have measurable effects on the emission and polarization properties of the ambient gas and might even be directly observable.

HYDROMAGNETIC EFFECTS

Production Mechanisms

Magnetic field effects have been implicated in the production of both extragalactic radio jets and galactic, molecular-cloud jets. One line of argument is based on a comparison of the dynamic thrust of the jets, as inferred from the parameters of the extended emission lobes, with the bolometric luminosity, L, of the central source. If the jets originate as a radiatively driven, optically thin wind, then the thrust, P, cannot exceed L/c. Typical values of P for radio galaxies lie in the range of 10^{31}–10^{35} dyne, whereas the correponding upper limits on L are 10^{41}–10^{44} erg sec^{-1}, which are too low to be consistent with this acceleration mechanism.[83] A similar conclusion can be drawn in the case of bipolar molecular sources: even for sources associated with relatively luminous central objects (L in the range of 10^2–10^5 $L_\odot$), the corresponding values of P (based on the inferred mass, size, and velocity of the emission-line lobes) are too large by factors of 10^2–10^3.[31] The thrust delivered by the radiation field at the origin could increase to $\tau L/c$ (where τ is the optical depth) if the acceleration region were optically thick.[99] In the limit that the jet velocity comes to exceed c/τ, the driving photons become "trapped" in the flow so that their energies (and not just their momenta) can be utilized in the acceleration process. In that limit, the central luminosity would be "supercritical" (i.e., for electron scattering opacity, L would exceed L_{Ed}, the Eddington luminosity); hence, this type of flow is unlikely to exist in double radio sources (which are generally associated with low-luminosity central objects), although it may be relevant in certain quasars and in SS433.[100] In the case of young stellar sources, the required values of τ (corresponding to continuum dust opacity) are simply too large to be consistent with the observed limits on the size of the jet production region. These considerations militate against radiative acceleration being the main driving mechanism in most radio galaxies and molecular-cloud sources, and point to hydromagnetic effects (which are in principle capable of giving rise to high-thrust flows) as the most natural alternative. One other conceivable mechanism for producing jets in AGN, namely, the formation of an electron-positron wind, is also subject to severe constraints.[101] On the one hand, it is necessary that the bulk Lorentz factor of such a wind be high in order for the travel time not to exceed the annihilation time of the e^+-e^- pairs; on the other hand, the Compton drag exerted on the pair plasma by any realistic ambient radiation field would (particularly in the case of quasars) prevent the attainment of highly relativistic speeds. Again, no such problems arise if the initial transport of power from the central source is mainly in the form of Poynting flux rather than particle kinetic energy flux.

One universal mechanism for the hydromagnetic production of jets involves centrifugally driven winds. Such winds could arise above Keplerian accretion disks with embedded open magnetic field lines. If the field lines make an angle of less than 60° with the surface of the disk, then material tied to the field can be flung out from the surface as a result of centrifugal forces, much as beads are pushed out along a rotating wire.[102] These MHD winds possess the usual three critical points that delineate different regions of the flow. Closest to the disk is the slow magnetosonic point, where the wind becomes sonic. This region is still magnetically dominated, and the balance between the magnetic pressure gradient and the magnetic tension causes the field lines to bend towards the rotation axis. It also produces the initial collimation of the flow. The wind acceleration continues to be primarily centrifugal all the way to the Alfvén critical point, where the poloidal components of the flow velocity and the Alfvén velocity become equal. Up until that point, the gas along a given magnetic flux tube is in approximate corotation with the disk material at the foot of the tube. The wind thus exerts a substantial back torque on the disk (with an effective lever arm corresponding to the Alfvén radius) and provides an effective mechanism for removing angular momentum from the accretion flow. Beyond the Alfvén critical point, the inertia of the matter causes the field lines to become progressively more toroidal. Magnetic pressure gradients (rather than centrifugal forces) now become the main acceleration mechanism; in addition, magnetic "hoop" stresses associated with the toroidal field component act to collimate the flow (see further discussion below). The acceleration process continues effectively to the fast magnetosonic point, where the poloidal flow velocity becomes equal to the total Alfvén velocity, $B/(4\pi\rho)^{1/2}$ (where B is the magnetic field magnitude and ρ is the gas density). The location of this last critical point depends on the initial temperature of the flow and on the shape of the field lines: for a cold, radial flow it lies at infinity, but for a nonzero initial temperature or a more rapidly divergent field topology it occurs at a finite distance above the disk. (Initially divergent cold flows that attain highly super-Alfvénic speeds would eventually, however, converge towards the axis to maintain transverse magnetic pressure support.[102]) The energy flux of winds that passes through the fast magnetosonic point at a finite distance from the disk is asymptotically dominated by the gas kinetic contribution. It is conceivable, however, that the jet is still Poynting-flux-dominated far away from the origin and that the transition to a matter-dominated flow occurs only as a result of its interaction with the ambient medium.[101]

The above mechanism is likely to operate in both AGN and molecular-cloud disks,[102,103] and in both cases it could conceivably control the evolution of the disk and the mass accretion rate through the efficient extraction of angular momentum. In fact, one of the attractive features of the centrifugally-driven-wind scenario is that it provides a possible resolution of the "angular momentum problem" in the theory of star formation, which is related in a natural way to the common occurrence of bipolar outflows in young stellar objects. In the case of a self-similar outflow from a Keplerian disk,[102] B scales with radius as $r^{-5/4}$, so most of the extracted angular momentum comes from the outer parts of the disk. (The thrust P in this solution is, however, dominated by the highest velocity matter driven from the inner edge of the disk.) The efficiency of angular-momentum removal by this mechanism depends on the degree of coupling between the magnetic field and the disk, which in turn is highly sensitive to the fractional ionization of the disk material. This question is particularly relevant in the case of molecular-cloud disks, where it is conceivable that the strength of the coupling (and hence the outflow and accretion processes) is regulated in a self-consistent manner by the ionizing radiation emitted by the accreting protostar itself.[103]

Centrifugally driven winds need not be associated only with disks, but could also emanate from the central objects themselves. For example, such winds might be driven by rapidly rotating protostars that retain sufficiently high ($B \gtrsim 10$ G) surface magnetic fields.[104] Another possibility is that, while the energy for the outflow comes from the rotation of the central object, most of the mass is provided by the surrounding cloud to which it remains magnetically linked.[105] When the central object is a rotating black hole, then an external mass supply is, in fact, mandatory. In that case, it is also necessary that an ambient mass distribution be present to support the field lines that thread the hole, but the basic angular-momentum (and energy) extraction mechanism should, in principle, still operate effectively.[83,106] Other effects may also be important. For example, the initial acceleration may be dominated by thermal, rather than centrifugal, effects. The heating, however, could itself be due to magnetic energy dissipation in the coronae of disks or stars.[107,108] In the case of protostars, this activity may be associated with dynamo action that is initiated when the stars become convectively unstable.[81] In addition, centrifugally driven outflows need not always be steady and they conceivably could also take the form of a sequence of MHD "twist waves."[109]

The applicability of the MHD approximation to the treatment of magnetized flows in AGN is still somewhat controversial. Since the rest-mass density of the ejected particles is expected to be much smaller than the magnetic energy density near the central black hole, some models have neglected the inertial terms in the particle equations of motion and adopted an electromagnetic (rather than a hydromagnetic) description of the energy extraction processes.[110,111] These models impose the condition of $\rho_e E + (J \times B)/c = 0$, corresponding to an electromagnetically force-free magnetosphere, instead of the perfect-MHD condition of $E + v \times B = 0$ (where ρ_e is the charge density and E, J, and v are the electric field, current density, and velocity vectors, respectively). The argument generally given in support of the MHD approximation is that the value of ρ_e that is required to short-out the vacuum electric field near a rapidly rotating, massive black hole (the analog of the Goldreich-Julian density in pulsar theory) is negligible in comparison with the actual density that is expected to exist in the vicinity of such a hole as a result of e^+-e^- pair-creation processes.[2,83] However, even if the MHD approximation is more appropriate for AGN outflows, it is still conceivable that pulsar-type winds could be associated with other astrophysical jets. In fact, it has recently been argued that pulsars themselves could give rise to well-collimated jets through the interaction of their charged particle beams with the surrounding supernova remnants.[112,113]

Magnetic Confinement and Collimation

As mentioned above, centrifugally driven winds would tend to be collimated both by the initial force-free topology of the magnetic field lines and, after inertial effects became important, by the pinch stress induced by the toroidal field component. The predominance of the toroidal component (or, more generally, of the component transverse to the direction of motion) sufficiently far downstream is, however, a general property of any highly conducting jet with a negligible amount of internal shear.[32] This follows from a simple flux-conservation argument, which implies that the stresses associated with the longitudinal component of a "frozen-in" magnetic field in a supersonic, nonrelativistic jet should decrease with increasing jet radius, R_j, as R_j^{-4}, but that the transverse stresses would decline only as R_j^{-2}. This argument receives observational support from radio polarization measurements of weak, extended, extragalactic jets, which indicate that most of these sources exhibit a transition from longitudinal to transverse orientation of the projected magnetic field vectors in the first 10% of their lengths.[1] Another consequence of the inferred scaling of the transverse magnetic field is that, for flows with an effective adiabatic index greater than one, the thermal pressure in the jet would eventually fall below the magnetic pressure even if the jet were gas-pressure dominated closer in. This implies that magnetic stresses could control the transverse dynamics of such jets at large distances from the origin (see further discussion below).

The collimation produced by toroidal magnetic fields is associated with the flow of net current along the jet, in accordance with the MHD relation $J = c\nabla \times B/4\pi$.[114,115] It is, however, somewhat misleading to describe the effect of the field as "self-confinement" since the magnetic stresses cannot contain themselves (the total pressure always exceeds the tension) and the field-carrying plasma must be confined by external forces.[116] A more accurate statement is that the return current need not flow through the observable regions of the jet, which could therefore carry a net current and experience enhanced collimation. In the context of the perfect-MHD approximation, the confining toroidal field lines outside the visible jet are frozen into jet material and therefore must have been transported to their present locations by the flow itself. The invisible jet material might, for example, be identified with the cocoon of shocked gas deposited at the advancing end of the jet. In this picture,[2] the return current flows near the outer boundary of the cocoon and there exists a large pressure difference between that boundary (which separates the jet material from the external medium) and the inner boundary (which abuts on the visible jet). This model may account for the apparent discrepancy between the pressure values inferred inside such sources as the quasar jet 4C32.69 or the radio lobes of the galactic X-ray source Sco X-1, and the much smaller values that are likely to characterize the ambient medium in each case.[117] These sources presumably correspond to relatively hot (or low-density), high-Mach-number jets, which are the types of flow expected to develop extensive cocoons,[95] although resistive field-redistribution effects behind the terminal shock[113,114] could modify this requirement. However, there is as yet no direct confirmation of this interpretation (such as might have been provided by a measurement of a sign reversal in the Faraday rotation measure across the jet) in any of these sources.

Lateral confinement could also be effected by an external magnetic field. In particular, a jet that propagates along the direction of a large-scale ambient field may be collimated by the transverse stresses of field lines that resist compression and bending by the flow. This possibility was first proposed in connection with molecular-cloud jets,[66] where it was motivated by the observed near alignment between the outflow directions and the projected orientations of the large-scale magnetic fields.[56-58] (As was noted above in connection with the formation of nozzles, the ambient magnetic field is likely to affect the collimation also through the anisotropic density distribution that it induces in the cloud.) Additional support for this possibility is provided by the inferred magnitudes of the embedded magnetic fields,[59-62] which indicate that in many cases the field may contribute significantly to the dynamical support of the cloud. An analogous idea has been explored in connection with a magnetized-neutron-star model for SS433;[100] there it was proposed that the tension of the dipolar field lines anchored in the star could stabilize a pair of de Laval nozzles and provide the initial collimation of the jets. An altogether different mechanism was suggested for beams composed of low-frequency electromagnetic waves, such as might be emitted by pulsars or spinars.[118] For sufficiently strong waves, the nonlinear dependence of the phase velocity on the wave amplitude could lead to a radial concentration (self-focusing) of the beam. However, under astrophysical circumstances, it is likely that such beams would be converted by various plasma instabilities into particle-energy (rather than Poynting-flux) dominated jets for which a fluid MHD description is more appropriate.[2]

Equilibria and Instabilities

As was noted above, simple flux-conservation arguments suggest that magnetized jets should become magnetic-pressure dominated sufficiently far away from the origin even if they are thermal-pressure dominated near the source. (In certain models of current-carrying jets,[115] the compression induced by the magnetic hoop stresses tends to keep the thermal pressure in rough equipartition with the magnetic pressure after the magnetic field becomes dynamically important; however, this approximate equality need not be maintained if the net current in the jet is small or if radiative cooling is efficient.) Under these circumstances, the jet may be expected to settle into a new, magnetically dominated configuration in which the effect of thermal pressure forces is negligible. By considering the hydrostatic balance equation in the reference frame of the jet and applying the perfect-MHD approximation, one finds that this condition is equivalent to having $(\nabla \times \boldsymbol{B}) \times \boldsymbol{B} = \boldsymbol{0}$. The first integral of this equation is $\nabla \times \boldsymbol{B} = \mu \boldsymbol{B}$, where μ is in general some scalar function of the coordinates that is constant along field lines ($\boldsymbol{B} \cdot \nabla \mu = 0$). This relation describes the so-called force-free field,[116] which has the property that the current density, $\boldsymbol{J}$, is everywhere parallel to the local direction of $\boldsymbol{B}$.

The transition to a force-free field configuration would in general be accompanied by a rearrangement of the field lines in the jet. This rearrangement, however, would be subject to various topological constraints resulting from the fact that, in ideal MHD, the field lines cannot be severed or reconnected. One such constraint is expressed by the conservation of the total magnetic helicity, K, given by the integral of $\boldsymbol{A} \cdot \boldsymbol{B}/8\pi$ over the volume of the jet (where $\boldsymbol{A}$ is the vector potential from which the field is derived: $\boldsymbol{B} = \nabla \times \boldsymbol{A}$). The total magnetic helicity is a measure of the global twist and knottedness of the field lines and, under perfect-MHD conditions, is a conserved and gauge-invariant quantity for any closed volume bounded by magnetic surfaces (surfaces that are everywhere parallel to the local direction of $\boldsymbol{B}$). The conservation of K has been a key concept in the interpretation of various laboratory plasma-confinement experiments, where it was hypothesized that it continues to hold even in the presence of a small, but finite resistivity (when other ideal-MHD constraints no longer apply).[119] The nonzero resistivity allows the plasma to dissipate energy and approach a minimum-energy state, which can be shown to correspond to a linear (i.e., $\mu = const.$) force-free field.

The hypothesis of magnetic energy dissipation taking place under the conditions of global helicity conservation has recently been applied to the modeling of magnetic-pressure-dominated jets.[120] It was argued that confined, super-Alfvénic jets that are also supersonic with respect to the external medium can be approximated at each point by a cylinder of radius R_j. Assuming, in addition, that the helicity per unit length is constant along the jet, it then follows that the minimum-energy equilibrium configuration corresponds to a force-free field that is locally linear (although μ can vary along the jet). Furthermore, as in the laboratory pinch models,[119] it turns out that this configuration is, in general, a linear superposition of only two modes: an axisymmetric mode that accounts for the net axial flux (and current) in the jet and a helical mode that varies along the jet with a wavelength $\lambda \approx 5 R_j$. The nonaxisymmetric mode becomes energetically favorable when the confining external pressure drops below a certain critical value that depends on the magnitudes of the conserved flux and helicity. This model therefore predicts that jets that propagate to regions of sufficiently low pressure would have nonaxisymmetric equilibrium configurations. The transition to nonaxisymmetric equilibria has been observed in laboratory experiments and is analogous to the transition of (axisymmetric) Maclaurin spheroids into (nonaxisymmetric) Jacobi ellipsoids in the theory of rotating fluid masses.

The nonaxisymmetric geometry of the helical mode (corresponding to two oppositely directed flux tubes wrapped around each other) should be reflected in the synchrotron emission pattern of relativistic electrons radiating in the equilibrium force-free field. In fact, it turns out that one can explain with this model a variety of nonaxisymmetric and oscillatory features exhibited by certain extended extragalactic radio jets such as NGC 6251.[21] In particular, one can account for the transverse oscillations of the ridge line that occur at various locations along that jet with a characteristic wavelength of $\sim 5 R_j$. However, in contrast to the Kelvin-Helmholtz kink-instability interpretation discussed above, this mechanism does not involve an actual transverse motion of the jet. In addition, one can account with this model for the quasi-periodic emission knots that are detected along extended radio jets. Again, this is merely an apparent effect that is due to the fact that the observed intensity of the synchrotron emission from the periodic, nonaxisymmetric field distribution in the jet depends only on the component of B normal to the line of sight. In this picture, the knots do not represent sites of actual compression of the jet material as in the Kelvin-Helmholtz pinch-instability interpretation. (It may be noted here that other hydrodynamic interpretations of emission knots, e.g., in terms of bow shocks attached to dense clumps accelerated by the flow[121] or of thermal instabilities in the jet,[122] generally also postulate actual pressure or density enhancements.) This picture thus provides an alternative explanation of knots and wiggles in jets, one that is based on the equilibrium structure of the flow rather than on the perturbations induced by various instabilities.

The minimum-energy force-free solution also has possible applications in compact extragalactic radio sources. In particular, it provides a natural interpretation of the large ($>180°$) polarization position-angle swings that have been measured in a number of BL Lacertae objects and highly variable quasars.[123] Such sources are often identified with relativistic jets that are observed at a small angle to the axis.[4] In this interpretation, the swings are attributed to the propagation of a shock wave along an unresolved, relativistic jet that has a nonaxisymmetric equilibrium field configuration: as the shock moves through successive transverse cross sections of the jet, it "illuminates" (by enhanced synchrotron emission) the progressively rotated magnetic field vectors associated with the nonaxisymmetric mode, thereby giving rise to a systematic variation of the apparent polarization position angle. Because of relativistic aberration, the apparent swing could exhibit a quasi-periodic, steplike behavior even for a constant-velocity shock. This behavior, which may have already been detected in one BL Lac object,[124] could help to distinguish this interpretation of the observed swings from certain alternative "random-walk" explanations.[125]

The force-free equilibria discussed in the preceding paragraphs represent minimum-energy states that should be immune to both ideal and resistive MHD instabilities on time scales short compared to the resistive diffusion time (i.e., on time scales over which the magnetic helicity remains approximately conserved). However, the jets under consideration would still be susceptible to the Kelvin-Helmholtz instability. Linear perturbation calculations[90] nevertheless indicate that strong magnetic fields parallel to the direction of the flow could have a stabilizing effect on the growth of large-scale modes, particularly those of the "ordinary" variety. Magnetic fields might also play a role in channeling some of the bulk kinetic energy that drives the instability into kinetic energy of relativistic electrons (which could then be liberated as synchrotron radiation). This might be accomplished by means of a turbulent cascade of MHD waves, which, for a fully turbulent flow, would take place throughout the entire volume of the jet.[126] The dissipation of the waves at the lower end of the cascade could, in principle, lead directly to particle acceleration, either by the Fermi mechanism[127] or through resonant interactions.[128] An alternative source of energy for the synchrotron emission in radio jets has, however, been suggested in the context of the force-free equilibrium model considered above.[129] Specifically, it was proposed that the requisite power could be provided by the dissipation of magnetic energy in an expanding or a contracting jet that would take place as the field continuously adjusted to maintain a minimum-energy configuration while conserving helicity. In this case, efficient particle acceleration might occur in magnetic field-reconnection sites (through induced DC electric fields) and need not be mediated by MHD waves. This energy dissipation mechanism is in many respects similar to that which is thought to operate in the heating of the solar corona and in the production of solar flares.[116,130]

SUMMARY

Several recent observational developments have provided qualitatively new data on cosmic jets. In particular, one can single out the multifrequency, high-dynamic-range, and high-resolution maps of extended extragalactic radio sources and the VLBI polarization measurements of compact radio sources. The study of galactic molecular-cloud jets has similarly been advanced by the development of millimeter-wavelength interferometry and through high-resolution CCD observations. These data have, for the first time, provided information on the detailed structure of the intrinsic magnetic field in extragalactic radio jets, as well as on the presence of disks and on the general physical conditions near the origin of molecular-cloud sources.

Despite the vast differences in scales, in physical parameters, and even in the detailed physical processes between bipolar stellar outflows and AGN jets, the striking morphological similarities between these two types of sources strongly suggest that they are different manifestations of the same basic astrophysical phenomenon. This, in turn, implies that certain universal dynamical mechanisms could be responsible for the production, collimation, and morphological properties of astrophysical jets; and, furthermore, that some of the basic dynamical processes that are more readily observable in one class of objects might also apply in other types of jets. In discussing the various possible mechanisms, it proved useful to distinguish between purely hydrodynamic and magnetic-field effects, even though these effects are probably intertwined in real sources. Thus, either nozzles or funnels could complement the action of magnetic stresses in channeling bipolar outflows from magnetized accretion disks. Similarly, both the ambient thermal pressure and the convected toroidal magnetic field lines could contribute to the confinement and collimation of observable jets, with each mechanism possibly dominating in a different region of the flow. In the same vein, both Kelvin-Helmholtz instabilities and minimum-energy MHD configurations might be involved in the production of the observed knots and wiggles in jets. Finally, to give one additional example, MHD waves could play a role in the dissipation of turbulent cascades that on the largest scale are driven by purely hydrodynamic instabilities.

Recent theoretical studies of jets have also been marked by qualitatively new developments. Sophisticated numerical codes have been applied to the simulation of axisymmetric flows, and fully three-dimensional experiments are apparently forthcoming. Together with linear perturbation analyses, these simulations have provided new insights into the expected evolution of various surface instabilities in jets. Another development has been the systematic introduction of viscosity and resistivity into the models, which should provide a more accurate representation of actual astrophysical conditions. This has already produced a number of promising results on the structure of turbulent as well as of magnetically dominated jets. Yet another example of a new approach can be found in the recent analytical and numerical attempts to explore in detail the connection between jets and accretion disks. In particular, these studies have raised the possibility that a self-regulating inflow-outflow pattern could be maintained by means of various feedback mechanisms involving a global network of mechanical, radiative, and magnetic couples. Further studies in this direction should provide additional insights into the universal nature of astrophysical jets.

REFERENCES

1. BRIDLE, A. H. & R. A. PERLEY. 1984. Annu. Rev. Astron. Astrophys. **22**: 319.
2. BEGELMAN, M. C., R. D. BLANDFORD & M. J. REES. 1984. Rev. Mod. Phys. **56**: 255.
3. MILEY, G. K. 1980. Annu. Rev. Astron. Astrophys. **18**: 165.
4. KELLERMANN, K. I. & I. I. K. PAULINY-TOTH. 1981. Annu. Rev. Astron. Astrophys. **19**: 373.
5. HEESCHEN, D. S. & C. M. WADE, EDS. 1982. IAU Symposium No. 97 on Extragalactic Radio Sources. D. Reidel Publishing Co. Dordrecht.
6. FANTI, R., K. KELLERMANN & G. SETTI, EDS. 1984. IAU Symposium No. 110 on VLBI and Compact Radio Sources. D. Reidel Publishing Co. Dordrecht.
7. FERRARI, A. & A. G. PACHOLCZYK, EDS. 1983. Proceedings of an International Workshop on Astrophysical Jets. Reidel. Dordrecht.
8. BRIDLE, A. H. & J. A. EILEK, EDS. 1985. Proc. NRAO Workshop No. 9 on the Physics of Energy Transport in Extragalactic Radio Sources. National Radio Astronomy Observatory. Green Bank, West Virginia.
9. MILEY, G. 1983. Reference 7, p. 99.
10. OSMER, P. S. 1978. Astrophys. J. **226**: L79.
11. GRAHAM, J. A. & R. M. PRICE. 1981. Astrophys. J. **247**: 813.
12. GRAHAM, J. A. 1983. Astrophys. J. **269**: 440.
13. FEIGELSON, E. D. et al. 1981. Astrophys. J. **251**: 31.
14. BURNS, J. O., E. D. FEIGELSON & E. J. SCHREIER. 1983. Astrophys. J. **273**: 128.
15. BRODIE, J., A. KÖNIGL & S. BOWYER. 1983. Astrophys. J. **273**: 154.
16. DREHER, J. W. & E. D. FEIGELSON. 1984. Nature (London) **308**: 43.
17. PERLEY, R. A., J. W. DREHER & J. J. COWAN. 1984. Astrophys. J. **285**: L35.
18. DREHER, J. W. 1981. Astron. J. **86**: 833.
19. LAING, R. 1981. Mon. Not. R. Astron. Soc. **195**: 261.
20. CORNWELL, T. J. & R. A. PERLEY. 1985. Reference 8, p. 39.
21. PERLEY, R. A., A. H. BRIDLE & A. G. WILLIS. 1984. Astrophys. J. Suppl. Ser. **54**: 291.
22. O'DEA, C. P. & F. N. OWEN. 1986. Astrophys. J. In press.
23. ROBERTS, D. H., R. I. POTASH, J. F. C. WARDLE, A. E. E. ROGERS & D. F. BURKE. 1984. Reference 6, p. 35.
24. COTTON, W. D. 1984. Astrophys. J. **286**: 503.
25. WALKER, R. C. 1985. Reference 8, p. 20.
26. CANTÓ, J. & E. E. MENDOZA, EDS. 1983. Proc. Symposium on Herbig-Haro Objects, T Tauri Stars, and Related Phenomena. Rev. Mex. Astron. Astrof. 7.
27. SCHWARTZ, R. D. 1983. Annu. Rev. Astron. Astrophys. **21**: 209.
28. BLACK, D. C. & M. S. MATTHEWS, EDS. 1985. Protostars and Planets II. University of Arizona Press. Tucson.

29. MUNDT, R. 1985. Reference 28, p. 414.
30. BIEGING, J. H., M. COHEN & P. R. SCHWARTZ. 1984. Astrophys. J. **282:** 699.
31. BALLY, J. & C. LADA. 1983. Astrophys. J. **265:** 824.
32. BLANDFORD, R. D. & M. J. REES. 1974. Mon. Not. R. Astron. Soc. **169:** 395.
33. BECKWITH, S., A. NATTA & E. E. SALPETER. 1983. Astrophys. J. **267:** 596.
34. KÖNIGL, A. 1983. Reference 26, p. 121.
35. HERBIG, G. H. & B. F. JONES. 1981. Astron. J. **86:** 1232.
36. CHOE, S-U., K-H. BÖHM & J. SOLF. 1985. Astrophys. J. **288:** 338.
37. NORMAN, C. A. & J. SILK. 1979. Astrophys. J. **228:** 197.
38. HERBIG, G. H. & B. F. JONES. 1983. Astron. J. **88:** 1040.
39. BIRETTA, J. A., F. N. OWEN & P. E. HARDEE. 1983. Astrophys. J. **274:** L27.
40. SNELL, R. L., R. B. LOREN & R. L. PLAMBECK. 1980. Astrophys. J. **239:** L17.
41. CUDWORTH, K. M. & G. HERBIG. 1979. Astron. J. **84:** 548.
42. MUNDT, R. & J. W. FRIED. 1983. Astrophys. J. **274:** L83.
43. STROM, K. M., S. E. STROM & J. STOCKE. 1983. Astrophys. J. **271:** L23.
44. SARCANDER, M., TH. NECKEL & H. ELSÄSSER. 1985. Astrophys. J. **288:** L51.
45. SNELL, R. L., J. BALLY, S. E. STROM & K. M. STROM. 1985. Astrophys. J. **290:** 587.
46. KAIFU, N. et al. 1984. Astron. Astrophys. **134:** 7.
47. SCHWARTZ, P. R., J. A. WAAK & H. A. SMITH. 1983. Astrophys. J. **267:** L109.
48. HASEGAWA, T. et al. 1984. Astrophys. J. **283:** 117.
49. BIEGING, J. H. 1984. Astrophys. J. **286:** 561.
50. DAVIDSON, J. A. & D. T. JAFFE. 1984. Astrophys. J. **277:** L13.
51. PLAMBECK, R. L., S. N. VOGEL, M. C. H. WRIGHT, J. H. BIEGING & W. J. WELCH. 1985. In Proc. URSI Conference on Millimeter- and Submillimeter-Wave Astronomy. J. Gomez-Gonzales, Ed.
52. PERSSON, S. E., T. R. GEBALLE, P. J. MCGREGOR, S. EDWARDS & C. J. LONSDALE. 1984. Astrophys. J. **286:** 289.
53. MCGREGOR, P. J., S. E. PERSSON & J. G. COHEN. 1984. Astrophys. J. **286:** 609.
54. SNELL, R. L. 1981. Astrophys. J. Suppl. Ser. **45:** 121.
55. SHU, F. H. 1977. Astrophys. J. **214:** 488.
56. VRBA, F. J., S. E. STROM & K. M. STROM. 1976. Astron. J. **81:** 958.
57. DYCK, H. M. & C. J. LONSDALE. 1979. Astron. J. **84:** 1339.
58. TURNSHEK, D. A., D. E. TURNSHEK & E. R. CRAINE. 1980. Astron. J. **85:** 1638.
59. HANSEN, S. S. 1982. Astrophys. J. **260:** 599.
60. WOUTERLOOT, J. G. A., H. J. HABING & J. HERMAN. 1980. Astron. Astrophys. **81:** L11.
61. DRAINE, B. T. & W. G. ROBERGE. 1982. Astrophys. J. **259:** L91.
62. CHERNOFF, D. F., D. J. HOLLENBACH & C. F. MCKEE. 1982. Astrophys. J. **259:** L97.
63. WIITA, P. J. 1978. Astrophys. J. **221:** 436.
64. NORMAN, M. L., L. SMARR, J. R. WILSON & M. D. SMITH. 1981. Astrophys. J. **247:** 52.
65. SMITH, M. D., L. SMARR, M. L. NORMAN & J. R. WILSON. 1983. Astrophys. J. **264:** 432.
66. KÖNIGL, A. 1982. Astrophys. J. **261:** 115.
67. ULRICH, R. K. 1976. Astrophys. J. **210:** 377.
68. MOUSCHOVIAS, T. CH. 1976. Astrophys. J. **207:** 141.
69. ABRAMOWICZ, M. A., M. JAROSZYŃSKI & M. SIKORA. 1978. Astron. Astrophys. **63:** 221.
70. SPARKE, L. S. 1982. Astrophys. J. **254:** 546.
71. PACZYŃSKI, B. & P. J. WIITA. 1980. Astron. Astrophys. **88:** 23.
72. REES, M. J., M. C. BEGELMAN, R. D. BLANDFORD & E. S. PHINNEY. 1982. Nature (London) **295:** 17.
73. FUKUE, J. 1982. Publ. Astron. Soc. Japan **34:** 163.
74. NARAYAN, R., R. NITYANDA & P. J. WIITA. 1983. Mon. Not. R. Astron. Soc. **205:** 1103.
75. HAWLEY, J. F. & L. L. SMARR. 1985. In Proc. Los Alamos Workshop on Magnetospheric Phenomena in Astrophysics (AIP).
76. ICKE, V. 1980. Astron. J. **85:** 329.
77. CHAKRABARTI, S. 1985. Astrophys. J. **288:** 7.
78. ELMEGREEN, B. G. 1978. Moon and Planets **19:** 261.
79. BEGELMAN, M. C., C. F. MCKEE & G. A. SHIELDS. 1983. Astrophys. J. **271:** 70.

80. MEIER, D. L. 1982. Astrophys. J. **256**: 706.
81. CASSEN, P., F. H. SHU & S. TEREBEY. 1985. Reference 28, p. 448.
82. SANDERS, R. H. 1983. Astrophys. J. **266**: 73.
83. PHINNEY, E. S. 1983. Ph.D. Thesis. University of Cambridge.
84. BICKNELL, G. V. 1985. Reference 8, p. 63.
85. BICKNELL, G. V. 1984. Astrophys. J. **286**: 68.
86. BAAN, W. A. 1980. Astrophys. J. **239**: 433.
87. HENRIKSEN, R. N. 1985. Reference 8, p. 211.
88. BEGELMAN, M. C. 1982. Reference 5, p. 223.
89. NORMAN, M. L., L. SMARR, K-H. A. WINKLER & M. D. SMITH. 1982. Astron. Astrophys. **113**: 285.
90. FERRARI, A., E. TRUSSONI & L. ZANINETTI. 1981. Mon. Not. R. Astron. Soc. **196**: 1051.
91. COHN, H. 1983. Astrophys. J. **269**: 500.
92. HARDEE, P. E. 1984. Astrophys. J. **277**: 106.
93. PAYNE, D. G. & H. COHN. 1985. Astrophys. J. **299**: 655.
94. NORMAN, M. L., K-H. A. WINKLER & L. L. SMARR. 1985. Reference 8, p. 150.
95. SMARR, L. L., M. L. NORMAN & K-H. A. WINKLER. 1984. Physica **12D**: 83.
96. WOODWARD, P. R. 1984. *In* Astrophysical Radiation Hydrodynamics. K-H. A. Winkler & M. L. Norman, Eds. Reidel. Dordrecht.
97. SUMI, D. M. & L. L. SMARR. 1985. Reference 8, p. 168.
98. BURNS, J. O., D. CLARKE, E. D. FEIGELSON & E. J. SCHREIER. 1985. Reference 8, p. 255.
99. PHILLIPS, J. P. & J. E. BECKMAN. 1980. Mon. Not. R. Astron. Soc. **193**: 245.
100. BEGELMAN, M. C. & M. J. REES. 1984. Mon. Not. R. Astron. Soc. **206**: 209.
101. REES, M. J. 1984. Reference 6, p. 207.
102. BLANDFORD, R. D. & D. G. PAYNE. 1982. Mon. Not. R. Astron. Soc. **199**: 883.
103. PUDRITZ, R. E. & C. A. NORMAN. 1983. Astrophys. J. **274**: 677.
104. HARTMANN, L. & K. B. MACGREGOR. 1982. Astrophys. J. **259**: 180.
105. DRAINE, B. T. 1983. Astrophys. J. **270**: 519.
106. PHINNEY, E. S. 1982. Reference 7, p. 201.
107. STURROCK, P. A. & C. BARNES. 1972. Astrophys. J. **176**: 31.
108. GALEEV, A. A., R. ROSNER & G. S. VAIANA. 1979. Astrophys. J. **229**: 318.
109. UCHIDA, Y. & K. SHIBATA. 1984. *In* IAU Symposium No. 107. M. Kundu, Ed. Reidel. Dordrecht.
110. BLANDFORD, R. D. & R. L. ZNAJEK. 1977. Mon. Not. R. Astron. Soc. **179**: 433.
111. BURNS, M. L. & R. V. E. LOVELACE. 1982. Astrophys. J. **262**: 87.
112. MICHEL, F. C. 1985. Astrophys. J. **288**: 138.
113. BENFORD, G. 1984. Astrophys. J. **282**: 154.
114. BENFORD, G. 1978. Mon. Not. R. Astron. Soc. **183**: 29.
115. CHAN, K. L. & R. N. HENRIKSEN. 1980. Astrophys. J. **241**: 534.
116. PARKER, E. N. 1979. Cosmical Magnetic Fields. Clarendon Press. Oxford.
117. ACHTERBERG, A., R. D. BLANDFORD & P. GOLDREICH. 1983. Nature (London) **304**: 607.
118. FERRARI, A., S. MASSAGLIA & M. DOBROWOLNY. 1980. Astron. Astrophys. **92**: 246.
119. TAYLOR, J. B. 1974. Phys. Rev. Lett. **33**: 1139.
120. KÖNIGL, A. & A. R. CHOUDHURI. 1985. Astrophys. J. **289**: 173.
121. BLANDFORD, R. D. & A. KÖNIGL. 1979. Astrophys. Lett. **20**: 15.
122. MARSCHER, A. P. 1980. Astrophys. J. **239**: 296.
123. KÖNIGL, A. & A. R. CHOUDHURI. 1985. Astrophys. J. **289**: 188.
124. ALLER, H. D., P. E. HODGE & M. F. ALLER. 1981. Astrophys. J. **248**: L5.
125. JONES, T. W. *et al.* 1985. Astrophys. J. **290**: 627.
126. FERRARI, A., E. TRUSSONI & L. ZANINETTI. 1979. Astron. Astrophys. **79**: 190.
127. BICKNELL, G. V. & D. B. MELROSE. 1982. Astrophys. J. **262**: 511.
128. EILEK, J. A. & R. N. HENRIKSEN. 1984. Astrophys. J. **277**: 820.
129. CHOUDHURI, A. R. & A. KÖNIGL. 1986. Astrophys. J. In press.
130. NORMAN, C. A. & J. HEYVAERTS. 1983. Astron. Astrophys. **124**: L1.

PART III. LARGE-SCALE STRUCTURE; CLUSTERING

Superclustering of Galaxy Clusters

NETA A. BAHCALL

Space Telescope Science Institute
Baltimore, Maryland 21218

THE SAMPLE

The sample used is Abell's (1958)[1] statistical sample of rich clusters of galaxies of distance class, $D \leq 4$ ($z \lesssim 0.1$), with redshifts for all but one of these clusters reported by Hoessel, Gunn, and Thuan (1980).[2] This sample includes all Abell clusters at $D \leq 4$ that are of richness class, $R \geq 1$, and are located at high galactic latitude (as specified in Abell's table 1, plus the requirement of $|b| \geq 30°$). A total of 104 clusters belong in this sample.

The area subtended by the above statistical sample is 4.26 steradians (2.64 in the north and 1.62 in the south). A summary of the sample properties and its division into distance and richness classes, as well as into hemispheres, is presented in Bahcall and Soneira, 1983a[3] (hereafter BS, 1983a). Also listed in the above reference are properties of the much larger and deeper $D = 5 + 6$ statistical sample ($|b| \geq 30°$) that includes 1547 clusters; while only a small fraction of the redshifts are measured for this sample, it is used, because of its much larger number of clusters, in various comparison tests to strengthen and confirm the results obtained from the $D \leq 4$ sample.

The distances to the clusters are calculated from the standard relation (e.g., Sandage, 1961[4]):

$$R = \frac{c}{H_0 q_0^2 (1+z)^2}\{q_0 z + (q_0 - 1)[(1 + 2q_0 z)^{1/2} - 1]\}. \tag{1}$$

A Hubble constant of $H_0 = 100\ h$ km s^{-1} Mpc^{-1} and $q_0 = \frac{1}{2}$ are used throughout the paper. (The distances increase by $\leq 2\%$ for $q_0 = 0$.) In equation (1), we also assume that peculiar velocities are small compared to the Hubble velocity.

The spatial density of clusters in the sample as a function of redshift is given in BS, 1983a.[3] As expected, the mean density remains approximately constant for small redshifts and then, due to the limited precision in estimating distances from m_{10} magnitudes, the density falls off, exhibiting a low density tail to $z = 0.14$. This observed variation of the Abell selection function with distance [i.e., $n(z)$ in figure 3 of BS, 1983a[3]] must be taken into account in any analysis requiring a distance-limited sample. We do so in our determination of the spatial correlation function by comparing the data with random catalogs constructed to have the same density selection function, $n(z)$, and the same boundaries as those of the real sample. We also use the observed $n(z)$ in our construction of the supercluster catalog (see below).

Selection effects that depend on position in the sky (e.g., due to absorption in the Galaxy or confusion with galactic stars) are also of importance in determining the correlation function or other distributions of clusters. We determined, as a function of galactic latitude and longitude, the ratio of observed number of clusters to that expected if the distribution on the sky was uniform. The results for both $D \leq 4$ and $D =$

5 + 6, presented in BS, 1983a,[3] show some decrease of cluster density with latitude. This dependence is corrected for in the determination of the cluster correlation function by including the same selection effect, $P(b)$, in the random catalogs (see below).

THE SPATIAL CORRELATION FUNCTION OF RICH CLUSTERS OF GALAXIES

The availability of a complete redshift sample of rich clusters of galaxies makes it possible, for the first time, to calculate relatively accurately cluster distances and separations and, in turn, to determine directly the spatial correlation function of rich clusters of galaxies. We do so in the present section. A more detailed description is given in BS, 1983a.[3]

The joint probability, dP, of finding two clusters separated by a distance, r, and within volume elements, dV_1 and dV_2, is written as:

$$dP = n(R_1)n(R_2)[1 + \xi(r)]dV_1 dV_2, \qquad (2)$$

where $\xi(r)$ is the two-point spatial correlation function and $n(R)$ is the space density of clusters at the distance, R, from the sun. (The space density in the present sample varies with the distance, R, due to the sample selection function discussed in the previous section.)

The frequency distribution, $F(r)$, of all pairs of clusters with separation, r, in the present sample was determined. Cluster separations were calculated using the redshift distances [equation (1)] and the angular separation on the sky [i.e., $r = (R_1^2 + R_2^2 - 2R_1 R_2 \cos\theta)^{1/2}$]; peculiar velocities were assumed to be small. In order to minimize the influence of selection effects on the determination of $\xi(r)$, we constructed a set of 1000 random catalogs, each containing 104 clusters randomly distributed within the angular boundaries of the survey region, but with the same selection functions in both redshift and angular position as the Abell redshift sample (see the previous section). The frequency distribution of cluster pairs was determined in both the real and random catalogs and the results were compared. This procedure ensures that the selection effects and boundary conditions will affect the data and random catalogs in the same manner, thus minimizing most of these effects.

The spatial correlation function was determined from the relation,

$$\xi(r) = F(r)/F^R(r) - 1, \qquad (3)$$

where $F(r)$ is the observed frequency of pairs in the Abell sample and $F^R(r)$ is the corresponding frequency of random pairs (as determined by the ensemble average frequency of the 1000 random catalogs). An ensemble average random frequency is used in order that $\xi(r)$ will not be affected by the fluctuations present in any particular realization of a single random sample. The correlation function was evaluated for various cases including: (a) no selection function in latitude [i.e., $P(b) = 1$]; (b) full selection function in latitude (discussed above); (c) northern and southern hemispheres treated separately; and (d) high and low latitude zones ($|b| > 50°$ and $|b| \leq 50°$) treated separately, each with its observed $n(z)$ [and $P(b)$] selection function.

The resulting correlation function is presented in FIGURE 1. Strong spatial

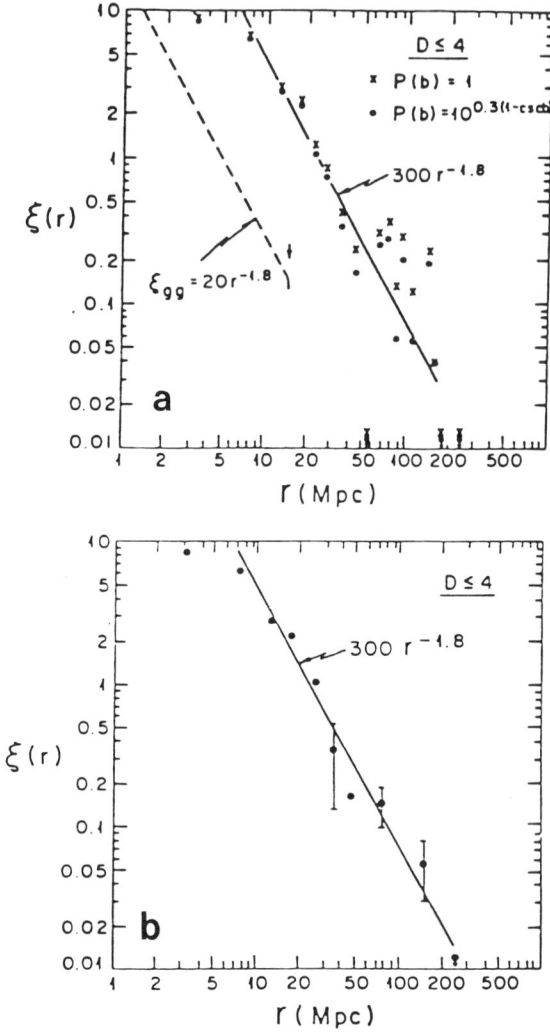

FIGURE 1. (a) The spatial correlation function of the $D \leq 4$ sample. [Crosses refer to no correction for latitude selection function; dots refer to the full correction of $P(b)$.] The solid line is the best-fit 1.8 power-law to the data. The dashed line is the galaxy-galaxy correlation function of Peebles and co-workers. (b) Same as (a), but plotted in larger bins at large separations.

correlations are observed at separations $\lesssim 25\ h^{-1}$ Mpc. Weaker correlations are observed to large separations of $\sim 100\ h^{-1}$ Mpc, where $\xi \sim 0.1$; beyond $150\ h^{-1}$ Mpc, no statistically significant correlations are observed in the present sample. The results for the correlation function do not differ greatly between the two extreme cases of zero and full correction to the latitude selection function, $P(b)$, thus implying that our main conclusions are not strongly influenced by the latitude selection. In addition, the overall results obtained using a two-zone random catalog (high and low latitude zones) show no significant difference from those obtained using the entire sample.

The correlation function in FIGURE 1 can be well approximated by a single power-law relation of the form,

$$\xi(r) = 300(rh)^{-1.8}, \quad 5 \lesssim r \lesssim 150 \; h^{-1} \; \text{Mpc}, \tag{4}$$

where r is in Mpc. The correlation function is smooth, with little scatter at $r \lesssim 50 \; h^{-1}$ Mpc. At $r > 50 \; h^{-1}$ Mpc, the scatter and uncertainties increase, but weak correlations of the order of 0.2 are still detected to these very large separations. These results, as expected, are consistent with the angular and redshift correlations observed at large angular and radial separations (BS, 1983a).[3] If real, the weak correlations at ~50–150 h^{-1} Mpc should also be found in future, larger (deeper) redshift samples of rich clusters. In the meantime, the agreement between the above derived correlation function and the angular correlation function of the much larger $D = 5 + 6$ sample (which includes 1547 clusters), both of which show correlation to projected separations of at least ~100 h^{-1} Mpc, suggests that the present results may indeed be representative of rich clusters in general.

At $1.5 < r \lesssim 5 \; h^{-1}$ Mpc, $\xi(r)$ appears to rise less steeply (only marginally) than expected from the 1.8 power-law relation. Some flattening is expected because of possible selection effects (although the first bin used at $r \geq 1.5 \; h^{-1}$ Mpc is outside the Abell radius), as well as real physical effects (possible tidal encounters and mergers of close clusters). A nonnegligible peculiar velocity among clusters will also decrease the number of close pairs estimated from a pure Hubble flow.

The cluster correlation function found above is remarkably similar in shape to the galaxy-galaxy correlation function discussed by Peebles (1974)[5] and Groth and Peebles (1977);[6] the latter is represented, for comparison, by the dashed curve in FIGURE 1. While similar in shape, the observed cluster correlation function has an amplitude that is larger than the galaxy correlation function amplitude by a factor of about 15 (at a given distance), or, equivalently, it has a distance scale shift of a factor of approximately 5 upward (for a given amplitude). The cluster correlation function is observed to extend much beyond the reported break in the galaxy spatial correlation function at $r \simeq 15 \; h^{-1}$ Mpc (Soneira and Peebles, 1978),[7] to at least 50 h^{-1} Mpc and possibly to ~150 h^{-1} Mpc. While the galaxy correlation function is unity at ~5 h^{-1} Mpc, the cluster correlation is unity at ~25 h^{-1} Mpc.

The present results rule out the possibility of a complete cutoff in the galaxy correlation function at $r \gtrsim 15 \; h^{-1}$ Mpc since galaxy correlations due to galaxies in rich clusters exist at $r \gtrsim 15 \; h^{-1}$ Mpc. The contribution to the galaxy correlation function from rich clusters can be estimated using the fraction of galaxies, f, that belong in rich ($R \geq 1$) clusters. The galaxy-galaxy correlation due to galaxies in such rich clusters (when normalized to the total galaxy density), ξ_{gg}^c, is given approximately by $\xi_{gg}^c \simeq f^2 \xi_{cc}$, where ξ_{cc} is the observed cluster-cluster correlation function. A lower limit to the total galaxy-galaxy correlation function, ξ_{gg}, is therefore given by

$$\xi_{gg} \geq \xi_{gg}^c \approx f^2 \xi_{cc}. \tag{5}$$

For $f \simeq 5\%$ [corresponding to galaxies within one Abell radius (Bahcall, 1979)[8]], the contribution to the galaxy correlation function from rich clusters, at any separation, is of the order of $\xi_{cc}/400$ ($\simeq \xi_{gg}/25$). Thus, beyond the reported break in ξ_{gg} (for example, at $r \simeq 20\ h^{-1}$ Mpc), a lower limit to the expected galaxy-galaxy correlation is: $\xi_{gg}(20\ h^{-1}$ Mpc) ≥ 0.005. Since galaxy-cluster cross-correlations indicate that a galaxy excess exists well beyond an Abell radius of a cluster, the fraction, f, is most likely larger than 5% and the galaxy correlation limit becomes higher [for $f = 10\%$, $\xi_{gg}(20h^{-1}) \geq 0.02$].

The correlation function of clusters of different richness classes ($R = 1$ and $R \geq 2$) were determined separately in order to test for possible dependence on richness. The large sample of 1547 $D = 5 + 6$ clusters was used for that purpose and the angular correlation functions of its 1125 $R = 1$ and 422 $R \geq 2$ clusters were determined separately. The amplitude of the angular correlation function is found to be strongly dependent on cluster richness, with richer clusters ($R \geq 2$) showing stronger correlations by a factor of about three as compared with the poorer ($R = 1$) clusters (FIGURE 2). (The $D \leq 4$ sample contains too few $R \geq 2$ clusters to yield statistically significant comparisons.) Both richness classes exhibit the same power-law shape correlation function as observed in the total sample. (See also BS, 1983a.[3])

In FIGURE 2b, we show the dependence of the correlation function on the richness of the system, from single galaxies to poor and rich clusters. It is apparent that the correlations become stronger with increasing richness (or luminosity) of the system. This implies that the chance, above random, of finding a rich system next to another rich system is considerably higher than that of finding two neighboring poorer systems.

In order to ensure that the above derived spatial correlation function is not due to some special peculiarities in the nearby $D \leq 4$ sample, the angular correlation function of the much larger and deeper $D = 5 + 6$ sample (1547 clusters) was determined and compared with that expected from the above spatial function. The agreement between the $D \leq 4$ and $D = 5 + 6$ functions is excellent (see figure 12 of BS, 1983a[3]), indicating that the derived spatial function indeed represents well the overall cluster distribution in the entire catalog. Additional checks are presented in BS, 1983a.[3]

The difference in the correlation strength between galaxies and rich clusters can be understood on the basis of the following model (Bahcall, 1986).[10] Assume that a given fraction of all galaxies belong to rich clusters and their low density extended envelopes (extending to a few tens of Mpc), and that the rest of the galaxies are only weakly correlated. In this case, the cluster correlation function will be the dominant function representing the large-scale structure and the galaxy correlation function will be mostly derived from the cluster correlation. The galaxy correlation will be weaker than the cluster correlation by a factor related to the fraction of galaxies in clusters. We therefore use the observed difference in the correlation strength between galaxies and clusters to determine the fraction of galaxies that belong in clusters and their extended surrounding. We find that if this fraction is

$$f \simeq 25\% \quad \text{(fraction of galaxies associated with rich clusters)}$$

and the rest of the galaxies are weakly correlated, then the galaxy correlation function is indeed about 16 times weaker than the cluster correlation function. It would imply that while rich clusters are preferentially formed in large-scale, high-density regions (superclusters), more galaxies are formed also in a lower-density, more diffuse component.

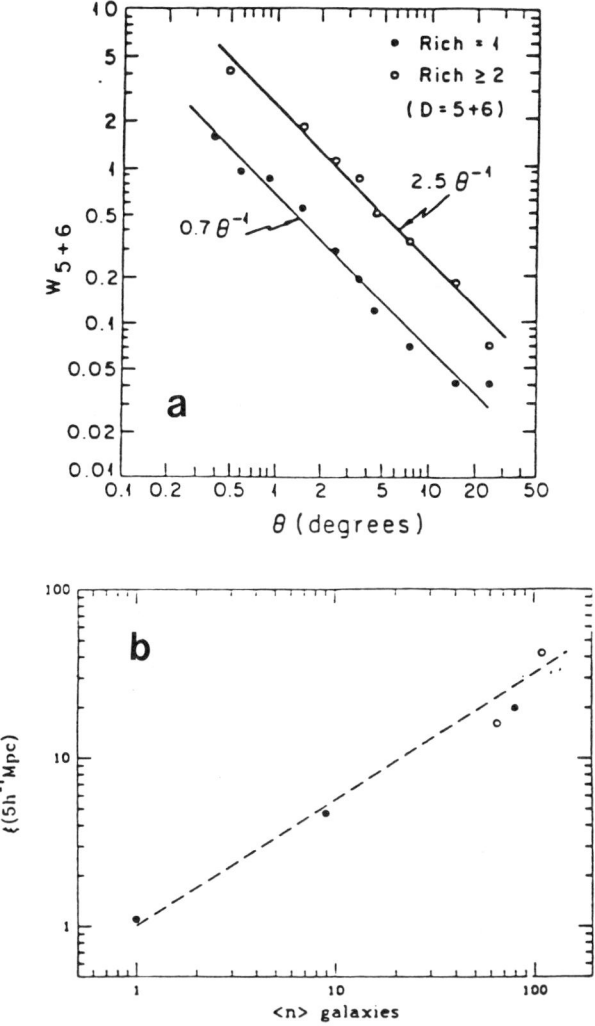

FIGURE 2. (a) The angular correlation function of richness 1 clusters (dots; 1125 clusters) and richness ≥ 2 clusters (circles; 422 clusters) in the $D = 5 + 6$ sample. (b) The dependence on richness of the two-point spatial correlation function. The spatial correlation function at $r = 5\,h^{-1}$ Mpc is plotted as a function of richness (roughly proportional to luminosity) for the galaxy-galaxy, galaxy-cluster, and cluster-cluster pairs ($R \geq 1$, as well as $R = 1$ and $R \geq 2$ independently; the latter are shown by open circles). Richness represents $\langle n \rangle = (n_1 n_2)^{1/2}$, where the subscripts "1" and "2" refer to the two members of the pair; n_i is the mean Abell galaxy richness count for the appropriate clusters; and $n_1 = 1$ is used for a galaxy. (The appropriate galaxy-cluster cross-correlation is from Peebles, 1980;[9] it excludes the term due to internal galaxy correlations within the cluster.) The dashed line, $\xi(n) \propto n^{0.8}$, is presented in order to guide the eye in the approximate richness dependence of the correlation.

The main conclusions reached from the above study are summarized below:

(1) Strong spatial correlations of rich Abell clusters are observed in both the $D \leq 4$ redshift sample and the larger and deeper $D = 5 + 6$ sample. Cluster correlations are detected to large separations of $r \lesssim 150 \ h^{-1}$ Mpc. The correlations are consistent in the angular, redshift, and spatial correlation functions of these samples (see also BS, 1983a[3]), with good agreement between the results obtained from the $D \leq 4$ and $D = 5 + 6$ samples. Various selection effects have been removed by comparing with appropriately synthesized random catalogs.

(2) The spatial correlation function of rich ($R \geq 1$) clusters is observed to fit a power-law relation of the form, $\xi(r) = 300(rh)^{-1.8}$, for $5 \lesssim r \lesssim 150 \ h^{-1}$ Mpc (r in Mpc). [When corrected for velocity broadening, the intrinsic function becomes $\xi(r) = 360(rh)^{-1.8}$, which best fits the observed angular and redshift correlation functions.] This correlation function has the same shape as the galaxy correlation function, but has an amplitude larger by a factor of ~ 18 and extends to greater distances than those observed for the galaxy correlation function. The cluster correlation function is unity at $r \simeq 25 \ h^{-1}$ Mpc, as compared with $r \simeq 5 \ h^{-1}$ Mpc in the galaxy correlation function. The extent of the rich cluster correlations beyond the reported 15 h^{-1} Mpc break in the galaxy correlation function yields a lower limit to the galaxy correlation at $r \gtrsim 15 \ h^{-1}$ Mpc.

(3) Integrating the intrinsic spatial correlation, $\xi(r) = 360(rh)^{-1.8}$ ($r \leq 150 \ h^{-1}$ Mpc), yields good agreement with the observed angular correlation functions of both the $D \leq 4$ and $D = 5 + 6$ samples. This agreement indicates that the $D \leq 4$ redshift sample is a fair sample of the much larger $D = 5 + 6$ sample and that the observed correlations represent real spatial correlations.

(4) The correlation function appears to depend strongly on cluster richness, with rich clusters ($R \geq 2$) showing stronger correlations by a factor of about three as compared with the poorer ($R = 1$) clusters (both are consistent with an $r^{-1.8}$ power law). This result, combined with the lower correlation amplitude of individual galaxies, indicates that progressively stronger correlations exist, at a given separation, for richer galaxy systems.

(5) A model that explains the large difference between the cluster and galaxy correlation functions suggests that approximately 25% of all galaxies belong to the rich clusters and their extended envelopes, and the rest of the galaxies are only weakly correlated (Bahcall, 1986).[10] The cluster correlation function would provide the dominant description of the existing structure on large scales.

A CATALOG OF SUPERCLUSTERS

Introduction

The strong correlation function among clusters of galaxies discussed in the preceding section arises from the strong tendency of clusters to cluster themselves, that is, to form superclusters. In order to investigate the properties of the large-scale clustering of clusters, a complete, well-defined catalog of superclusters—defined as clusters of clusters of galaxies—is required. We have recently constructed such a catalog (Bahcall and Soneira, 1983b,[11] hereafter BS, 1983b) using the complete redshift sample of Abell clusters (discussed above) and an objective selection criterion of a spatial density enhancement. We report on these results in the present section.

Method of Selection

The following selection procedure is adopted. For each cluster in the sample, a sphere of maximum radius, r_m, centered on the cluster, is determined. The radius is chosen so that the density of clusters within r_m, $n(r \leq r_m) = N(\leq r_m)/[4(\pi/3)r_m^3]$, satisfies:

$$n(r \leq r_m) \geq fn_0(R). \tag{6}$$

Here, $N(\leq r_m)$ is the number of sample clusters whose centers fall within the sphere of radius, r_m, $n_0(R)$ is the mean space density of clusters at the cluster distance, $R(z)$, from us, and f is the density enhancement factor. The radius, r_m, is the maximum radius around the given cluster at which the cluster density exceeds the mean level by a factor of f. The density of clusters in the sample, especially at high z, is observed to depend on the distance, $R(z)$, from the Galaxy (discussed above and in BS, 1983a[3]) since the sample is magnitude-limited (with some precision) rather than redshift-limited (as defined in the Abell catalog); therefore, for any given value of f, equation (6) is calculated using the sample's actual value of $n_0(R)$ for each cluster at its appropriate distance, R. For each of the 104 clusters in the sample, a sphere of radius r_m is determined, resulting in a series of 104 spheres of various sizes. The spheres typically fall into two categories: (1) single, isolated spheres and (2) distinct, nonoverlapping clumps of spheres. The exact division into these categories depends on the assumed value of f. The outside boundary of each clump of spheres approximates an isodensity enhancement contour with an average volume density of clusters within the boundary of $fn_0(R)$. Each of the distinct clumps of spheres at a given f is identified as a separate group of clusters, that is, a supercluster. Any cluster lying outside the clump boundaries (i.e., all the isolated spheres) are considered single clusters for that value of f.

The sphere size for any value of f depends on the density of neighboring clusters. We have also carried out a somewhat different analysis using a constant sphere size. Both of these methods are similar to the percolation-analysis technique (e.g., Zeldovich et al., 1982),[12] where the percolation-size parameter is r_m (or, approximately, $\propto f^{-1/3}$).

The groups and singles selected as described above, as well as the boundaries and richnesses of the groups (i.e., the number of clusters per group), depend, by definition, on the value adopted for the density enhancement, f (or, equivalently, the size of the selecting sphere). Low values of f will tend to select groups of low density and high separation, some of which may not be real physical associations. Higher values of f will select tighter, more compact groups. Many of these tight groups constitute the dense cores of larger groups selected with lower f values (see below). The group boundaries clearly do not define strict physical limits to the superclusters, but rather define volumes of various levels of overdensities.

In order to quantitatively study how the selection of superclusters is affected by the assumed value of f, we performed the analysis for various values of f, ranging from f = 10 to 400. The typical maximum radii, r_m, around single clusters are 17.5, 13.8, 10.9, 8.1, 6.4, and 5.1 h^{-1} Mpc, respectively, for f = 10, 20, 40, 100, 200, and 400 (at a distance of R = 200 h^{-1} Mpc). Due to the procedure by which connected (i.e., "percolated") spheres are selected, the maximum separation between any binary cluster is $2r_m$.

The Superclusters

The results for f = 20, 40, 100, 200, and 400 are summarized in TABLE 1 (for the complete sample of 104 $R \geq 1$ clusters). In this table, we present a catalog of superclusters, complete to $z \leq 0.08$, for each of the various values of f. The mean position and redshift of the superclusters as determined from the members in the f = 20 catalog are also listed. These values change somewhat as f increases and the groups break up into more compact subgroups (or vice versa).

The total number of superclusters decreases from 16 at f = 20 to 7 at f = 400. This factor of two decrease is small considering the factor of twenty increase in the overdensity. (Random catalogs yield a factor of fourteen decrease in the number of random "superclusters.") The multiplicity, or number of clusters per supercluster, $N_{cl/sc}$, varies from 2 to 15 for the f = 20 superclusters and reduces to a value of 2 to 3 clusters per f = 400 supercluster. The superclusters contain a large fraction of all clusters; this fraction, $F_{cl}(sc)$, is 54% at f = 20 and reduces to 16% at f = 400 (see also FIGURE 4 below). The fractional volume of space occupied by the superclusters is, however, small; it varies from ~3% at f = 20 to 0.04% at f = 400. The f = 400 superclusters have, as expected, the smallest cluster separations observed; all cluster pairs in these superclusters are separated by $\lesssim 13\ h^{-1}$ Mpc. These high density superclusters are either identical to or constitute the densest part (or parts) of the large superclusters selected with the lower f values. The effect of increasing f on the breakup of large groups into dense, compact subgroups and eventually into cores of tight binaries and triplets is apparent in TABLE 1.

The linear size of the largest observed superclusters increases rapidly from 13 h^{-1} Mpc for the highest density clumps (f = 40) to over 100 h^{-1} Mpc at f = 20. This increase is considerably steeper than observed in the random catalogs and indicates the stronger connections among clusters in the real data. A comparison of the mean separation of all cluster pairs in superclusters as seen both projected on the sky and in the redshift direction appears to suggest a velocity broadening component superimposed on the Hubble flow (see BS, 1983a,[3] 1983b[11]).

A map of all superclusters selected at f = 20, 40, 100, 200, and 400 is presented in FIGURE 3 for both the northern and southern galactic hemispheres. The density enhancements of the superclusters for f = 20 to 400 are indicated by the density contours. The supercluster numbers refer to those listed in TABLE 1.

Some specific superclusters are noteworthy. Supercluster BS8 is a rich Ursa Major supercluster. BS10 is the Coma supercluster (which contains A1656 and A1367 as its $R \geq 1$ members). BS12B is a dense Corona Borealis supercluster (that may extend to 12A and 12C, as shown, and possibly even further; it is located in the region of highest

TABLE 1. Supercluster Catalog[a]

BS	α, δ (1950) ($f = 20$)	z ($f = 20$)	Abell Cluster Members ($R \geq 1$)				
			$f = 20$	$f = 40$	$f = 100$	$f = 200$	$f = 400$
	h m ° ′						
1	0053 − 1238	0.0541	85, 151	—	—	—	—
2	0112 + 0204	0.0433	⎡ 119, 168	119, 168	119, 168	119, 168	—
	0103 − 0047[b]	0.0452[b]	⎣ 189, 193				
3	0122 + 1759	0.0652	154, 225	—	—	—	—
4	0258 + 1307	0.0738	399, 401	399, 401	399, 401	399, 401	399, 401
5	0442 − 2122	0.0682	500, 514	500, 514	500, 514	500, 514	—
6	1049 + 4012	0.0795	1035, 1187	—	—	—	—
7	1113 + 3151	0.0347	1185, 1228	1185, 1228	1185, 1228	1185, 1228	—
8	1142 + 5547	0.0581	⎡1291, 1318 1377, 1383 1436	⎡1291, 1318 1377, 1383 1436	⎡1291, 1318 1377, 1383 1436	⎡1291, 1318 1383	⎡1291, 1318 1383
9	1145 − 0210	0.0992	1364, 1399	—	—	—	—
10	1220 + 2415	0.0218	1367, 1656	—	—	—	—
11	1345 + 0356	0.0782	1773, 1809	1773, 1809	1773, 1809	1773, 1809	—
12A	1348 + 2714[b]	0.0699[b]	⎡1775, 1793 1795, 1831 1927, 2022	⎡1775, 1795 1831	1775, 1831	—	—
12B	1527 + 3040[b]	0.0710[b]	2061, 2065 2067, 2079 2089, 2092 2124	⎡2061, 2065 2067, 2079 2089, 2092 2124	⎡2061, 2065 2067, 2079 2089, 2092 2124	⎡2061, 2065 2067, 2089 2079, 2092	⎡2065, 2067 2089 2079, 2092
12C	1607 + 2842[b]	0.0945[b]	2142, 2175	2142, 2175	—	—	—
13	1449 + 1506	0.0509	⎡1913, 1983 ⎣1991, 2040	—	—	—	—
14	1509 + 0605 1508 + 0650[b]	0.0831 0.0774[b]	⎡2028, 2029 ⎣2048	2028, 2029	2028, 2029	2028, 2029	2028, 2029
15	1549 + 1615 1602 + 1651[c]	0.0388 0.0373[c]	⎡2063, 2107 2147, 2151 2152	⎡2107, 2147 2151, 2152	⎡2147, 2151 2152[d]	⎡2147, 2151 2152[d]	⎡2147, 2151 2152[d]
16	1627 + 4021	0.0308	2197, 2199	2197, 2199	2197, 2199	2197, 2199	2197, 2199

[a]Supercluster catalog for $f = 20$ through $f = 400$ for the $D \leq 4$, $R \geq 1$ complete redshift sample [using HGT redshifts, except z(A1318) = 0.0586; see text]. Column headings: (1) BS supercluster number; (2–3) mean supercluster position and redshift for $f = 20$ (average of the $f = 20$ members, except where otherwise indicated); (4–8) a list of the Abell cluster members ($R \geq 1$) in the superclusters for each of the overdensities, $f = 20$ to 400.
[b]Mean position and redshifts of the $f \geq 40$ members.
[c]Mean position and redshifts of the $f \geq 100$ members.
[d]Included in the $f \geq 100$ superclusters if z(A2152) = 0.0383 is used (as measured from 22 galaxy redshifts; SRS).

density of rich clusters in the sample). BS15 is the Hercules supercluster, which, from galaxy redshift surveys (Chincarini, Rood, and Thompson, 1981[13]) as well as from our extended $R \geq 0$ cluster sample distribution (BS, 1983b[11]), appears to extend at least $\sim 50\ h^{-1}$ Mpc north to connect with BS16. It is apparent from the maps that a large concentration of superclusters (i.e., BS12 through 16) is present in one quadrant of the

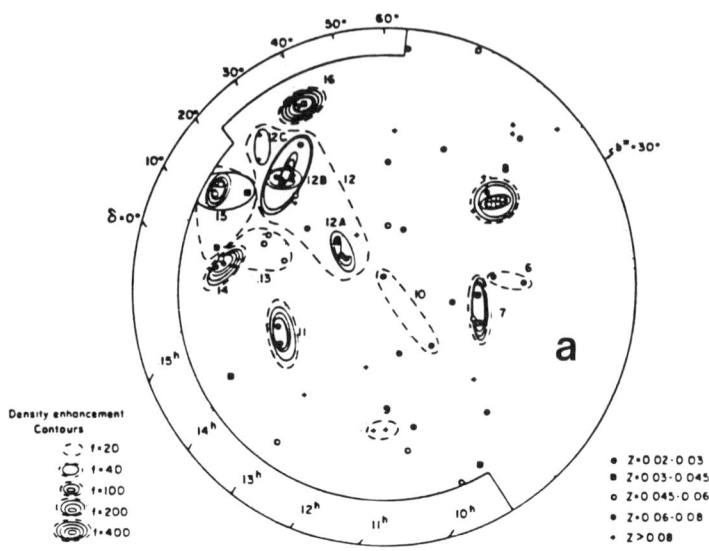

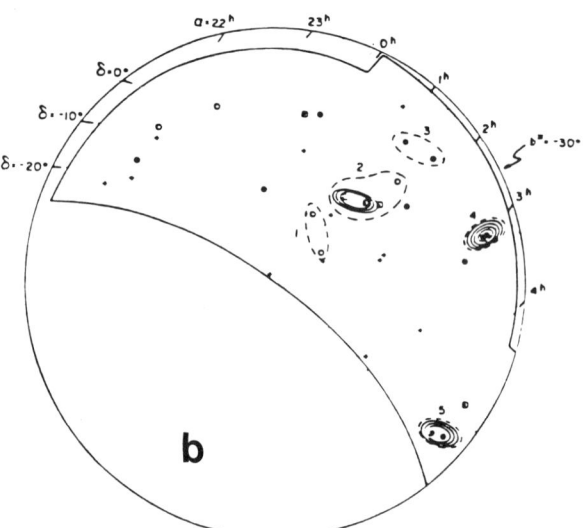

FIGURE 3. Projected equal-area map of the northern (a) and southern (b) clusters and superclusters in the $R \geq 1$ sample. Outer contour is $|b| = 30°$; inner contour is the completeness limit of the sample. The galactic poles are at the respective centers of these polar maps. Longitude 1^{II} is $0°$ at the west and increases clockwise. Clusters at different redshift ranges are marked by different symbols. The density contours represent the three-dimensional density enhancement of the superclusters from $f = 20$ to $f = 400$. The superclusters are numbered by their BS catalog numbers for $f = 20$ (TABLE 1). At $f = 10$, superclusters BS12 + 13 + 15 + 16 coalesce into a single large-scale supercluster.

northern hemisphere. The large galaxy void in Bootes (Kirshner et al., 1981[14]) is located close-by and is apparently related to this high concentration of rich galaxy clusters and superclusters (Bahcall and Soneira, 1982).[15,16]

Superclusters were selected also from a larger sample of 175 clusters that included all $R = 0$ clusters with measured redshifts ($D \leq 4$). The superclusters identified in this $R \geq 0$ sample are generally consistent with and strengthen those found in the smaller $R \geq 1$ sample (see BS, 1983b[11]).

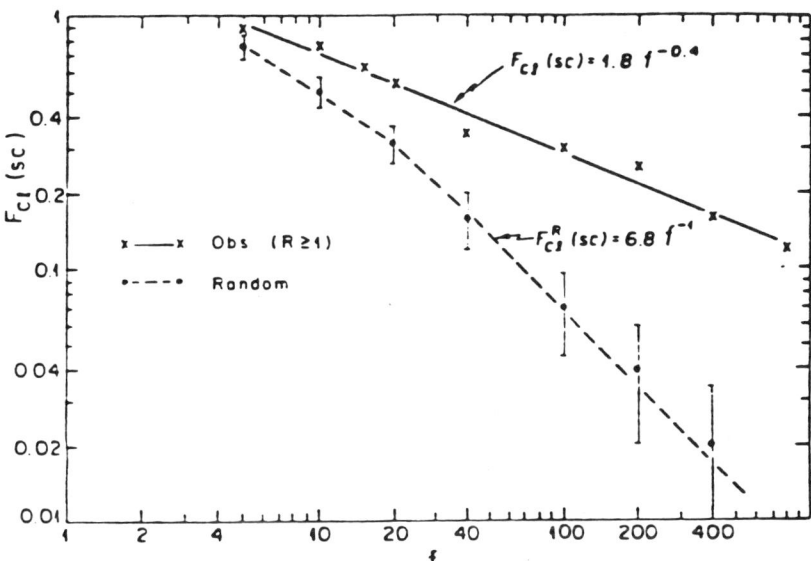

FIGURE 4. The fraction of all $R \geq 1$ clusters that are supercluster members as a function of the supercluster density enhancement, f. Crosses represent the data; dots represent the ensemble average of 100 random catalogs. $1 - \sigma$ error-bars are shown for the random points. The curves are power-law fits for $f \gg 1$.

Tests and Comparisons with Random Catalogs

In order to evaluate the statistical significance of the superclusters in TABLE 1, we used 100 random catalogs that were generated with the same selection functions as the real sample (see BS, 1983a[3]) and analyzed with an identical procedure to that used on the real clusters. The number of superclusters as a function of richness, $n_{cl/sc}$ (i.e., number of cluster members per supercluster), was determined for both the observed and the average of the 100 random catalogs. For all the values of f that were studied, the observed catalog yields a larger number of superclusters with high richnesses than do the random catalogs.

The fraction of rich clusters that belong in the superclusters, $F_{cl}(sc)$, is presented as a function of f in FIGURE 4 for both the observed and random catalogs. As expected, $F_{cl}(sc)$ decreases smoothly with f in both the observed and random catalogs. However, a larger fraction of clusters are supercluster members in the real data than in the random catalogs for $f \gtrsim 10$. In addition, the decrease in $F_{cl}(sc)$ with f is considerably less steep in the real data than in the random catalogs. The observed fraction of rich ($R \geq 1$) clusters that are in superclusters can be approximated by $F_{cl}(sc) \simeq 1.8 f^{-0.4}$ (for $f \gg 1$). The fractional volume of space occupied by the superclusters is very small (<10% for $f \gtrsim 10$) and decreases rapidly with increasing density enhancement.

The superclusters selected by using a constant-size sphere around each cluster are also found to be similar to those identified using a density enhancement criterion (see BS, 1983b[11]).

By repeating the selection procedure using a somewhat different redshift catalog than that of HGT (i.e., those of Sarazin, Rood, and Struble, 1982;[17] Fetisova, 1981[18]), we conclude that the main features of the superclusters identified in this paper are insensitive to the precise determination of the cluster redshifts within the typical expected uncertainties.

THE MULTIPLICITY FUNCTION

The frequency distribution of clusters among superclusters of different richnesses, $n_{cl/sc}$ (where $n_{cl/sc}$ is the number of clusters in a supercluster) is plotted in FIGURE 5. This figure shows the fraction of clusters that are members of superclusters at any given richness; the distribution $F_{cl}(n_{cl/sc})$ is the multiplicity function of the clusters.

The frequency distributions are plotted in FIGURE 5 for different values of f, from $f = 10$ to 400, for both the observed and the average of 100 random catalogs. FIGURE 5 shows that the observed and random catalogs yield different distributions. The fraction, F_{cl}, for the random catalog falls off smoothly and steeply with increasing $n_{cl/sc}$ (for $f \gtrsim 10$); thus the random catalogs have essentially no power at large $n_{cl/sc}$. The observed superclusters have systems with more members than seen in the random catalogs for all $f \gtrsim 10$. The observed high richness superclusters appear to grow rapidly (in richness and size) as f decreases. As these richest, largest-scale structures grow, a gap of medium richness superclusters appears to be rapidly forming (see FIGURE 5). Neither the gap nor the related largest-scale structure exist in the random catalogs.

LARGE-SCALE SUPERCLUSTERS SURROUNDING THE GIANT GALAXY VOID IN BOÖTES

The overdensity of galaxies observed by Kirshner et al. (1981)[14] on both redshift sides of the $z \simeq 0.04$–0.06 galaxy void in Boötes is found to coincide in redshift space with similar overdensities of clusters and superclusters of galaxies (see Bahcall and Soneira, 1982a).[15] The main contributors to these overdensities are superclusters BS12 and BS15–16 from our catalog (FIGURE 3). These dense, large-scale superclusters appear to surround the giant galaxy void and extend to scales of $\gtrsim 100 \ h^{-1}$ Mpc in the tail of their galaxy distribution. Previous observational evidence, together with these results, suggest that galaxy voids may generally be associated with surrounding galaxy excesses; moreover, the bigger the void, the stronger is the related excess (see Bahcall and Soneira, 1982a[15] for more details).

AN ~300-Mpc VOID OF RICH CLUSTERS OF GALAXIES

A huge void of cataloged nearby rich clusters of galaxies is observed in the complete $D \leq 4$ Abell sample discussed above (see Bahcall and Soneira, 1982b[13]). The void is in the approximate redshift range of $z \simeq 0.03$–0.08 and it extends ~100° across the sky (i.e., ~300 h^{-1} Mpc). Its projected area is completely devoid of nearby (but not distant) rich clusters ($R \geq 1$). The void does not appear to be caused by absorption in the Galaxy. If this apparent void in nearby rich clusters is real, it subtends a volume of more than $10^6 \, h^{-3}$ Mpc^3.

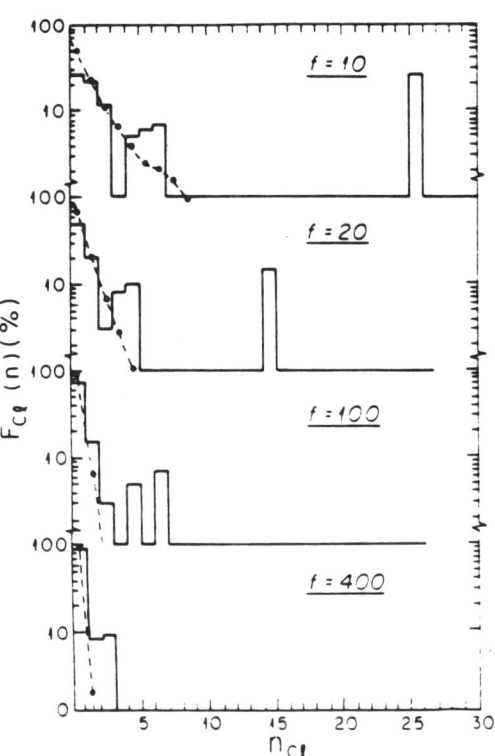

FIGURE 5. The frequency distribution of $R \geq 1$ clusters that are members of superclusters of various richnesses, $n_{cl/sc}$. (The first bin, $n_{cl/sc} = 1$, refers to single clusters; $n_{cl/sc} = 2, 3$, etc., represents superclusters with two, three, etc., cluster members.) The distributions are given for four different density enhancements (f). The histogram represents the observed catalog; the dashed line is the average of 100 random catalogs.

All the findings described above appear to indicate the existence of structure in the universe on scales much larger (~100 h^{-1} Mpc) than previously seen or expected. This provides strong limitations to models of the formation of galaxies, clusters, and the large-scale structure of the universe.

REFERENCES

1. ABELL, G. O. 1958. Astrophys. J. Suppl. Ser. **3:** 211.
2. HOESSEL, J. G., J. E. GUNN & T. X. THUAN. 1980. Astrophys. J. **241:** 486.

3. BAHCALL, N. A. & R. M. SONEIRA. 1983a. Astrophys. J. **270**: 20.
4. SANDAGE, A. 1961. Astrophys. J. **133**: 355.
5. PEEBLES, P. J. E. 1974. Astrophys. J. Suppl. Ser. **28**: 37.
6. GROTH, E. & J. P. E. PEEBLES. 1977. Astrophys. J. **217**: 385.
7. SONIERA, R. M. & J. P. E. PEEBLES. 1978. Astron. J. **83**: 845.
8. BAHCALL, N. A. 1979. Astrophys. J. **232**: 689.
9. PEEBLES, P. J. E. 1980. *In* Proceedings of Les Houches Summer School, session XXXII, Physical Cosmology.
10. BAHCALL, N. A. 1986. Astrophys. J. Lett. **300**. In press.
11. BAHCALL, N. A. & R. M. SONEIRA. 1983b. Astrophys. J. **277**: 27.
12. ZELDOVICH, YA. B., J. EINASTO & S. F. SHANDARIN. 1982. Nature **300**: 407.
13. CHINCARINI, G., H. J. HOOD & L. A. THOMPSON. 1981. Astrophys. J. Lett. **249**: L47.
14. KIRSHNER, R. P., A. OEMLER, JR., P. L. SCHECHTER & S. A. SHECTMAN. 1981. Astrophys. J. Lett. **248**: L57.
15. BAHCALL, N. A. & R. M. SONEIRA. 1982a. Astrophys. J. Lett. **258**: L17.
16. BAHCALL, N. A. & R. M. SONEIRA. 1982b. Astrophys. J. **262**: 419.
17. SARAZIN, C. L., H. J. HOOD & M. F. STRUBLE. 1982. Astron. Astrophys. Lett. **108**: L7.
18. FETISOVA, T. S. 1981. Astron. Zh. **58**: 1137.

The Galaxy Distribution and the Large-Scale Structure of the Universe[a]

MARGARET J. GELLER,[b] VALÉRIE DE LAPPARENT,[b,c]
AND MICHAEL J. KURTZ[b]

[b]*Harvard-Smithsonian Center for Astrophysics
Cambridge, Massachusetts 02138*

[c]*Ecole Normale Supérieure de Jeunes Filles
and
Université Paris 7
Paris, France*

INTRODUCTION

Over the past few years, there has been a tremendous increase in our knowledge of the distribution of galaxies [Seldner et al., 1977;[1] Peebles, 1980;[2] Davis and Peebles, 1983;[3] Bean et al., 1983;[4] Kirshner, Oemler, and Schechter (KOS hereafter), 1978[5] and 1979[6]]. Before the redshift surveys of Huchra et al. (1982; CfA hereafter),[7] Bean et al. (1983; AAT hereafter),[4] and Kirshner et al. (1983),[8] knowledge of the distribution was based primarily on the distribution projected on the sky. In spite of the advances made by the new redshift surveys, many fundamental questions remain unresolved.

One of these unresolved issues is the statistical description of galaxy clustering on scales $\gtrsim 10$ Mpc (where the Hubble constant, $H_0 = 100$ km s^{-1}Mpc^{-1}). Because of the need for large samples, our knowledge of the distribution on these scales is still heavily dependent upon observation of the distribution projected on the sky. In particular, the appearance and detailed statistics of the map derived from the Shane-Wirtanen counts (1967)[9] have attracted much attention. The filamentary structure (Moody, Turner, and Gott, 1983;[10] Kuhn and Uson, 1982[11]) and the break in the two-point correlation function derived from the map (Groth and Peebles, 1977[12]) have been important for reconciling theoretical models with observation (Davis and Peebles, 1977;[13] Davis, Groth, and Peebles, 1977;[14] Gott, Turner, and Aarseth, 1979;[15] Zeldovich, Einasto, and Shandarin, 1982;[16] Dekel and West, 1985[17]).

Here, we begin with a brief review of the current status of knowledge about the large-scale galaxy distribution. The bulk of this paper is a critical analysis of the Shane-Wirtanen counts; we emphasize the effects of residual systematic errors on measures of the galaxy distribution. Although the Shane-Wirtanen counts have been important for defining the issues in the study of large-scale structure, they are inadequate to answer the questions. We conclude by evaluating the kind of new data required to measure the galaxy correlation function on scales $\gtrsim 10$ Mpc.

---

[a]This research was supported in part by NASA grant no. NAGW-201.

CURRENT STATUS

On scales $\lesssim 5$ Mpc, there is general agreement that the two-point spatial correlation function is well approximated by a power law,

$$\xi_g(r) = (r_o/r)^\gamma, \quad (1)$$

where $r_o = 5.4 \pm 0.3$ Mpc (Davis and Peebles, 1983[3]) and $\gamma = 1.8$.

The analysis of the Shane-Wirtanen counts by Groth and Peebles (1977)[12] yielded the first indication of a large-scale feature in the projected two-point correlation function. Groth and Peebles commented that this feature at 2.5° or a projected scale of ~10 Mpc is close to the noise and is therefore "tentative" (Peebles, 1980).[18] Subsequent studies show that the value of the mean mass density, Ω, the transition between the linear and nonlinear clustering regime, and the initial spectrum of fluctuations can all affect the shape and position of such a "break" in the correlation function (Peebles, 1980;[2] Davis and Peebles, 1977[13]).

The break in the angular correlation function for the Shane-Wirtanen counts could, however, be of local origin. Seldner and Uson (1983)[19] demonstrate that the interpretation of the break is complicated by the possible role of galactic obscuration. We show below that the break is indistinguishable from an artifact introduced by residual systematic variations in the effective depth of the Shane-Wirtanen survey (see also references 20 and 21).

Soneira and Peebles (1978)[22] show that a spatial correlation function that steepens sharply at ~15 Mpc reproduces the projected correlation function derived from the Shane-Wirtanen counts. Three recent redshift surveys yield more direct measures of the large-scale behavior of the spatial correlation function, but no two of these are convincingly consistent. The AAT surveys indicate that a break in the correlation function occurs at 5 Mpc (Shanks et al., 1983).[23] The correlation function derived by Kirshner, Oemler, and Schechter (1979)[6] remains significant at a scale of 30 Mpc. The CfA redshift survey, which contains about ten times as many galaxies as the AAT or KOS surveys, yields a correlation function with a break at ~10 Mpc in reasonable agreement with the Shane-Wirtanen result. However, on large scales, the CfA survey results may be affected the sample depth of ~60 Mpc, which is somewhat shallow compared with the scales of interest here.

Studies of the correlation function for the distribution of Abell clusters (1958)[24] provide another measure of the large-scale galaxy distribution (see the paper by N. Bahcall in this volume). The rich cluster correlation function of approximate form, $\xi_c(r) \propto r^{-2}$, extends to scales of ~100 Mpc. This result is inconsistent with a very steeply falling galaxy correlation function on scales $\gtrsim 15$ Mpc (Kaiser, 1984;[25] Bahcall and Soneira, 1983[26]).

Observations of individual very large systems are an important complement to the statistical studies. It is now clear that there are individual very large-scale features in the galaxy distribution like the void in Boötes (Kirshner et al., 1981)[27] and the Perseus-Pisces and Lynx-Ursa Major chains (Giovanelli and Haynes, 1982;[28] Einasto et al., 1980;[29] Giovanelli, 1983[30]). The existence of these structures poses serious challenges for models of large-scale structure formation. However, a more reliable comparison of models with the data requires a knowledge of the "typical" distribution on these large scales.

The Shane-Wirtanen counts are sufficiently deep that the volume surveyed could include many large structures. The filamentary appearance of the Shane-Wirtanen map has been taken by some as evidence that long, thin structures like the Perseus-Pisces chain are common. Although the data are certainly suggestive to the eye, statistical tests of the filamentary nature of the Shane-Wirtanen counts yield, at best, ambiguous results. Kuhn and Uson (1982)[11] discuss a statistic that distinguishes between the Shane-Wirtanen map and model galaxy catalogs with the same low order statistics, but with no added chainlike structure. Their technique does not map the filament positions and it is therefore somewhat difficult to evaluate the features of the distribution to which the statistic responds. Moody, Turner, and Gott demonstrate an algorithm for locating filaments. They find that long, high-contrast filaments occur significantly more frequently in the Shane-Wirtanen map than in simulations of hierarchical clustering. Further analysis using the same statistical technique shows that some of the most impressive filamentary structures in the Shane-Wirtanen map are caused by residual systematic errors in the data (de Lapparent, Geller, Kurtz, and Turner, 1985).[31]

In order to begin to sort out some of the discrepancies in the analysis and interpretation of data on large-scale structure, we reexamined the Shane-Wirtanen counts. We have been able to identify some of the sources of systematic bias in the SSGP map. Perhaps the most important result of this work is the derivation of the remarkably stringent limits on systematic errors that must be met in order to obtain a reliable estimate of the statistics of structures on scales $\gtrsim 10$ Mpc.

THE SHANE-WIRTANEN DATA

The Shane-Wirtanen counts are one of the most impressive achievements in extragalactic astronomy. Over a twelve-year period, Shane and Wirtanen counted more than 1.2 million galaxies. The counts were made in $10' \times 10'$ cells on 1246 plates, each covering a $6° \times 6°$ region of the sky. The plates overlap by at least $1°$ in both right ascension and declination.

Shane and Wirtanen were conservative in their analysis of the counts; they published the counts in $1° \times 1°$ cells and smoothed their maps even further. They concluded: "The distribution of galaxies in space is characterized by a hierarchy of structural features ranging upward from double and multiple galaxies, through groups, clusters, and clouds. The latter may be several tens of megaparsecs in diameter. The possibility of even larger irregularities measured in hundreds of megaparsecs is suggested."[9] The conservative tone of the last statement may be based in part on a comparison that Shane and Wirtanen made of two counts of the same region (see figure 7 in Shane and Wirtanen, 1967[9]). These two maps show that the highest contrast structures are stable, but there is a significant variation in the low level (large-scale) galaxy surface density contours due, at least in part, to systematic errors in the counts. Both the quality of the plates and the details of the counting can contribute to these variations.

In 1977, Seldner, Siebers, Groth, and Peebles (SSGP; 1977)[1] published a new reduction of the Shane-Wirtanen counts based on the original $10' \times 10'$ data. The widely reproduced map has been a driver for much important research in the field. The algorithm for producing the map is described in SSGP. In this algorithm, each pixel in the map is multiplied by a correction factor that is the product of "plate" and "pixel" correction factors. The "plate" correction factors compensate for variation from plate to plate and are determined from the overlapping portions of adjacent plates. The "pixel" correction factors account for variations in the mean count across the "average" plate. The pixel correction factors are applied before matching the overlaps to calculate the plate correction factors. Tapes of the original counts, the processed counts listed plate by plate, and the resulting map were kindly provided to us by the Princeton group.

In order to examine the systematics in the counts, we constructed maps of the region, $10^h10^m < \alpha < 15^h30^m$ and $-22.5° < \delta < 37.5°$, projected onto a Cartesian grid. The rows of pixels are laid down along lines of constant declination. This approximation is sufficiently accurate for $\delta \leq 40°$. At $\delta = 0°$, the image pixels are exactly the data pixels. As $|\delta|$ increases, the projection of the data pixels is stretched along the R.A. axis. The mapping takes this stretching into account by adding the fractional contribution from each overlapping data pixel to the appropriate image pixel. There is no averaging in regions of plate overlap; we follow the procedure used by SSGP and take data pixels from the plate with the closest center.

For ease of reproduction, FIGURE 1 shows a smoothed version of this region of the sky in the SSGP reduction. The data have been smoothed with a Gaussian of 47' ($\sim$ five

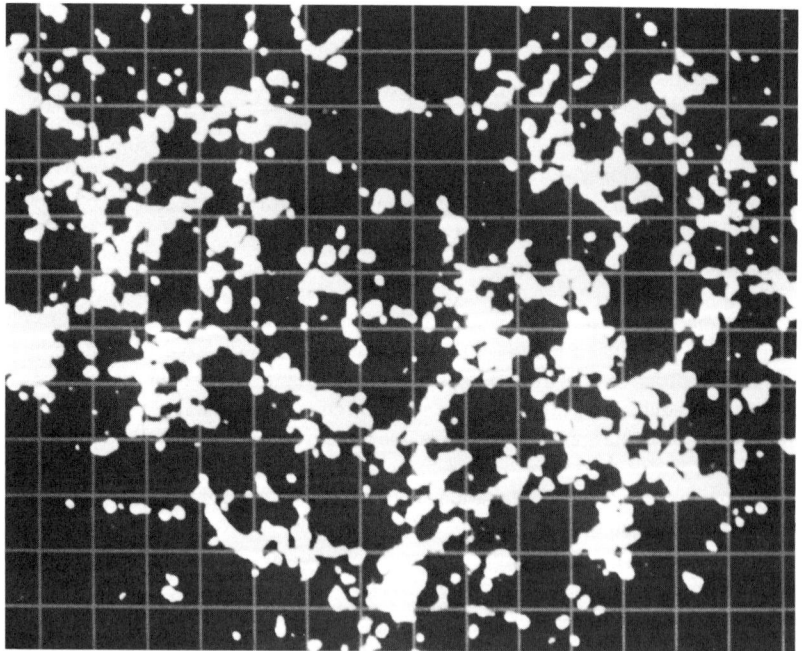

FIGURE 1. The smoothed SSGP map in the region, $10^h10^m < \alpha < 15^h30^m$ and $-22.5° < \delta < 37.5°$.

FIGURE 2. The distribution of zero (black) pixels in the raw image of FIGURE 1. The patterns at top and bottom are dislocations.

pixel) FWHM. The white regions have a galaxy surface density ≥ 2.3 per pixel; the mean in the region is 1.5 galaxies per pixel. The white region is about 18% of the area of the map; we display the upper 18% of the pixels in white.

The Cartesian grid in right ascension and declination makes it easier to see features related to plate boundaries and plate-to-plate variations. The grid of plate boundaries is overlaid on the map. One odd feature of the figure is the common appearance of white regions that either run along plate boundaries or are truncated at or very near the boundaries. The SSGP procedure suppresses the appearance of plate edges, but individual plates are still identifiable in the map. Many of the plates in this region are either light or dark.

The residual significance of the plate-to-plate variations and the fundamental problems in removing them can be demonstrated by processing of the unsmoothed image. FIGURE 2 shows the distribution of zero (black) and nonzero pixels in the map of FIGURE 1. The patterns at the top and bottom of the image are dislocations that occur because a data pixel is a map pixel in this image. Individual plates are apparent in this image because the fraction of zero pixels varies substantially from plate to plate. These problems are obviously in the data and not in the sky.

The plates that are most apparent in FIGURE 2 have markedly different means. For example, in the upper part of FIGURE 1 there is a dark plate immediately northwest of the Coma cluster. [If the plates are numbered (x, y), where x runs from left to right and y runs from top to bottom, this plate is (9,2).] In FIGURE 2, this plate is also anomalous because it contains many zero pixels. In the raw data, the mean count per pixel on this plate is 1.0; the means on the plates to the left and right are 3.2 and 2.8, respectively. In the SSGP reduction, the plate remains anomalous with a mean count per pixel of 1.3; for the plates to the left and right, the means are 1.9. The low plate was counted by Shane; the other two were counted by Wirtanen. [FIGURES 4A and 4B show the effect that this variation has on the SSGP reduction of the counts in the region around Coma (discussed further below).]

A quantitative test demonstrates the relationship between the pattern of zero pixels and the plate positions. For a set of 64 contiguous plates with $10^h10^m < \alpha < 15^h30^m$ and $17.5° < \delta < 37.5°$, we sample the data with a 30 × 30 pixel grid displaced from the original plate positions by a fixed number of columns of pixels. The shifts of the grid range from -15 to $+15$ columns. For each position of the grid, we calculate the

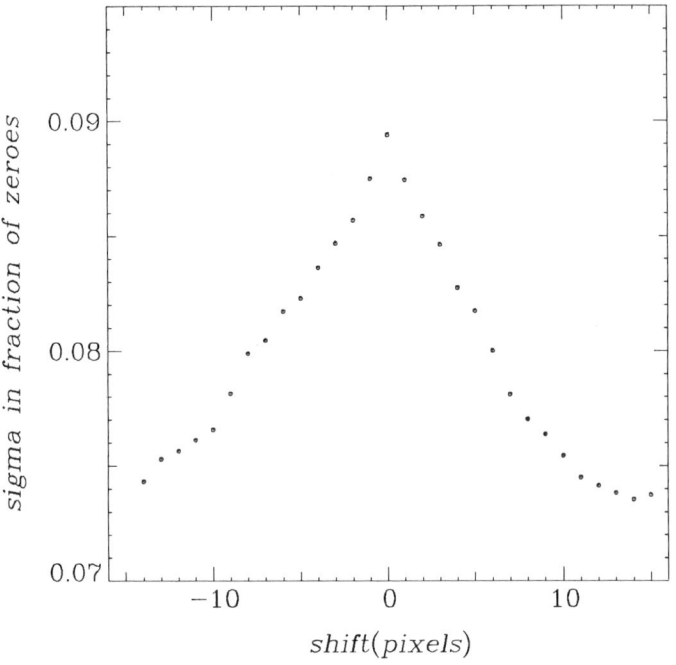

FIGURE 3. Sigma in the fraction of zeroes as a function of grid position.

statistics of the distribution of zeroes on the 30 × 30 pixel scale. For example, FIGURE 3 shows the σ in the fraction of zeroes as a function of the shift. Regardless of the sample of plates, the peak is steep and it occurs at zero shift. The fractional error in σ is $\delta\sigma/\sigma \sim 0.015$. If there were no relationship between the pattern of zeroes and the plate positions, the peak could occur at any grid position. Obviously, this pattern in the zeroes cannot be removed from the data by any multiplicative correction.

The pattern of zeroes introduces a systematic error in the determination of the mean for any plate. The fraction of zeroes due to systematic errors, including variations in the magnitude limit, counting, and plate quality, is large; from simulated data, we estimate that the r.m.s. number of "excess" zero pixels in a plate is ~15%. These large variations can be correlated over large scales because the counting pattern and the plate quality are not random functions of position on the sky.

Plate-to-plate variations may introduce large-scale features in the galaxy distribution. It is, for example, instructive to look at the relationship between features in the map and the pattern of plates counted by Shane and Wirtanen. FIGURE 4A shows a smoothed map of the original counts in the region of $10^h10^m < \alpha < 15^h30^m$ and $-22.5° < \delta < 37.5°$ superposed on the plate counting pattern. The darker plates were counted by Wirtanen; the rest were counted by Shane. FIGURE 4B shows the smoothed SSGP reduction of the same region. The white region in FIGURE 4A has a galaxy surface density $\gtrsim 2.1$ per pixel; the mean count in the region is 1.7 galaxies per pixel. In FIGURE 4B, the white region has galaxy surface density $\gtrsim 2.3$ and the mean count is 1.5 galaxies per pixel. In both maps, we display the highest 18% of the pixels in white. In the SSGP map, the contrast is stretched relative to that for the raw data (see reference 21 for a detailed discussion).

There are at least two notable differences between the maps in FIGURES 4A and 4B. First consider the region near the Coma cluster (the brightest region near the top center of the map). In FIGURE 4A, the cluster itself is more extended than in FIGURE 4B. In FIGURE 4B, there is a striking large dark region to the south of the Coma cluster that is far less impressive in FIGURE 4A. This "hole" in FIGURE 4B is probably due at least in part to the propagation of the effects of correcting for the bad plate discussed above.

In the lower regions of the map, Wirtanen generally counted plates along diagonals. These diagonals are more apparent and the filamentary structures appear finer in FIGURE 4B than in FIGURE 4A. Some of these apparent long filamentary structures run along the diagonals, through plates counted by Shane. The relationship between the large-scale structures and the counting pattern is worrisome.

The counting efficiency is a function of position on the plate and this function is different for the two observers. This difference in the average "shape" of the plates counted by the two observers contributes to the filamentary pattern. SSGP take pixel correction factors that are observer independent. In fact, these factors are observer and time dependent (de Lapparent, Kurtz, and Geller, 1985).[21] The variations are often as large as 10%.

In the region of FIGURE 4A, the gradient in Wirtanen's counting efficiency is generally smaller than in Shane's: in the raw counts, the difference between the mean count for pixels near the plate center and for those near the plate edge is smaller for the plates counted by Wirtanen. After correction (FIGURE 4B), most of the Wirtanen plates are slightly shallower at the center and deeper near the edge because SSGP apply a single set of pixel correction factors. The edge-matching procedure of SSGP therefore introduces an artificial contrast between the central regions of the plates counted by the two observers; the bias suppresses the plates counted by Wirtanen and enhances the plates counted by Shane. In the region shown in the figures, this artificial contrast stretching explains why nearly all the plates counted by Wirtanen are dark in FIGURE 4B and why the ladderlike filamentary pattern mimics the plate counting pattern (de Lapparent, Kurtz, and Geller, 1985).[21]

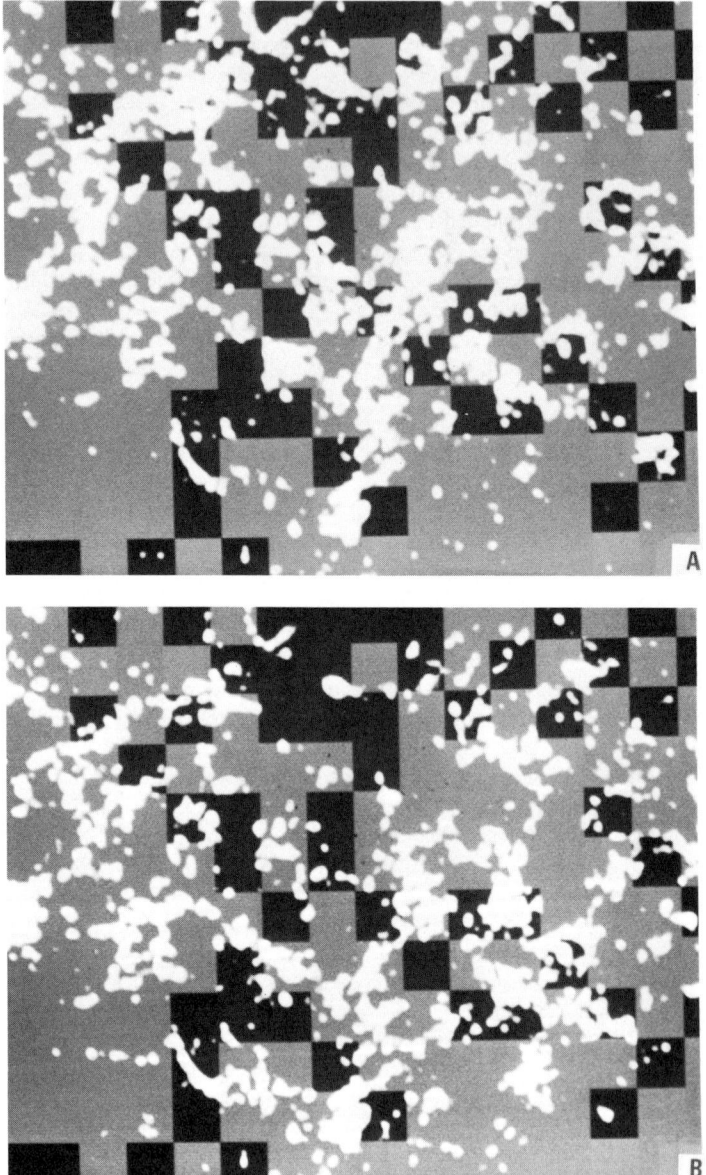

FIGURE 4. (A) The smoothed raw data superposed on the plate counting pattern. The black plates were counted by Wirtanen and the grey by Shane. (B) The smoothed SSGP map superposed on the plate counting pattern. The color scheme is the same as in FIGURE 4A.

MEASURING THE GALAXY CORRELATION FUNCTION ON LARGE SCALES

The plate-to-plate variations present in the original data and in the SSGP reduction can have a marked effect on the galaxy correlation function derived from the data. For example, FIGURE 5 shows the correlation function for a strip of ten Shane-Wirtanen plates at a declination of 10° and centered at a right ascension of

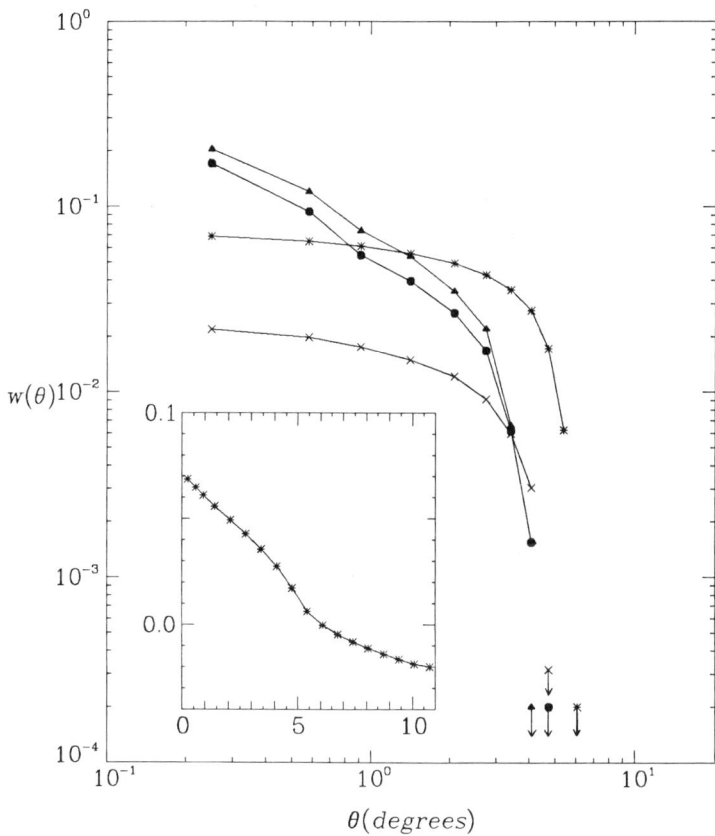

FIGURE 5. The correlation functions, $w_R(\theta)$ [▲], $w_S(\theta)$ [●], and $w_P(\theta)$ [×], for the 10° strip and $w_P(\theta)$ [∗] for the 0° strip.

12^h50^m. FIGURE 5 shows the correlation functions for both the original Shane-Wirtanen data, $w_R(\theta)$, and the correlation function, $w_S(\theta)$, for the SSGP reduction. Recall that in the SSGP reduction every pixel in the map is multiplied by a correction factor that is the product of a plate and a pixel correction factor. The plate correction factors are supposed to account for variations from plate to plate. The pixel correction factors account for variations across a single plate.

In FIGURE 5, we also plot the correlation function, $w_P(\theta)$, for the correction factors applied by SSGP in strips at 0° and 10° declination; in other words, the value we assign to each pixel in the strip is just the plate correction factor. The strip becomes a series of uniform 30 × 30 pixel plates, each with an amplitude specified by the plate correction factor. The amplitude of $w_P(\theta)$ is the normalized variance in the plate correction factors. The position of the break in $w_P(\theta)$ is determined by the plate scale and is affected by the detailed sequence of plate correction factors. For the sequence of plates at 10°, the break is very near the angular scale of the break in the galaxy correlation function derived from the counts.

We use $w_P(\theta)$ as a model for the residual plate-to-plate variations in the SSGP reduction. The amplitude of the residual systematics should be smaller than the amplitude of the correction factors, but the range of shapes of the $w_P(\theta)$ for the residual systematics can be well represented by the range for the correction factors. The systematic correlation between the means on adjacent plates is important in determining the behavior of $w_P(\theta)$ at scales $\gtrsim 2.5°$.

We next model the effect of unremoved plate-to-plate variations on the correlation function derived from the data. If we let $w_I(\theta)$ be the intrinsic correlation function of the galaxy distribution and $w_P(\theta)$ be the correlation function for the plate correction factors, the observed correlation function, $w_O(\theta)$, is

$$w_O(\theta) = w_I(\theta)w_P(\theta) + w_I(\theta) + w_P(\theta), \tag{2}$$

where we have assumed that the correction factor for any pixel is a multiplicative factor uncorrelated with the count in that pixel. This expression is approximately correct if the plate-to-plate variations are due to variations in the limiting apparent magnitude (see references 19 and 21). In fact, the systematic errors in the data are complex; the two observers counted differently as a function of the local density of galaxies on the plates (see the discussion of FIGURES 4A and 4B above; see also reference 21).

The inset of FIGURE 5 shows the behavior of $w_P(\theta)$ at large angular scale. For $\theta \gtrsim 6°$, $w_P(\theta)$ is negative. In other words, systematic errors can introduce anticorrelation on some scales larger than an individual plate. It is this anticorrelation that introduces a break in $w_O(\theta)$. Although we do not, of course, know the true $w_P(\theta)$, the pattern of plates counted by Shane and Wirtanen coupled with the difference in the shape of the plates counted by the two observers introduces such behavior (see reference 21).

Equation (2) provides a basis for exploring the relationship between the observed and intrinsic correlation functions. If the intrinsic correlation function is

$$w_I(\theta) = (\theta_0)/(\theta) \tag{3}$$

with $\theta_0 = 0.04$, we can reproduce the observed correlation function with a realizable sequence of residual plate-to-plate variations with an r.m.s. amplitude of only 15% (FIGURE 6). The observed correlation function is biased by the systematic errors. In this example, the slope of the correlation function at scales larger than the break is largely determined by the details of the systematics along with the binning of the data.

This model gives an indication of the systematic errors that can be tolerated in a measurement of the large-scale behavior of the galaxy correlation function. To avoid the introduction of biases in the correlation function, we require

$$w_P(\theta_B) \ll w_I(\theta_B), \quad (4)$$

where θ_B is the angular scale where we wish to measure the correlation function. If we assume that the plate-to-plate variations result from variations in the limiting magnitude (an oversimplification), the rule of thumb for avoiding artifacts in $w_O(\theta)$

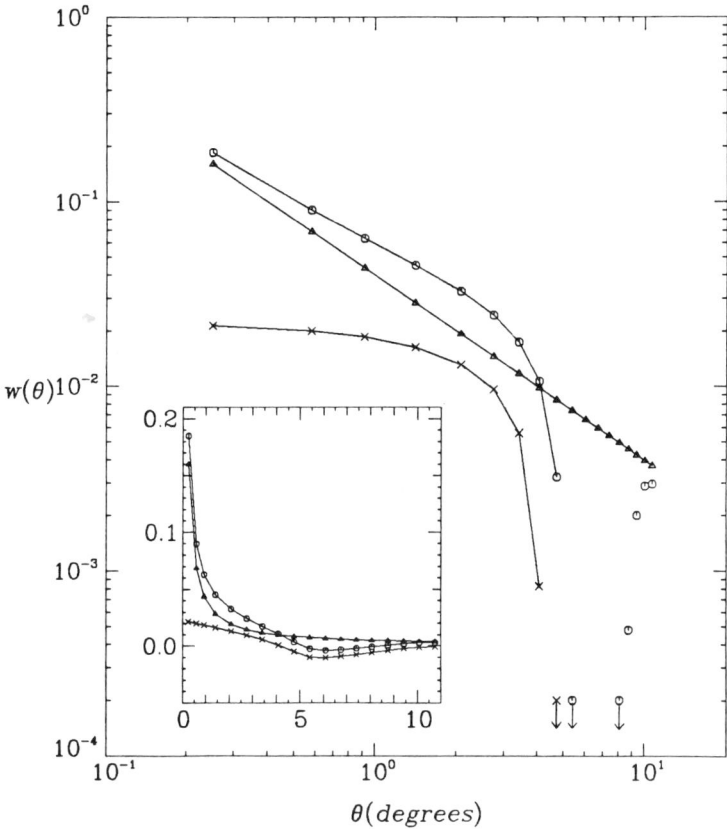

FIGURE 6. The correlation functions, $w_I(\theta)$ [△], $w_P(\theta)$ [×], and $w_O(\theta)$ [○], in the model of equation (3).

is

$$\Delta m \ll (w_I(\theta_B))^{1/2}, \quad (5)$$

where Δm is the systematic variation in the magnitude limit on the scale, θ_B (Geller, de Lapparent, and Kurtz, 1984).[20] For the Shane-Wirtanen counts, we require $\Delta m \lesssim 0.^m05$ from plate to plate. This requirement is remarkably stringent.

Both the break in the correlation function and the filamentary appearance of the Shane-Wirtanen map are suspect because of residual systematic errors in the SSGP map. The systematic errors introduced by differences in the counting efficiency of the two observers are alone sufficient to introduce both of these features.

There is a clear need for new, carefully calibrated surveys designed to measure the large-scale behavior of the correlation function and to examine the filamentariness of the galaxy distribution. At least two kinds of surveys are now feasible: (1) deep samples in regions of small angular scale and (2) shallower samples in regions of larger angular scale. For the angular correlation function, the amplitude of the angular correlation function at a fixed projected spatial scale is inversely proportional to the depth of the survey. Thus, equation (5) requires smaller systematic variations in the magnitude limit in deeper surveys. A shallower large angular scale survey coupled with redshift measurements is probably the best method of attack.

ACKNOWLEDGMENTS

We thank the National Science Foundation and the Smithsonian Institution Research Opportunities Fund for providing partial support for one of us (MJG) to present this paper at the Twelfth Texas Symposium. V. de Lapparent is partially supported by an Amelia Earhart Fellowship from Zonta International.

REFERENCES

1. SELDNER, M., B. SIEBERS, E. J. GROTH & P. J. E. PEEBLES. 1977. Astron. J. **82**: 249.
2. PEEBLES, P. J. E. 1980. The Large-Scale Structure of the Universe. Princeton University Press. Princeton.
3. DAVIS, M. & P. J. E. PEEBLES. 1983. Astrophys. J. **267**: 465.
4. BEAN, A. J., G. EFSTATHIOU, R. S. ELLIS, B. A. PETERSON & T. SHANKS. 1983. Mon. Not. R. Astron. Soc. **205**: 605.
5. KIRSHNER, R. P., A. OEMLER, JR. & P. L. SCHECHTER. 1978. Astron. J. **83**: 1549.
6. KIRSHNER, R. P., A. OEMLER, JR. & P. L. SCHECHTER. 1979. Astron. J. **84**: 951.
7. HUCHRA, J., M. DAVIS, D. LATHAM & J. TONRY. 1982. Astrophys. J. Suppl. Ser. **52**: 89.
8. KIRSHNER, R. P., A. OEMLER, JR., P. L. SCHECHTER & S. A. SHECTMAN. 1983. Astron. J. **88**: 1285.
9. SHANE, C. D. & C. A. WIRTANEN. 1967. Publ. Lick Obs. **XXII** (part 1).
10. MOODY, J. E., E. L. TURNER, & J. R. GOTT III. 1983. Astrophys. J. **273**: 16.
11. KUHN, J. R. & J. M. USON. 1982. Astrophys. J. **263**: L47.
12. GROTH, E. J. & P. J. E. PEEBLES. 1977. Astrophys. J. **217**: 385.
13. DAVIS, M. & P. J. E. PEEBLES. 1977. Astrophys. J. Suppl. Ser. **34**: 425.
14. DAVIS, M., E. J. GROTH & P. J. E. PEEBLES. 1977. Astrophys. J. Lett. **212**: L107.
15. GOTT, J. R., E. L. TURNER & S. J. AARSETH. 1979. Astrophys. J. **234**: 13.
16. ZELDOVICH, YA. B., J. EINASTO & S. F. SHANDARIN. 1982. Nature **300**: 407.
17. DEKEL, A. & M. J. WEST. 1985. Astrophys. J. **288**: 411.
18. PEEBLES, P. J. E. 1980. Ninth Texas Symposium on Relativistic Astrophysics. Ehlers, Perry & Walker, Eds: 161–171. N.Y. Academy of Sciences. New York.
19. SELDNER, M. & J. M. USON. 1983. Astrophys. J. **264**: 1.
20. GELLER, M. J., V. DE LAPPARENT & M. J. KURTZ. 1984. Astrophys. J. Lett. **287**: L55.
21. DE LAPPARENT, V., M. J. KURTZ & M. J. GELLER. 1986. Astrophys. J. In press.
22. SONEIRA, R. M. & P. J. E. PEEBLES. 1978. Astron. J. **83**: 7.
23. SHANKS, T., A. J. BEAN, G. EFSTATHIOU, R. S. ELLIS, R. FONG & B. A. PETERSON. 1983. Astrophys. J. **274**: 529.

24. ABELL, G. O. 1958. Astrophys. J. Suppl. Ser. **3:** 211.
25. KAISER, N. 1985. Astrophys. J. In press.
26. BAHCALL, N. A. & R. M. SONEIRA. 1983. Astrophys. J. **270:** 20.
27. KIRSHNER, R. P., A. OEMLER, JR., P. L. SCHECHTER & S. A. SHECTMAN. 1981. Astrophys. J. Lett. **248:** L57.
28. GIOVANELLI, R. & M. P. HAYNES. 1982. Astron. J. **87:** 1355.
29. EINASTO, J., M. JÔEVEER & E. SAAR. 1980. Mon. Not. R. Astron. Soc. **193:** 353.
30. GIOVANELLI, R. 1983. *In* IAU Symposium No. 104. G.O. Abell & G. Chincarini, Eds.: 273–280. Reidel. Dordrecht.
31. DE LAPPARENT, V., M. J. GELLER, M. J. KURTZ & E. L. TURNER. 1985. In preparation.

PART IV. GENERAL RELATIVITY

Energy and Its Definition in General Relativity

ROGER PENROSE

Mathematical Institute
University of Oxford
Oxford OX1 3LB, England

BACKGROUND

There is an essential difficulty—and it has been with us since Einstein first put forward his general theory of relativity—concerning the definition, within that theory, of the fundamental concept: energy. The difficulty is present also for the related concepts of momentum and angular momentum, but I shall phrase most of my preliminary discussion in terms of energy (or, equivalently, mass). This difficulty arises because in Einstein's field equation,

$$R_{ab} - \frac{1}{2} R g_{ab} + \lambda g_{ab} = -8\pi G T_{ab} \tag{1}$$

(with a cosmological constant, λ, being allowed for), the energy tensor, T_{ab}, refers to the energy of matter fields only and not to the energy of the gravitational field itself. It is clear on physical grounds that the gravitational field must in fact contribute to the total mass-energy of a physical system. There are two particular ways that this can come about. First, there is the (negative) Newtonian potential energy contribution to the total energy of two (or more) masses. The closer the masses are to each other, the greater will be the amount by which the total mass-energy falls short of the sum of their individual mass-energies. Second, there is the (positive) contribution to the total mass-energy from gravitational waves. These are "ripples" in empty space, so $T_{ab} = 0$ there. Nevertheless, the positivity of gravitational wave energy is a clear implication of the Bondi-Sachs[1] expression for the mass measured at null infinity ($\mathcal{J}^+$).

In mathematical terms, the reason that energy can arise other than from direct contributions from T_{ab} is related to the fact that the covariant divergence law,

$$\nabla_a T^{ab} = 0, \tag{2}$$

does not give rise to an integral conservation law (essentially because of the extra index, "b"), which is in contrast with the case of charge conservation, where $\nabla_a J^a = 0$ does provide us with an integral conservation law for charge. The total mass-energy concept thus must involve contributions not simply arising from "adding up" T_{ab}. Einstein's original proposal for coping with this problem was, of course, to add a "pseudotensor" quantity to T_{ab} that was supposed to represent the gravitational contribution to the energy. However, this quantity has no local coordinate-independent meaning and Einstein regarded it as providing merely a means to calculating the total energy of a system. Even this could be achieved only when the coordinates were specialized to have a very particular form at infinity. Locally, the pseudotensor can always be reduced to zero, emphasizing the fact that the gravitational energy concept is necessarily a nonlocal one.

Physicists seem to have gradually become accustomed to the idea that a meaningful measure of energy for a physical system that incorporates both the gravitational and matter contributions would be too much to ask for, except in the case of the total energy of an entire isolated physical system. For this total energy (for an asymptotically flat space-time), successful expressions do indeed exist: at spatial infinity (i^0), there is the ADM expression[2] and at null infinity ($\mathcal{J}^+$), the Bondi-Sachs expression[1]— the latter having the advantage that by examining its value at different retarded times, the contribution from gravitational or electromagnetic waves can be separated out. These expressions apply to the standard Einstein theory with $\lambda = 0$. With negative λ, for asymptotically anti–de Sitter space-times, a corresponding expression has been introduced by Abbot and Deser,[3] which has been somewhat clarified by Ashtekar and Magnon.[4]

QUASI-LOCAL EXPRESSIONS

It would clearly be physically useful to be able to say something more detailed about the distribution of energy in a physical space-time. In fact, on various occasions, expressions applicable at a more local level have been suggested. I shall refer to an expression that purports to measure the energy (or other physical quantities) surrounded (i.e., "linked") by a closed 2-surface $\mathcal{S}$ as a *quasi-local* expression if it involves only the geometry (or other gauge invariant physical fields) and only in the immediate neighborhood of $\mathcal{S}$. I shall indicate various suggestions that have been made for expressions of this kind.

Historically, the most important attempt at a quasi-local energy expression is that due to Komar.[5] Here one assumes that the space-time, $\mathcal{M}$, is stationary, i.e., it contains a timelike Killing vector, k^a. We assume that k^a is normalized to be a unit vector at infinity, generating the standard time-translations. The suggested energy expression is then

$$(8\pi G)^{-1} \oint e_{abpq} \nabla^a k^b \, dx^p \wedge dx^q \tag{3}$$

(where e_{abpq} is the alternating tensor). This has the significant property that its value is unchanged as $\mathcal{S}$ is moved continuously through a region of vacuum ($T_{ab} = 0$). Moreover, if $\mathcal{S}$ is taken to infinity, expression (3) agrees with the ADM value (at i^0) and the Bondi-Sachs value (at $\mathcal{J}^+$). However, it has the drawback that the "source-term" for expression (3) is, in effect, the Ricci tensor rather than its trace-reverse—which it should be, according to equation (1) (with $\lambda = 0$)—and this is related to the fact that certain anomalies concerning factors of two tend to arise. (One of these concerns the angular momentum and the other concerns the contributions when matter fields are present, which I shall remark upon shortly.) One further slight drawback is that expression (3) is not, strictly speaking, a completely quasi-local expression since a normalization for k^a at infinity is required.

Winicour and Tamburino[6] adapted Komar's expression to define what they called an energy "linkage," for which k^a need not be assumed actually to be a Killing vector. For the integral to have physical significance, it is necessary to place suitable restrictions on k^a. When this is done at $\mathcal{J}^+$, the Bondi-Sachs definition can be recovered. However, in the general case, considerable ambiguities remain.

An expression due to Hawking,[7]

$$(A/16\pi G^2)^{1/2} \left(1 - \frac{1}{2\pi} \oint \rho \rho' d\mathcal{S}\right), \tag{4}$$

is also worthy of note. Here, A is the area of the 2-surface $\mathcal{S}$, assumed spacelike and with the topology of a sphere, and ρ and ρ' are the spin-coefficients[8,9] describing the convergence of the null normals to $\mathcal{S}$. When $\mathcal{S}$ is taken to $\mathcal{J}^+$ in a suitable way, the Hawking expression [expression (4)] also reproduces the Bondi-Sachs definition. However, in a general way, expression (4) cannot be used just as it stands and it must be supplemented by some condition such as of "maximal roundness." (Otherwise unreasonable answers would arise, notably nonzero values for flat space-time.)

A different definition has been more recently proposed by Ludvigsen and Vickers.[10] This arose out of their work developing Witten's argument for the positivity of the ADM mass, which was done in order to provide one of the proofs of the somewhat stronger result that the Bondi-Sachs mass is likewise positive-definite. The link with this positivity theorem gives their approach some physical plausibility, but their definition is not really a quasi-local one in the sense described above since it depends upon information carried in from infinity on a null hypersurface through $\mathcal{S}$.

Finally, there is a quasi-local definition that I myself proposed a few years ago[11,12] for the mass, momentum, and angular momentum, surrounded by a spacelike topological 2-sphere. The remainder of this account will be concerned with this expression: its motivation, definition, and latest results.

MOTIVATION

The paradigm for a quasi-local expression would appear to be provided by Maxwell theory. Here, one has the expressions,

$$Q = \frac{1}{8\pi} \oint_\mathcal{S} {}^*F_{ab} \, dx^a \wedge dx^b \tag{5}$$

and

$$0 = \frac{1}{8\pi} \oint_\mathcal{S} F_{ab} \, dx^a \wedge dx^b, \tag{6}$$

respectively, for the electric and magnetic charges surrounded by $\mathcal{S}$. Here

$${}^*F_{ab} = \frac{1}{2} e_{abcd} F^{cd} \tag{7}$$

and the Maxwell equations,

$$\nabla_{[a} F_{bc]} = 0, \quad \nabla^a F_{ab} = 4\pi J_b, \tag{8}$$

are being assumed. Gauss's law, generalized to a general-relativistic setting, enables equation (5) to be converted to a 3-volume integral,

$$Q = \frac{1}{6} \int_\Sigma J^a \, e_{apqr} \, dx^p \wedge dx^q \wedge dx^r, \qquad (9)$$

of charge density over the compact 3-region, Σ, where $\mathcal{S} = \partial \Sigma$.

It should be pointed out that the mere existence of a "quasi-local" expression for charge, such as equation (5), in itself implies a weak form of charge conservation. For it makes no difference to the value of Q how the 2-surface $\mathcal{S}$ is spanned. Any two compact 3-regions, Σ, Σ', whose boundaries are the same surface,

$$\partial \Sigma = \mathcal{S} = \partial \Sigma', \qquad (10)$$

would be deemed to intercept the same total charge, namely Q, even in the absence of some explicit formula such as equation (9). One needs only some convincing reason to believe in the physical appropriateness of the proposed quasi-local expression [here equation (5)]. Its conservation, in this weak sense, is a necessary consequence.

There is, however, a stronger form of conservation that does not follow from the mere existence of a quasi-local expression. Strong conservation arises when the "charge" value remains constant whenever the 2-surface $\mathcal{S}$ is moved continuously through a source-free region. It is clear that this requirement of strong conservation is indeed satisfied in Maxwell theory. In physical terms, we say that the field is itself "uncharged." We do not, however, expect that a corresponding property would hold for Yang-Mills theory—nor, indeed, for general relativity since, as was remarked above, the gravitational field can itself physically possess mass-energy, albeit in some nonlocal form. The existence of a quasi-local energy expression in general relativity is perfectly consistent with these physical requirements.

To motivate my own suggestion for such a quasi-local expression, consider first the linearized theory. In this case, we do expect a strong conservation law, with the energy of gravitation being quadratic in the field (as can be seen, for example, in the Bondi-Sachs flux at $\mathcal{I}^+$). Thus, it can be ignored in the linearized limit. The idea is to take the (weak field) curvature tensor, R_{abcd}, which is a spin-2 object, and reduce it to a spin-1 object—which behaves formally like a Maxwell field—by contracting off two of its indices with a suitably defined object, W^{ab} ($= -W^{ba}$):

$$R_{abcd} W^{cd}. \qquad (11)$$

The equation to be satisfied by W^{ab} turns out to be

$$\nabla^{(a} W^{b)c} - \nabla^{(a} W^{c)b} + g^{a[b} \nabla_d W^{c]d} = 0 \qquad (12)$$

(round parentheses denoting symmetrization and square brackets denoting antisymmetrization). Putting

$$k^a = \frac{2}{3} e^{abcd} \nabla_b W_{cd}, \qquad (13)$$

we find that k^a is automatically a Killing vector of the flat background space-time of the linearized theory. Computing the charge integral for the "Maxwell field" in equation (11), we actually obtain[11,12] the energy, momentum, or angular momentum component of the sources surrounded by $\mathcal{S}$, which correspond to the Killing vector defined by equation (13). (The results agree with those of Sachs and Bergmann,[13] though their procedures were completely different.)

As stated, the reason this works is somewhat obscure. Things are rather more transparent if a spinor formalism is used. Using the abstract index notation,[9] we have, in the vacuum region,

$$R_{abcd} = \Psi_{ABCD}\, \epsilon_{A'B'}\epsilon_{C'D'} + \epsilon_{AB}\epsilon_{CD}\overline{\Psi}_{A'B'C'D'}, \tag{14}$$

where Ψ_{ABCD} is symmetric and satisfies

$$\nabla^{AA'}\Psi_{ABCD} = 0 \tag{15}$$

(from the linearized Bianchi identity); we can also set

$$W^{ab} = \omega^{AB}\epsilon^{A'B'} + \epsilon^{AB}\overline{\omega}^{A'B'}, \tag{16}$$

where ω^{AB} is symmetric and satisfies

$$\nabla^{(A}_{A'}\omega^{BC)} = 0 \tag{17}$$

[which is the spinor form of equation (12)]. We now find that equation (11) has the spinor expression,

$$\chi_{AB}\,\epsilon_{A'B'} + \epsilon_{AB}\overline{\chi}_{A'B'}, \tag{18}$$

where

$$\chi_{AB} = \psi_{ABCD}\omega^{CD} \tag{19}$$

is symmetric and satisfies the spinor form of Maxwell's free-space equation,[9]

$$\nabla^{AA'}\chi_{AB} = 0. \tag{20}$$

This equation is an immediate consequence of equations (15) and (17). Indeed, equations (15) and (20) are particular cases of the massless free-field equation for spin $n/2$:

$$\nabla^{AA'}\phi_{AB...L} = 0, \tag{21}$$

where $\phi_{AB...L}$ has n symmetric indices. As a consequence of equations (17) and (21), the quantity,

$$\phi_{AB...JKL}\omega^{KL}, \tag{22}$$

which is symmetric with $n - 2$ indices, satisfies the massless free-field equation for spin $(n - 2)/2$. Thus, the quantity, ω^{AB}, subject to equation (17), provides a procedure in the general case of an arbitrary massless field for lowering the spin by one unit.

In Minkowski space-time, equation (12) has twenty linearly independent real solutions [i.e., equation (17) has ten independent complex solutions]. However, for ten independent real solutions, the Killing vector of equation (13) vanishes and the integrals of equation (11) must therefore also vanish. The remaining ten independent real solutions of equation (12) provide integrals that do not generally vanish and they give the ten components of energy, momentum, and angular momentum for the sources surrounded by $\mathcal{S}$. Indeed, one obtains a strong conservation law for these quantities.

In general relativity proper, however, equation (12) [or, equivalently, equation (17)] will in general have no solutions (other than zero). This is not surprising since equation (12) represents sixteen independent equations on only six independent unknowns [or, in complex terms, equation (17) represents eight equations on three unknowns]. Indeed, it would be physically undesirable for equation (12) to have nontrivial solutions in the general case since these would lead to strong conversation laws in general relativity for the corresponding energy, momentum, or angular momentum components. All we seek (at this stage) is a reasonable quasi-local *definition* of such quantities so that weak, rather than strong, conservation will hold. We require something that "looks like" W^{ab} (or ω^{AB}) as far as $\mathscr{S}$ is concerned, but which need not extend into the ambient space-time, $\mathscr{M}$, in any specific way. It is natural to demand, therefore, merely that the "tangential part" of equation (12) [or (17)] be satisfied. However, we now find that the resulting equations are underdetermined since there are only four equations for the six unknowns (or two equations for three unknowns, in the complex description).

The way around this impasse turns out to be to lower the spin in two steps of one-half each (rather than in a single step). This is achieved, in flat space, by use of a spinor quantity, ω^A, which satisfies

$$\nabla_{A'}^{(A} \omega^{B)} = 0. \tag{23}$$

For, by the same reasoning as before, if $\phi_{AB...L}$ (symmetric) satisfies the massless free-field equation (21) for spin $n/2$, then

$$\phi_{AB...KL} \omega^L \tag{24}$$

satisfies the same equation for spin $(n-1)/2$. Equation (23) is called the twistor equation[14,15] and its solutions, in Minkowski space-time, constitute a four-complex-dimensional vector space, T, referred to as twistor space. (This is just one of many possible definitions of twistor space. It suffices for our purposes here. The conformal supersymmetry generators of Wess and Zumino[16] are also effectively twistors in this sense.)

Again, equation (23) does not generally have (nontrivial) solutions in a curved space-time, $\mathscr{M}$, since there are six complex equations for two complex unknowns. However, restricted to $\mathscr{S}$, we now obtain two equations for two unknowns. Elementary index theory shows us that indeed there is, at least in the generic case, a four-complex-dimensional space of solutions of these tangential equations, provided that $\mathscr{S}$ is spacelike and topologically a 2-sphere. This space is called the 2-surface twistor space and is denoted by $\mathsf{T}(\mathscr{S})$. The required analogues of ω^{AB}, needed for defining the analogues of W^{ab} used in equation (11), are now the elements of the symmetric tensor product,

$$\mathsf{T}(\mathscr{S}) \odot \mathsf{T}(\mathscr{S}). \tag{25}$$

It is remarkable that these objects do not seem to be definable directly, and it appears to be necessary to employ "spin-½" objects in order to define what would appear to be essentially "integral spin" concepts, namely, energy-momentum and angular momentum.

TECHNICAL DEFINITIONS

The tangential part of equation (23) is conveniently expressed in terms of the compacted spin-coefficient formalism.[9,17] The equations become

$$\eth'\omega^0 = \sigma'\omega^1, \quad \eth\omega^1 = \sigma\omega^0, \tag{26}$$

where $\eth'$ and $\eth$ are suitably defined (covariant) derivative operators within $\mathcal{S}$, with the spin-frame being chosen to have its "flagpole" directions orthogonal to $\mathcal{S}$. A solution of equation (26) on $\mathcal{S}$ [element of $\mathsf{T}(\mathcal{S})$] may be denoted by Z^α (a 2-surface twistor), where the four-dimensional abstract index, α, refers to the 4-space $\mathsf{T}(\mathcal{S})$. Taking Z^α and $\tilde{Z}^\alpha$ to be two arbitrary elements of $\mathsf{T}(\mathcal{S})$ [given, respectively, by the solutions, ω^0, ω^1 and $\tilde{\omega}^0$, $\tilde{\omega}^1$, of equation (26)], we can tentatively define the element, $A_{\alpha\beta}$, of $\{\mathsf{T}(\mathcal{S}) \odot \mathsf{T}(\mathcal{S})\}^*$ by the relation,

$$A_{\alpha\beta} Z^\alpha \tilde{Z}^\beta = (4\pi i G)^{-1} \oint_{\mathcal{S}} (P\omega^0\tilde{\omega}^0 + Q(\omega^0\tilde{\omega}^1 + \omega^1\tilde{\omega}^0) + R\omega^1\tilde{\omega}^1)d\mathcal{S}. \tag{27}$$

Here

$$P = \Psi_1 - \Phi_{0,1} = R_{abcd}\ell^a m^b \overline{m}^c m^d, \tag{28}$$

$$Q = \Psi_2 - \Phi_{1,1} - \Lambda + \lambda/6 = \frac{1}{2} R_{abcd}(\ell^a n^b + \overline{m}^a m^b)\overline{m}^c m^d + \lambda/6, \tag{29}$$

and

$$R = \Psi_3 - \Phi_{2,1} = R_{abcd}\overline{m}^a n^b \overline{m}^c m^d \tag{30}$$

[with a cosmological constant, λ, being allowed for, as in equation (1)]. The null tetrad $(\ell^a, m^a, \overline{m}^a, n^a)$ is that defined by the spin-frame.

This (tentatively) defines the total angular momentum twistor, $A_{\alpha\beta}$, for the matter + gravitation surrounded by $\mathcal{S}$. In linearized theory (flat space), this takes the form,[15]

$$A_{\alpha\beta} = \left[\begin{array}{c|c} 0 & \text{energy-momentum} \\ \hline \text{energy-momentum} & \text{angular momentum} \end{array}\right], \tag{31}$$

and a Hermitian (squared) norm, $\|Z^\alpha\|$, of signature, $(++--)$, is defined on T, whose natural extension to $\{\mathsf{T} \odot \mathsf{T}\}^*$ allows us to define the squared rest-mass as

$$m^2 = -\frac{1}{2}\|A_{\alpha\beta}\|. \tag{32}$$

The particular form of equation (31) entails that only ten of the possible twenty independent components of $A_{\alpha\beta}$ survive in the linearized theory (as noted earlier). It is not yet known whether an analogous reduction occurs in the full theory in all cases. Moreover, there is, as yet, no clear agreement as to the best definition of the Hermitian norm on $\mathsf{T}(\mathcal{S})$ in all cases, although in many circumstances the choice is clear.

SPECIFIC EXAMPLES

At null infinity, $\mathcal{J}^+$, this definition provides an energy-momentum 4-vector that is in complete agreement[11] with that of Bondi-Sachs. It also provides a definition of angular momentum that has some significant advantages[15] over those that had been suggested earlier,[18] notably proper agreement with linearized theory (for "bad cuts" of $\mathcal{J}^+$). At spacelike infinity, i^0, agreement is obtained with the ADM energy-momentum[20] and the Ashtekar-Hansen angular momentum.[20] (There seems also to be agreement with the Abbot-Deser-Ashtekar-Magnon definition for the negative cosmological constant.[21])

Several striking results have been obtained by Tod[22] for finite surfaces $\mathcal{S}$. For example, if $\mathcal{S}$ is drawn arbitrarily in a constant-time hypersurface in the Schwarzschild solution (or, indeed, in any rotationally symmetric hypersurface, $t = f(r)$, in standard coordinates), then the Schwarzschild mass is obtained for the m of equation (32) whenever $\mathcal{S}$ links the source (once) and the value of $m = 0$ is obtained if $\mathcal{S}$ does not link the source. For the Reissner-Nordstrom solution, there are somewhat analogous results, but now there is the (expected) positive contribution from the energy density of the electromagnetic field. Thus, the value of m depends on its particular location in addition to whether or not it links the source. [It is worth remarking here that if the Komar expression in expression (3) is used, then the contribution from the electromagnetic energy density comes out wrong by a factor of two.]

Surfaces $\mathcal{S}$ drawn in $t = f(r)$ hypersurfaces in these solutions have the special property that they can be embedded in conformally flat space-time without changing their first or second fundamental forms. Such surfaces we refer to as *uncontorted;* they are *contorted* if they *cannot* be so embedded. It is natural to conjecture that for a general (contorted) $\mathcal{S}$ in the Schwarzschild solution, it should continue to be the case that the m obtained is the Schwarzschild mass when $\mathcal{S}$ links the source (once) and zero when $\mathcal{S}$ does not link the source. We shall see the status of this conjecture shortly.

Tod[22] also examined Friedmann-Robertson-Walker models and found, in particular, that for a spherically symmetrical surface, m has the same value as would be the case in Euclidean space with a sphere of the same area that was immersed in a fluid of the same density as that of the model. Thus, for example, in the closed models, $\mathcal{S}$ reaches its maximum for an "equatorial" 2-sphere and reduces back to zero when the entire universe is encompassed. One can also say quite generally, with this definition, that the total mass-energy of any closed universe is zero (in agreement with expectations). Moreover, in any such universe, the mass m on one side of any $\mathcal{S}$ is always equal to that on the other.

One of the most striking of Tod's[22] results concerns 2-surfaces $\mathcal{S}$ in a conformally flat 3-surface $\mathcal{H}$ of time-symmetry. If $\mathcal{S}$ is continuously deformed throughout a vacuum region of $\mathcal{H}$, then it turns out that m does not change. Furthermore, one may envisage $\mathcal{S}$ linking one or the other of two source regions (yielding the results m_1, m_2, respectively) or else linking both sources together (yielding the result $m_{1,2}$). Tod finds the physically expected inequality,

$$m_{1,2} < m_1 + m_2, \qquad (33)$$

and he also finds the result that for large separations, the difference agrees with the correct Newtonian potential-energy contribution. This result is particularly striking because potential-energy effects did not directly enter into the motivational considerations for the definition of m given here.

A more recent result obtained by Tod[23] concerns a static "black hole in a cage." The cage is not to prevent the hole from "escaping," but merely to distort it from spherical shape. For given surface area (which would not be changed if the cage is altered "adiabatically"), one would anticipate that the hole configuration of least energy would be that of spherical shape. Thus, indeed, Tod finds the inequality,

$$m^2 \geq A/16\pi G^2, \tag{34}$$

with equality only in the spherically symmetrical Schwarzschild case.

THE η-FACTOR MODIFICATION

These results are all very gratifying, but difficulties have arisen from another calculation performed by Tod[24] and checked by R. M. Kelly. The light cone, $\mathcal{C}$, of some arbitrary point, p, in a space-time, $\mathcal{M}$, is considered. A timelike vector, t^a, is selected at p and cross sections, $\mathcal{S}_r$, of constant affine distance, r, from p are drawn, where the various affine parameters along the rays of $\mathcal{C}$ are normalized against t^a at p. The question is: what is the value of m (and the structure of $A_{\alpha\beta}$) for $\mathcal{S}_r$ at small r? When $\mathcal{M}$ is not vacuum at p, the expected energy-momentum,

$$\frac{4}{3}\pi r^3 T_{ab} t^b, \tag{35}$$

is indeed found at order, r^3 (as could be inferred in any case from agreement with the linearized limit). When T_{ab} vanishes, the first possibility for a contribution occurs at order, r^5. It is found that there is such a contribution for the definition in equation (**27**) and that if $\mathcal{M}$ is the Schwarzschild solution, then this contribution is nonvanishing when t^a does not lie in a (t, r)-plane. This contradicts the aforementioned conjecture, as it is stated, but it is worse than that: the m^2 obtained turns out to be negative (i.e., m is imaginary). These are cases when the surfaces $\mathcal{S}_r$ are contorted.

In the meantime, I had by myself obtained a procedure from attempting to prove the opposite result, that, in a suitable sense, this conjecture ought to be true. I was able to show that if a certain "Property K" were true for the Schwarzschild solution, then any $\mathcal{S}$ not linking the source, whether contorted or not, should yield zero for m (and for $A_{\alpha\beta}$) if a certain modification is made to the integral in equation (**27**). This consists of including a factor, η, that multiplies the entire integrand, where η is proportional to the determinant,

$$\begin{vmatrix} \omega_1^0 & \omega_2^0 & \omega_3^0 & \omega_4^0 \\ \omega_1^1 & \omega_2^1 & \omega_3^1 & \omega_4^1 \\ \eth\omega_1^0 & \eth\omega_2^0 & \eth\omega_3^0 & \omega_4^0 \\ \eth'\omega_1^1 & \eth'\omega_2^1 & \eth'\omega_3^1 & \eth'\omega_4^1 \end{vmatrix}, \tag{36}$$

with ω_1^A, ω_2^A, ω_3^A, and ω_4^A providing four independent solutions of equation (**26**). When $\mathcal{S}$

is uncontorted (and in certain other cases, such as when $\mathcal{S}$ is an arbitrary "cut" of $\mathcal{I}^+$), expression (36) is constant over $\mathcal{S}$ and so η can be chosen to be unity. In the general case, we need some normalization for η, with one suggested possibility being

$$\oint_\mathcal{S} \log \eta \, d\mathcal{S} = 0. \tag{37}$$

When this modification is made to equation (27), Tod and Kelly find that m now vanishes to order, r^5. However, Property K itself remains unestablished. This property asserts that the Killing spinor[15,25] [which is a particular solution of equation (17) arising in any type {22}—i.e., type D—space-time] always belongs to $\mathsf{T}(\mathcal{S}) \odot \mathsf{T}(\mathcal{S})$. This property must be false for the C-metrics[26] and for the Kerr solution,[26] but the circumstances for the Schwarzschild solution seem to be considerably more favorable.[a]

Shaw[27] has recently performed calculations of a reverse nature to those of Tod and Kelly, where expansions in terms of negative powers of an affine parameter, r, are obtained ("large spheres"). Physically reasonable results are so far obtained for the definition of equation (27) (with or without η-factor), though the Hawking and Ludvigsen-Vickers definitions contain some anomalies.

REFERENCES

1. BONDI, H. 1960. Nature **186:** 535; BONDI, H., M. G. J. VAN DER BURG & A. W. K. METZNER. 1962. Proc. R. Soc. London **A269:** 21; SACHS, R. K. 1962. Proc. R. Soc. London **A270:** 103.
2. ARNOWITT, R., S. DESER & C. W. MISNER. 1962. In Gravitation: An Introduction to Current Research. L. Witten, Ed. Wiley. New York.
3. ABBOT, L. F. & S. DESER. 1982. Nucl. Phys. **B195:** 76.
4. ASHTEKAR, A. & A. MAGNON. 1984. Class. Quantum Grav. **1:** L30.
5. KOMAR, A. 1959. Phys. Rev. **113:** 934.
6. TAMBURINO, L. & J. WINICOUR. 1966. Phys. Rev. **150:** 1039.
7. HAWKING, S. W. 1968. J. Math. Phys. **9:** 598.
8. NEWMAN, E. T. & R. PENROSE. 1962. J. Math. Phys. **3:** 566; 1963. J. Math. Phys. **4:** 998.
9. PENROSE, R. & W. RINDLER. 1984. Spinors and Space-Time, vol. 1: Two-Spinor Calculus and Relativistic Fields. Cambridge University Press. Cambridge.
10. LUDVIGSEN, M. & J. A. G. VICKERS. 1982. Bondi Momentum and its Quasi-Local Null Surface Extension. University of Canterbury Preprint. Christchurch, New Zealand.
11. PENROSE, R. 1982. Proc. R. Soc. London **A381:** 53.
12. PENROSE, R. 1984. J. C. Maxwell, the Sesquicentennial Symposium. M.S. Berger, Ed. Elsevier. Amsterdam/New York.
13. SACHS, R. K. & P. G. BERGMANN. 1958. Phys. Rev. **112:** 674.
14. PENROSE, R. 1967. J. Math. Phys. **8:** 345.
15. PENROSE, R. & W. RINDLER. 1986. Spinors and Space-Time, vol. 2: Spinor and Twistor Methods in Space-Time Geometry. Cambridge University Press. Cambridge.
16. WESS, J. & B. ZUMINO. 1974. Nucl. Phys. **70:** 39.
17. GEROCH, R., A. HELD & R. PENROSE. 1973. J. Math. Phys. **14:** 874.
18. WINICOUR, J. 1980. In General Relativity and Gravitation, vol. 2. A. Held, Ed. Plenum. New York.

---

[a]After this lecture was given, some results of N.M.J. Woodhouse,[28] using an original calculational procedure, have confirmed the Tod-Kelly "small spheres" calculation, but also seemingly yield a new anomaly, not removed by the η-factor, for "small ellipsoids." [**Note added in proof:** This new anomaly has subsequently disappeared.]

19. ASHTEKAR, A. & R. O. HANSEN. 1978. J. Math. Phys. **19:** 1542.
20. SHAW, W. T. 1983. Proc. R. Soc. London **A390:** 191.
21. KELLY, R. M. Private communication.
22. TOD, K. P. 1983. Proc. R. Soc. London **A388:** 457.
23. TOD, K. P., R. M. KELLY & N. M. J. WOODHOUSE. To appear.
24. TOD, K. P. 1984. Twistor Newsletter **18**. Math. Inst. Oxford.
25. WALKER, M. & R. PENROSE. 1970. Commun. Math. Phys. **18:** 265.
26. KRAMER, D., H. STEPHANI, M. MACCALLUM & E. HERLT. 1980. Exact Solutions of Einstein's Field Equations. Deut. Verlag Wissenschaften. Berlin.
27. SHAW, W. T. 1985. Twistor Newsletter **19**. Math. Inst. Oxford.
28. WOODHOUSE, N. M. J. 1985. Twistor Newsletter **19**. Math. Inst. Oxford.

Gravitational Collapse

DEMETRIOS CHRISTODOULOU

Departments of Physics and Mathematics
Syracuse University
Syracuse, New York 13210

This lecture is a report of my recent work on the global initial value problem for Einstein's equations in the spherically symmetric case with a massless scalar field as the material model. This work is a mathematical study of the dynamics of gravitational collapse and the formation of black holes.

PART I

By virtue of spherical symmetry, with introducing coordinates, u and r, where u is the retarded time and r is the luminosity distance, the metric is of the form,

$$ds^2 = -e^{2\nu} du^2 - 2 e^{\nu+\lambda} du\, dr + r^2\, d\Sigma^2,$$

where $d\Sigma^2$ is the metric of the unit 2-sphere. The Einstein equations are

$$R_{\mu\nu} - \frac{1}{2} g_{\mu\nu} R = 8\pi T_{\mu\nu}$$

and

$$T_{\mu\nu} = \partial_\mu \phi\, \partial_\nu \phi - \frac{1}{2} g_{\mu\nu} g^{\alpha\beta} \partial_\alpha \phi\, \partial_\beta \phi.$$

The integrability condition is the conservation law of the energy momentum tensor, which is equivalent to

$$\Box_g \phi = 0.$$

In the present case, the equations reduce to a single nonlinear evolution equation. For any function, $f(u, r)$, let $\bar f(u, r)$ denote the mean value function of f with respect to r:

$$\bar f(u, r) = \frac{1}{r} \int_0^r f(u, r')\, dr'.$$

I denote by h the principal unknown function. Let

$$g = \exp\left[-4\pi \int_r^\infty (h - \bar h)^2\, \frac{dr}{r} \right]$$

and

$$D = \frac{\partial}{\partial u} - \frac{1}{2} g \frac{\partial}{\partial r}.$$

Then the nonlinear evolution equation is

$$Dh = \frac{1}{2r}(g - \bar{g})(h - \bar{h}).$$

Let h be a solution of this equation that is regular on the central line. Then ν, λ, ϕ, defined by

$$e^{\nu+\lambda} = g, \quad e^{\nu-\lambda} = \bar{g}, \quad \phi = \bar{h},$$

satisfies the Einstein equations. Conversely, if ν, λ, ϕ is a spherically symmetric solution of the Einstein equations, then h defined by

$$h = \frac{\partial}{\partial r}(r\phi)$$

satisfies the nonlinear evolution equation. If the assumption of regularity on the central line is dropped, then the equivalence with the Einstein equations holds if and only if the mass equation (see below) is imposed as an additional condition.

The evolution law of $\bar{h}$ along the integral curves of D is

$$D\bar{h} = \frac{\xi}{2r},$$

where

$$\xi = \int_0^r \bar{g}(h - \bar{h}) \frac{dr}{r}$$

represents the local radiative amplitude. We define the local mass, m, by

$$m = \frac{r}{2}(1 - \bar{g}/g).$$

m is a nonnegative increasing function of r at each u representing the mass, which at retarded time, u, is enclosed within the sphere of radius, r. The mass equation is

$$Dm = -\frac{4\pi r^2}{g}(D\bar{h})^2 \quad \text{or} \quad Dm = -\frac{\pi}{g}\xi^2.$$

The total (Bondi) mass, M, is

$$M = \lim_{r \to \infty} m(u, r)$$

and we have

$$M = \frac{1}{2}\int_0^\infty (1 - g)\,dr.$$

The problem is covariant under scaling group, $(u, r) \to (u/a, r/a), a > 0$; that is, if h is a solution, then h' defined by $h'(u, r) = h(u/a, r/a)$ is also a solution and $M'(u) = aM(u/a)$.

The characteristics of the problem are the integral curves of D and they satisfy the ordinary differential equation,

$$\frac{dr}{du} = -\frac{1}{2}\bar{g}(u, r).$$

The characteristics have the property that they converge toward the future:

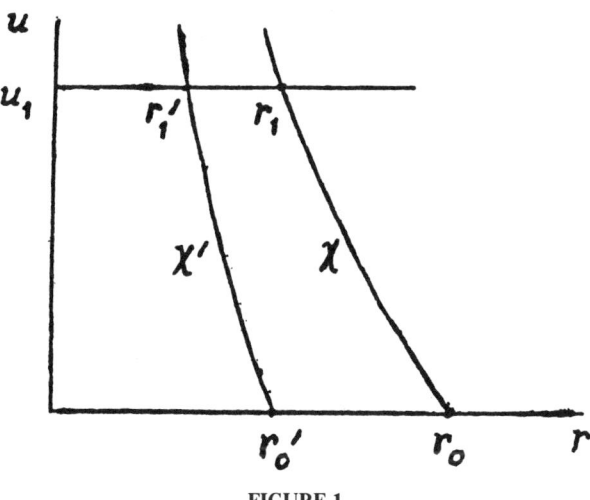

FIGURE 1

The convergence factor is

$$\zeta(u_1, r_1) = \lim_{r_1' \to r_1} \frac{r_1 - r_1'}{r_0 - r_0'} = \exp\left\{-\frac{1}{2}\int_0^{u_1}\left[\frac{1}{r}(g - \bar{g})\right]_\chi du\right\}.$$

Integrating the mass equation along a characteristic χ, we obtain the mass-flux relation:

$$m(u_1, r_1) + \mu \int_0^{u_1}\left[\frac{\xi^2}{g}\right]_\chi du = m(0, r_0)$$

Let us denote by M_1 the final total mass:

$$M_1 = \lim_{u \to \infty} M(u).$$

Proposition: For $r_0 > 2M_1$, the timelike lines of $r = r_0$ are complete toward the future.

I take $u = 0$ as the initial hypersurface and I consider initial data, $h(0, r) \in C^1[0,\infty[$, of finite initial total mass, M_0. In part, I consider classical solutions. A classical solution is a differentiable function, $h(u, r)$, whose derivatives are continuous even on $r = 0$. I first show the local, in retarded time, existence and global uniqueness of classical solutions and then I prove:

Theorem (global existence and asymptotic behavior of classical solutions for small initial data): We consider initial data, $h(0, r) \in C^1[0, \infty[$, such that $h(0, r) = O(r^{-3})$ and $\partial h/\partial r = O(r^{-4})$. We denote:

$$d_0 = \inf_{a>0} \sup_{r \geq 0} \left\{ \left(1 + \frac{r}{a}\right)^3 |h(0,r)| + \left(1 + \frac{r}{a}\right)^4 \left|a \frac{\partial h}{\partial r}(0, r)\right| \right\}.$$

Then there exists a $\delta > 0$ such that if $d_0 < \delta$, there exists a global classical solution, $h(u, r) \in C^1([0, \infty[\times [0, \infty[)$, of the nonlinear evolution equation taking at $u = 0$ the given data. This solution has the decay property of

$$|h(u, r)| \leq C(1 + u + r)^{-3}, \qquad \left|\frac{\partial h}{\partial r}(u, r)\right| \leq C(1 + u + r)^{-4}.$$

The corresponding space-time is timelike and null geodesically complete toward the future and $M_1 = 0$.

PART II

I then proceed to study the global problem for arbitrarily large initial data. For such data, there may not exist a classical solution for all retarded times. I therefore introduce an appropriate concept of generalized solution. Let Q denote the complement of the central line:

$$Q = \{(u, r) \mid 0 \leq u < \infty, 0 < r < \infty\}.$$

Definition: A global generalized solution of the problem is a function, $h \in C^1(Q)$, such that at each u, h belongs to $L^2(0, \infty)$, and $\int_0^\infty h^2 \, dr$ is bounded by a continuous function of u, having the following properties: h satisfies the nonlinear evolution equation,

$$Dh = \frac{1}{2r}(g - \bar{g})(h - \bar{h})$$

in Q, with $\bar{h}, g, \bar{g}$ being continuous in Q, and

$$h(0, r) = h_0(r) : \text{given initial data}.$$

Also, at each u, $g/\bar{g}$ belongs to $L^1(0, r_0)$, r_0 arbitrary. Furthermore, for almost all u,

$$\xi = \lim_{\delta \to 0} \int_\delta^r \bar{g}(h - \bar{h}) \frac{dr}{r}$$

exists and $g^{1/2}\xi/\bar{g}r^{1/2} \in L^2((0, u_0) \times (0, r_0))$, u_0, r_0 arbitary. In addition, $\bar{h}$ is weakly differentiable in Q and

$$D\bar{h} = \frac{\xi}{2r}$$

and m is weakly differentiable in Q and

$$Dm = -\frac{\pi}{g}\xi^2.$$

Finally, for each $(u_1, r_1) \in Q$, the main integral identity,

$$\int_0^{r_1} \frac{g}{\bar{g}}(u_1, r)\, dr + 2\pi \iint_{Q(u_1, r_1)} \frac{g\xi^2}{\bar{g}^2 r}\, drdu + \frac{1}{2}\int_0^{u_1} g(u, 0)\, du = \int_0^{r_0} \frac{g}{\bar{g}}(0, r)\, dr,$$

holds where:

$$Q(u_1, r_1) = \{(u, r) | 0 < r < \chi_{u_1}(u; r_1), 0 < u < u_1\}, \qquad r_0 = \chi_{u_1}(0; r_1).$$

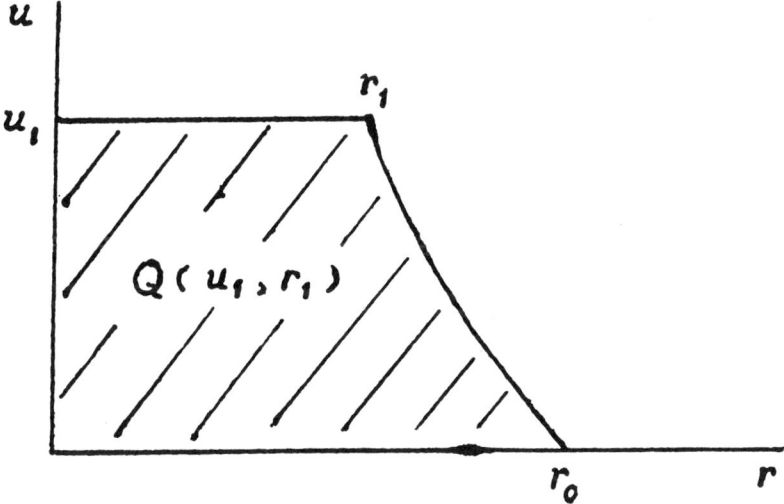

FIGURE 2

(Through each $(u_1, r_1) \in Q$, there passes a unique characteristic.) I prove:

Theorem: For each initial data, $h_0 \in C^1 [0, \infty[$, of finite initial total mass, M_0, there exists a global generalized solution of the problem.

I then study the structure of the solutions. The metric function, g, is continuous with respect to r even at $r = 0$. $g(u, 0)$ is a measurable function of u, with $0 \leq g(u, 0) \leq 1$. We can define the measure, $g(u, 0)du$, which represents proper time duration on the central line. The function, $g(u, 0)$, may also vanish for a set of values of u of nonzero du measure. Such a set will correspond to a set of events on the central line whose duration as observed from infinity is nonzero, but whose proper time duration is zero. The tangent of each characteristic is absolutely continuous in the parameter, u.

If at a certain value of u, $g^{1/2} \xi/\bar{g} r^{1/2} \in L^2(0, r_0)$ for some (and therefore for all) $r_0 > 0$, I call the corresponding light cone, "regular." On the other hand, if the contrary is true, I call the corresponding light cone, "singular."

Corollary: The set of singular cones is of zero measure.

Proposition: On each regular cone, $\xi/\bar{g}^{1/2}$ is a continuous and uniformly bounded function of r, such that $\xi/\bar{g}^{1/2} \to 0$ as $r \to 0$. Furthermore,

$$\sup_{r \geq 0} \left| \frac{\xi}{\bar{g}^{1/2}} (u, r) \right| \in L^2(0, u_0), u_0 \text{ arbitrary.}$$

Proposition: On a regular cone, $\bar{g}^{1/2} \bar{h}$ tends to a limit, f, as $r \to 0$. The function, $f(u)$, which is thus defined for almost all u, belongs to $L^2(0, u_0)$, u_0 arbitrary.

The sectional curvatures of planes tangential to the light cones are continuous everywhere in the complement of $r = 0$. On the other hand, the sectional curvatures of planes transversal to the light cones may blow up at the singular cones. However, at each $r > 0$, some belong to $L^2(0, u_0)$ and some to $L^1(0, u_0)$, u_0 arbitrary, as functions of u.

For each regular cone, we can define the total radiative amplitude,

$$\Xi = \lim_{r \to \infty} \xi = \lim_{\delta \to 0} \int_\delta^\infty \bar{g} (h - \bar{h}) \frac{dr}{r},$$

where we have $\Xi \in L^2(0, \infty)$. Let also

$$N = \int_0^\infty h \, dr,$$

so we have

$$\frac{dN}{du} = \frac{1}{2} \Xi$$

and

$$\frac{dM}{du} = -\pi \Xi^2.$$

Thus, the total (Bondi) mass is an absolutely continuous function of u.

I also prove the following uniqueness theorem:

Theorem: A generalized solution having the same data as a classical solution coincides with it in the domain of existence of the latter.

PART III

In the final part, I study the asymptotic behavior of the global generalized solutions as the retarded time $u \to \infty$. I demonstrate that when the final Bondi mass, M_1, is different from zero, as $u \to \infty$, a black hole forms of mass, M_1, surrounded by vacuum. I first prove:

Theorem: For $r > 2 M_1$, $M(u) - m(u, r) \to 0$ as $u \to \infty$; that is, the mass remaining outside each sphere of radius greater than $2M_1$ tends to zero as u tends to infinity. The main steps in establishing this theorem are the following lemmas:

Lemma: For each $r_0 > 2M_1$ and $\epsilon > 0$,

$$\sup_{r \geq r_0} \left\{ r^{3-\epsilon} \left| \frac{\partial h}{\partial u}(u, r) \right| \right\} \to 0 \quad \text{as} \quad u \to \infty;$$

Lemma: $N(u) \to 0$ as $u \to \infty$.

As a consequence of the above theorem, I obtain:

Corollary: At each $r \neq 2M_1$,

$$g \to g_1 = \begin{cases} 1 & \text{for} \quad r > 2M_1 \\ 0 & \text{for} \quad r < 2M_1 \end{cases} \quad \text{as } u \to \infty$$

pointwise, uniformly in each $[0, r_1] \cup [r_2, \infty[$, $r_1 < 2M_1$, $r_2 > 2M_1$. Also

$$\bar{g} \to \bar{g}_1 = \begin{cases} 1 - 2M_1/r & \text{for} \quad r > 2M_1 \\ 0 & \text{for} \quad r \leq 2M_1 \end{cases} \quad \text{as } u \to \infty$$

uniformly in r:

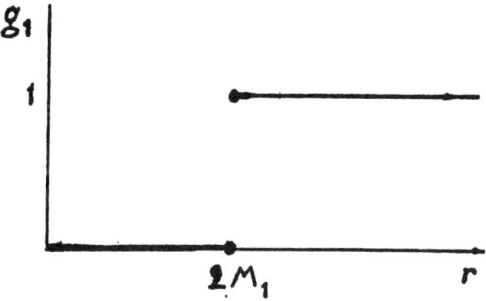

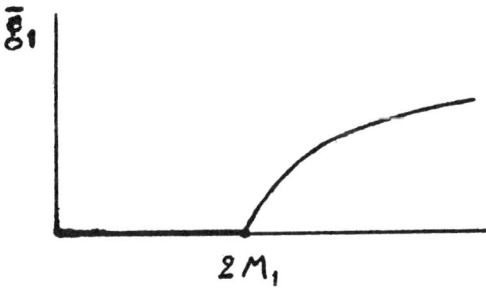

FIGURE 3

I then establish the formation of the event horizon:

Theorem: In the interval, $[0, 2M_1[$, there is a continuous increasing function, $u_0(r)$, such that in the region,

$$\{(u, r) \mid u \geq u_0(r), r \in [0, 2M_1[\},$$

we have:

$$g(u, r) \leq e^{-[u-u_0(r)]/32M_1}.$$

For each $r_1 \in [0, 2M_1[$, the timelike lines of $r = r_1$ are incomplete and their proper length, $T(r_1)$, is a continuous increasing function in $[0, 2M_1[$. Also for each $r_1 \in]0, 2M_1[$, there is a unique characteristic, χ_{r_1}, asymptotic to the line, $r = r_1$, as $u \to \infty$. (As $r_1 \to 2M_1$, $u_0(r_1)$ and $T(r_1) \to \infty$.)

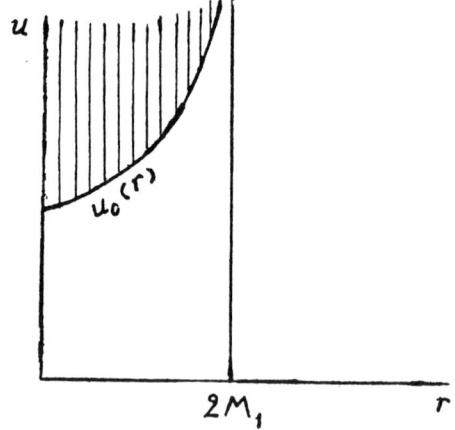

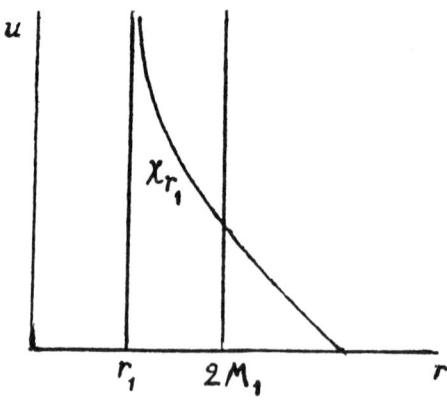

FIGURE 4

The limiting hypersurface, $u = \infty$, has therefore the following meaning: The part, $r > 2M_1$, represents future timelike infinity. The part, $r < 2M_1$, is the future event horizon. The point, $r = 2M_1$, is the point at infinity on the horizon.

I finally study the behavior of the scalar field on the horizon:

Theorem: At each $r \in\;]0, 2M_1[$,

$$\bar{h} \to \bar{h}_1, \quad h \to h_1, \quad \partial h/\partial r \to \partial h_1/\partial r \quad \text{as } u \to \infty$$

pointwise, uniformly in each compact subinterval of the interval, $]0, 2M_1[$. h_1 is a continuously differentiable function on the interval, $]0, 2M_1[$, and $h_1 \in L^2(0, r_1)$ for each $r_1 < 2 M_1$, but $h_1 \notin L^2(0, 2M_1)$. $\bar{h}_1$ is the mean value function of h_1:

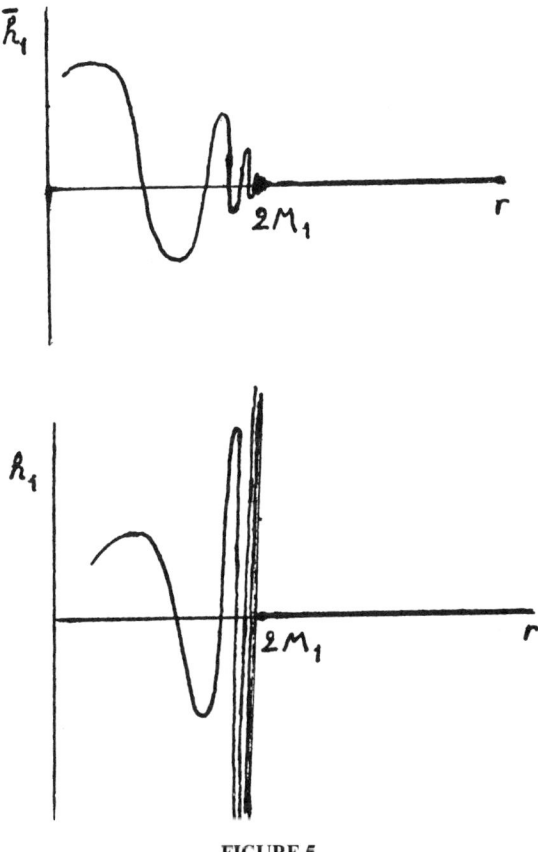

FIGURE 5

ACKNOWLEDGMENTS

The derivations of the results reported here are contained in a series of three articles to appear in *Communications in Mathematical Physics*.

PART V. DATA FROM SATELLITES

The IRAS View of the Extragalactic Sky[a]

B. T. SOIFER

Division of Physics, Mathematics, and Astronomy
California Institute of Technology
Pasadena, California 91125

INTRODUCTION

The Infrared Astronomy Satellite (IRAS) performed the first large-scale all-sky survey at wavelengths from 12 µm to 100 µm with sufficient sensitivity to detect significant numbers of extragalactic objects. The overall mission is described by Neugebauer *et al.* (1984)[1] and the first scientific results from IRAS were reported in the accompanying papers. Since then, a significant amount of new work has been done to further explore the IRAS data. In this paper, I will attempt to summarize the study of the extragalactic sky as viewed by IRAS. It should be emphasized at the outset that the exploration of the IRAS data has really just begun and the work that is described here has barely scratched the surface of the IRAS data.

OBSERVATIONS OF NORMAL GALAXIES

In the closest galaxies, the IRAS spatial resolution of ~1' can be used to identify the locations of the infrared emission. FIGURE 1 (from Habing *et al.*, 1984)[2] shows the 60-µm image of the nearby giant spiral galaxy, M31, from the IRAS data and it also shows images of this quite normal galaxy at other wavelengths. There are two significant structures in the infrared, corresponding to the nucleus and the large dust ring at about 9 kiloparsecs from the center. The nuclear emission is a very small fraction (~1%) of the optical emission from the same volume and is due to the heating of a small amount of interstellar dust by the intense stellar radiation field of the central bulge of the galaxy.

The infrared ring corresponds to a similar ring in atomic hydrogen, in nonthermal radio emission, and in interstellar dust. This emission is associated with the active star-forming regions in this galaxy. It appears that in M31 the bulk of the infrared emission is localized to the sites where young stars are now forming or have recently formed. This is not surprising and it reflects the colocation of the gas available to become stars, the dust that converts stellar radiation into infrared radiation, and the young, luminous stars that are the most recent generation of stars in this galaxy.

In other galaxies where the IRAS spatial resolution is sufficient to locate the sites of the infrared emission, e.g., M33, it has been found that the infrared emission is also localized to regions of active ongoing star formation (Rice *et al.*, 1985).[3]

[a]This work was supported through the Jet Propulsion Laboratory under contract with NASA.

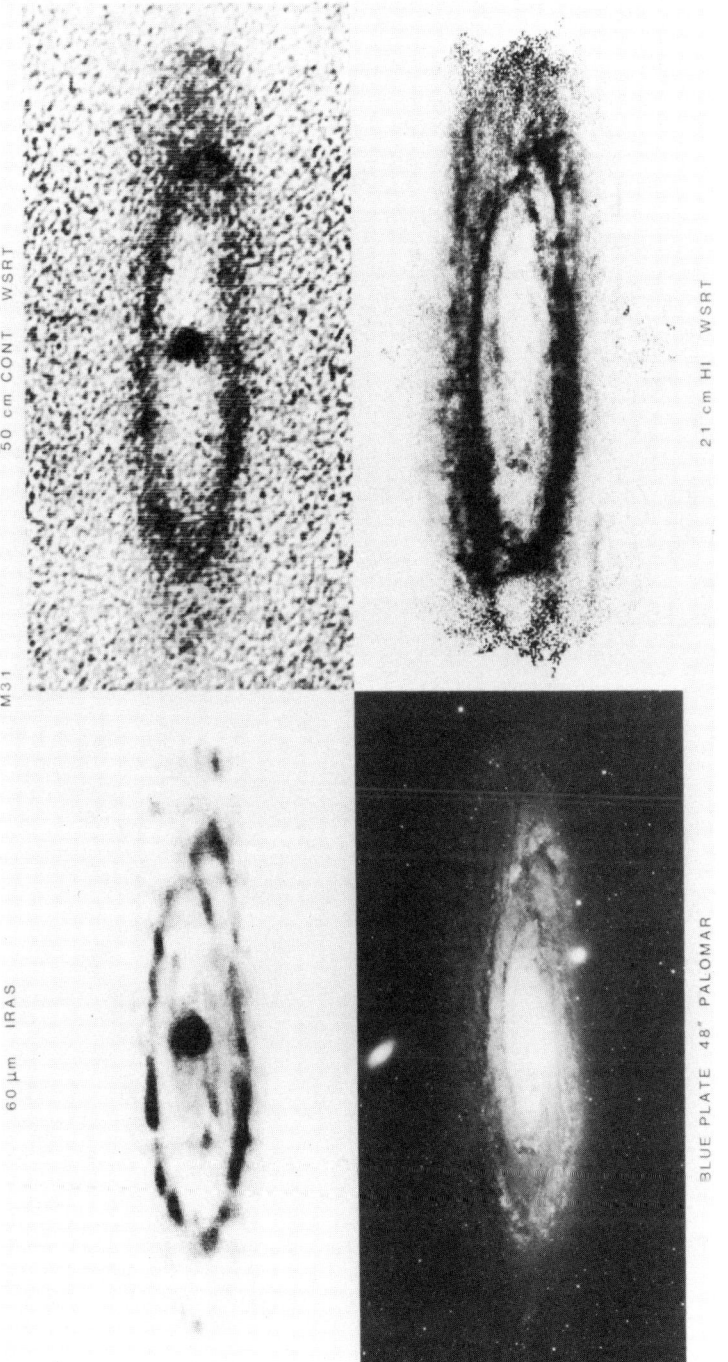

FIGURE 1. Images of M31 at four wavelengths from Habing et al., 1984.[1] The upper left image is the IRAS 60-μm image; the upper right is an image of the 50-cm continuum radiation that traces the synchrotron emission in the galaxy. The lower left image is a photograph in blue light, while the image in the lower right is in the 21-cm line of atomic hydrogen.

GLOBAL PROPERTIES OF INFRARED EMISSION IN GALAXIES

The study of the global properties of the infrared emission from galaxies is consistent with the picture developed from the closest galaxies. FIGURE 2 (from de Jong et al., 1984)[4] shows a histogram of the IRAS detections of galaxies broken down by galaxy morphological type. As is consistent with the observations of the closest galaxies, the brighter infrared galaxies are the dustier, i.e., later type, galaxies.

While the observations of the optically bright galaxies by IRAS has confirmed our previous prejudices regarding the origin of infrared radiation in normal galaxies, the study of large samples of galaxies has shown that the infrared activity of galaxies is quite varied. This is illustrated in FIGURE 3 where the fraction of galaxies having a measured value of far infrared to blue luminosity ratio is plotted. The ratio of far

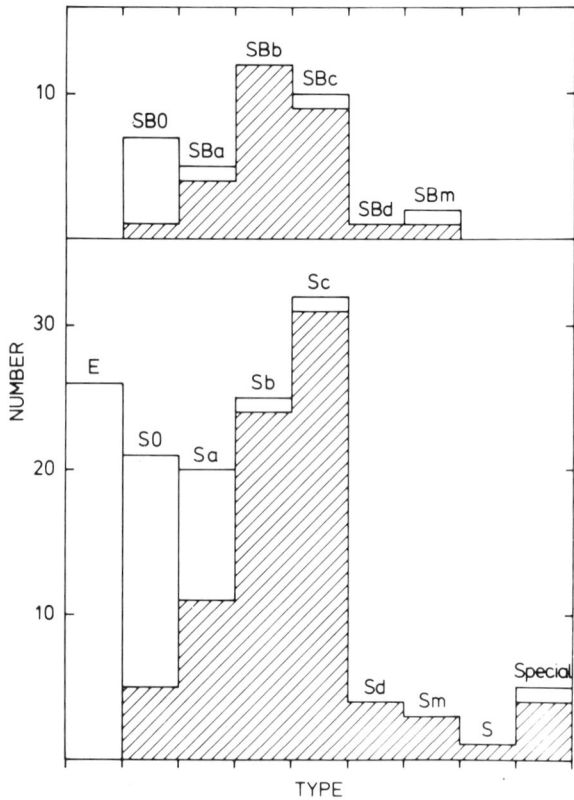

FIGURE 2. The distribution of morphological types of galaxies in the sample of de Jong et al. (1984).[4] The white areas represent the galaxies scanned by IRAS, while the hatched areas are those galaxies detected by IRAS. The histograms are separated for barred (top) and unbarred (bottom) galaxies. Note the successively higher detection rate for galaxies by IRAS with later galaxy type.

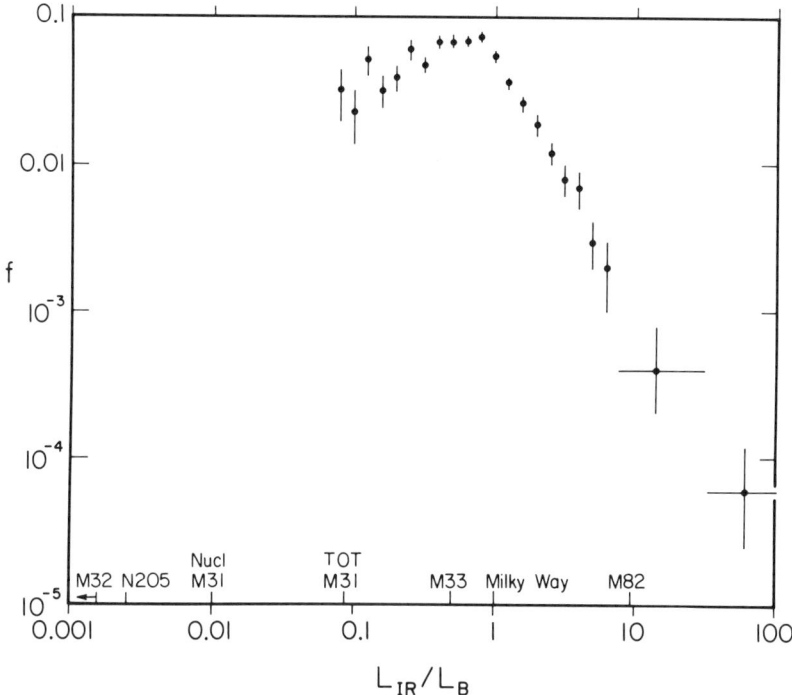

FIGURE 3. The distribution of the fraction of galaxies having a measured value of infrared to blue luminosity ratio given by the abscissa for the sample of galaxies brighter than $m_b = 14.5$ mag in the UGC catalog (Nilson, 1973).[5] For reference, the measured values of this ratio for several nearby, well-known galaxies are also indicated.

infrared to blue luminosity (hereafter denoted as IR/B) is a useful measure of the infrared activity of a galaxy since it normalizes the infrared luminosity to the stellar luminosity of the galaxy and is independent of distance. This plot was generated from the data for galaxies in the UGC catalog (Nilson, 1973)[5] that were brighter than $m_b = 14.5$ mag. In addition to the fractional distribution, the locations of certain individual galaxies are also indicated in the figure.

The first point of significance from this figure is the large range spanned in IR/B ratio. For galaxies shown in this plot, the range in this parameter is almost five orders of magnitude, i.e., ranging from 10^{-3} to 10^2. Thus, we see that galaxies span the entire range from producing virtually no infrared luminosity to producing all their luminosity in the infrared.

The "normal" galaxies that are typified by spiral galaxies occupy the range of IR/B ratio of ~0.1 to ~1. The large majority of the galaxies fall in this range. The lack of detected galaxies at IR/B less than ~0.1 is simply an indication of the sensitivity limits of the IRAS survey. The total fraction of the galaxies detected from this complete optical sample, ~80% when corrected for the IRAS sensitivity limits, is consistent with the previous conclusion that IRAS is detecting the dusty galaxies, and those galaxies devoid of dust, i.e., the elliptical galaxies, are not detected in the IRAS survey.

The galaxies with values of IR/B > 1 are seen to be increasingly rare, but are quite interesting galaxies. The fraction of galaxies with a given IR/B ratio, $f(IR/B)$, drops with a power law, $n \propto f^{-1.7}$, for $f > 1$. The galaxies having $f > 3$ are quite rare, comprising only 2% of the galaxies brighter than $m_b = 14.5$ mag. These are perhaps the most interesting of the galaxies detected in the IRAS survey.

FIGURE 4 shows a plot of the IR/B ratio for galaxies as a function of the 100-μm to 60-μm flux density ratio (or effectively the color temperature of the far infrared radiation emergent from the galaxy). While the scatter in these data is quite large, there is a trend, first pointed out by de Jong et al. (1984),[4] for the galaxies that are more active in the infrared (as measured by larger IR/B ratios) to have warmer color temperatures. Since the infrared emission is undoubtedly optically thin, this suggests that the more active infrared galaxies are producing more luminosity per unit volume

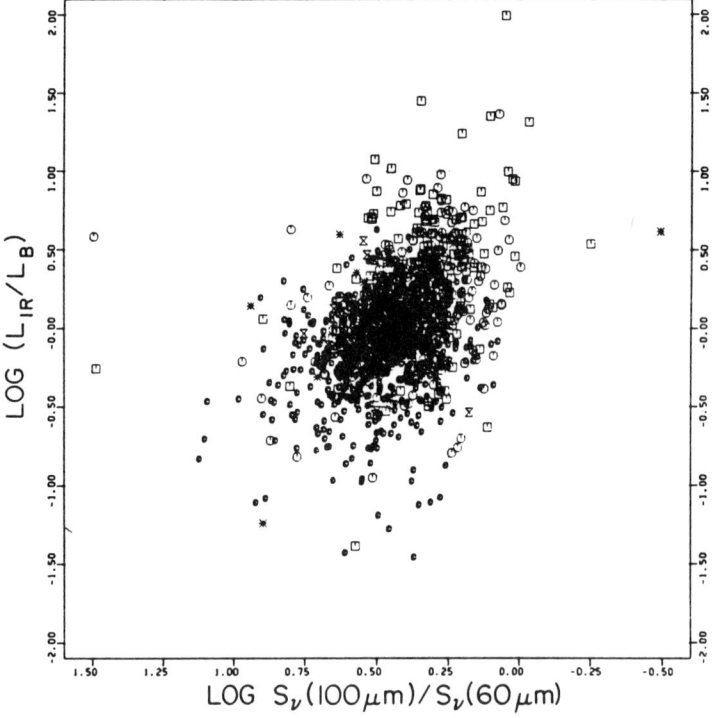

FIGURE 4. The value of the infrared to blue luminosity ratio is plotted versus the 100-μm to 60-μm flux density ratio for the galaxies detected from the IRAS survey and brighter than $m_b = 14.5$ mag in the UGC catalog (Nilson, 1973).[5] Note that decreasing the 100-μm to 60-μm flux density ratio corresponds to increasing the radiation color temperature, with the value of ten for this ratio corresponding to a temperature of ~23K and the value of one corresponding to a temperature of ~60K.

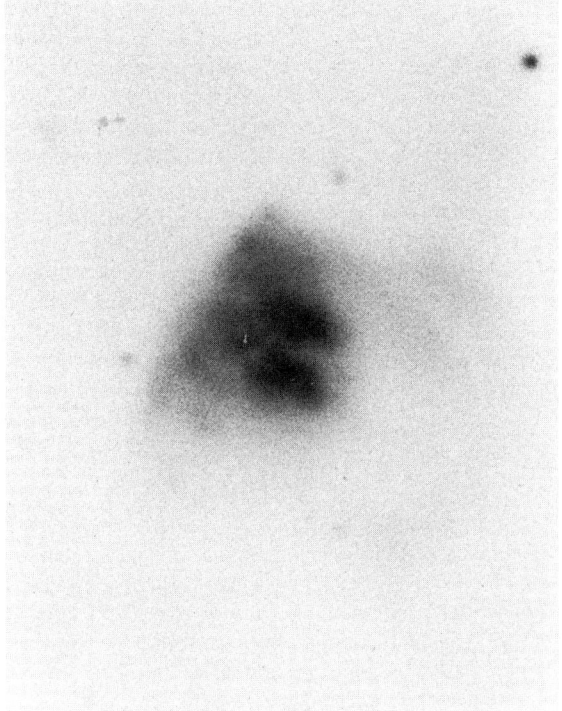

FIGURE 5. An optical photograph of the interacting galaxy Arp220 (c. 1966, California Institute of Technology). This galaxy was found to be one of the brightest galaxies in the sky in the IRAS survey.

in order to increase the energy density of the absorbed radiation field and hence the temperature of the emitting grains. Whether this is a result of more active star formation in these galaxies, or the increasing dominance of a central luminosity source in these galaxies, is a question that can only be resolved by observations of the galaxies that are among the most extreme infrared active galaxies.

The extremely active infrared galaxies, where IR/B > 10, are certainly among the most interesting objects found in the IRAS survey. FIGURE 5 shows an optical photograph of Arp 220, one of the extreme infrared active galaxies discovered in the IRAS survey and one of the brightest infrared emitting galaxies in the sky (Soifer *et al.*, 1984).[6] This galaxy illustrates many of the features of the extreme infrared galaxies. First is its enormous luminosity. Its total luminosity of $2 \times 10^{12} L_\odot$ places it in the range of the quasars' energy output. Further, the faint wisps seen in the figure are quite suggestive of the remnant of a merger between two galaxies. Many of the more active infrared galaxies discovered in the IRAS survey are apparently such interacting galaxies (Lonsdale *et al.*, 1984).[7] In addition, while the image shown here is suggestive of a double nucleus, more recent CCD imagery of Arp 220 (Danielson, Lonsdale, and Soifer, 1985)[8] suggests rather an extensive dust lane cutting across the galaxy, which explains why the luminosity is emerging in the infrared, i.e., the luminosity source is behind a large column density of dust that is optically virtually opaque.

While Arp 220 is by no means a commonly occurring galaxy, it is also not unique in the IRAS survey. Houck et al. (1984)[9] reported the discovery of a class of objects in the IRAS survey that had the infrared properties of galaxies, but had no optical counterparts on the Palomar photographic sky survey. Deep imaging and spectroscopy reported by Houck et al. (1985)[10] and Aaronson and Olszewski (1984)[11] have demonstrated that these sources are indeed galaxies having luminosities as great or greater than that of Arp 220, but with infrared to blue luminosity ratios in excess of 100, ranging up to values of 300. These galaxies appear to have characteristics both of enormously active star-forming galaxies and galaxies having at least some Seyfert-like activity. The relation between the active nuclei and the star formation activity, along with the ultimate origin of the activity in these galaxies, will await the detailed study of larger samples of such galaxies that are available with the recent release of the IRAS catalog.

CONCLUSIONS

The unbiased, high sensitivity survey of the sky performed by the Infrared Astronomy Satellite has truly opened a new window on the universe. As has been the case with all previous major astronomical surveys, many new discoveries have been made in this survey. Among the most exciting of these discoveries has been the identification of a new class of extremely luminous galaxies that produce 99% or more of their luminosity in the infrared. The understanding of these galaxies will await full and complete observations using all the tools currently available to observational astronomers, and perhaps even platforms not yet available, such as ST and SIRTF.

ACKNOWLEDGMENTS

It is a pleasure to thank my many colleagues on the IRAS science team whose work I have freely quoted in this review. I want to particularly thank the many workers at the IRAS Science Data Analysis System, without whose diligent and dedicated efforts the IRAS catalogs and atlases would still be but a gleam in the eye of the IRAS science team.

REFERENCES

1. NEUGEBAUER, G. et al. 1984. Astrophys. J. Lett. **278**: L1
2. HABING, H. J. et al. 1984. Astrophys. J. Lett. **278**: L59.
3. RICE, W. J. et al. 1985. In preparation.
4. DE JONG, T. et al. 1984. Astrophys. J. Lett. **278**: L67.
5. NILSON, P. Uppsala General Catalog of Galaxies. Act Universitatis Upsaliensis, Nova Regial Societatis Upsaliensis, Senes, V: A, vol. 1.
6. SOIFER, B. T. et al. 1984. Astrophys. J. Lett. **283**: L1.
7. LONSDALE, C. J., G. NEUGEBAUER & B. T. SOIFER. 1984. Bull. Am. Astron. Soc. **16**: 470.
8. DANIELSON, G. E., C. J. LONSDALE & B. T. SOIFER. 1985. In preparation.
9. HOUCK, J. R. et al. 1984. Astrophys. J. Lett. **278**: L63.
10. HOUCK, J. R. et al. 1985. Astrophys. J. Lett. In press.
11. AARONSON, M. & OLSZEWSKI. 1984. Nature **309**: 414.

Recent Results from the Japanese X-Ray Astronomy Satellites

Y. TANAKA

Institute of Space and Astronautical Science
4-6-1 Komaba, Meguro-ku
Tokyo 153, Japan

The Institute of Space and Astronautical Science (ISAS) has thus far launched two X-ray astronomy satellites. The first one, Hakucho, is a small 97-kg satellite launched in 1979.[1] The wide field of view of the rotating modulation collimator of Hakucho provided us with a systematic survey of X-ray burst sources and a long-term monitor of X-ray pulsars as well as other sources of interest. The second one, Tenma, is a 220-kg satellite launched in 1983.[2] The main instrument of Tenma is a large area array of gas scintillation proportional counters (GSPC), which comprises ten units of each 80 cm^2 effective area.[3] GSPC possesses twice the energy resolution of ordinary proportional counters used in X-ray astronomy observations so far. Therefore, this is the first substantial improvement of the energy resolution over a wide energy range in the 20 year history of X-ray astronomy. The third X-ray astronomy mission, Astro C, is presently under preparation for the launch in early 1987.

We have observed various classes of objects from Hakucho and Tenma, and have obtained many interesting results. It is hardly possible to describe these results in any comprehensive manner. Instead, I shall attempt in this paper to outline (i) the investigation of neutron stars and their environment in low-mass binary X-ray sources and (ii) the emission and absorption of iron, based on the observations from Hakucho and Tenma.

OBSERVATION OF NEUTRON STARS THROUGH X-RAY BURSTS

Many galactic X-ray sources exhibit a violent phenomenon called X-ray bursts (for a general review of X-ray bursts, see Lewin and Joss, 1983[4]). An example of a typical X-ray burst is shown in FIGURE 1. Presently, there are about 30 burst sources known in our galaxy, of which 9 were discovered and 4 were established by Hakucho. The burst sources are highly concentrated towards the galactic center, with 50% of them located within only ten degrees of the galactic center. There is enough evidence for these sources to be low-mass binary systems with an accreting neutron star.

These X-ray bursts are currently interpreted as nuclear shell flashes on the neutron star surface (for early works, see Woosley and Taam, 1976;[5] Maraschi and Cavaliere, 1977;[6] Joss, 1978[7]). The spectrum of burst emission is of a blackbody gradually cooling with time and the apparent blackbody radius is found to remain essentially constant at about 10 km during the burst decay.[8–10] Important enough, this fact suggests that the size of a neutron star can be measured through X-ray bursts. As a matter of fact, X-ray

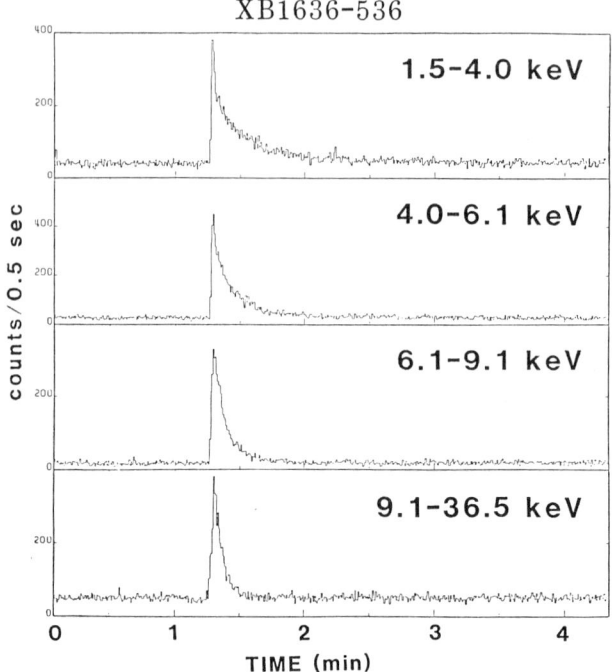

FIGURE 1. Time profile of an X-ray burst from X1636-53 (Tenma). Count rates are plotted in four energy bands. A softening of spectrum during the decay (higher energy part decays faster) evidently shows gradual cooling with time.

burst has been the subject of intensive study with Hakucho and Tenma. In particular, the large area GSPC of Tenma enables us to study the spectral features of burst in much more detail than before.

FIGURE 2 shows examples of burst spectrum measured with the GSPC of Tenma at two temperatures. The observed spectra are in very good agreement with blackbody spectra except for a significant excess above 10 keV, probably indicating the Comptonization by a hotter plasma. With the blackbody temperature, T_b, and the bolometric flux, F_x, observed, one defines the apparent blackbody radius, r_b, by

$$L = 4\pi D^2 F_x = 4\pi r_b^2 \sigma T_b^4, \tag{1a}$$

where D is the distance to the source. It is important to note that T_b is the color temperature and not the effective temperature. For an electron-scattering dominant atmosphere during a burst, the color temperature can be significantly higher than the effective temperature and the flux is reduced by an emissivity factor, ϵ, due to electron scatterings (e.g., see Ebisuzaki et al., 1984[11]). Taking into account the general relativistic effect,

$$4\pi D^2 F_x = 4\pi r_o^2 g^{-2} \sigma T_b^4 \epsilon, \quad (r_b = r_o g^{-1} \epsilon^{1/2}), \tag{1b}$$

where r_o is the true neutron star radius and

$$g^2 = 1 - \frac{2GM}{c^2 r_o}. \quad (2)$$

FIGURE 3 shows the relation between r_b and T_b for a number of bursts observed from X1636-53.[12] This plot clearly reflects the temperature dependence of ϵ, which increases with decreasing temperature.

If ϵ and g are known, one can determine the mass and radius of the neutron star from equations (**1a**), (**1b**), and (**2**), provided the source distance, D, is known. The emissivity factor, ϵ, is to be evaluated by solving the radiative transfer during a burst (e.g., van Paradijs, 1982;[13] Czerny and Sztajno, 1982.[14] These calculations do not take into account the Compton effect, however.) Effort of this line with inclusion of the Compton effect is in progress.

Since X-ray bursts are considered to be the phenomenon on the neutron star surface, a significant general relativistic effect is expected. If a gravitational redshift were observed in some spectral feature, the g-value can be obtained from the redshift factor, $(1 + z) = 1/g$. From equation (**2**), the g-value also gives the mass to radius ratio, M/r_o, independently of the source distance. In fact, we discovered remarkable features in the burst spectra of X1636-53 observed from Tenma in 1983, which were attributable to a gravitationally redshifted absorption line.[15] As shown in FIGURE 4, the three largest bursts and one medium-size burst among a total of twelve observed commonly reveal an absorption line at 4.1 keV during the decay.

If the absorption were caused by iron, which is the most abundant heavy element in

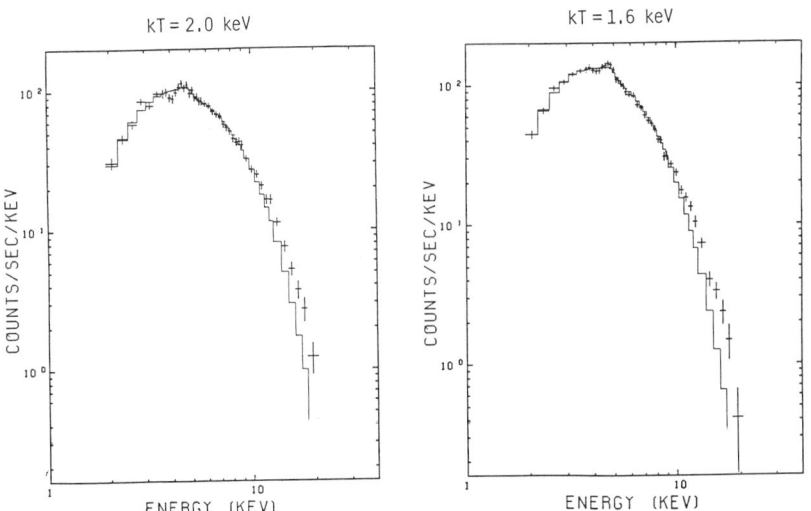

FIGURE 2. Burst spectra observed from X1636-53 at two blackbody temperatures, $kT = 1.6$ keV and 2.0 keV (Tenma). Histograms are the best-fit blackbody spectra. Note a significant excess above 10 keV.

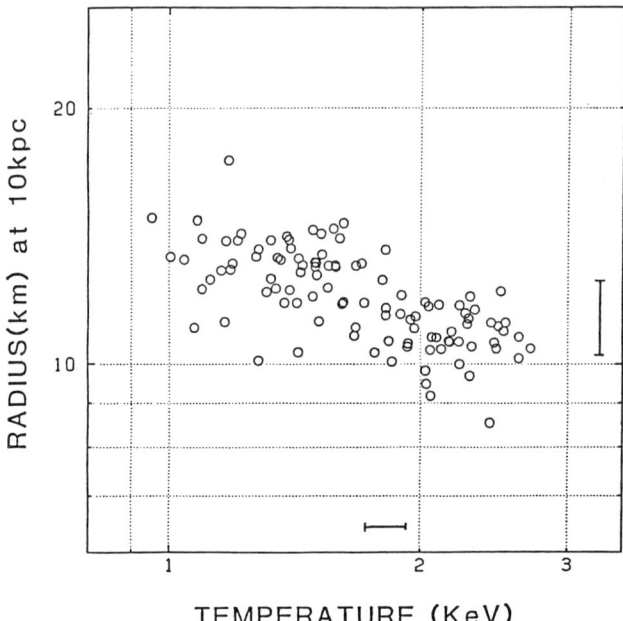

FIGURE 3. Relation between the apparent blackbody radius and the blackbody temperature during the burst decay, plotted for twelve bursts observed from X1636-53.[12] Typical 90% confidence errors are indicated in the figure.

the accreted matter, the gravitational redshift factor would be $(1 + z) = 1.61 \pm 0.04$ or $g = 0.62 \pm 0.02$. On the other hand, as shown in FIGURE 5, this g-value leads to the serious problem that the resulting large value of M/r_o is really at the limit for most of the current theoretical models of stable neutron stars. While the presence of the absorption line is beyond doubt, more observational as well as theoretical examinations are desired in view of the important impact of this result.

Since the rise time of a burst, typically one second, is orders of magnitude longer than the dynamical time scale on the neutron star, the emission during a burst can be regarded as quasi-stationary. Then, the burst peak luminosity would saturate at the Eddington limit. It has recently become evident that there is an upper bound to the burst peak luminosity. Ohashi et al. (1982)[17] noted that the highest five of the peak fluxes among the observed X1636-53 bursts from Hakucho were the same within the errors. Basinska et al. (1984)[18] showed that the peak flux of bursts from X1728-34 qualitatively increases with size, but saturates at a certain value.

The Tenma result shows more detailed feature when the upper bound is reached.[12] FIGURE 6 shows three bursts from X1636-53 with the largest peak flux among the rest. They exhibit a flat top all at the same value for a few seconds during which the blackbody temperature and the apparent blackbody radius undergo a large excursion. This feature is fully consistent with the expected phenomenon when the luminosity reaches the Eddington limit. When it occurs, the radiation pressure will expand the neutron star atmosphere and the temperature will drop. As the radiation pressure starts to decrease, the atmosphere gradually contracts back and the temperature rises accordingly (while maintaining the luminosity constant at the Eddington limit). We therefore believe that such a flat-topped burst peak associated with the atmospheric expansion is the manifestation of the Eddington limit. Very long bursts with a precursor are interpreted likewise, in which the luminosity is kept constant for more than 100 sec.[19,20]

The Eddington limit luminosity, L_E, for a distant observer is given by

$$L_E = \frac{4\pi cGM}{\kappa_o (1 + X)} g = 1.28 \times 10^{38} \left(\frac{M}{1.4 M_\odot}\right)\left(\frac{1.7}{1 + X}\right)\left(\frac{g}{0.62}\right) \text{erg cm}^{-2} \text{sec}^{-1}, \quad (3)$$

where $\kappa_o (1 + X)$ is the Thomson scattering opacity and X denotes the mass fraction of hydrogen. For many burst sources, however, the burst peak luminosity largely exceeds the Eddington limit for a 1.4 solar-mass neutron star, if the distances are about 8 kpc. FIGURE 7 shows the distribution of the burst peak fluxes for five sources all within six degrees of the galactic center, as observed from Hakucho.[21] From the sharp concentra-

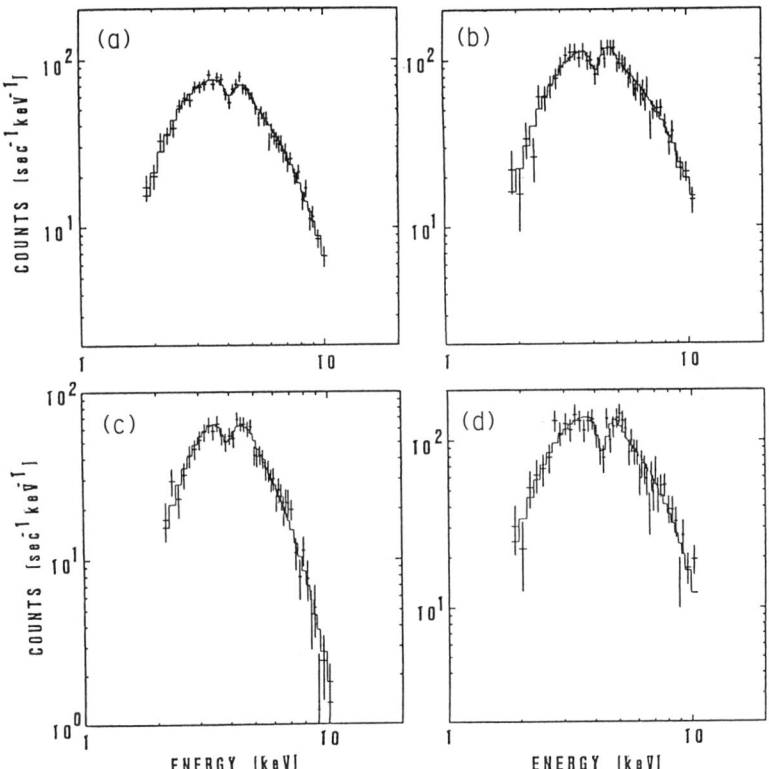

FIGURE 4. Absorption lines observed in the burst spectra during the decay, for four different bursts from X1636-53.[15] The histograms are the best-fit time-averaged blackbody spectra including an absorption line.

tion of burst sources, most of them should be within 1 kpc of the galactic center. The observed maximum peak fluxes all turned out to be in excess of that expected from the Eddington limit by a factor ranging from three to six. Similar results have been obtained for the bursts from the globular cluster sources in NGC 6624[22] and Trz 2.[23]

This large discrepancy is still an unresolved issue. Even a helium-rich atmosphere ($X = 0$) does not remove the discrepancy. One way to resolve it is to consider that the burst peak luminosity can indeed largely exceed the Eddington limit ("super-Eddington" luminosity). For instance, Melia and Joss (1984)[24] suggest that the stellar wind of the neutron star carries the burst energy away and converts it to radiation in the accretion disk. This model assumes the kinetic energy of the wind to be at least 30

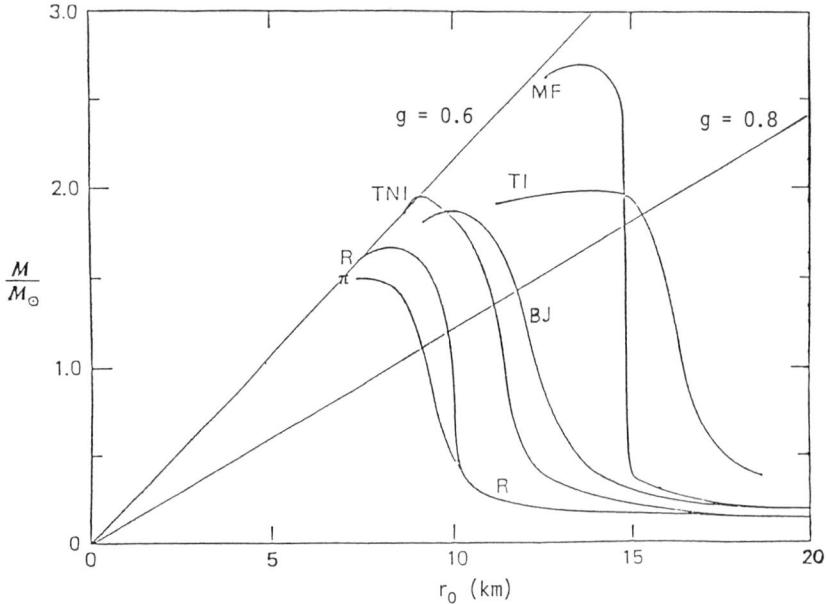

FIGURE 5. Mass versus radius relations for various neutron star models (see Baym and Pethick, 1979[16]). Two lines for $g = 0.6$ and $g = 0.8$ are indicated in the figure.

times the Eddington limit. However, this would cause a severe problem on the amount of energy release in a burst. Apart from this, there is already a concern about the amount of energy release from a nuclear shell flash. Sometimes, the time-averaged burst luminosity becomes as high as 5% of the persistent luminosity as compared to about 1% expected for a helium flash, which is the ratio of nuclear energy release to gravitational energy release. Furthermore, several cases have been observed in which two bursts occurred in succession within ten minutes.[25-27] These correspond to the luminosity ratio of the order 100%, which is hard to explain unless some nuclear fuel reservoir is invoked.

Another way is to consider that the galactic center is substantially closer than 8 kpc. Ebisuzaki et al. (1984)[11] argue from the theoretical basis that the burst peak

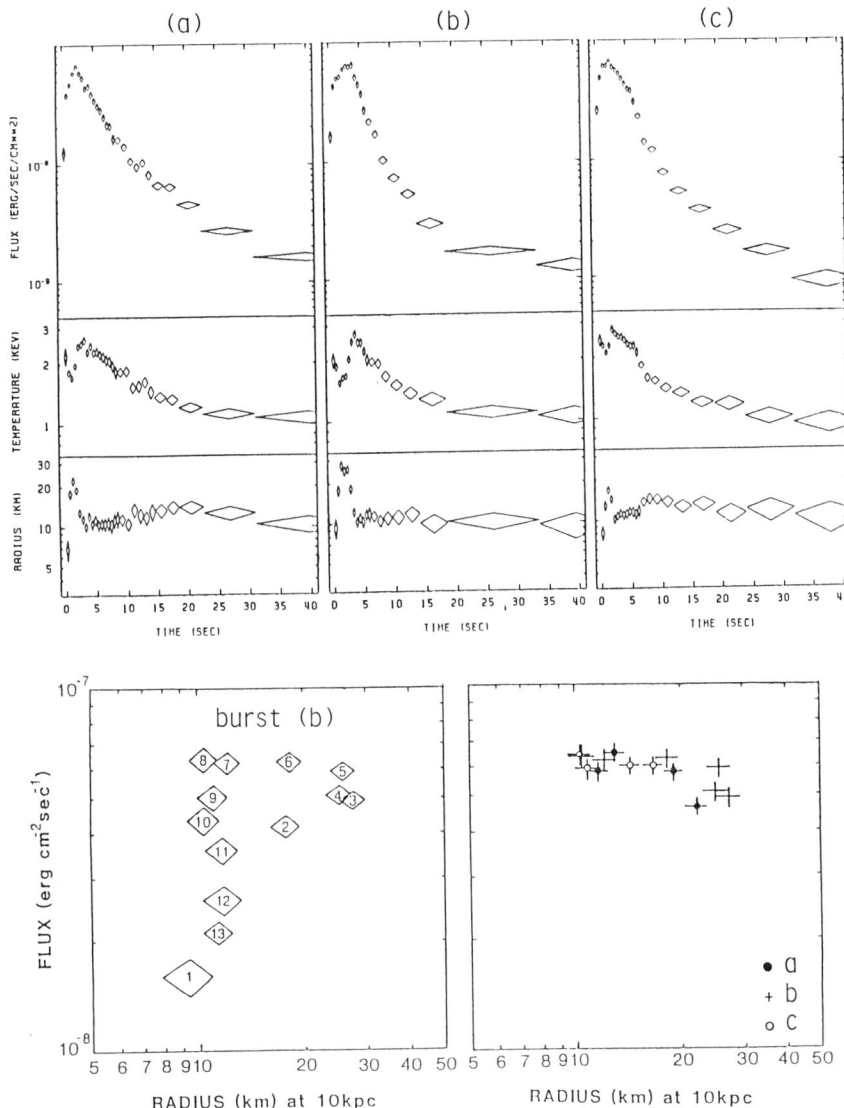

FIGURE 6. Upper panels: Three bursts with the largest peak flux among twelve bursts observed from X1636-53. The bolometric flux, the blackbody temperature, and the apparent blackbody radius are shown respectively as functions of time. Lower panels: The bolometric flux versus the apparent blackbody radius plotted every 0.5 sec, numbered according to the sequence of time (left). The bolometric flux versus the apparent blackbody radius for the peak portions of the three bursts (right). Within statistical uncertainties, the peak luminosities of these bursts are the same and remain constant for a few seconds, while the apparent blackbody radius decreases.[12]

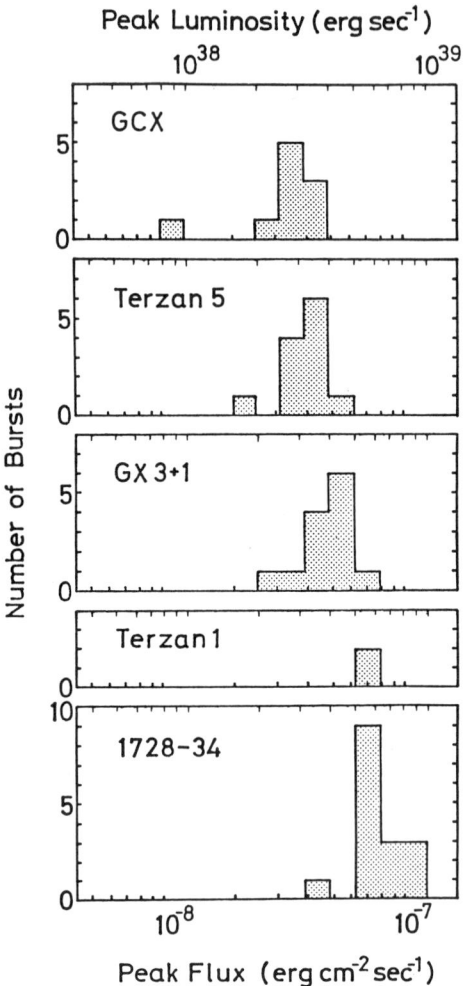

FIGURE 7. Peak flux distributions for the bursts from five burst sources within six degrees of the galactic center, observed from Hakucho.[21] The upper scale indicates the corresponding luminosity for 8 kpc distance.

luminosity for nuclear shell flashes could exceed the Eddington limit only very slightly and they instead suggest that the galactic center is not much farther than 6 kpc.

OBSERVATION OF NEUTRON STARS AND THE ACCRETION DISKS

Nature of the spectrum of nonpulsating low-mass binary X-ray sources (bulge sources) has not been well understood yet. Accreting matter is considered to form an accretion disk that extends close to the neutron star surface because of the lack of an intense magnetic field. In the standard accretion disk model (Hoshi, 1984,[28] and references therein), the accretion disk will be optically thick except for the innermost region wherein the radiation pressure dominates and the disk becomes optically thin.

Matter circulating along the Keplerian orbit gradually falls inward and the gravitational energy released will be equipartitioned to thermal and rotational energies. From the optically thick disk, the thermal energy is efficiently radiated away. On the other hand, though, the emissivity of the innermost optically thin region is very low and the thermal energy released in this region will be transported together with the rotational energy onto the neutron star surface. Therefore, one expects two separate emission regions: (i) the optically thick accretion disk and (ii) the neutron star surface.

The spectrum from the optically thick accretion disk, $F_d(E)$, will be given by the following "multicolor" blackbody spectrum:

$$F_d(E) = (\cos \theta / D^2) \int_{r_{in}}^{\infty} 2\pi r B(E, T(r)) \, dr, \tag{4a}$$

where $B(E,T)$ is the Planckian distribution with temperature, T, θ is the inclination angle of the disk, D is the source distance, and r_{in} is the inner boundary radius of the optically thick disk. Since T is proportional to $r^{-3/4}$ in the optically thick disk, the above equation is rewritten as

$$F_d(E) = (8\pi \cos \theta / 3D^2) \, r_{in}^2 \int_0^{T_{in}} (T/T_{in})^{-11/3} B(E,T) \, dT/T_{in}, \tag{4b}$$

where T_{in} is the temperature at the inner boundary, r_{in}.

On the other hand, the spectrum from the neutron star surface, $F_s(E)$, is expected to be a blackbody spectrum and is expressed by

$$F_b(E) = (S'/D^2) \, B(E, T_b), \tag{5}$$

where S' is the projected area of the emitting surface.

We analyzed the spectra of four low-mass binary sources observed from Tenma: Sco X-1, X1608-52, GX5-1 and GX349+2.[29] These sources showed intensity variations by a factor of two to three on a time scale of a few hours. They all exhibit a common relation that the spectrum becomes harder as the intensity increases. The spectra when the intensity is high are compared with those in the adjacent periods of a lower intensity. For all the four sources, we find that the difference between the high- and low-intensity spectra is invariably a blackbody spectrum with kT of approximately 2 keV, as shown in FIGURE 8. This fact suggests that the spectra of these sources contain a 2-keV blackbody component in themselves and that the intensity variation of this component explains the change of the spectrum.

In fact, the observed spectra of these sources are shown to be decomposed into two spectral components: a 2-keV blackbody component and a softer component, as shown in FIGURE 9. The softer component is found to be best expressed by the "multicolor" blackbody spectrum given by equations (4a) and (4b) with kT_{in} of 1.3 to 1.4 keV. It is striking that the temperatures for these two components are not only constant with time, but also almost the same for different sources. FIGURE 10 shows the intensities of the two components as functions of time. We clearly see that the "multicolor" component is remarkably stable, while the 2-keV component is highly variable.

Based on the accretion disk model, it is very natural to interpret that the "multicolor" component and the 2-keV blackbody component respectively correspond to the emission from the optically thick accretion disk and that from the neutron star surface. The parameter values obtained appear to be consistent with this interpreta-

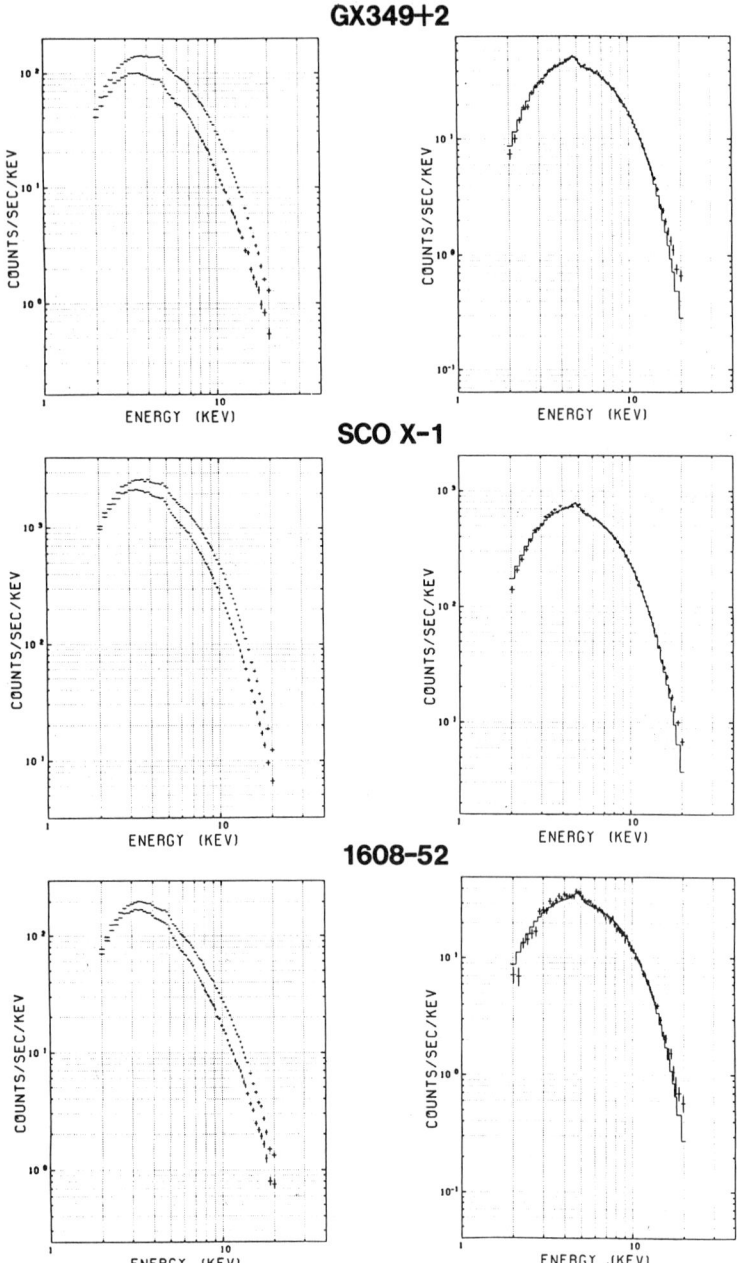

FIGURE 8. Left panels: Examples of the observed pulse-height spectra for three low-mass binary sources, GX349+2, Sco X-1, and X1608-52, at high- and low-intensity levels.[20] Right panels: The difference between the high- and low-intensity spectra for each of the three sources, which is invariably best-fit to a blackbody spectrum with kT of approximately 2 keV (histograms). A slight excess above 15 keV is noted.

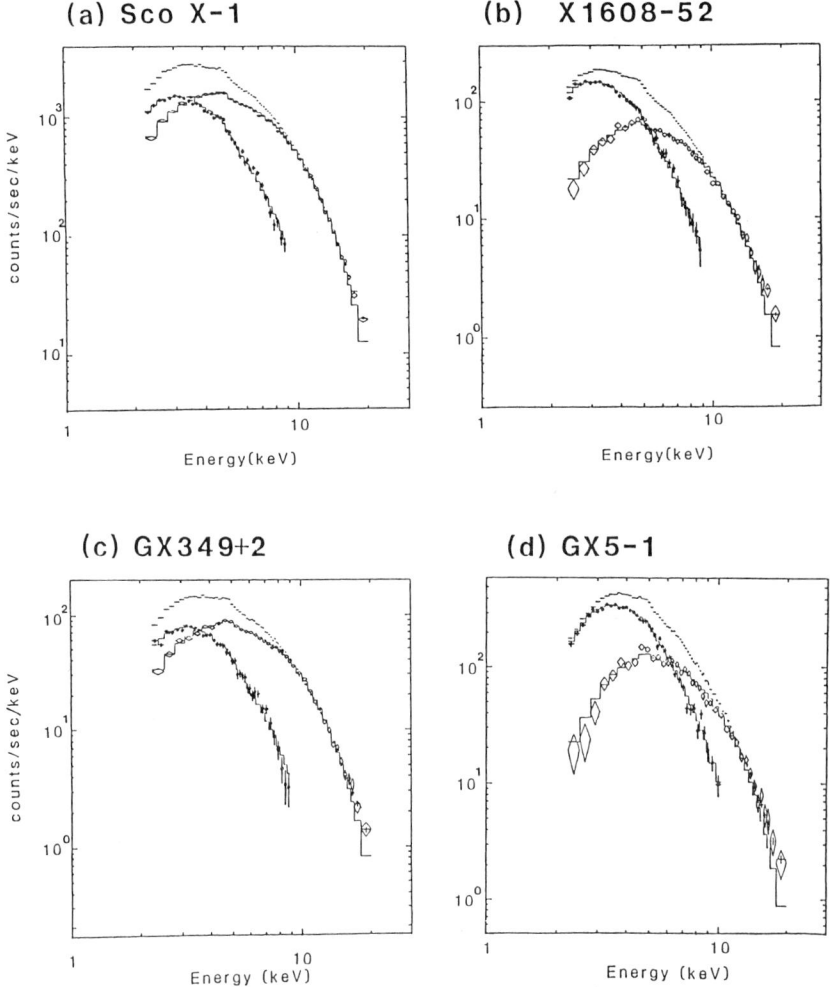

FIGURE 9. Decomposition of the observed spectra for (a) Sco X-1, (b) X1608-52, (c) GX349+2, and (d) GX5-1 into hard and soft components.[29] The hard component is the 2-keV blackbody spectrum as shown in FIGURE 8, whereas the soft component is best expressed by the "multicolor" spectrum expected from an optically thick accretion disk (see text).

tion. For instance, X-ray bursts from X1608-52 give the apparent radius, r_b, of the neutron star at 2 keV. For this r_b, the required inequality relation is indeed satisfied:

$$r' < r_b \simeq r'_{in} < r_{in}, \qquad (6)$$

where

$$r' = (S'/\pi)^{1/2} \text{ and } r'_{in} = r_{in} (\cos \theta)^{1/2}. \qquad (7)$$

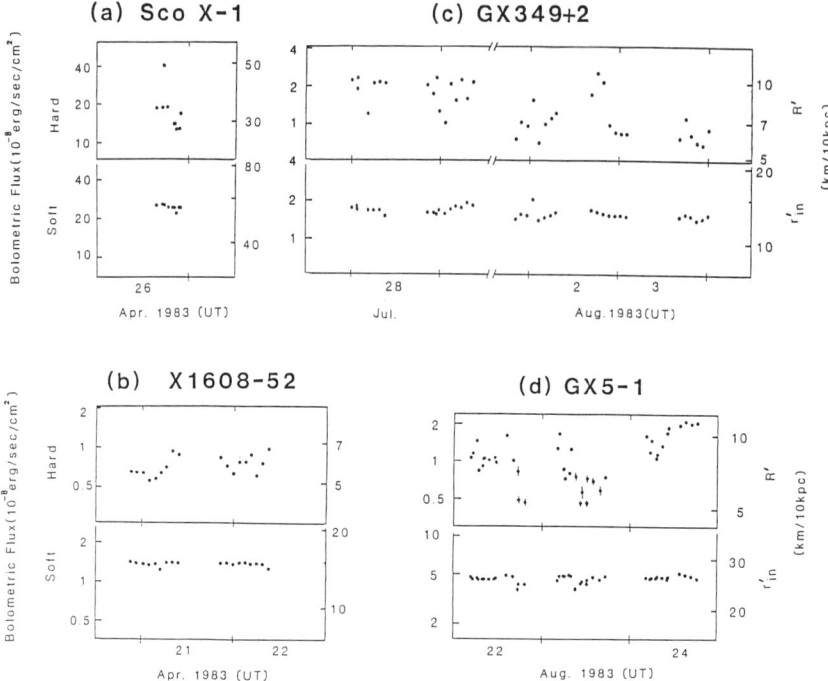

FIGURE 10. Time variations of the 2-keV component and the soft "multicolor" component for (a) Sco X-1, (b) X1608-52, (c) GX349+2, and (d) GX5-1, respectively.

According to the above result, we are now able to observe the emissions from the accretion disk and the neutron star surface separately. The emission from the optically thick accretion disk appears to be remarkably stable, which would indicate a constant accretion rate. On the other hand, though, the large changes in the flux from the neutron star surface manifest a varying flow inside r_{in}.

An interesting case is the transient dips observed from GX5-1.[29] The light curves in FIGURE 11a show some of these dips, which last about a minute. The spectrum during the dip, as shown in FIGURE 11b, is expressed solely by the "multicolor" component. No significant 2-keV component is present. This implies that the flow onto the neutron star surface stopped for a while.

The above results suggest that some kind of instability is disturbing the flow inside r_{in} (and even stops the flow sometimes), while a constant accretion continues. Consequently, the disk could temporarily store matter and serve for a reservoir. This implication might have relevance to the Rapid Burster problem that is discussed later.

AN EXOTIC ACCRETION/THE RAPID BURSTER

The Rapid Burster is a unique source for its peculiar behavior, which is unlike any other source in our galaxy. The Rapid Burster was discovered by Lewin *et al.* (1976)[30]

and so designated because it produces rapidly repetitive bursts. Unlike those bursts discussed earlier, the rapid bursts do not show a significant cooling during the decay. The rapid bursts are interpreted to be the gravitational energy release and to be produced as the result of chopped accretion flow by some instabilities.[31] The Rapid Burster also produces the type of bursts discussed in the first section of this paper [designated by Hoffman et al. (1978)[31] as Type I bursts, while the rapid bursts are designated as Type II bursts], from which the Rapid Burster is also considered to be an accreting neutron star.

The Rapid Burster is located within the globular cluster, Liller 1. The Rapid Burster is a recurrent transient with a period of about six months (though its recurrence is not reliable). When it appears, the rapid burst activity lasts for a few weeks. We observed the rapid burst activity in August 1979 from Hakucho. Since then, in spite of our effort, we did not detect its activity for four years until we found it active in August 1983 and again in July 1984 from Tenma.

One of the striking features of the Rapid Burster is its nature of a relaxation oscillator. The integrated flux (size) of a burst varies largely from burst to burst, but the time interval to the next burst is in proportion to the size of the burst.[30] This distinct characteristic would imply the presence of a reservoir with a fixed capacity somewhere outside the neutron star. However, what serves for the reservoir and what mechanism triggers and stops the flow are yet unknown. From the discussions in the previous section, the accretion disk could store matter and might act as the reservoir.

The mode of the rapid burst activity is also various. One extreme is a fairly periodic repetition of spiky bursts. Another extreme is a train of long flat-topped bursts. Various profiles of the rapid bursts are shown in FIGURE 12. There is a clear tendency that the peak flux of a large-size burst saturates. This effect causes a flat top of burst, sometimes lasting as long as ten minutes. However, this saturation level varies from

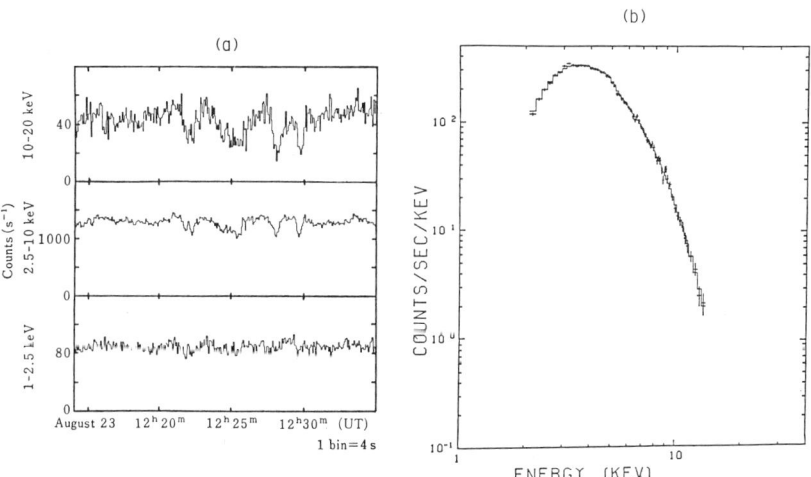

FIGURE 11. (a) Count-rate histograms of GX5-1 in three energy bands observed from Tenma showing transient dips. (b) Pulse-height spectrum during a dip, which is best expressed by the "multicolor" spectrum.[29]

time to time. In 1979, a long train of flat-topped bursts were observed from Hakucho[32] and the saturation level varied by as large as a factor of four.

The persistent component of the Rapid Burster is usually very weak. It shows up in a peculiar way when the interval between bursts is long enough. It begins to grow about a minute after the end of a relatively large burst. Then it starts to decline halfway and disappears about a minute before the onset of the next burst. The maximum intensity of this component is usually of the order of one-tenth the saturation level. However, in the observation from Tenma in August 1983,[33] the Rapid Burster started off as a

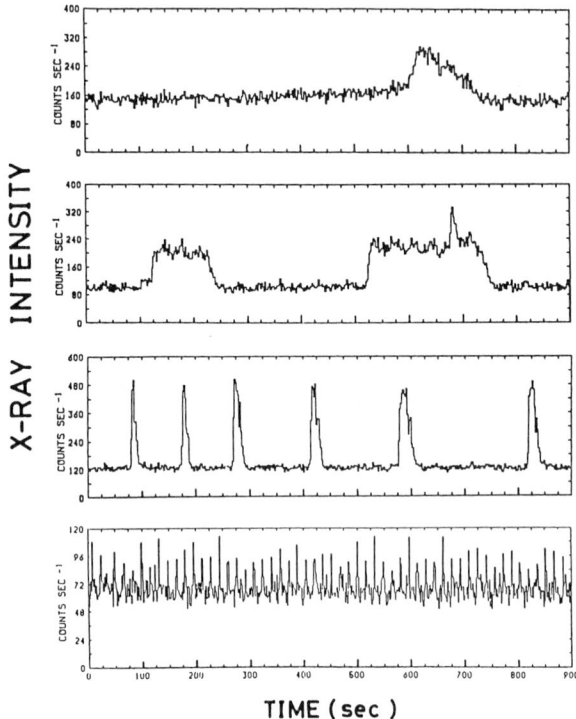

FIGURE 12. Various appearances of Type II bursts from the Rapid Burster during the activity in August 1983.[33]

persistent source and remained persistently bright at about 50 mCrab intensity for at least ten days, which is new for this source. During this period, Type I bursts occurred fairly periodically. Later on when the rapid burst activity started, this persistent component disappeared.

The GSPC on board Tenma can provide good quality spectra. The energy spectra of the rapid bursts are best represented by blackbody spectra with a significant hard tail. An example is shown in FIGURE 13. The blackbody temperature and the apparent blackbody radius are not constant for individual bursts. The blackbody temperature

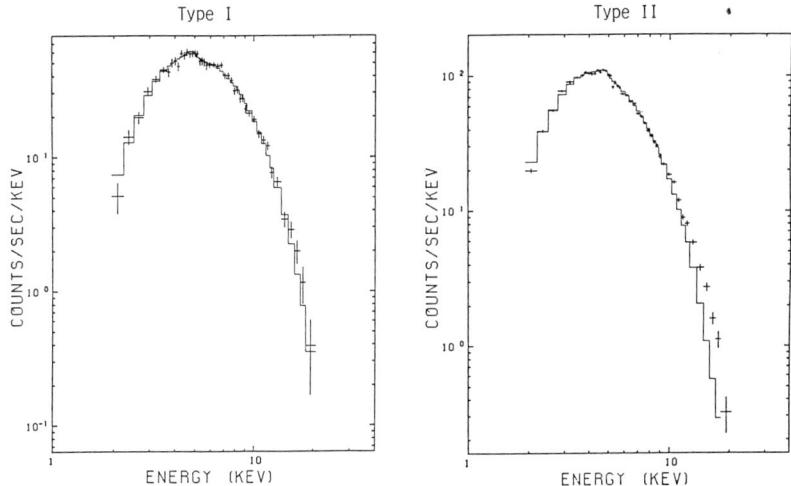

FIGURE 13. Pulse-height spectra of Type I bursts (left) and Type II bursts (right) observed from Tenma. Histograms are the best-fit blackbody spectra.

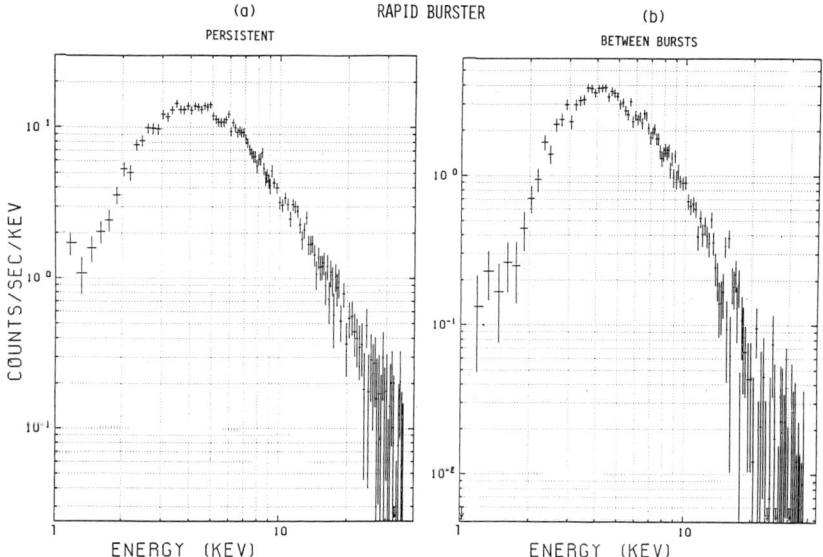

FIGURE 14. Spectra of the persistent emission from the Rapid Burster observed from Tenma (a) during the first half of the August 1983 activity when the Rapid Burster was persistently bright and (b) during long pauses after large Type II bursts in the July 1984 activity.

varies from burst to burst in the range of 1.6–2.0 keV and the apparent blackbody radius changes between 8 and 15 km for 10 kpc distance. On the other hand, Type I bursts from the same source invariably give an apparent blackbody radius of 7 km in the same temperature range, which is regarded to represent the size of the neutron star. This indicates an important fact that the rapid bursts are emitted from an extended yet optically thick region, and not from the neutron star surface. It is also worth mentioning that, for flat-topped bursts, the temperature as well as the radius remain constant through the interval of the flat top. This fact implies that a steady accretion flow is maintained during this period of time.

The spectrum of the persistent component also has the blackbody nature. FIGURE 14 shows two spectra of the persistent component. Although the spectrum for the long persistent phase in August 1983 is not entirely free from the ambiguity in eliminating the contribution of the neighboring X1728-34, both spectra are represented by a blackbody spectrum at about 1.6 keV of kT with a pronounced hard excess.

Thus, we see that the emission from the Rapid Burster, whether or not it is the rapid burst or the persistent emission, is always of blackbody nature. This is qualitatively different from other low-mass binary sources discussed in the previous section. If an optically thick accretion disk were extending close to the neutron star surface, one would expect a soft "multicolor" component to be present, as shown in the preceding section. Since we do not see it significantly, the optically thick disk would be stopped at some distance farther than a few times 10E6 cm from the neutron star and thus the emission therefrom falls below the X-ray energy band. This might suggest the presence of a magnetic field of the order of 10E8 gauss or greater, but weak enough not to make it a pulsar.

Another interesting finding is a significant 0.5-sec oscillation in two long flat-topped bursts observed from Hakucho.[34] This periodicity does not seem to be due to the neutron star rotation because the periods in two bursts are slightly different from each other.

Apart from regular oscillations, a kind of ringing is often noted in the decay part of large-size bursts. A very striking characteristic emerged from the Tenma observation. FIGURE 15 shows these bursts displayed in the order of size and duration. The saturation levels of these bursts are nearly the same. One clearly notices that the larger the burst size, the slower is the ringing. Moreover, it exhibits a fine structure and does not seem to be a simple damping oscillation.

To our surprise, as seen in FIGURE 16, the decay part of every burst reproduces a nearly identical pattern to a fine detail, if the time unit is properly scaled. In other words, the pattern is time-scale invariant; the pattern is described by a single function, $F(t/\tau)$, with τ being a characteristic time for each burst. If one adopts the period for the fundamental frequency of ringing as τ, this time-scale invariant pattern can be verified over the range of τ from 10 sec down to 0.3 sec, which is as far as the employed time resolution allows us to follow. More detailed analysis is presently in progress.

The X-ray intensity is considered to follow the mass flow rate. Therefore, the flow rate appears to be finely controlled during a burst by some subtle mechanism, according to the function, $F(t/\tau)$, regardless of τ. It is also difficult to understand that the characteristic time scale is so long as ten seconds. The above results make the Rapid Burster even more puzzling than it already was.

WHAT WOULD A BLACK HOLE SOURCE LOOK LIKE?

Some X-ray sources reveal an ultrasoft spectrum, which is significantly softer than those of most low-mass binary sources discussed in the second section of this paper. White (1983)[35] pointed out that these sources should be examined for potential black hole candidates since all black hole candidates known to date exhibit a state with an ultrasoft spectrum. Cyg X-1 has been the best black hole candidate because of its mass lower limit of 8 $M_\odot$ for the compact object. Cyg X-1 exhibits two distinct spectral states, one of which is an ultrasoft spectrum. Recently, Cowley et al. (1983)[36] gave a mass lower limit of 7 $M_\odot$ for the compact object in the LMC X-3 system. Hutchings et al. (1983)[37] reported the compact object in LMC X-1 to be more massive than 3 $M_\odot$.

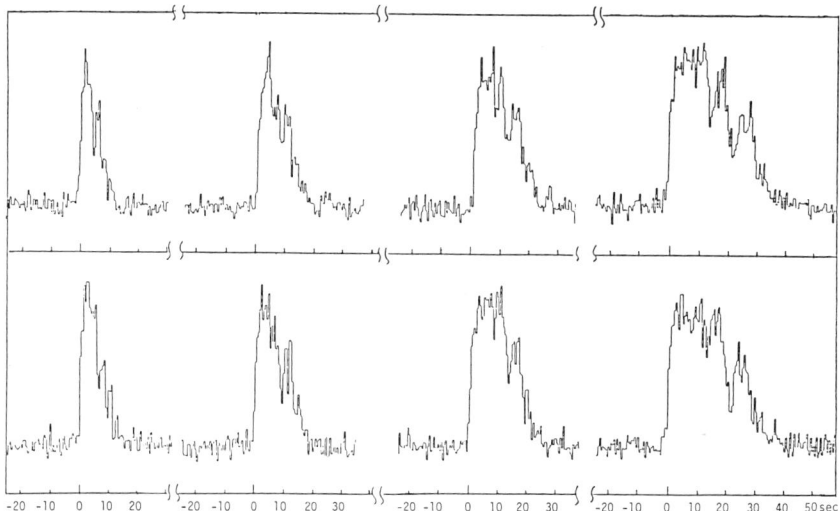

FIGURE 15. Detailed profiles of Type II bursts from the Rapid Burster observed from Tenma, displayed in the order of duration. Note the ringing and the striking similarity in the fine structures in the decay part of these bursts.

These values are above the current theoretical limit for the mass of a stable neutron star. White and Marshall (1984)[38] showed that the spectra of both LMC X-1 and LMC X-3 are ultrasoft, similar to the soft state spectrum of Cyg X-1. In addition to these sources, GX339-4 and Cir X-1 also exhibit an ultrasoft spectrum sometimes. These sources are also suspected of black hole candidates because of their flickering, which is characteristic of Cyg X-1. FIGURE 17 shows the ultrasoft spectra of GX339-4 and Cir X-1 observed from Tenma.

Since a black hole is generally considered to possess no magnetic field, the accretion disk will extend close to the Schwarzschild radius. The essential difference in the accretion flow to a black hole as compared to a neutron star is the absence of a solid surface. Hence, the spectrum from an accreting black hole would lack the 2-keV

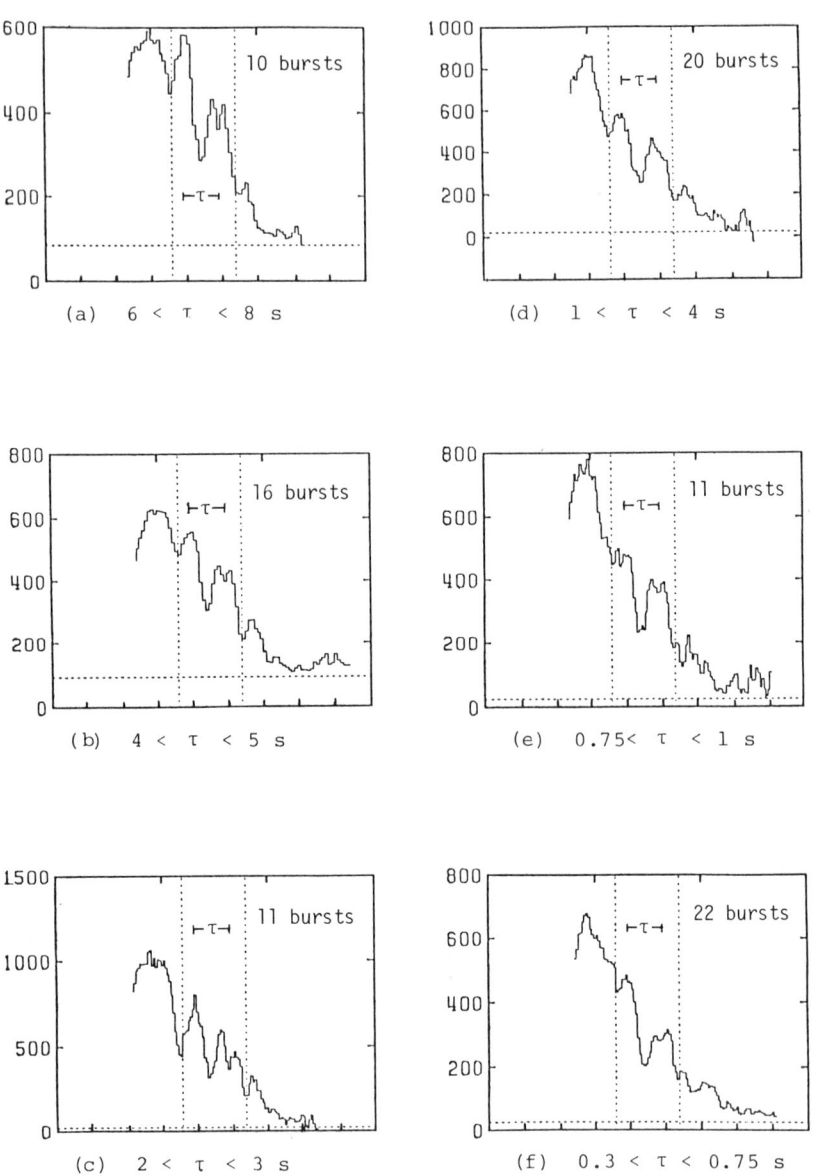

FIGURE 16. Composite profiles of Type II bursts from the Rapid Burster are shown for different ranges of the characteristic time, constructed by scaling time for each burst according to the characteristic time (time scale is expressed in units of the characteristic time). The number of bursts superposed is given in each diagram.

blackbody component expected from the neutron star surface (see the second section of this paper), which is shown schematically in FIGURE 18. As a matter of fact, the observed "multicolor" spectrum expected from the accretion disk of low-mass binary sources is very similar to the ultrasoft spectrum of the black hole candidates. This picture would explain, at least qualitatively, the nature of the ultrasoft spectrum for black hole sources.

THE EMISSION LINE AND ABSORPTION EDGE OF IRON

A good energy resolution of the GSPC of Tenma enables us to determine the intensities of emission lines (if present) and the ionization states significantly better

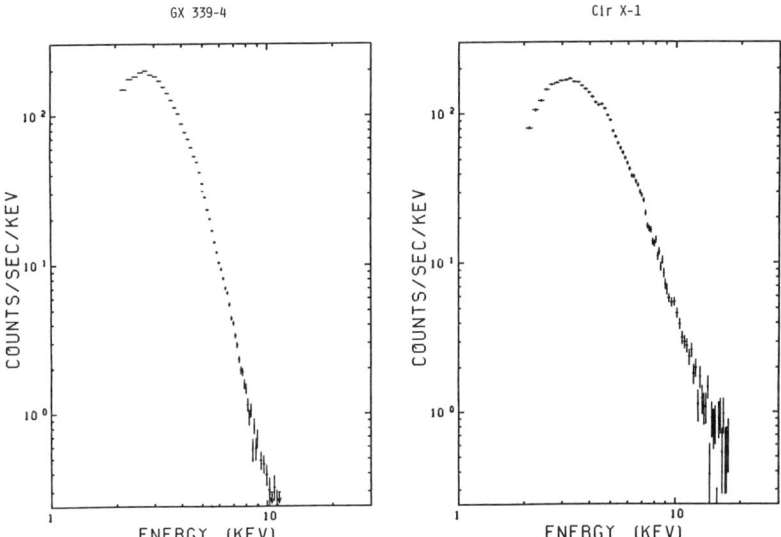

FIGURE 17. High-(soft-)state spectra of two black hole candidates, GX339-4 and Cir X-1, observed from Tenma.

than before. In particular, the iron emission lines can be studied to a much lower intensity than previously detectable and are found to be rich resources of information. The K-absorption edge of iron also gives the ionization state as well as the abundance of iron.

X-Ray Pulsars

The X-ray pulsars that we observed exhibit the iron emission line invariably at 6.4 keV, except for the low-mass binary pulsar X1626-67, which shows no significant iron line. Examples of the observed spectra of Vel X-1 and GX301-2 are shown in FIGURE

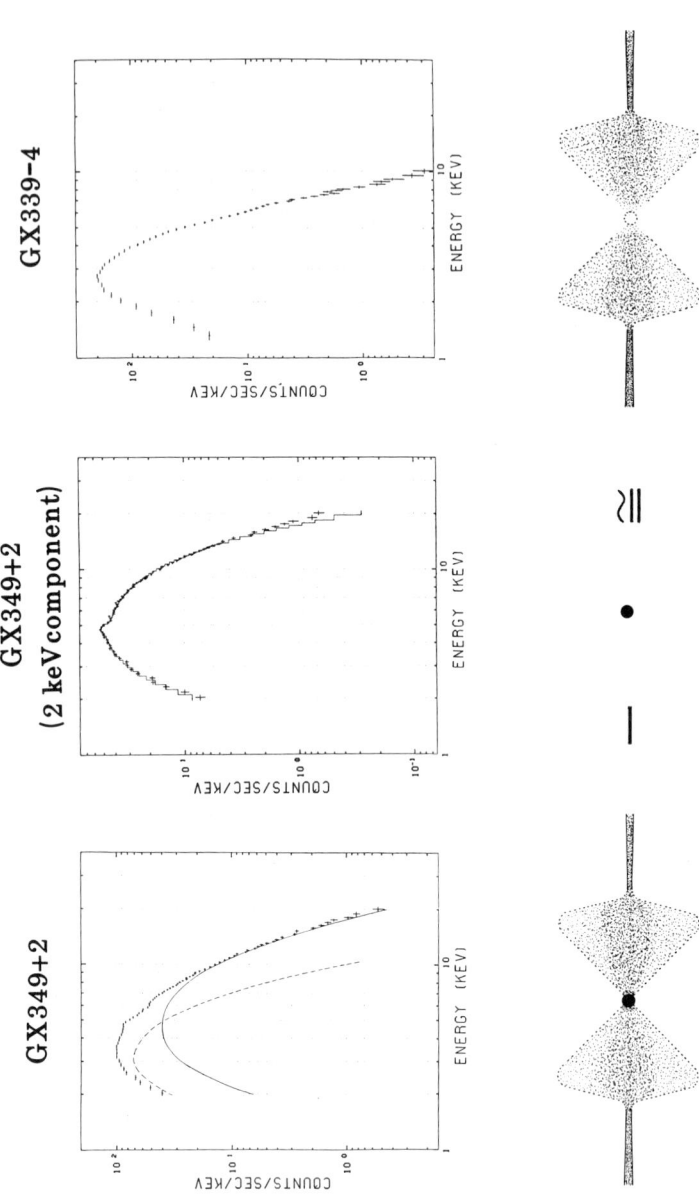

FIGURE 18. Schematic diagram illustrating a qualitative difference between an X-ray source of a neutron star and that of a black hole. The absence of a solid surface may explain the lack of the blackbody component in the spectrum of a black hole source.

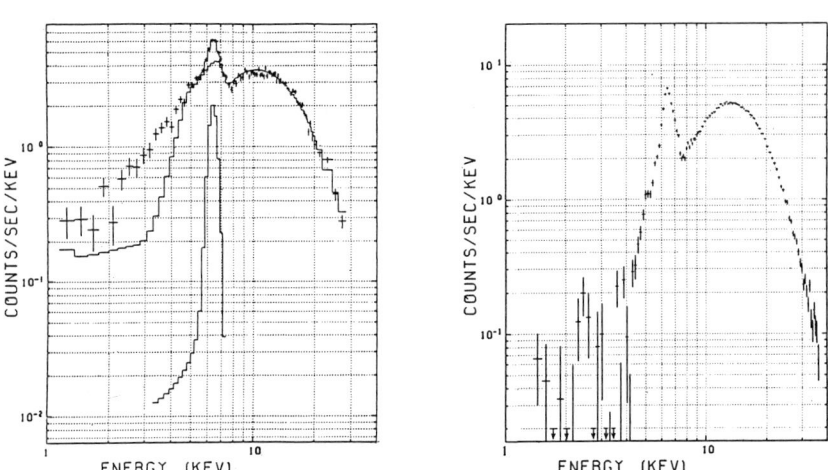

FIGURE 19. Examples of the pulse-height spectra of two X-ray pulsars, Vela X-1 and GX301-2, observed from Tenma in the period of heavy absorption.

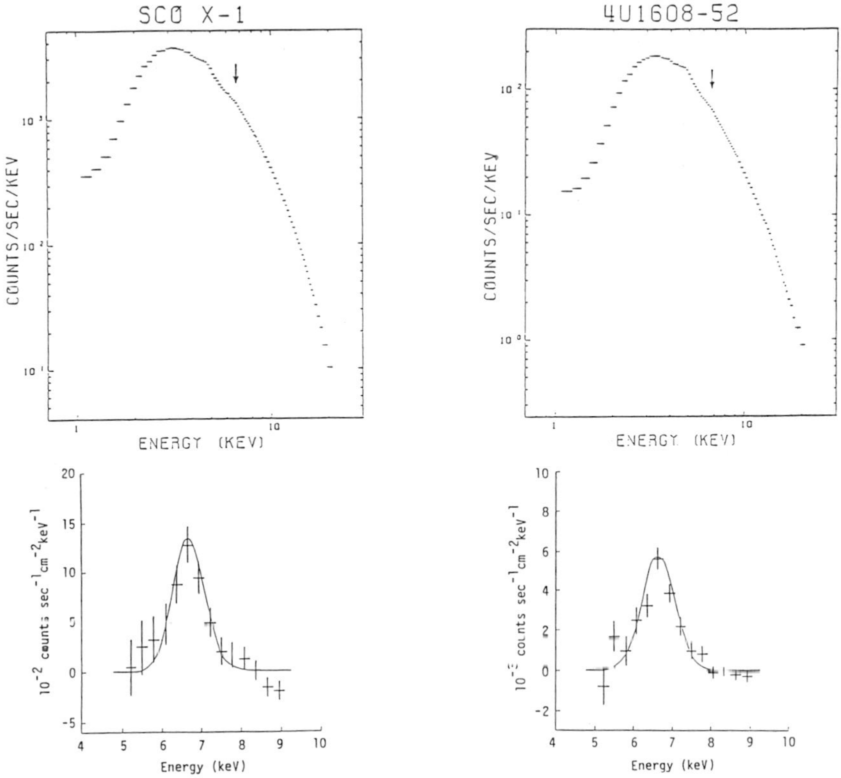

FIGURE 20. The emission line of iron observed from two low-mass binary sources, Sco X-1 and X1608-52.[40] Upper panels show the overall pulse-height spectra, in which the iron lines are marked by the arrows. The line profiles after subtracting the underlying continua are shown in the lower panels.

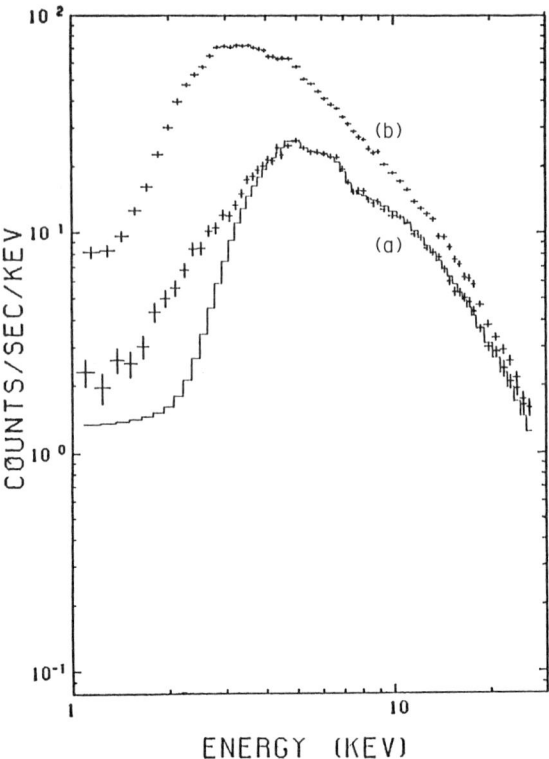

FIGURE 21. Observed pulse-height spectra of Cyg X-1 (a) during a transient intensity dip and (b) after the dip.[41] The histogram shows the expected spectrum when the spectrum (b) is subjected to an absorption by a certain amount of neutral column. Note the K-absorption edge of iron and a marked low-energy excess in the spectrum (a).

19. This energy of 6.4 keV implies the fluorescent origin from relatively cool matter surrounding the neutron star. The energy of the observed K-absorption edge also indicates a relatively low ionization state. An interesting finding is that the iron line intensity of these sources does not pulsate. The result for Vel X-1 and its implications are discussed by Ohashi *et al.* (1984).[39]

Low-Mass Binary Sources

Several nonpulsating low-mass binary sources are found to emit the iron line. These lines are found at 6.7 keV, indicating the emission from the helium-like state of iron. This is in contrast to the 6.4-keV line for X-ray pulsars. The iron lines from Sco X-1 and X1608-52 are shown in FIGURE 20.[40] The 6.7-keV line is interpreted to be emitted from the outer accretion disk, due to X-ray irradiation.

Transient Absorption

For an example, the case of Cyg X-1 is shown here. Cyg X-1 frequently exhibits intensity dips near superior conjunction. FIGURE 21 shows the spectra observed during and after such a dip.[41] The spectrum during the dip clearly indicates that the intensity dip of Cyg X-1 is a transient absorption phenomenon. The absorption edge at 7.2 keV implies a relatively cool matter. The depth of absorption yields the iron abundance, which is consistent with being cosmical. The absence of any significant fluorescent iron line associated with the heavy absorption suggests that some cloud(s), originating from the main star, crossed the line of sight rather than wrapping up the compact source. A pronounced low-energy excess over the expected absorption is noted, which can be interpreted as due to partial obscuration of the emission region by these clouds.

Extragalactic Sources

Clusters of galaxies are known commonly to emit the iron lines, from which the iron abundance can be estimated. Also, some of the active galaxies are found to emit fairly intense iron lines. FIGURE 22 shows the observed spectra of the Seyfert galaxy NGC 4151 and the radio galaxy Cen A. The iron lines from both galaxies are found to be at 6.4 keV, suggesting the fluorescent origin from relatively cool matter. However, the observed iron line from NGC 4151 appears too intense to be accounted for in terms of the fluorescence from a sphere of gas with the cosmical iron abundance surrounding the central source. The iron line from Cen A is apparently broadened. These features

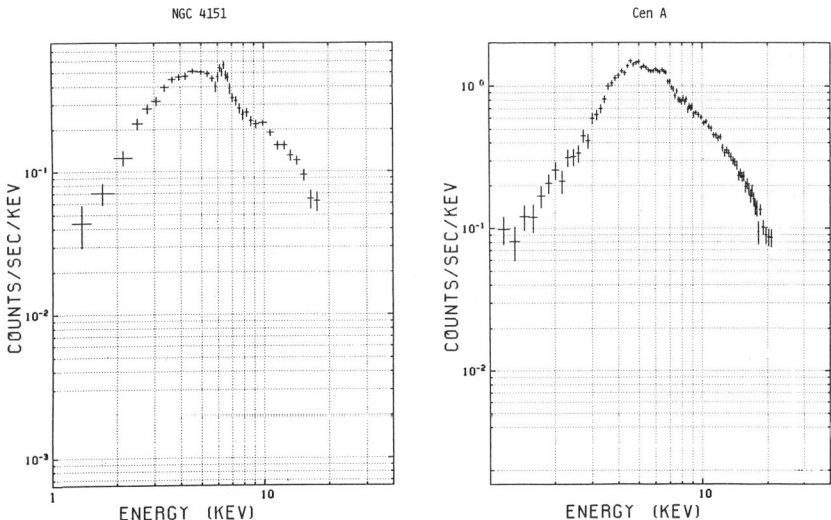

FIGURE 22. Observed pulse-height spectra of the Seyfert galaxy NGC4151 and the radio galaxy Cen A.

are discussed in more detail elsewhere. Obviously, these new results add important information for the study of active galactic nuclei.

Thermal Emission along the Galactic Plane

The presence of an excess emission along the galactic plane was noted previously from the HEAO-1 survey.[42] However, the nature of this emission has remained unknown. The Tenma observation revealed an intense line of iron at 6.7 keV associated with this excess emission. The observed spectra are in agreement with thermal emission from thin hot plasma of kT in the range of 3–8 keV, varying in intensity and temperature from region to region.[43] Examples of the observed spectra are shown in FIGURE 23. Presence of such hot plasma along the galactic plane is quite unexpected. A diffuse plasma of these temperatures extended to the galactic scale is unlikely because

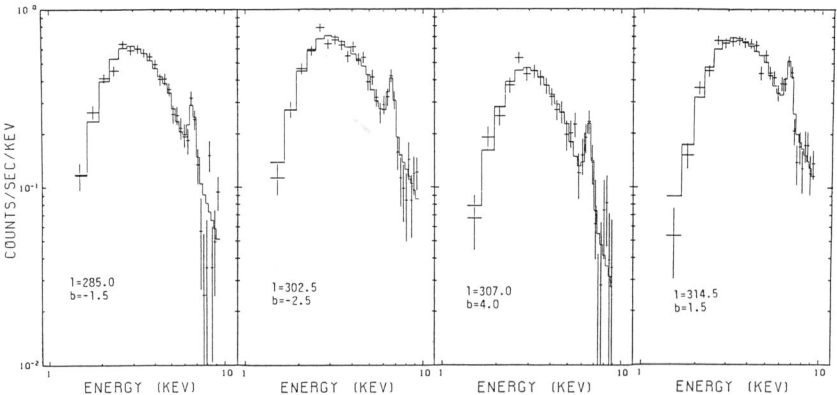

FIGURE 23. Examples of the pulse-height spectra observed from selected source-free regions along the galactic plane.[43] Position of the center of the field of view is given in each diagram.

of too high of a pressure to be bound. Stellar origin would also be difficult since stellar coronae of such high temperature are rarely observed. The observed spectra are reminiscent of young supernova remnants. However, the measured X-ray flux requires very many supernovae to be present, which escaped detection. Thus, the origin of this hot plasma is a big issue at present.

REFERENCES

1. KONDO, I., H. INOUE, K. KOYAMA et al. 1981. Space Sci. Instrum. **5:** 211.
2. TANAKA, Y., M. FUJII, H. INOUE et al. 1984. Publ. Astron. Soc. Japan **36:** 641.
3. KOYAMA, K., T. IKEGAMI, H. INOUE et al. 1984. Publ. Astron. Soc. Japan **36:** 659.
4. LEWIN, W. H. G. & P. C. JOSS. 1983. Accretion-Driven Stellar X-Ray Sources. Lewin & van den Heuvel, Eds.: 41. Cambridge University Press. London/New York.
5. WOOSLEY, S. E. & R. E. TAAM. 1976. Nature **263:** 101.

6. MARASCHI, L. & A. CAVALIERE. 1977. Highlights in Astronomy, vol. 4, part 1, p. 127.
7. JOSS, P. C. 1978. Astrophys. J. **225**: L23.
8. SWANK, J. H., R. H. BECKER, E. A. BOLDT *et al.* 1977. Astrophys. J. **212**: L73.
9. HOFFMAN, J. A., W. H. G. LEWIN & J. DOTY. 1977. Astrophys. J. **217**: L23; Mon. Not. R. Astron. Soc. **179**: 57.
10. VAN PARADIJS, J. 1978. Nature **274**: 650.
11. EBISUZAKI, T., T. HANAWA & D. SUGIMOTO. 1984. Publ. Astron. Soc. Japan **36**: 551.
12. INOUE, H., I. WAKI, K. KOYAMA *et al.* 1984. Publ. Astron. Soc. Japan **36**: 831.
13. VAN PARADIJS, J. 1982. Astron. Astrophys. **107**: 51.
14. CZERNY, M. & M. SZTAJNO. 1983. Acta Astron. **33**: 213.
15. WAKI, I., H. INOUE, K. KOYAMA *et al.* 1984. Publ. Astron. Soc. Japan **36**: 819.
16. BAYM, G. & C. PETHICK. 1979. Annu. Rev. Astron. Astrophys. **17**: 415.
17. OHASHI, T., H. INOUE, K. KOYAMA *et al.* Astrophys. J. **258**: 254.
18. BASINSKA, E. M., W. H. G. LEWIN, M. SZTAJNO *et al.* 1984. Astrophys. J. **281**: 337.
19. LEWIN, W. H. G., W. D. VACCA & E. M. BASINSKA. 1983. Astrophys. J. **277**: L57.
20. TAWARA, Y., S. HAYAKAWA, T. KII *et al.* 1984. Publ. Astron. Soc. Japan **36**: 845.
21. INOUE, H., K. KOYAMA, K. MAKISHIMA *et al.* 1981. Astrophys. J. **250**: L71.
22. CLARK, G. W., J. G. JERNIGAN, H. BRADT *et al.* 1976. Astrophys. J. **207**: L105.
23. GRINDLAY, J. E., H. L. MARSHALL, P. HERTZ *et al.* 1980. Astrophys. J. **240**: L121.
24. MELIA, F. & P. C. JOSS. 1984. Submitted to Astrophys. J.
25. LEWIN, W. H. G., J. A. HOFFMAN, J. DOTY *et al.* 1976. Mon. Not. R. Astron. Soc. **177**: 83.
26. MURAKAMI, T., H. INOUE, K. KOYAMA *et al.* 1980. Publ. Astron. Soc. Japan **32**: 543.
27. INOUE, H., K. KOYAMA, F. MAKINO *et al.* 1984. Publ. Astron. Soc. Japan **36**: 855.
28. HOSHI, R. 1984. Publ. Astron. Soc. Japan **36**: 785.
29. MITSUDA, K., H. INOUE, K. KOYAMA *et al.* 1984. Publ. Astron. Soc. Japan **36**: 741.
30. LEWIN, W. H. G., J. DOTY, G. W. CLARK *et al.* 1976. Astrophys. J. **207**: L95.
31. HOFFMAN, J. A., H. L. MARSHALL & W. H. G. LEWIN. 1978. Nature **271**: 630.
32. INOUE, H., K. KOYAMA, K. MAKISHIMA *et al.* 1980. Nature **283**: 358.
33. KUNIEDA, H., Y. TAWARA, S. HAYAKAWA *et al.* 1984. Publ. Astron. Soc. Japan **36**: 807.
34. TAWARA, Y., S. HAYAKAWA, H. KUNIEDA *et al.* 1982. Nature **299**: 38.
35. WHITE, N. E. 1983. COSPAR/IAU Symposium on Advances in High-Energy Astrophysics and Cosmology, Rojen, Bulgaria, 1983.
36. COWLEY, A. P., D. CRAMPTON, J. B. HUTCHINGS *et al.* 1983. Astrophys. J. **272**: 118.
37. HUTCHINGS, J. B., D. CRAMPTON & A. P. COWLEY. 1983. Astrophys. J. **275**: L43.
38. WHITE, N. E. & F. E. MARSHALL. 1984. Astrophys. J. **281**: 354.
39. OHASI, T., H. INOUE, K. KOYAMA *et al.* 1984. Publ. Astron. Soc. Japan **36**: 699.
40. SUZUKI, K., M. MATSUOKA, H. INOUE *et al.* 1984. Publ. Astron. Soc. Japan **36**: 761.
41. KITAMODO, S., S. MIYAMOTO, Y. TANAKA *et al.* 1984. Publ. Astron. Soc. Japan **36**: 731.
42. WORRALL, D. M., F. E. MARSHALL, E. A. BOLT *et al.* 1982. Astrophys. J. **255**: 111.
43. KOYAMA, K. 1984. X-Ray Astronomy '84. Bologna, Italy.

PART VI. COSMIC RAYS; GAMMA RAYS

Abundances in Cosmic Rays[a]

MARTIN H. ISRAEL
Department of Physics
and
McDonnell Center for the Space Sciences
Washington University
St. Louis, Missouri 63130

INTRODUCTION

This paper gives an overview of cosmic ray abundances. It is intended to summarize the observations and to organize these results in terms of models of cosmic ray sources. The following paper by David Eichler complements this one by dealing with the theory of cosmic ray acceleration.

Three types of data will be considered here. First are abundances of elements, in particular the recently measured abundances of the very rare elements with atomic number (Z) greater than 30, which are called ultraheavy (UH) elements in cosmic ray literature. Abundances of individual elements have been measured for every element with $Z < 33$ and for even-Z elements with $Z < 60$. In addition, abundances have been measured with resolution of approximately plus-or-minus one unit of atomic number at higher Z. It is clear from these observations that the cosmic ray source differs in composition from the solar system in significant ways, while the general trend of abundances is the same. I will discuss the significance of the differences both with respect to the nucleosynthesis of the material and the cosmic ray acceleration process.

The second type of data are isotopic abundances. Individual isotopes have been resolved for the most abundant elements in the cosmic rays. The most abundant isotope in the cosmic ray source is the same as in the solar system for each of the elements, $_1$H, $_2$He, $_6$C, $_8$O, $_{10}$Ne, $_{12}$Mg, $_{14}$Si, and $_{26}$Fe. However, for all three elements where the cosmic ray source abundances of the other isotopes have been determined (namely, Ne, Mg, and Si), there are distinct enrichments of the neutron-rich isotopes in the cosmic rays, giving clear indication of a different nucleosynthesis history.

Finally, I will describe three observations that have been made of cosmic ray antiprotons. The observed antiproton flux is distinctly higher than that predicted from standard models of secondary production resulting from cosmic ray proton collisions with nuclei of the interstellar gas. Two of these observations can be interpreted as requiring some changes in the standard models of cosmic ray propagation in the galaxy, while the other observation, at lower energies, is more difficult to reconcile with secondary antiproton origin.

In all discussions of cosmic ray composition, it is important to distinguish between

---

[a]My preparation of this paper was supported in part by NASA grant nos. NGR 26-008-001 and NAG8-498, and my participation in the conference was supported in part by an NSF Travel Grant, in part by the McDonnell Center for the Space Sciences of Washington University, and in part by the Israeli conference organizers.

the "observed" composition and the "source" composition. The observed composition reflects (1) the composition of ambient material at the site of cosmic ray acceleration—its nucleosynthesis history and elemental fractionation that occurred in the formation of the region; (2) the acceleration process, which could give further elemental fractionation depending, for example, on the charge state of the atoms; and (3) the propagation in the interstellar medium, where primary accelerated nuclei can fragment in nuclear collisions with the interstellar gas.

The cosmic ray "source" composition is inferred from the observed composition by calculations that correct for the fragmentation in the interstellar medium. The elemental source composition reflects both nucleosynthesis processes and fractionation effects, while the isotopic composition is expected to reflect principally the nucleosynthesis since the fractionation is expected to be dominated by atomic properties.

Throughout this discussion, it is useful to remember that the abundances of elements and isotopes in the solar system reflect the composition of the local interstellar medium about 5×10^9 years ago. The cosmic ray source may represent a different region of the galaxy and it samples that region relatively recently, about 10^7 years ago.

OBSERVED ELEMENTAL ABUNDANCES

Cosmic ray abundances observed near earth are compared with the abundances of elements in the solar system in FIGURES 1[1] and 2.[2-7] The solar system abundances are intended to describe the composition of the cloud of interstellar material that formed the solar system. For most elements, the solar system abundances are derived from measurements in type-1 carbonaceous chondrite (C1) meteorites and are expected to be representative of the solar photosphere.

Two features of these figures are at once apparent. First, there is good general agreement between solar system and cosmic ray abundances of the most abundant elements; the peaks of the two sets of abundances generally agree within a factor of three, even though the overall abundances vary over twelve decades. Second, the abundances differ widely for elements that are in the valleys of the solar system distribution, such as the elements with atomic numbers of 3–5, 21–25, 44–48, and 60–74.

The broad agreement in abundances of the more abundant elements is evidence that the bulk of the cosmic rays have a nucleosynthesis history that is not radically different from that of the solar system. However, the differences of factors of three or so are significant, and these will be discussed further below.

The filling in of the valleys in the solar system abundances is well understood as the result of fragmentation of heavier nuclei in collisions with interstellar gas. The fragments of any particular nuclide have a broad distribution of atomic number spreading over all lighter elements. Where the solar system abundance of an element is much lower than that of a heavier element, the cosmic rays observed near earth are almost entirely secondaries and the observations give no information about the abundance of that element in the cosmic ray source. The standard model of cosmic ray propagation in the galaxy has been constructed to fit the abundances of these secondary elements. That model can then be used to extrapolate observed abundances

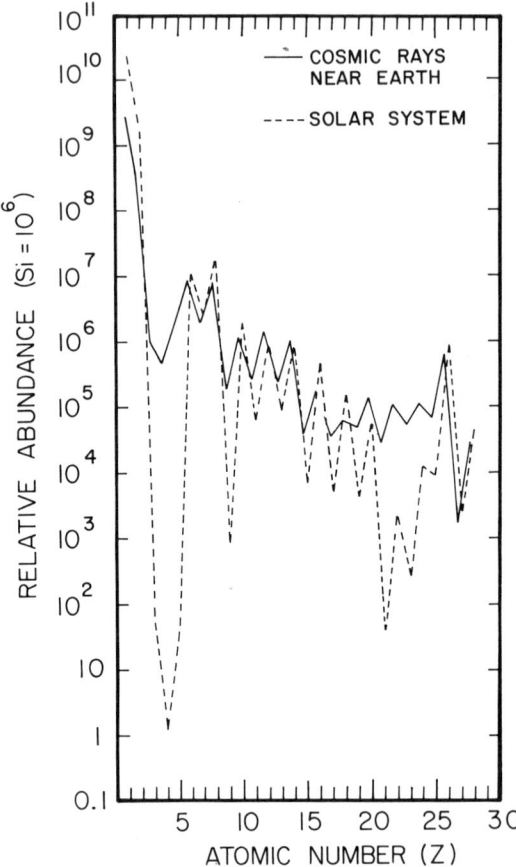

FIGURE 1. Relative abundances of elements in the solar system and in the arriving cosmic rays for elements with $Z \leq 28$. Figure adapted from reference 1.

of the other elements back to the source. For the more abundant elements, most of the observed nuclei are primary and the extrapolation back to the source is relatively insensitive to the exact form of the propagation model. In this paper, I will principally be concerned with the source abundances and so I will confine the discussion to those elements for which the source abundance can be derived with confidence from the observations.

ULTRAHEAVY ELEMENTS AND NUCLEOSYNTHESIS

No observations of isotopic composition of ultraheavy (UH) elements ($Z > 30$) have yet been possible, but the elemental abundances of these elements can display the signature of their nucleosynthesis. The solar system abundances of these elements can be explained, in a somewhat simplified manner, as the superposition of two principal processes of neutron capture, namely, the slow s-process and the rapid r-process. The solar system abundances can be decomposed, isotope by isotope, into r-process and s-process components. FIGURE 3 displays the r and s components of the solar system, element by element.[8-10] (In order to display the trends more clearly, only even-Z elements are plotted in this figure.)

It is apparent from this figure that the elemental abundances of UH cosmic rays can be expected to indicate the relative contribution of r-process and s-process elements. The charge interval, $50 \leq Z \leq 58$, is especially important in this regard since here the solar system has about equal contributions from r and s, with r contributing most of the $_{52}$Te and $_{54}$Xe and s contributing most of the $_{50}$Sn, $_{56}$Ba, and $_{58}$Ce. If the cosmic rays were dominated by either the r- or the s-process, one would expect to see a distinct lack of either Te and Xe or of Sn and Ba; however, as shown in FIGURE 2, there are significant quantities of all five elements observed. The Ba and Sn abundances are somewhat higher than those of Te and Xe, suggesting an enrichment of s-process elements, but, as will be discussed in the next section, this enhancement is consistent

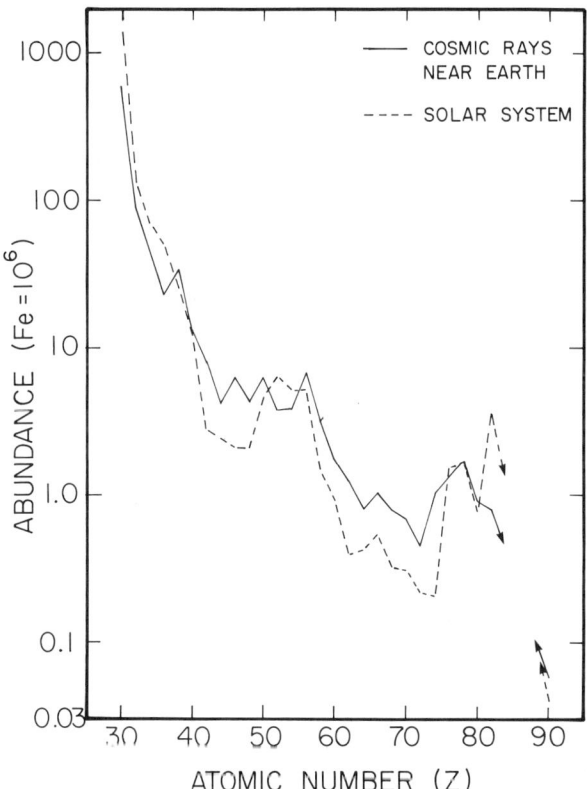

FIGURE 2. Relative abundances of elements in the solar system[17] and in the arriving cosmic rays[2-7] for even-Z elements with $Z \geq 30$. For $Z \leq 42$, plotted is the abundance for each even-Z element only; for $Z \geq 44$, plotted is the abundance of element Z plus the abundance of element $Z + 1$; the point plotted at $Z = 90$ is for $Z \geq 90$, i.e., for Th plus U in the solar system and for all actinide elements in the cosmic rays. For $Z \geq 44$, normalization of abundance to Fe is correct to about ±20%.

with a source fractionation effect that has been observed for other atomic numbers. When source abundances are inferred from the observations and a source fractionation governed by first ionization potential is included, it has been shown that the cosmic rays in this range of atomic numbers have significant contributions from both the r- and the s-process in a mixture similar to that found in the solar system.[2,11]

On the other hand, the lack of $_{82}$Pb observed in the cosmic rays relative to $_{76}$Os, $_{77}$Ir,

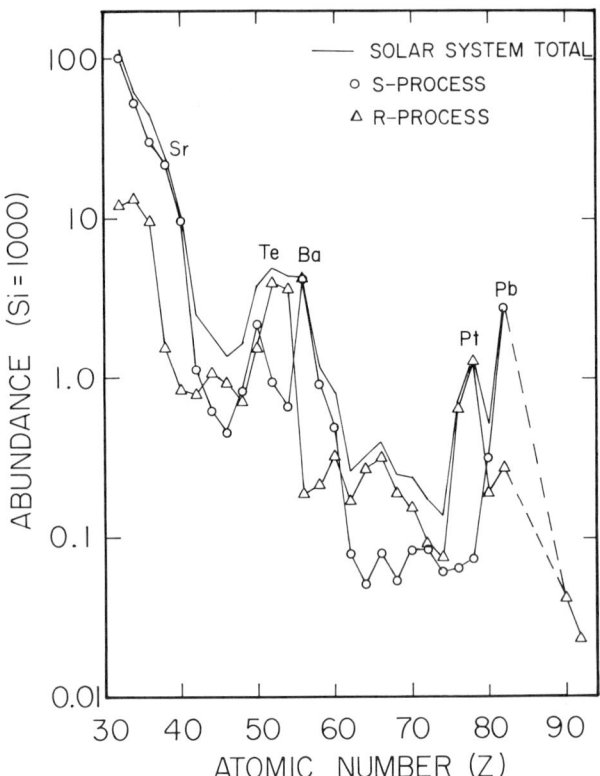

FIGURE 3. Contributions of r-process and s-process nucleosynthesis to the solar system abundances of even-Z elements. (Abundances of odd-Z elements are generally significantly less than those of their even-Z neighbors.) For $Z \leq 36$, from reference 8; for $Z = 82$, from reference 10, which includes the cyclic s-process; for all other elements, from reference 9.

and $_{78}$Pt suggests a cosmic ray source with less s-process component than in the solar system. However, it is also quite plausible that the underabundance of Pb in the cosmic rays is the result of either an error in the difficult determination of the solar system Pb abundance or of a source-fractionation effect.[6] (This point will be discussed below.)

The abundance of the actinide elements, $_{90}$Th and $_{92}$U, in the cosmic rays was previously thought to be exceptionally high compared to the solar system. If true, this would have been an indication of r-process enhancement in the cosmic rays since these elements are not made at all in the s-process. In fact, observations[7] have demonstrated that the observed cosmic ray actinide abundance is approximately what would be expected if the cosmic ray source had the same composition as the solar system.

While the UH results discussed above have all been based on observations from the Heavy Nuclei Experiment on the third High Energy Astronomy Observatory (HEAO-3), generally consistent results have been obtained by the Bristol University experiment on the Ariel-6 satellite,[12,13] although that instrument had poorer charge resolution than the HEAO-3 instrument and so did not resolve individual elements.

Thus the observations of the UH cosmic rays are consistent with a source composition with a similar mix of r-process and s-process nuclei as is found in the solar system. However, a precise decomposition of the cosmic ray abundances into r and s components is not possible because of uncertainties of the effects of elemental fractionation effects. Enhancement by a factor of two to three of either the r or s component of the cosmic rays compared to the solar system cannot be excluded. Better understanding of source fractionation effects or measurements of the isotopic composition will be required for a more definitive analysis of the nucleosynthesis history of these UH cosmic rays.

ELEMENTAL ABUNDANCES AND SOURCE FRACTIONATION

When elemental abundances in the cosmic ray source (CRS) are compared with those in the solar system (SS), one finds (FIGURE 4) that the ratio, CRS/SS, for the various elements is organized by the first ionization potential (FIP). (Cosmic ray source abundances in this and later figures are derived from results of several experiments. For H and He, they come from balloon observations;[14,15] for $_6$C through $_{27}$Co and for $_{29}$Cu and $_{31}$Ga, the results are from the Danish-French experiment on HEAO-3;[16] and for $_{28}$Ni, $_{30}$Zn, and the elements with $Z > 31$, the results are from the HEAO-3 Heavy Nuclei Experiment.[5] The solar system abundances[17] are mainly from type C1 meteorites, except H, C, N, and O are from photospheric measurements, He is from the solar wind H/He ratio, $_{10}$Ne is from the solar wind Ne/Ar ratio and from astronomical measurements of extra-solar-system nebulae, and the heavier noble gases, $_{18}$Ar, $_{36}$Kr, and $_{54}$Xe, are simply interpolated from nearby elements. The error bars plotted are the quadratic sums of the cosmic ray and solar system errors.)

Elements with higher FIP generally are less abundant in the cosmic rays, as if atoms that are more difficult to ionize are less likely to be accelerated. Qualitatively, this anticorrelation suggests acceleration out of a gas at a temperature of about 10,000 K. However, solar energetic particles (SEP), which are accelerated in flares in the million degree solar corona, display a similar anticorrelation with FIP, as shown in FIGURE 5, which gives results from the Voyager spacecraft averaged over many flares.[18]

The similarity between FIGURES 4 and 5 suggests a similar origin for the source fractionation of the galactic cosmic rays and the solar energetic particles. One plausible explanation is that the FIP-dependent effects for SEP occur at the interface between the photosphere and the corona, where the temperatures are about right for the first ionization potential to matter. In this case, the solar energetic particles would

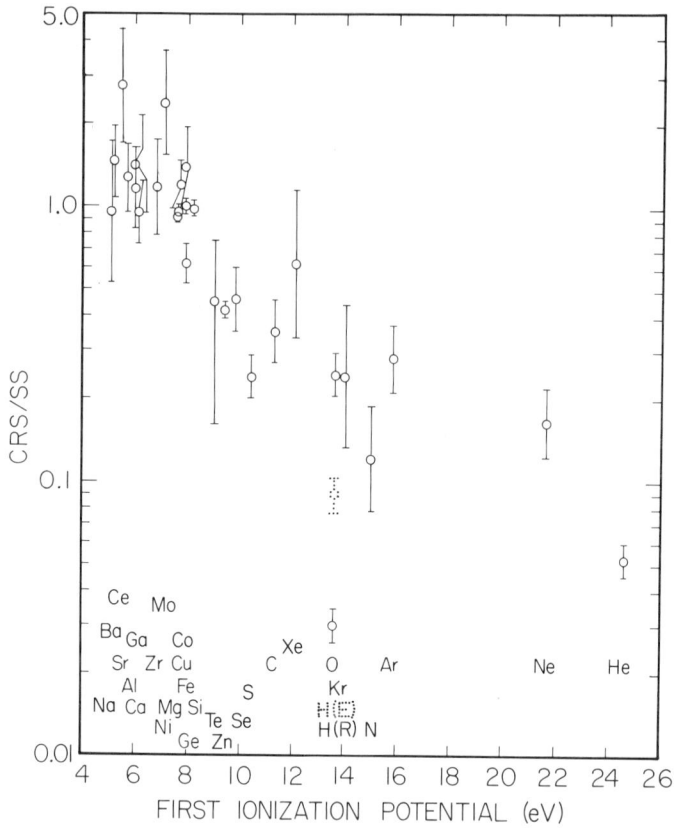

FIGURE 4. The ratio of cosmic ray source[5,16] to solar system[17] abundances, normalized to unity for Fe, as a function of the first ionization potential. Error bars indicate the quadratic sum of the error on solar system abundance and the error on cosmic ray source abundance.

be representative of the corona, and there is some evidence[18] that the coronal composition does differ from that of the photosphere in a similar manner to the SEP. Then the similar correlation with FIP in the galactic cosmic rays could be the result of galactic cosmic ray acceleration occurring in the hot (million-degree) component of the interstellar medium, which could have a composition differing from the cool bulk of the interstellar gas because of similar effects at the interface. On the other hand, the FIP correlation could be the result of injection into the acceleration process from stars with similar flare processes as the sun.[19] Alternatively, the similarity between FIGURES 4 and 5 could be the result of similar fractionation effects in the acceleration process itself. Such a process will be discussed shortly.

The correlation of CRS/SS with FIP as shown in FIGURE 4 is clearly imperfect. For example, $_{32}$Ge and $_{26}$Fe have the same value of FIP, yet the Ge point in this figure falls at about 60% of the Fe point and the difference is well beyond the error bars. One possible explanation is that the use of C1 meteorites gives an incorrect estimate of the solar system composition. There has been discussion of the possibility that C2 meteorites are more nearly representative of the solar system and it has been pointed out[20] that replacing the C1 values in the standard solar system abundances with C2 values[21] removes the Ge discrepancy and extends up to at least 10 eV the flat region where CRS/"SS" is roughly unity (FIGURE 6). We also note that the C2 abundance of Pb is about half the C1 value, so use of C2 meteorites for determining the solar system abundances would remove most of the disrepancy between cosmic ray and solar system abundances in the Pt-Pb region. While there is evidence that the C2 abundances are in fact not as good representatives of the solar system as C1,[22] it is important to bear in mind that there may be systematic errors in the standard solar system abundances.

In considering either FIGURE 4 or 6, one cannot ignore the serious discrepancy of the most abundant element, H. Two H points are plotted—the lower one compares H with the other elements at the same rigidity (momentum per unit charge), while the upper one compares H with the other elements at the same energy per nucleon (or the same velocity). The rigidity correlation is more likely appropriate since the H/He ratio at the same rigidity is nearly independent of rigidity, while the ratio at the same energy per nucleon is quite energy dependent.[15] [The H(E) point plotted here uses the high-energy limit, which is the highest H/He value observed.] Using the H(R) point, H is discrepant by about a factor of ten compared with oxygen, which has the same first ionization potential.

Since there is no good quantitative model to explain the FIP correlation and since the FIP correlation is imperfect (especially when H is included), it is useful to look for other correlations. Eichler[23] has proposed a model for shock acceleration of cosmic rays in which energetic particles are drawn directly from the hot interstellar medium. (See

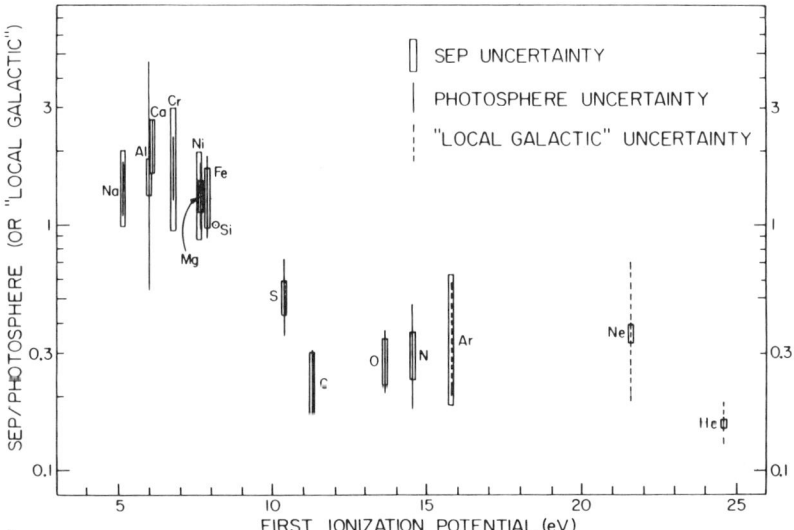

FIGURE 5. The ratio of abundance in solar energetic particles to solar system abundances, normalized to unity for Si, as a function of first ionization potential. (From reference 18.)

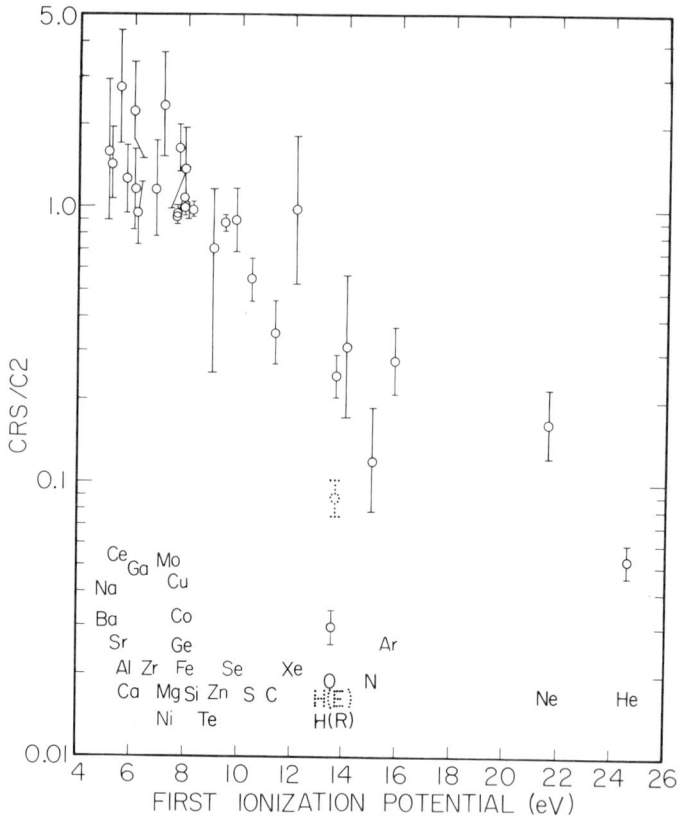

FIGURE 6. Same as FIGURE 4, except that for those elements where reference 17 used C1 meteorite abundances, we have here substituted C2 meteorite abundances.[21]

also Eichler's paper in this conference.) In the hot interstellar medium, the lighter nuclei are fully ionized, while the heavier nuclei are not. Eichler and Hainebach[24] have shown that in this model the acceleration should be more effective for the partially ionized heavy nuclei because they have a higher rigidity at the same velocity, which results in a higher diffusion coefficient; consequently, they see a sharper velocity contrast across the shock. According to this idea, if the hot interstellar gas had the same composition as the solar system, the values of CRS/SS for the various elements would be ordered by the mass-to-charge ratio of those elements in a million-degree gas. As an estimate of the charge state in such a gas, Eichler suggested use of Q_{120}, which is the state the element would have after removal of all electrons whose ionization potential is less than 120 eV.

FIGURE 7 has the same ordinate as FIGURE 4 (namely, the ratio, CRS/SS), but the abscissa is A/Q_{120} (where A is the atomic weight). With significant scatter, this plot does show a general correlation in the expected direction. In particular, it provides a natural explanation for the very low hydrogen abundance, especially when the constant-rigidity H(R) point is used. The ordering of the data when plotted against A/Q_{120} is improved significantly in FIGURE 8, where the C2 meteorites are substituted for the C1 in determining the solar system abundances (as was done previously for FIGURE 6). Here the only elements that fall significantly off the smooth trend in A/Q_{120} are the noble gases, Ne, Ar, and Kr, for which the solar system abundances are essentially unmeasured.

Thus, the elemental abundances of the cosmic ray source show significant differences from the standard solar system abundances. Since the isotope results, which will be discussed in the next section, show unmistakable evidence of a difference between the nucleosynthesis history of cosmic rays and of the solar system, the elemental abundance differences must be affected by nucleosynthesis to some extent. However, it is clear that the cosmic ray elemental abundances also reflect some combination of fractionation effects in the acceleration process, fractionation in the

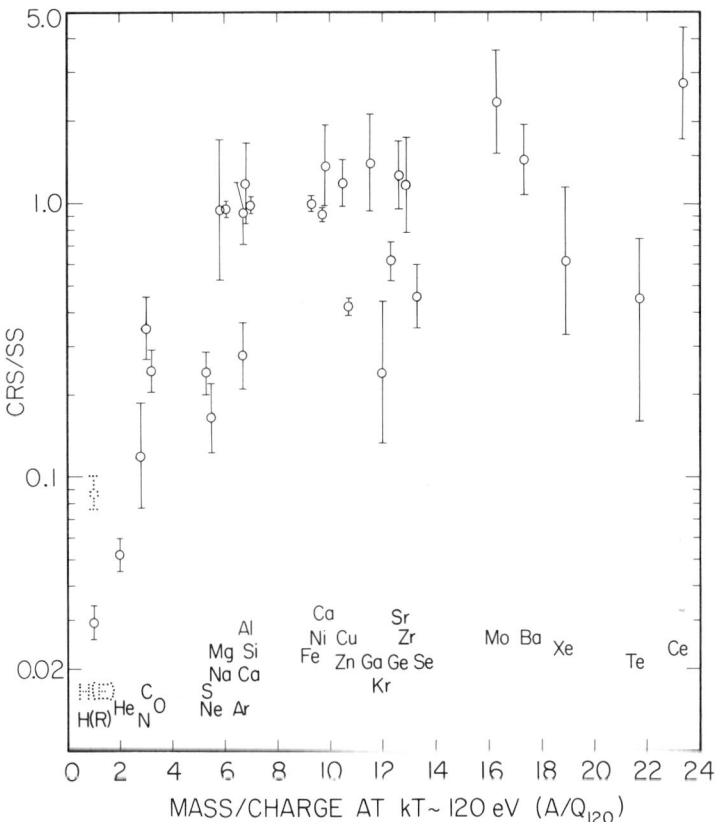

FIGURE 7. The ratio of cosmic ray source to solar system abundances (same as in FIGURE 4), as a function of A/Q_{120}, where A is the atomic mass and Q_{120} is the charge state the element would have after removal of all electrons with ionization potential less than 120 eV. Q_{120} is an estimate for the charge state in the hot interstellar medium.

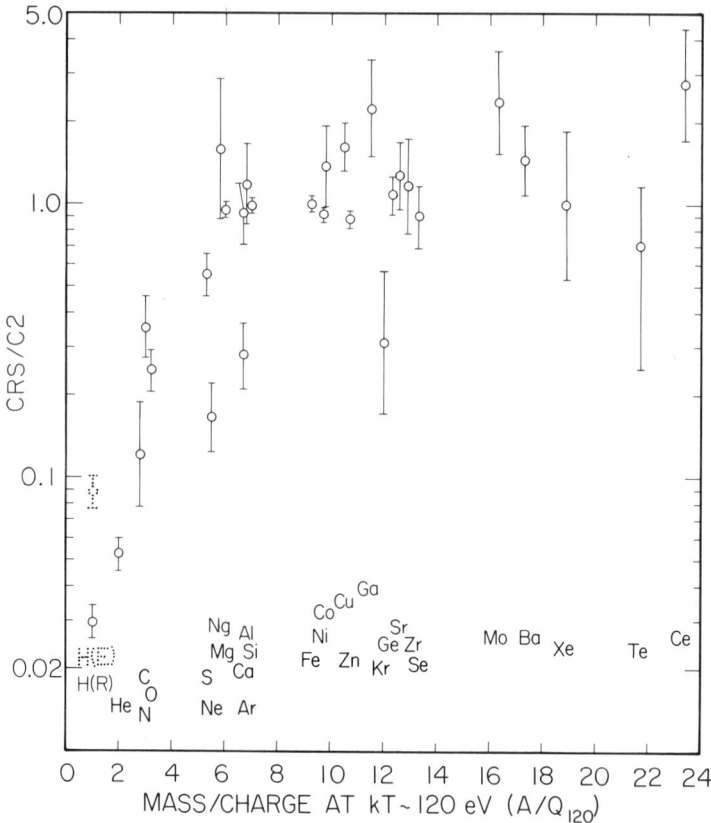

FIGURE 8. Same ordinate as FIGURE 6, with same abscissa as FIGURE 7.

production of the medium from which the acceleration takes place, and/or errors in our standard set of solar system abundances.

ISOTOPIC COMPOSITION

While the elemental composition of cosmic rays can be significantly affected by various atomic processes, it is unlikely that major effects on the isotopic composition would be seen from these processes. The electronic properties of the different isotopes of an element are, of course, the same and the mass-to-charge ratios of different isotopes of the elements to be considered here differ by 10% or less.

The elements, $_{10}$Ne, $_{12}$Mg, and $_{14}$Si, are the only ones for which individual isotopes have been clearly resolved and for which the observed abundances of the rarer isotopes can be extrapolated back to the source with reasonable precision. Like all other elements for which isotopic measurements have been made, the most abundant isotope of each of these elements is the same as the most abundant isotope in the solar system, namely, ^{20}Ne, ^{24}Mg, and ^{28}Si. However, for all three of these elements, the neutron-rich isotopes are significantly enriched in the cosmic rays. The ratio, ^{22}Ne/^{20}Ne, is four times larger in the cosmic rays than in the solar system, while the ratios, ^{25}Mg/^{24}Mg, ^{26}Mg/^{24}Mg, ^{29}Si/^{28}Si, and ^{30}Si/^{28}Si, are each about 1.6 times larger in the cosmic rays than in the solar system.[25,26] For other elements, the source abundances of the rarer isotopes have not been determined with enough precision to see enhancements of a factor of 1.6, either because of small numbers of observed events and thus very poor statistical precision or because secondary fragments of heavier elements contribute a major fraction of the observed flux and so extrapolation back to the source introduces significant uncertainty.

The data points in FIGURE 9 for C through S come from the ISEE-3 spacecraft;[25,26] the Fe data are from a compilation of observations.[26] The ordinate in this figure is the ratio of two abundance ratios; for example, the Ne point gives the ^{22}Ne/^{20}Ne abundance ratio in the cosmic ray source divided by the same ratio in the solar system. If the cosmic ray source and the solar system had the same composition, then all points on the figure would be at unity. The error bars include the effects of the statistical precision of the measurements and of the uncertainties in the extrapolation back to the source.

These isotope observations indicate clearly that the material at the cosmic ray source has had a different nucleosynthesis history than did the material of which the solar system was made. The difference must be due to the fact that the cosmic rays come from a different region of the galaxy than did the solar system and/or that the cosmic rays are a sample of the interstellar medium as it was fairly recently, only ten million years ago, as compared to the solar system, which was formed from the interstellar medium almost five billion years ago.

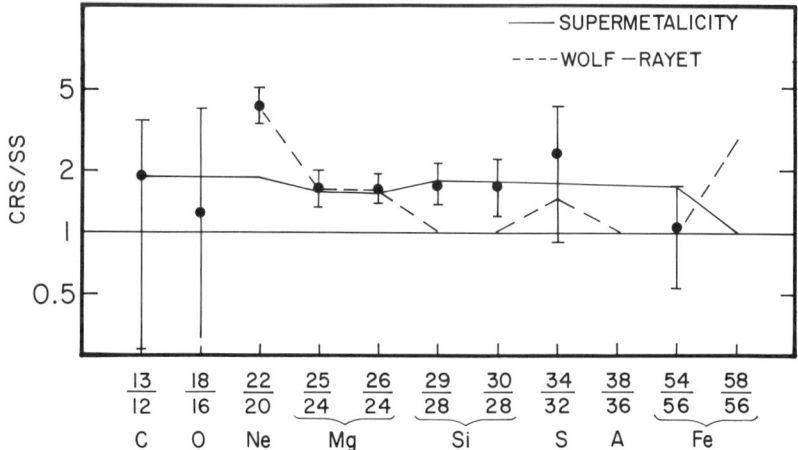

FIGURE 9. Abundance of rare isotopes in the cosmic ray source compared to abundance in the solar system.[25,26] Lines indicate results of calculations using the supermetallicity[27] or the Wolf-Rayet[28-30] model.

Several specific models have been proposed to explain all or some of the five isotope enrichments. These models have the merit of making predictions about the isotopic composition of other elements for which we do not yet have precise enough data. Predictions of two of these models are drawn onto FIGURE 9.

An explanation of the enhancements of ^{25,26}Mg and of ^{29,30}Si, but only part of the enhancement of ^{22}Ne, comes from the supermetallicity model.[27] In this picture, cosmic rays come from a region in which stars have a higher metallicity (i.e., higher abundance of elements heavier than He, principally C, N, and O) than did the stars whose nucleosynthesis produced the material of the solar system. This could be due either to the cosmic rays having contributions from regions closer to the galactic center where a galactic composition gradient gives higher metallicity or simply to chemical evolution of the galaxy during the nearly five billion years since the formation of the solar system.

The supermetallicity model depends on the fact that the hydrogen-burning CNO cycle converts most of the CNO to ^{14}N. The neutron-rich isotopes of Ne, Mg, and Si are made by the series of reactions, ^{14}N(α, γ) ^{18}F$(e^+\nu)$ ^{18}O(α, γ) ^{22}Ne(α, n) ^{25}Mg(n, γ) ^{26}Mg.... A factor of 1.6 metallicity enhancement leads to about a factor of 1.6 enrichment of the neutron-rich isotopes on this chain. This model predicts a similar enrichment of the isotope ^{54}Fe, but no enrichment of ^{58}Fe, which is not made in this process. (Although ^{54}Fe is not neutron-rich compared to the more common ^{56}Fe, it has two more neutrons than protons, as does ^{22}Ne, ^{26}Mg, and ^{30}Si, and ^{56}Fe is created as ^{56}Ni, which has equal numbers of protons and neutrons.) Thus with one free parameter, the metallicity enhancement, this model fits four of the five neutron enrichments and it makes predictions of enrichments of several other isotopes.

Another model explains the full enrichment of ^{22}Ne and of the two isotopes of Mg (but not the Si) by assuming that the cosmic rays are enriched compared to the solar system in material flowing out from Wolf-Rayet stars.[28-30] These are stars that are observed to have anomalous surface composition and strong stellar winds. This model, however, does not account for the enrichment of the heavy isotopes of Si. The predictions of this model for the Fe isotopes are just the opposite of that of the supermetallicity model. With the Wolf-Rayet model one expects no enhancement of ^{54}Fe, but a factor of about three enrichment of ^{58}Fe.

All five of the isotope enrichments can be explained by a suitable combination of the supermetallicity and the Wolf-Rayet models.[31] This combination is plausible because both metal-enriched and Wolf-Rayet stars are more abundant closer to the galactic center and because a plausible model of cosmic ray propagation can have a significant fraction of the cosmic rays we observe originating a few kiloparsecs closer to the galactic center.

A totally different explanation for the cosmic ray isotope enrichments is the suggestion[32] that it may be the solar system rather than the cosmic rays whose abundance is anomalous. In this picture, a last minute injection of alpha-particle nuclei from a nearby supernova enriched the solar system, relative to the average interstellar matter, in the isotopes, ^{20}Ne, ^{24}Mg, and ^{28}Si.

It is clear that isotope measurements have just begun to give us important new information about the cosmic ray source and measurements of other rare isotopes will be necessary to distinguish among the various models that have been suggested.

ANTIPROTONS

We have noted above that the cosmic rays we observe include nuclei that originated as secondary fragments from collisions of heavier cosmic ray nuclei with interstellar gas. In fact, for some elements that are extremely rare in most astrophysical settings, such as $_3$Li, $_4$Be, and $_5$B, the cosmic ray nuclei we observe are almost entirely secondaries. Similarly, it is expected that secondary antiprotons should be present in the cosmic rays, resulting principally from collisions of cosmic ray protons with interstellar hydrogen.

Using the standard propagation model that fits the observations of secondary nuclei, the antiproton production cross sections measured at accelerators, and the observed flux of protons and other nuclei, one can predict the flux of antiprotons that

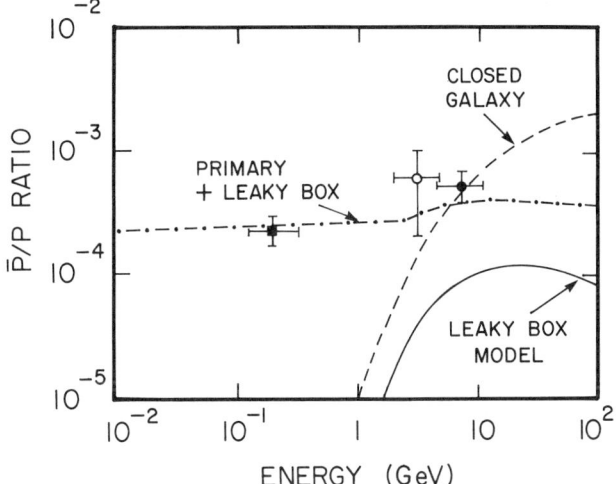

FIGURE 10. Observations of the antiproton-to-proton ratio at various energies[33,34,38] and predictions based on various models.[35,41]

should be observed. Three independent observations of antiprotons have now been reported and all three give significantly greater antiproton fluxes than predicted from this standard model.

The highest energy data point in FIGURE 10 is a measurement made with a balloon-borne superconducting magnetic spectrometer that identifies negative particles by their curvature in the magnetic field. It distinguishes antiprotons from other negative particles—electrons and atmospheric secondary muons—by comparing their rigidity, which is given by their magnetic curvature, with their velocity, given by a gas Cherenkov counter.[33] In this balloon flight,[28] antiprotons were clearly identified, leading to an antiproton to proton ratio of 5×10^{-4} in the energy interval of 5 to 12 GeV. A second balloon-borne magnetic spectrometer experiment reports detection of two antiprotons in the 2 to 5 GeV interval, resulting in an antiproton to proton ratio[34] in agreement with that found above 5 GeV.

The curve labeled "leaky box model" in FIGURE 10 gives the predicted antiproton abundance in the standard propagation model that pictures the galaxy as a leaky box from which the cosmic rays escape after traversing on average about 5 g/cm^2 of interstellar hydrogen.[35] Various modifications of the standard propagation model have been suggested to account for the higher observed antiproton flux. One model suggests that some fraction of the cosmic ray sources are embedded in dense regions of the galaxy from which cosmic rays can emerge after traversing about 50 g/cm^2 of material.[36] From such thick sources almost none of the primary or secondary heavier nuclei would emerge, but these sources would contribute to the observed flux of protons and would be very significant sources of antiprotons.

Another explanation of the high antiproton flux is derived from an alternative model of cosmic ray propagation, the closed galaxy model,[37] in which cosmic rays do not escape the galaxy and both protons and antiprotons are produced in significant numbers by fragmentation over a long time. A closed galaxy model curve[35] with parameters selected that are consistent with observations of secondary nuclei is plotted in FIGURE 10 and is consistent with the high-energy antiproton observations. This model predicts an increasing antiproton fraction with increasing energy and it would obviously be very interesting to make antiproton observations at higher energies.

The lowest point in FIGURE 10 results from a balloon-borne experiment that identified antiprotons as single charged particles for which a Cherenkov counter shows a velocity less than that of a 320-MeV proton and a spark chamber photograph shows an emission of several minimum-ionizing particles characteristic of a nuclear interaction in which about 2 GeV of energy are released.[38] Since the incident particle had much less than 2 GeV of kinetic energy, release of this amount of energy can only be attributed to annihilation of an incident antiproton with a proton of the detector. In this experiment, 14 such events were identified as antiprotons with incident energy between 100 and 300 MeV. It is fair to say that there has been some controversy about whether these 14 events could have been caused by normal particles that might have a very small, but finite chance of simulating an antiproton event in this detector system. Careful calculations by the experimenters have shown that each of the suggested modes by which such a false event could be caused has a probability much too low to account for any significant part of the 14 events. Nevertheless, it would be extremely important for an independent experiment to confirm this observation.

The importance of this observation has been noted by many people; a good review of the situation has been written by Gaisser.[39] The basic point is that the kinematics of antiproton production makes it very difficult to produce antiprotons with energy as low as those observed in this experiment because production of a proton-antiproton pair requires nearly 2 GeV in the center of mass, which translates to nearly 6 GeV in the laboratory frame for a cosmic ray proton incident on an interstellar gas nucleus at rest. This kinematic constraint accounts for the rapid fall with decreasing energy below about 1 GeV of both the leaky box and the closed galaxy curves in FIGURE 10. One possible way to overcome this kinematic constraint would be to have the antiprotons produced and trapped in an expanding region so that their energy is reduced by adiabatic cooling;[40] however, no quantitative prediction of antiproton spectrum has been made with this idea.

The excitement caused by these low-energy antiproton observations is because of the third curve in this figure, which fits the data by making the ad hoc assumption of a primary antiproton flux of about 10^{-4} of the proton flux, nearly independent of energy, superposed on the secondary flux of the standard leakybox model.[41] Of course, the detection of primary antimatter in the cosmic rays would have profound astrophysical and cosmological implications; however, no antinuclei heavier than protons have yet been detected and some experiments have reached sensitivity levels for antihelium that would have detected antiparticle fractions of several times 10^{-4}, comparable to the observed antiproton flux.

Thus, the antiproton observations could turn out to be of great cosmological significance and they are at least telling us about the locations of cosmic ray sources and the subsequent propagation of cosmic rays in the galaxy. Further independent observations of antiprotons at several energies are clearly needed and further searches for heavier antinuclei may also shed light on the subject.

ACKNOWLEDGMENTS

Like any review paper, this one depends on the results of many scientists and thanks are due to all of those whose work is quoted here. I particularly want to thank my coinvestigators on the HEAO-3 Heavy Nuclei Experiment, namely, W. R. Binns and J. Klarmann at Washington University, E. C. Stone at California Institute of Technology, and C. J. Waddington at University of Minnesota; the results on ultraheavy cosmic rays described here are the result of close collaboration among all of us and many of the ideas for synthesizing the results emerged from discussions with this group. Thanks to David Eichler for discussing his work prior to the conference and for suggesting I display the element abundances in an A/Q plot. Thanks to Jonathan Ormes for urging me to include antiprotons in this paper and for providing me with an unpublished proposal that provided an excellent review of this subject.

REFERENCES

1. SIMPSON, J. A. 1983. Annu. Rev. Nucl. Part. Sci. **33**: 323.
2. BINNS, W. R., R. K. FICKLE, T. L. GARRARD, M. H. ISRAEL, J. KLARMANN, E. C. STONE & C. J. WADDINGTON. 1981. Astrophys. J. **247**: L115.
3. BINNS, W. R., D. P. GROSSMAN, M. H. ISRAEL, M. D. JONES, J. KLARMANN, T. L. GARRARD, E. C. STONE, R. K. FICKLE & C. J. WADDINGTON. 1983. 18th International Cosmic Ray Conference, Bangalore (18th ICRC) **9**: 106.
4. STONE, E. C., T. L. GARRARD, K. E. KROMBEL, W. R. BINNS, M. H. ISRAEL, J. KLARMANN, N. R. BREWSTER, R. K. FICKLE & C. J. WADDINGTON. 1983. 18th ICRC **9**: 115.
5. ISRAEL, M. H., W. R. BINNS, D. P. GROSSMAN, J. KLARMANN, S. H. MARGOLIS, E. C. STONE, T. L. GARRARD, K. E. KROMBEL, N. R. BREWSTER, R. K. FICKLE & C. J. WADDINGTON. 1983. 18th ICRC **9**: 305.
6. BINNS, W. R., N. R. BREWSTER, D. J. FIXSEN, T. L. GARRARD, M. H. ISRAEL, J. KLARMANN, B. J. NEWPORT, E. C. STONE & C. J. WADDINGTON. 1985. Astrophys. J. **297**: 111.
7. BINNS, W. R., R. K. FICKLE, T. L. GARRARD, M. H. ISRAEL, J. KLARMANN, E. C. STONE & C. J. WADDINGTON. 1982. Astrophys. J. **261**: L117.
8. CAMERON, A. G. W. 1982. Astrophys. Space Sci. **82**: 123.

9. FIXSEN, D. J. 1985. University of Minnesota Cosmic Ray Report, no. CR-195, and appendix to reference 6.
10. MARGOLIS, S. H. & J. B. BLAKE. 1985. Astrophys. J. **299**: 334.
11. BINNS, W. R., R. K. FICKLE, T. L. GARRARD, M. H. ISRAEL, J. KLARMANN, K. E. KROMBEL, E. C. STONE & C. J. WADDINGTON. 1983. Astrophys. J. **267**: L93.
12. FOWLER, P. H., R. N. F. WALKER, M. R. W. MASHEDER, R. T. MOSES & A. WORLEY. 1981. Nature **291**: 45.
13. FOWLER, P. H., M. R. W. MASHEDER, R. T. MOSES, R. N. F. WALKER & A. WORLEY. 1984. Ninth European Cosmic Ray Symposium, Kosice.
14. WEBBER, W. R. 1982. Astrophys. J. **255**: 329.
15. WEBBER, W. R. & J. A. LEZNIAK. Astrophys. Space Sci. **30**: 361.
16. LUND, N. 1984. Adv. Space Res. **4**(2–3): 5.
17. ANDERS, E. & M. EBIHARA. 1982. Geochim Cosmochim. Acta **46**: 2363.
18. COOK, W. R., E. C. STONE, R. E. VOGT, J. H. TRAINOR & W. R. WEBBER. 1981. 16th ICRC, Kyoto **12**: 265.
19. MEYER, J-P. 1985. Astrophys. J. Suppl. Ser. **57**: 173.
20. BINNS, W. R., D. J. FIXSEN, T. L. GARRARD, M. H. ISRAEL, J. KLARMANN, E. C. STONE & C. J. WADDINGTON. 1984. Adv. Space Res. **4**(2–3): 225.
21. MASON, B. 1979. Geological Survey Professional Paper 440-B-1.
22. MEYER, J-P. 1979. Les Elements et leurs Isotopes dans l'Univers, Liege: 153.
23. EICHLER, D. 1979. Astrophys. J. **329**: 419.
24. EICHLER, D. & K. HAINEBACH. 1981. Phys. Rev. Lett. **47**: 1560.
25. WIEDENBECK, M. E. & D. E. GREINER. 1981. Phys. Rev. Lett. **46**: 682.
26. WIEDENBECK, M. E. 1984. Adv. Space Res. **4**(2–3): 15.
27. WOOSLEY, S. E. & T. A. WEAVER. 1983. Astrophys. J. **243**: 651.
28. CASSE, M. & J. A. PAUL. 1982. Astrophys. J. **258**: 860.
29. MAEDER, A. 1983. Astron. Astrophys. **120**: 130.
30. BLAKE, J. B. & D. S. P. DEARBORN. 1984. Adv. Space Res. **4**(2–3): 89.
31. MAEDER, A. 1984. Adv. Space Res. **4**(2–3): 55.
32. OLIVE, K. A. & D. N. SCHRAMM. 1982. Astrophys. J. **257**: 276.
33. GOLDEN, R. L., S. HORAN, B. G. MAUGER, G. D. BADHWAR, J. L. LACY, S. A. STEPHENS, R. R. DANIEL & J. E. ZIPSE. 1979. Phys. Rev. Lett. **43**: 1196.
34. BOGOMOLOV, E. A., N. D. LUBYANAYA, V. A. ROMANOV, S. V. STEPANOV & M. S. SHULAKOVA. 1979. 16th ICRC **1**: 330.
35. PROTHEROE, R. J. 1981. Astrophys. J. **251**: 387.
36. CESARSKY, C. J. & T. M. MONTMERLE. 1981. 17th ICRC, Paris **9**: 207.
37. PETERS, B. & N. J. WESTERGAARD. 1977. Astrophys. Space Sci. **48**: 21.
38. BUFFINGTON, A., S. M. SCHINDLER & C. R. PENNYPACKER. 1981. Astrophys. J. **248**: 1179.
39. GAISSER, T. K. 1982. Second Moriond Astrophysics Meeting, Les Arcs: 347.
40. EICHLER, D. 1982. Nature **295**: 391.
41. STECKER, F. W., R. J. PROTHEROE & D. KAZANAS. 1983. Astrophys. Space Sci. **96**: 171.

The Origin of Cosmic Rays[a]

DAVID EICHLER

Astronomy Program
University of Maryland
College Park, Maryland 20742
and
Department of Physics
Ben Gurion University
Beer Shera, Israel

INTRODUCTION

Perhaps the most important observation of cosmic rays that a theory of their origin must account for is that they exist. While many other fascinating astrophysical phenomena were anticipated long before their discovery, cosmic rays have precisely the opposite history. It seems improbable, in fact, that theorists would have ever predicted the existence of cosmic rays had they not been observed first. Many theories of cosmic ray acceleration that have appeared in the literature through the years are consistent with an intensity of zero and thus they do not really explain why cosmic rays exist.

Moreover, cosmic rays appear almost wherever there are violent processes in tenuous, ionized astrophysical plasmas. That nature produces them so routinely is itself an important clue: it suggests that there is some basic physical process at work that, for some reason, distributes energy among particles in a highly undemocratic fashion, giving a small minority of them much more energy than most.

Cosmic rays account for much of the nonthermal radiation that has fertilized new branches of astronomy in the past several decades. In order to fully interpret such radiation, we, as astronomers, need to understand the necessary and sufficient physical conditions for cosmic ray production.

The quantitative observational features that a successful theory of cosmic ray origin should account for include: (a) The universal power law spectrum: In a wide variety of production sites, the differential energy spectrum of cosmic rays has the form, $E^{-\alpha}$, where α is typically a bit more than two; (b) High efficiency somewhere between 10 and 100%, with the latter fraction being a safe upper limit; (c) A relatively undistorted composition: In cases where both can be reliably measured, the composition of cosmic rays is remarkably similar to that of the ambient plasma in which they are produced. There may be a slight enhancement of heavy elements relative to hydrogen. There has also been a correlation observed with first ionization potential, in that the elements that are easier to ionize have a higher relative abundance than in the solar photosphere, but this may reflect the relative abundances in the hot phase of the thermal plasma rather than the acceleration mechanism.

It has been shown within the past decade that shocklike solutions to the steady state

---

[a]This work was supported in part by NSF grant no. AST-83-17755 and a NASA STTP grant.

Boltzmann equation of

$$-\vec{v}_L \cdot \frac{\partial f}{\partial \underline{x}} = \frac{f - \langle f \rangle}{\tau} \tag{1}$$

account for all of these observations, including the inevitability of cosmic ray production, given that it is a reasonable description of particle scattering by the plasma turbulence in the vicinity of a shock. Here, v_L is the particle velocity in the frame of the shock, $f(x, p)$ is the phase-space distribution function, the brackets denote an angle average in the frame of scattering centers, which are assumed to have a shocklike velocity profile of $u_s(x)$, and τ is the scattering time, which is in principle a function of momentum and rigidity. Note that the collision operator on the right-hand side of equation (1) represents isotropic elastic scattering in the frame of the scattering centers (pitch angle diffusion by small increments gives all of the same results and could have been used as well). It is believed that this is a reasonable description of particle scattering by hydromagnetic turbulence. It will be seen that cosmic ray production occurs as long as τ does not decrease too rapidly with momentum, p. If the scattering is due to strong magnetic turbulence, τ should be on the order of several gyroperiods and should not decrease at all with p.

More specifically, if u_s is assumed to be piecewise discontinuous, say u_- at $x < 0$ (upstream of the shock) and u_+ at $x > 0$, then for $v \gg u_s$, the postshock spectrum, $f(p)$, can be shown to be proportional to $p^{-3r/(r-1)}$, where $r = u_-/u_+$.

If one applies equation (1) to all particles in the fluid and chooses $u_s(x)$ self-consistently so that the scatterers absorb no net momentum or energy from the particles, one finds, for sufficiently strong shocks and very general $\tau(p)$, the following results: (a) Cosmic rays must be produced; (b) The relativistic cosmic ray spectrum is close to p^{-4} for shock parameters that would be typical of supernova-generated shocks in the hot phase of the interstellar medium; (c) The cosmic rays are produced with high efficiency (of order unity); (d) Given that τ is proportional to gyroperiod and given ionization states of a 10^6 K plasma, the predicted composition is consistent with what is observed to within the accuracy of the observations. All of these statements are made quantitative by the mathematical theory.

Equation (1) is perhaps the simplest imaginable Boltzmann equation. If it fails to include all of the details of particle motion in the vicinity of shocks, it at least illustrates the simple physics behind cosmic ray production in nature. Its success suggests that those simple aspects of the physics prevail over the remaining details.

The following section reviews the work over the past several decades that has led to the above conclusions. After that, I will give a brief discussion of the plasma-physical justification, to the extent that one exists, for using equation (1) to describe particle motion at a strong astrophysical shock. The final section reviews recent observations, some of which support the collisionless shock-wave theory of cosmic ray origin and some of which remain intriguing.

THEORETICAL RESULTS

Fermi, recognizing the importance of the power-law spectrum, proposed[1] that cosmic rays are accelerated by events in which they gain energy geometrically, such as collisions with moving clouds. Given that the escape rate from the acceleration process is a constant, say R, the integral spectrum, Fermi showed, is proportional to $E^{-(R/g)}$, where g is the acceleration rate. However, this demanded an escape rate that is virtually independent of energy, which is unreasonable unless there is some deep geometrical constraint enforcing such an independence.

Addressing the question of how particles are lifted out of the thermal pool to become cosmic rays, Fermi proposed that cosmic rays "reproduced" by knock-on collisions, which would impart several hundred MeV to each (previously thermal) proton, enough to permit it to be picked up by the acceleration mechanism. However, as he himself noted, heavy elements had just been discovered[2] to be a component of cosmic rays and the knock-on scheme could not account for them. This illustrates how measurements of cosmic ray composition can clearly rule out otherwise ingenious theories of cosmic ray origin.

More than a decade later, Schatzmann[3] proposed that cosmic rays could be accelerated into a power-law spectrum at shock fronts. For each round-trip shock crossing, a cosmic ray gains a fixed fraction of its original energy (as in Fermi's scheme). Now for completely laminar trajectories, an individual cosmic ray will make a completely predictable number of shock crossings so that its energy will be multiplied by a well-determined factor, which is uninteresting. Theories of shock acceleration that invoked laminar trajectories traditionally floundered for this reason. Schatzmann, on the other hand, proposed that the cosmic rays undergo scattering while making their way through the shock, so that they do so via a somewhat random trajectory. The actual number of shock crossings is, in this case, a matter of chance. Assuming the influence of the scattering to be energy independent (which is *ad hoc* in the geometry considered by Schatzmann), he was able to obtain power-law spectra whose index, however, depends on the scattering rate.

The modern theory of shock acceleration avoids the *ad hoc* assumptions of Schatzmann and is thus more compelling. Schatzmann had assumed the magnetic field to lie nearly within the shock plane so that the $\vec{v} \times \vec{B}$ Lorentz force would turn a cosmic ray back toward the shock soon after it had crossed it. The actual spectrum then depends on the competition between the scattering and the timescale for the laminar trajectory. The modern theory assumes that the magnetic field has a significant component along the shock normal, and it relies purely on the scattering inhomogeneities to keep the cosmic rays near the shock and to eventually sweep them through; the spectrum then depends purely on simple kinematic considerations. (The picture of Schatzmann, of course, may be more correct in some circumstances, such as in quasi-perpendicular interplanetary shocks, where the scattering apparently does not dominate the interaction of particles with the shock.[4]) Undoubtedly, the work of Lerche,[5] Wentzel,[6] and Kulsrud and Pierce[7] on instabilities that result from anisotropy and of Parker[8] and Jokipii[9] on cosmic ray diffusion in the presence of scattering centers was crucial in bringing about a shift to the modern picture of shock acceleration, which treats the cosmic rays as a second fluid coupled to the thermal fluid via the scatterers. Previously, they had been viewed as individual particles making helical trajectories against the background of the magnetized fluid.

The modern theory of shock acceleration uses equation (1) to describe the motion of particles. When the particle velocity, v, greatly exceeds the flow velocity, u_s, of the scatterers, the distribution function is almost isotropic and one may expand in the anisotropy to obtain from equation (1) the differential question,

$$\frac{\partial f}{\partial t} = 0 = -\vec{\nabla} \cdot (\vec{u}_s f) + \frac{1}{3}(\vec{\nabla} \cdot \vec{u}_s)\left(\frac{1}{p^2}\frac{\partial}{\partial p}p^3 f\right) + \vec{\nabla} \cdot \overleftrightarrow{D} \cdot \vec{\nabla} f. \quad (2)$$

Here the first term describes convection, the second term, compression, and the third, diffusion, where $\overleftrightarrow{D}$ is the diffusion tensor. This equation was introduced by Parker[8] to describe cosmic ray modulation by the solar wind. Axford and Fisk[11,12] applied it to particle motion at a discontinuous shock front. Exploiting the fact that a power law in momentum is an eigenfunction of the compression operator, $1/p^2\, \partial/\partial p\, p^3$, they could reduce the problem to a purely spatial diffusion problem by using an input spectrum that was a power law in momentum.

Several years later, Axford, Leer, and Skadron,[13] Krymsky,[14] Bell,[15] and Blandford and Ostriker[16] derived steady state solutions that showed that the shock acceleration mechanism would convert a delta function spectrum into a power law and that the spectral index was fixed at $3r/(r - 1)$, where r is the shock compression ratio. This showed explicitly that a single shock could accelerate particles by many orders of magnitude in energy and it accounted for the universal power-law spectrum of cosmic rays with much weaker, more plausible assumptions than any previous theory. The astrophysical significance of this result needs no elaboration.

Later, Eichler[17] suggested that if collisionless shocks formed via the hydromagnetic firehose instability, as several models had proposed,[18-21] then a high Alfven Mach number collisionless shock could be reasonably described using the same collision operator that the above theories of shock acceleration had applied to cosmic rays. In this model of a shock, all the particles are assumed to obey equation (1) and the velocity of the scatterers is calculated self-consistently so that conservation of momentum and energy is obeyed. Thus, what had formerly been considered a mechanism for accelerating cosmic rays in the vicinity of a shock front was taken to be the shock mechanism itself. A cosmic ray population, it was argued, would spontaneously arise out of the thermal plasma during the shock process given that the diffusion coefficient is an increasing function of energy in the hydromagnetic turbulence. For then the highest energy particles, at whatever energy they happened to be, would be accelerated roughly according to the linear picture, while the less energetic ones would absorb energy from the bulk flow only after it had been mediated by the higher energy ones. Though all particles are subjected to the mechanism, the available energy is not squandered on bulk heating, as would be the case for many other mechanisms. Rather, the highest energy particles plutocratically control the acceleration of the lower energy ones, thus restricting the entry of the latter into their ranks enough to maintain the privilege of having extremely high energies. The distribution of energy among the particles is as undemocratic as the observations of cosmic rays demand.

We thus have a complete, if not universally believed, description of how particles rise out of a "thermal" pool to attain cosmic ray energies. Moreover, cosmic ray production is unavoidable in strong shocks described by equation (1) and must occur in nature if and wherever such shocks exist. High production efficiency is automatic because it is only the dynamical significance of the cosmic rays that inhibits their own production. That their composition is so close to that of the thermal plasma also follows automatically since they come directly from it. Heavy, partially ionized elements, because they have higher diffusion coefficients, are preferentially accelerated relative to protons (in accordance with the above discussion) by an amount that is consistent with observations. (Actually, the simplest theory predicts this effect to be about a factor of two higher than the observational mean. However, neither theory nor observation should be expected to be more accurate than that.)

The nonlinear theory, in its present state, includes a wide variety of effects, including an explicit description of the waves that couple the particles to the fluid. The full set of equations is:

$$\rho u = c_1 \quad (3)$$

$$\rho u^2 + P_e + P_w + P_g = c_2 \quad (4)$$

$$\frac{d\mathcal{F}_w}{dx} = \frac{udP_w}{dx} + v_{ph}\frac{dP_e}{dx} - RP_w \quad (5)$$

$$RP_w = c_1 T ds \quad (6)$$

$$T ds = dh - \frac{dP_g}{\rho} \quad (7)$$

$$\frac{-\partial}{\partial x}u_s F + \frac{1}{3}\frac{du_s}{dx}p\frac{\partial F}{\partial p} + \frac{\partial}{\partial x}D\frac{\partial}{\partial x}F = 0. \quad (8)$$

Here ρ is the fluid density, u is the fluid velocity, P_e is the pressure in energetic particles, P_w is the wave pressure, P_g is the pressure in gas that has not yet been shocked, $\mathcal{F}_w$ is the energy flux due to the waves, v_{ph} is the phase velocity of the scattering waves relative to the background medium, $u - u_s$, R is the wave damping rate, Tds is the heat per unit mass that enters the fluid via the wave damping, h is the enthalpy per unit mass, and $F = 4\pi p^3 f$. Equations (3) and (4) express mass and momentum conservation, respectively. Equation (5) is an energy inventory equation for the waves with the terms on the right-hand side representing, in order of appearance, amplification of existing waves by compression, growth due to the cyclotron unstable cosmic ray gradient, and damping. It was derived by McKenzie and Volk,[22] who used the results of several earlier contributions.[23–25] Equations (6) and (7) describe the energy deposition into the thermal fluid by the wave damping and equation (8) is the one-dimensional form of equation (2). These equations can all be solved using a nonperturbative nonseparable analytic technique that proves to be extremely accurate in comparison with numerical solutions.[28] The numerical solutions referred to are possible only over a limited dynamical energy range. The analytic technique exploits the fact that in a real situation there is a very large dynamic energy range—ten decades for galactic cosmic rays—so it should be even more accurate in the latter case than in the comparison with the numerical results.

For parameters that are reasonable for supernova remnants in the hot phase of the interstellar medium, the relativistic particle spectra in the nonlinear theory are found[29] to be rather close to the E^{-2} power law yielded by the test-particle picture, though the compression ratios in the former tend to be higher. While the tail of the nonlinear solutions at the highest energies is considerably harder, the overall appearance of the solutions appears closer in a visual inspection to the test-particle solution than either of them are to the observed $E^{-2.7}$ spectrum, which is assumed to be influenced by an energy dependent escape rate. The efficiency of relativistic particle production exceeds 30%.

Another interesting result from the nonlinear theory[27] is that if wave damping is assumed to be negligible, the magnetic field and the energetic particles come into rough equipartition at shocks of modest Alfven Mach numbers ($4 \gtrsim M_A \gtrsim 6$).

COMMENTS ON THE PLASMA PHYSICAL BASIS OF THE THEORY

If cosmic rays are scattered mainly by hydromagnetic turbulence (namely, Alfven waves), then it is reasonable to suppose that they are isotropized in the frame of the waves and that the collision operator of equation (1) is valid.

How might the theory fail? One possibility is that, as noted earlier, the trajectories of the particles may not be completely stochastic, which is a basic assumption of equation (1). For a quasi-perpendicular field geometry, the Lorentz force due to the background magnetic field certainly introduces order and memory into the trajectory of the particles, particularly with regard to their motion along the shock normal. Indeed, shock acceleration appears to be more effective in quasi-parallel shocks where (a) scattering is needed more to impede particle motion along the shock normal and (b) streaming instabilities, which make the hydromagnetic turbulence, are more likely to arise.[30,31] On the other hand, particle trajectories can retain order even in quasi-parallel configurations; simulations by Ghosh[32] show that the waves driven unstable by the energetic particles can take on the appearance of large amplitude solitonlike structures and undergo various limit cycles, depending on the parameter regime. Shock acceleration theory, which invokes the random phase approximation in the description of wave-particle interaction, cannot, in its present state, handle such complications. For oblique shocks, the component of the magnetic field perpendicular to the shock normal is compressed and this introduces a coherent, spatially varying background electromagnetic field into the problem that could significantly affect thermal ions passing through it. Such a structure is not expected for exactly parallel shocks, but it is not yet clear how close to parallel the shocks must be to avoid the effect.

Another limitation of the theory is that it treats the velocity of the scatterers, u_s, only as a function of position. In reality, particles of different rigidity scatter off different waves, which could have different velocities at a given position.

It is also possible that energetic particles interact with the instreaming fluid at the shock via something other than hydromagnetic waves. In the case of superthermal particles streaming through a background fluid, this seems unlikely; Lerche, in his original consideration of the Alfven wave streaming instability, investigated[33] many others and could find none that compete with it. Recent systematic numerical solutions[34] to the dispersion relations for such a system corroborate his conclusions. Though the analogous question for thermal particles is more complex, simulations[35-37] have shown that high Mach number interpenetrating beams of particles stream freely through each other on scales shorter than an ion gyroradius. This suggests that the interaction between the beams is left to instabilities that generate hydromagnetic waves, which, by definition, are of the order of one gyroradius or longer. Simulations of parallel shocks,[38-40] though still in a preliminary state at the time of this writing, have thus far proven to be consistent with the view that they are generated by hydromagnetic instabilities (namely, the firehose instability and its resonant counterpart, the ion-cyclotron instability, which governs the coupling of the superthermal particles to the fluid). Much work in this area remains to be done and is eagerly awaited by the astrophysical community.

THE OBSERVATIONS

Besides being able to account reasonably for the "universal" observations of cosmic rays reviewed in the introduction, the collisionless shock-wave theory of cosmic ray origin has received much support from detailed observations in the solar system. First, energetic particles have been observed to coincide with solar flare-induced interplanetary shocks since the 1960s,[41,42] and the earth's bow shock was eventually detected to be a rather steady source of high energy particles.[43] As the mathematical theory of shock acceleration has developed, increasingly quantitative predictions have become available. At present, the spacecraft observations at the earth's bow shock show agreement with a large number of theoretical predictions, including: (a) the correlation between energetic particle intensities and the magnetic field orientation of the solar wind;[44] (b) the time dependence of the intensity when the field changes;[45] (c) the spectra of all the ion species in steady state;[46] (d) the spatial profile of the energetic particles;[47–49] (e) their composition and absolute intensity;[50] and (f) their velocity anisotropy.[47–49] Except for an occasional fudge factor of order unity, these observations are all accounted for quantitatively with essentially no free parameters. Similar, though less detailed, observational tests for interplanetary shocks prove successful.[51]

The earth's bow shock is really the only astrophysical laboratory available to us. Consequently, the theory of shock acceleration is one of the best tested astrophysical theories.

On the other hand, there still remains much that is poorly understood. The acceleration of electrons, for example, is poorly understood at the plasma physical level. Recently, there have been reports of ultrahigh energy gamma ray emission from several binary X-ray sources.[52–56] They may raise some qualitatively new questions about particle acceleration. While "ultrahigh energy" used to mean above 10^{12} eV (when gamma rays from these sources had been detected by mirrors sensitive to these energies), it currently means $\sim 10^{16}$ eV now that these sources have been detected by air shower arrays. The UHE flux from these objects implies that the systems are putting a large fraction of their energy into particles at $E > 3 \times 10^{16}$ eV. This is truly incredible, even to one who believes the claims made here about the efficiency of particle acceleration in nature.

The UHE gamma radiation typically shows orbital periodicity peaking, in the cases of Cyg X3 and LMC X4, when the neutron star is just emerging from (or going into) eclipse. This can be interpreted as being due to the interaction of the UHE primaries with the atmospheres of the companion star or accretion disk that eclipses them.[57] However, this interpretation implies that the UHE primaries must be produced near the neutron star and must travel to the intervening atmosphere in roughly a straight line. In the case of Her X1, the gamma rays have the 1.237 second spin period of the neutron star,[56,58] lending further support to the picture that the acceleration takes place in a very compact region. How this acceleration takes place though is still a difficult question. Attributing such acceleration to pulsars suffers from the difficulty that many of the neutron stars, such as in LMC X4 and Vela X1, are rotating too slowly to generate a potential of 10^{17} eV. The suggestion that the pulsar mechanism is actually taking place in the magnetosphere of the accretion disk[59] suffers from the difficulties (a) that in the case of Her X1, it is hard to understand why the gamma rays have the 1.237 second periodicity and (b) that in order to establish potential drops of 10^{17} volts, the magnetic fields of the neutron stars must be weak enough to allow Keplerian velocities of the order of $c/3$ before trapping the accretion flow; this implies field strengths of only 10^9 gauss or so and implies that the accretion disks would spin up the neutron stars well beyond their observed periods. In the case of Her X1, the surface field is deduced to be several times 10^{12} gauss,[60] implying that the inner radius of the accretion disk is about 3×10^8 cm; thus, any energy released outside of this radius due ultimately to accretion must be only about 0.3% of the total.

Finally, shock acceleration, though not clearly ruled out, faces formidable limitations from the fact that at large enough energies, synchrotron radiation or particle diffusion away from the region suppresses further acceleration. This maximum energy attainable, in a region of size, R, can be shown to be

$$E_{max} \approx \beta_s \, (R/10^8 \text{cm})^{1/3} \times 10^{17} \text{ eV} \qquad (9)$$

for protons,[61] where $\beta_s c$ is the shock velocity. It is clearly difficult to achieve an energy of $\sim 10^{17}$ eV if the shock is in an accretion flow of a neutron star. Relativistic outflows, such as the one from SS433, might prove to be more promising if they can be shocked near the neutron star.

At the time of this writing, the observations are being reported at a rapid pace and it is difficult to say anything more conclusive. If they prove to be correct, particularly if all of them prove to be, they will pose a great challenge.

SUMMARY AND FURTHER REMARKS

Enormous progress has been made in understanding the origin of cosmic rays within the past decade. The success of equation (**1**) in accounting for the observed properties of cosmic rays at both the general and the detailed level is a striking illustration that nature can do marvelous things with simple equations.

Also illustrated here is the important role of detailed, systematic spacecraft observations in the heliosphere in testing theories of relevance to the distant mysterious phenomena that pique the curiosity of astronomers.

Tracing the origin of cosmic rays back to collisionless shocks has reminded plasma astrophysicists of how much remains to be understood about the physics of such shocks, which account for much of the radiation that high energy astrophysics is based upon. The X-ray emission from shock-heated electrons, for example, cannot be fully interpreted until the physics of the shocks is understood. It is hoped that plasma simulations of shocks combined with intensive studies of the relevant microphysics will eventually lead us to such an understanding.

Finally, the recent observations of UHE gamma radiation from binary X-ray rources, each one more baffling than the previous one, show that nature remains more clever than theorists in its remarkable ability to accelerate particles.

ACKNOWLEDGMENTS

I am indebted to D. Ellison for vital help in preparing this review.

REFERENCES

1. FERMI, E. 1949. Phys. Rev. **75**: 7169.
2. FREIER, P., N. LOFGREN & R. OPPENHEIMER. 1948. Phys. Rev. **74**: 449.
3. SCHATZMANN, E. 1962. Ann. Astrophys. **137**: 135.
4. PESSES, M. E. 1979. Ph.D. thesis (University of Iowa) and references therein.
5. LERCHE, I. 1967. Astrophys. J. **147**: 689.
6. WENTZEL, D. G. 1968. Astrophys. J. **152**: 987.
7. KULSRUD, R. & W. F. PEARCE. 1969. Astrophys. J. **156**: 445.
8. PARKER, E. N. 1965. Planet. Space Sci. **13**: 9.
9. JOKIPII, J. R. 1966. Astrophys. J. Lett. **146**: L80.
10. JOKIPII, J. R. 1971. Rev. Geophys. Space Sci. **9**: 27.
11. FISK, L. A. & W. A. AXFORD. 1969. Trans. Am. Geophys. Union **50**: 307.
12. FISK, L. A. 1971. J. Geophys. Res. **76**: 1662.
13. AXFORD, W. A., E. LEER & C. SKADRON. 1977.
14. KRYMSKY, G. F. 1977. Dok. Akad. Nauk. SSSR **234**: 1306.
15. BELL, A. R. 1978. Mon. Not. Astron. Soc. **182**: 147.
16. BLANDFORD, R. D. & J. P. OSTRIKER. 1978. Astrophys. J. Lett. **221**: L19.
17. EICHLER, D. 1979. Astrophys. J. **229**: 419.
18. PARKER, D. 1979. Astrophys. J. **229**: 419.
19. KENNEL, C. F. & R. Z. SAGDEEV. 1967. J. Geophys. Res. **72**: 3302.
20. JACKSON, R. W. 1983. J. Geophys. Res. **88**: 9981.
21. AUER, R. D. & H. J. VOLK. 1973. Astrophys. Space Sci. **22**: 243.
22. MCKENZIE, J. F. & H. J. VOLK. 1982. Astron. Astrophys. **116**: 191.
23. DEWAR, R. L. 1970. Phys. Fluids **13**: 2710.
24. SCHWARZ, S. J. & J. SKILLING. 1978. Astron. Astrophys. **70**: 607.
25. ACHTERBERG, A. & C. A. NORMAN. 1980. Astron. Astrophys. **89**: 353.
26. EICHLER, D. 1984. Astrophys. J. **277**: 429.
27. EICHLER, D. 1985. Astrophys. J. In press.
28. ELLISON, D. C. & D. EICHLER. 1984. Astrophys. J. **286**: 691.
29. ELLISON, D. C. & D. EICHLER. In preparation.
30. IPAVICH, F. M., M. SCHOLER & G. GLOECKLER. 1981. J. Geophys. Res. **86**: 11,153.
31. KENNEL, C. F., F. L. SCARF, F. V. CORONITI, E. J. SMITH & D. A. GURNETT. 1982. J. Geophys. Res. **72**: 3303.
32. GHOSH, S. 1985. Ph.D. thesis, University of Maryland.
33. LERCHE, I. Private communication.
34. GARY, S. P. 1981. J. Geophys. Res. **86**: 4331.
35. MCKEE, C. F. 1970. Phys. Rev. Lett. **24**: 990.
36. FORSLUND, D. W. & C. R. SHONK. 1970. Phys. Rev. Lett. **25**: 1699.
37. PAPADAPOULOUS, K. 1972. Unpublished.
38. KAN, J. R. & D. W. SWIFT. 1983. J. Geophys. Res. **88**: 9981.
39. MANDT, M. E. & J. R. KAN. J. Geophys. Res. In press.
40. QUEST, K. B. 1985. Proc. of the Chapman Conference on Collisionless Shocks.
41. AXFORD, W. I. & G. C. REID. 1962. J. Geophys. Res. **67**: 1692.
42. AXFORD, W. I. & G. C. REID. 1963. J. Geophys. Res. **68**: 1793.
43. ASBRIDGE, J. R., S. J. BAME & I. B. STRONG. 1968. J. Geophys. Res. **73**: 5777.
44. SCHOLER, M. J., F. M. IPAVICH & G. GLOECKLER. 1981. J. Geophys. Res. **86**: 4374.

45. IPAVICH, F. M., M. SCHOLER, & G. GLOECKLER. 1981. J. Geophys. Res. **86:** 11,153.
46. EICHLER, D. 1981. Astrophys. J. **244:** 711.
47. ELLISON, D. C. 1981. Geophys. Res. Lett. **8:** 991.
48. LEE, M. A., G. SKADRON & L. A. FISK. 1981. Geophys. Res. Lett. **8:** 401.
49. FORMAN, M. A. 1981. Proc. 17th International Cosmic Ray Conference **3:** 467.
50. ELLISON, D. C. 1981. Ph.D. thesis, Catholic University.
51. FISK, L. & M. A. LEE. 1980. Astrophys. J. **237:** 620.
52. SAMORSKI, M. & W. STAMM. 1983. Astrophys. J. Lett. **268:** L17.
53. LLOYD-EVANS, J. et al. 1983. Nature **305:** 784.
54. PROTHEROE, R. J. et al. 1984. Astrophys. J. Lett. **280:** L47.
55. PROTHEROE, R. J. et al. 1984. Nature. Submitted.
56. BALTRUSAITUS, R. M. et al. 1985. Preprint.
57. VESTRAND, W. T. & D. EICHLER. 1982. Astrophys. J. **261:** 251.
58. DOWTHWAITE, J. C. et al. 1984. Nature **309:** 691.
59. CHANMUGAM, G. & K. BRECHER. 1985. Nature. In press.
60. TRUMPER, J. et al. 1978. Astrophys. J. Lett. **219:** L105.
61. EICHLER, D. & W. T. VESTRAND. In preparation.

Topics in Gamma Ray Astronomy

R. RAMATY[a] AND R. E. LINGENFELTER[b]

[a]*Laboratory for High Energy Astrophysics*
NASA/Goddard Space Flight Center
Greenbelt, Maryland 20771

[b]*Center for Astrophysics and Space Science*
University of California, San Diego
La Jolla, California 92093

INTRODUCTION

Gamma ray emission has been observed from a rich variety of astronomical sites, ranging from solar flares to active galaxies. These observations, carried out with detectors on the ground, on balloons, on satellites, and on space probes, provide unique insights into a variety of astrophysical problems. We have reviewed these in considerable detail previously.[1] Since the completion of this review, new observations have become available: Gamma ray spectra from solar flares have been observed[2] in much greater detail by a spectrometer on the Solar Maximum Mission (SMM), allowing several new studies including the determination of chemical abundances in the solar atmosphere. A gamma ray line from radioactive ^{26}Al was seen[3] from the interstellar medium by a high-resolution spectrometer on the Third High Energy Astronomical Observatory (HEAO-3), providing new information on processes of explosive nucleosynthesis in the Galaxy. Gamma ray lines have been reported[4] from the compact galactic object SS433, also with this HEAO-3 instrument, possibly providing clues for the understanding of the acceleration of the jets that are revealed by optical and radio observations. Gamma rays up to $\sim 10^{16}$ eV have been observed[5] from the galactic binary compact object Cygnus X-3 and very high-energy gamma rays have been detected from several other compact objects as well (TABLE 1), demonstrating that particles can be accelerated in compact sources to ultrahigh energies.

The known sources of astronomical gamma rays are listed in TABLE 1. In the present paper, we review the observations and implications of gamma rays from solar flares, gamma ray bursts, the Galactic Center, galactic nucleosynthesis, SS433, and Cygnus X-3. The galactic point sources not discussed in the present paper were reviewed in reference 7, diffuse galactic continuum emission was reviewed in reference 1 (see also reference 14), extragalactic point sources were reviewed in reference 1, and diffuse extragalactic emission was reviewed in references 1 and 17.

SOLAR FLARES

Recent observational and theoretical studies of gamma rays and neutrons from solar flares have provided new insights into the problem of particle acceleration and confinement, and they have given new information on the composition of the solar atmosphere. These results have been discussed in a number of recent papers (e.g., see references 2, 18–21). The gamma ray lines and neutrons result from nuclear interactions of accelerated protons and heavier nuclei, while the continuum is due to relativistic electron bremsstrahlung and the superposition of Doppler-broadened gamma ray lines.

Theoretical studies predicted[22] that the principal gamma ray lines should be those at 2.223 MeV from neutron capture on ^{1}H, at 0.511 MeV from positron annihilation,

TABLE 1. Astronomical Gamma Ray Sources

Source	Observational Characteristic	Source Characteristic
Solar Flares (second section in this paper)	0.4–7 MeV lines $\lesssim$100 MeV continuum	Particle acceleration Nuclear interactions
GBS 0526-66 (third section)	$\lesssim$2 MeV continuum ~0.4 MeV line 8 sec period Repeating outbursts	Neutron star localized to SNR N49 in LMC
Other Gamma Ray Transients (third section and ref. 1 for review)	$\lesssim$10 MeV continuum <0.1 MeV lines ~0.4–6 MeV lines 0.1–1000 sec durations	Magnetic neutron stars
Galactic Center (fourth section)	0.511 MeV line $\lesssim$3 MeV continuum <0.5 yr variability	Black hole Relativistic plasma
SS433 (sixth section)	Doppler-shifted lines	Relativistic jets
Crab Pulsar (ref. 6 and ref. 1 for review)	$\lesssim 10^3$ GeV continuum ~75 keV line 33 msec period	Magnetic neutron star
Vela Pulsar (ref. 7 for review)	$\lesssim$ GeV continuum 89 msec period	Magnetic neutron star
Geminga (refs. 8, 9 and ref. 7 for review)	$\lesssim 10^3$ GeV continuum 59 sec period	Compact object
Hercules X-1 (refs. 10, 11, 12)	$\lesssim 10^3$ GeV continuum 1.24 sec period $\lesssim$50 keV line	Magnetic neutron star in binary system
Cygnus X-3 (seventh section)	$\lesssim 10^{16}$ eV continuum 4.8 hour period	Compact object in binary system
Vela X-1 (ref. 13)	$\gtrsim 3 \times 10^{15}$ eV continuum 8.9 days period	Compact object in binary system
Rho Oph (ref. 1 for review)	$\lesssim$ GeV continuum	Interstellar cloud cosmic ray interactions
Other Galactic Sources (ref. 7 for review)	$\lesssim$ GeV continuum localized emission in the COS B survey	Unknown

TABLE 1. Continued

Source	Observational Characteristic	Source Characteristic
Interstellar Medium (fifth section and ref. 1 for review)	Line at 1.809 MeV Continuum ≲ GeV	Explosive nucleosynthesis. Cosmic ray interactions
LMC X-1 (ref. 15)	≳10^{16} eV continuum 1.4 day period	Compact object in binary system
NGC 4151 (ref. 1 for review)	≲ MeV continuum	Seyfert Galaxy
MCG 8-11-11 (ref. 16)	≲ MeV continuum	Seyfert Galaxy
Cen A (ref. 1 for review)	≲10^2 GeV continuum	Radio Galaxy
3C273 (ref. 1 for review)	≲200 MeV continuum	Quasar
Unresolved Extragalactic Emission (ref. 1 for review)	Continuum ≲200 MeV	Unknown

and at 4.438 and 6.129 MeV from de-excitation of nuclear levels in ^{12}C and ^{16}O, respectively. These predictions were confirmed when gamma rays were first observed[23] with a detector on OSO-7 from the solar flare of 4 August 1972. These and other weaker lines have since been observed from more than 30 flares by detectors on HEAO-1,[24] HEAO-3,[25] HINOTORI,[26] and, most extensively, SMM.[2,19,27] Neutrons from solar flares have also been observed, thus confirming earlier predictions.[28] The neutron observations consist of direct spacecraft[29,30] and ground-based detections,[31,32] as well as of the measurement[33] of the protons resulting from the decay of the neutrons in interplanetary space.

Energetic particles from solar flares have been observed in interplanetary space on numerous occasions, but there is clear evidence that the nuclear interactions that produce the gamma rays and neutrons are caused by accelerated particles that remain trapped in the magnetic fields of the flare region and interact as they slow down in the solar atmosphere. This is most clearly seen (e.g., reference 34) by the fact that, if the escaping particles were responsible for the observed gamma ray emission, they should also show great enrichments in spallation products, such as ^{2}H, ^{3}H, Li, Be, and B, which are not observed.[35]

Further evidence for this trapping comes from the comparison of the number of particles required to produce the observed gamma rays and neutrons with the number of escaping particles, and from the comparison of the number of positrons produced at the sun with the observed flux in the 0.511-MeV line. We now proceed to discuss these comparisons in more detail.

The number of gamma ray producing particles can be derived from measurements of the neutron-capture line at 2.2 MeV and the photon flux in the 4 to 7 MeV band,

TABLE 2. Energetic Particle Parameters in Solar Flares[a]

	In Solar Atmosphere				In Interplanetary Space	
	Bessel Function		Power Law			
Flare	αT	N_p (>30 MeV)	s	N_p (>30 MeV)	Spectral Index	N_p (>30 MeV)
	Determined from Gamma Ray Line Measurements					
4 Aug. 1972	0.029 ± 0.004	1.0×10^{33}	3.3 ± 0.2	7.2×10^{32}	—	4.3×10^{34}
11 Jul. 1978	~0.032	1.6×10^{33}	~3.1	1.3×10^{33}	—	—
9 Nov. 1979	0.018 ± 0.003	3.6×10^{32}	3.7 ± 0.2	2.6×10^{32}	—	—
7 Jun. 1980	0.021 ± 0.003	9.3×10^{31}	3.5 ± 0.2	6.6×10^{31}	$\alpha T \simeq 0.015$	8×10^{29}
1 Jul. 1980	0.025 ± 0.006	2.8×10^{31}	3.4 ± 0.2	1.9×10^{31}	—	$<4 \times 10^{28}$
6 Nov. 1980	0.025 ± 0.003	1.3×10^{32}	3.3 ± 0.2	1.0×10^{32}	—	3×10^{29}
10 Apr. 1981	0.019 ± 0.003	1.4×10^{32}	3.6 ± 0.2	1.0×10^{32}	—	—
	Determined from Neutron and Gamma Ray Line Measurements					
21 Jun. 1980	0.025 ± 0.005	7.2×10^{32}	inconsistent		$\alpha T \simeq 0.025$	1.5×10^{31}
3 Jun. 1982	0.034 ± 0.005	2.9×10^{33}	inconsistent		$s \simeq 1.7$	3.6×10^{32}

[a]From reference 20.

which is dominated[36,37] by C and O de-excitation lines. Since the effective threshold for neutron production is significantly higher than that for C and O excitations, the 2.2-MeV line and the 4 to 7 MeV band sample different portions of the accelerated particle spectrum. The ratio of the fluxes in the 2.2-MeV line and in the 4 to 7 MeV band therefore constrains the particle spectrum, while the 4 to 7 MeV flux determines the particle number. Results for several flares from which gamma rays were observed are summarized in TABLE 2. The spectral indexes and total proton numbers at the sun are given for two possible forms for the accelerated particle energy spectra; namely, a power law in kinetic energy and a Bessel function. For the former, the number of accelerated particles per unit kinetic energy is proportional to E^{-s}, where E is particle kinetic energy. For the latter, this number is proportional to $K_2[2(3p/m_p c \alpha T)^{1/2}]$, where p is the particle momentum per nucleon and αT is an index characterizing the hardness of the spectrum. A power law in kinetic energy is the nonrelativistic approximation of a power law in momentum, which is the spectral form expected (e.g., reference 38) from first order shock acceleration at a planar and infinite shock. The Bessel-function spectrum is the nonrelativistic approximation to the spectrum expected from stochastic acceleration.[39] Nonrelativistic approximations are adequate for calculations involving protons and nuclei since the bulk of the nuclear reactions in flares occur at energies much lower than $m_p c^2$.

The number of neutron-producing particles and their energy spectrum can be derived from observations of the time-dependent neutron flux at Earth. For consistency, this number and this spectrum must be the same as those derived from the gamma ray observations. Observations[2] of a time-dependent neutron flux are shown in FIGURE 1, together with calculated[20] fluxes. These fluxes are normalized such that the calculated 4 to 7 MeV flux agrees with the observed flux in this energy band, namely, ~76 photons/cm². It is evident that the combined neutron and gamma ray emission cannot result from particles with a power-law spectrum. For, as we see from FIGURE 1B, none of the combinations of power-law spectra and total particle numbers that could produce the observed 4 to 7 MeV flux can also produce a neutron flux consistent with that which was measured. As can be seen in FIGURE 1A, however, both observations are quite consistent with accelerated particles having a Bessel-function spectrum with $\alpha T \sim 0.025$ and a total number of 7×10^{32} protons >30 MeV. Qualitatively, the difference between this Bessel-function spectrum and a power law in kinetic energy amounts to the gradual steepening of the former as the energy increases. Shock acceleration can also produce[38] such a steepening (or high-energy cutoff) if the shock is of finite size and the acceleration is of finite duration. Thus, while these results cannot differentiate between the acceleration mechanisms, they demonstrate that a consistent interaction model can be set up involving either one of them.

Comparing these results with those inferred from the direct particle observations (TABLE 2), we see that independent of the spectral form, the number of particles that produce the observed gamma rays and neutrons are generally much higher than the number of interplanetary particles from flares that produce detectable gamma rays. This implies that the gamma rays and neutrons are produced predominantly in closed magnetic configurations from which very few charged particles escape. As mentioned above, the absence of spallation products in the escaping particles indicates that this latter population is not involved in significant gamma ray and neutron production. We discuss separately the exceptional case of the 4 August 1972 flare for which the number of particles observed in interplanetary space was much larger than the number of trapped particles (TABLE 2).

Further evidence that the gamma rays are generally produced in closed magnetic configurations comes from the analysis of the time-dependent flux of the 0.511-MeV

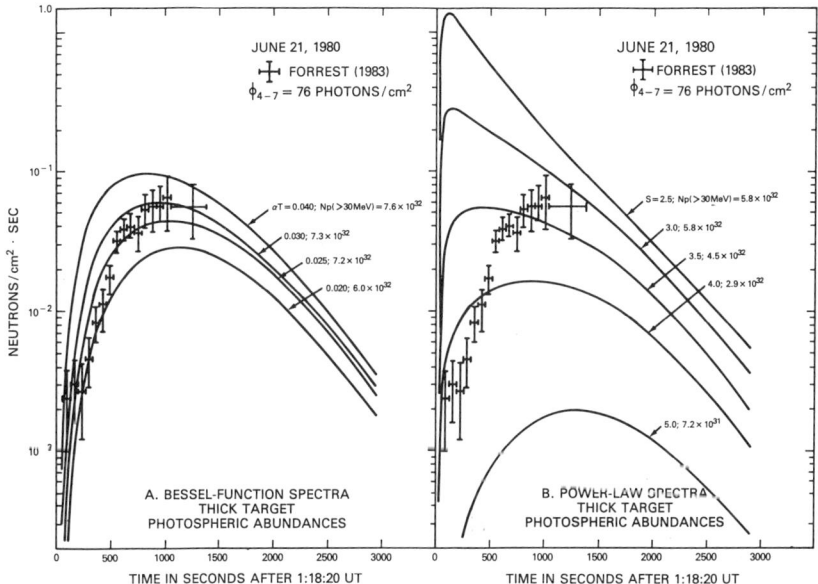

FIGURE 1. Determination[20] of the number and spectrum of flare-accelerated protons at the sun from observations[2] of the time-dependent neutron flux and the gamma ray emission in the 4–7 MeV range.

line from positron annihilation. This is shown in FIGURE 2, where observations[40] of the 21 June 1980 flare are compared with the calculated[20] 0.511-MeV flux. In these calculations, the radioactive β^+ emitters and π^+ mesons were produced by accelerated particles with the same spectrum and total number as determined from the neutron and 4 to 7 MeV observations, and it was assumed that the positrons remain trapped at the sun and that they annihilate essentially instantaneously. The agreement with the observations seen in FIGURE 2 is an indication that these assumptions are valid. The trapping of the positrons is further evidence for the trapping of all the gamma ray producing charged particles, while their short annihilation time (<10 sec) implies that the ambient density is greater than $>10^{11}$ cm^{-3}. This suggests that the annihilation site, and hence probably also the site of the nuclear interactions, is in the chromosphere below the transition layer.

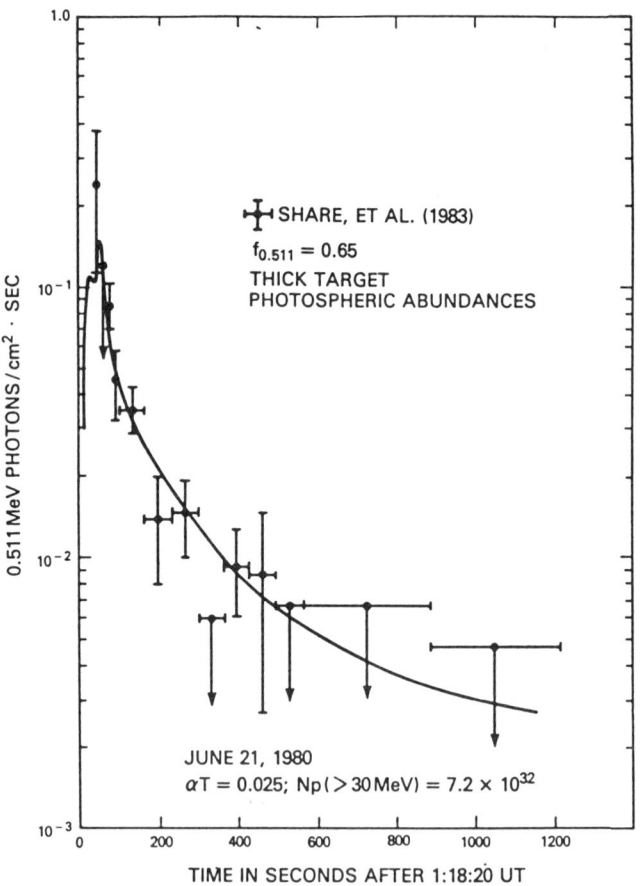

FIGURE 2. Observed[40] 0.511-MeV line flux from the 21 June 1980 flare compared with that expected[20] from the number and spectrum of accelerated particles determined in FIGURE 1.

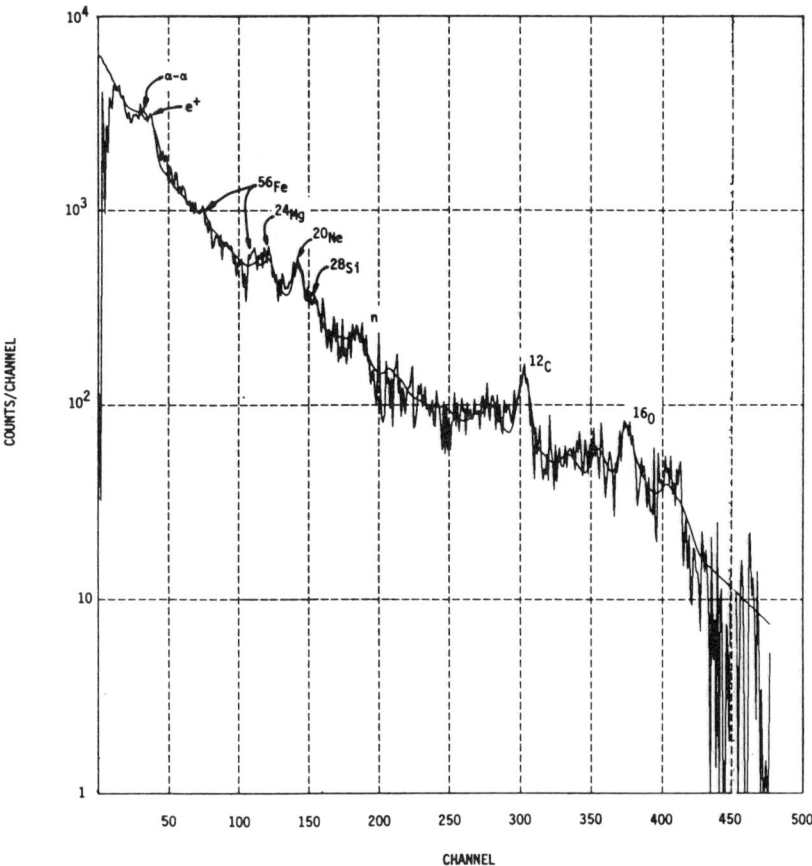

FIGURE 3. Observed[2,19] and calculated[42] spectra of the 27 April 1981 flare.

In addition to the 4 August 1972 flare, for which the number of interplanetary particles was much larger than that involved in gamma ray production, there are many other flares[41] that produce large fluxes of interplanetary particles without producing detectable gamma rays. These particles are devoid of spallation products and are most likely accelerated at coronal sites with ready access to interplanetary space.

We turn now to the question of the composition of the solar atmosphere in the flare region as determined from comparisons of the various de-excitation line intensities. The spectrum shown in FIGURE 3 was observed[2,19] from the 27 April 1981 flare by the gamma ray spectrometer on SMM. As already mentioned, nuclear reactions of accelerated protons and α particles with heavier nuclei in the ambient gas produce narrow lines, such as those seen at 6.129 MeV from de-excitation of $^{16}O^*$, at 4.438 MeV from $^{12}C^*$, at 1.779 MeV from $^{28}Si^*$, at 1.634 MeV from $^{20}Ne^*$, at 1.369 MeV from $^{24}Mg^*$, and at 0.847 MeV from $^{56}Fe^*$. The inverse reactions, between accelerated heavy nuclei and ambient H and He, produce broad lines that effectively merge into a continuum. Also evident are the lines at 2.223 and 0.511 MeV. The feature just below the positron annihilation line results from reactions between accelerated α particles and ambient He nuclei leading to $^7\text{Li}^{*0.478\text{MeV}}$ and $^7\text{Be}^{*0.431\text{MeV}}$. The continuum, upon which the narrow lines are superimposed, is due to both relativistic electron bremsstrahlung and the Doppler-broadened de-excitation lines of the accelerated heavy nuclei.

The relative intensities of the narrow nuclear de-excitation lines depend on several factors, such as the energy spectrum of the accelerated particles, but they are most sensitive to the elemental abundances of the ambient gas in the interaction region. Even though the location of this region cannot be determined by direct gamma ray imaging, indirect arguments, such as the time dependence of the 0.511-MeV line discussed above, indicate that most of the nuclear reactions take place in the chromosphere. The observed gamma ray spectrum, therefore, can be used to infer the chromospheric abundances. The most direct evaluation[21] consists of a theoretical calculation of the spectrum with variations of the abundances until the best fit to the data is achieved. The resultant best-fitting spectrum[42] is shown by the smooth curve in FIGURE 3. The implied abundances are shown in FIGURE 4 relative to local galactic abundances,[43] which are believed to be similar to photospheric abundances. The two abundance sets can be normalized arbitrarily, but for the result shown in FIGURE 4, the normalizations were chosen so that the discrepancy between the sets is minimal.

With this normalization, the principal difference between the gamma ray and local galactic abundances is the underabundance of C and O in the gamma ray set. The Fe, Si, Mg, and Ne abundances are in good agreement, but the statistical errors for Ca, S, Al, and N and the systematic errors for H and He are too large to permit any quantitative conclusion (see reference 21). A similar suppression of C and O in the coronal abundances relative to local galactic abundances has been pointed out in reference 43, where it was suggested that the suppression may be caused by charge-dependent mass transport from the photosphere to the corona. Since the photosphere is collisionally ionized at a relatively low temperature, the transport could depend on the first ionization potentials of the elements. Mass transport to the chromosphere could be influenced by similar fractionation effects. However, if the Ne abundance in the photosphere (where it cannot be measured) is the same as in the local galactic set, then the mechanism that produces differences between the gamma ray and photospheric abundances must include additional effects because correlation with first ionization potential alone would predict a Ne abundance at least as low as the O abundance, which is contrary to the results shown in FIGURE 4.

Independent of the mechanism responsible for the fractionation, it is quite obvious that significant abundance differences exist between various sites in the solar atmosphere. It seems inevitable that similar fractionation phenomena could affect the abundance determinations of objects other than the sun.

GAMMA RAY BURSTS

Gamma ray bursts were discovered[44] accidentally in 1967 by detectors on board the Vela satellites whose primary purpose was to monitor artificial nuclear detonations in space. The observational properties of the bursts and current theoretical ideas about their origin have been extensively reviewed in recent workshop proceedings.[45,46]

Gamma ray bursts are generally observed in the photon energy range from a few tens of keV to several MeV, with event durations ranging from about 0.1 to 100 sec. The observed burst energy fluences (>30 keV) range from about 10^{-7} to 10^{-3} erg/cm^2 and the frequency of occurrence of bursts ranges from about ten per year with fluences

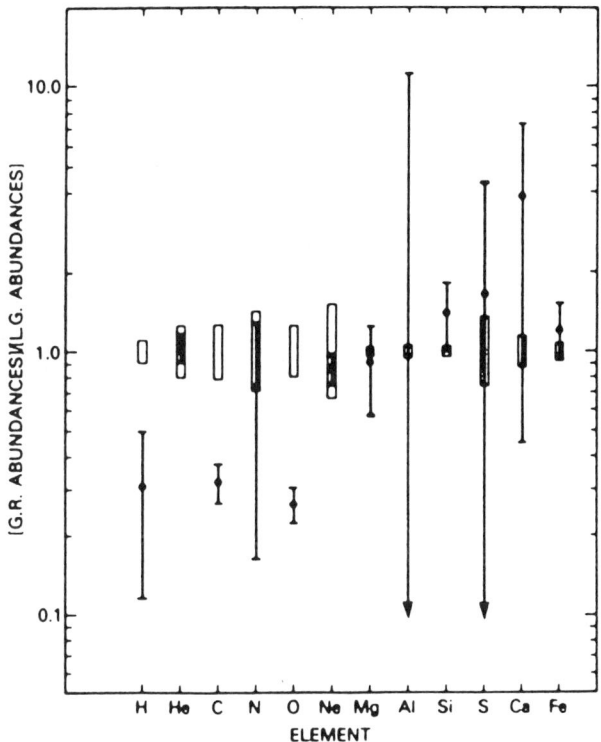

FIGURE 4. Ratios of chemical abundances.[21] The solid dots are ratios of the solar atmospheric abundances determined from the gamma ray spectrum shown in FIGURE 3 to the mean local galactic abundances,[43] while the bars represent the local galactic abundances divided by their mean.

>10^{-4} erg/cm^2 to a few thousand per year with fluences > 10^{-7} erg/cm^2. At fluences less than 10^{-5} erg/cm^2, the frequency of bursts falls below that which might be expected from an unbounded, isotropic, and homogeneous distribution of sources.[47,48] Although it has been suggested that this results from the finite galactic distribution of sources and is thus evidence for a galactic origin, recent studies[49,50] have shown that this deviation can be explained entirely by temporal and spectral selection biases in the detectors.

The distribution of gamma ray burst source directions on the sky is essentially isotropic,[47] which suggests that if they are galactic, then the sources typically lie within a scale-height of the disk ≤ 1 kpc and release energies of $<10^{39}$ ergs.

The determination[51] of several very precise source positions, however, has not led to the identification of any burst sources with known objects, except for one case. That exception is the source of the 5 March 1979 burst, GBS 0526-66, whose positional error box[52] of size 0.1 arcmin^2 lies within the supernova remnant N49 in the Large Magellanic Cloud at a distance of 55 kpc. If the burst source is at this distance, the total radiated energy is $\sim 10^{44}$ ergs, which is about five orders of magnitude larger than that inferred for a typical galactic gamma ray burst. However, the March 5 burst exhibited a number of remarkable and possibly unique observational characteristics, including[53,54] the extremely rapid rise time ($<2 \times 10^{-4}$ sec) of the impulsive emission spike, the relatively short duration (~ 0.15 sec) and high luminosity of this spike, the 8-sec pulsed emission following the impulsive spike, and 15 subsequent,[55] apparently nonrandom,[56] outbursts of lower intensity from the same source direction over the last several years. Thus, this burster appears[53,57] to belong to a different class of transients than that of most other galactic bursts.

Although searches (e.g., reference 57) of other positional error boxes have not produced any likely source objects, a search[59,60] of archival optical plates has revealed evidence of possible optical flashes from a couple of the burst sources in the past. Very recently, optical flashes have also been detected[61] from the direction of the repeating 5 March 1979 source direction. This appears to open a new window for monitoring such bursts, but simultaneous optical and gamma ray observations are still needed before it can be established that gamma ray bursts are in fact accompanied by detectable optical flashes.

The best insight into the nature of gamma ray burst sources has come from the discovery[62] of absorption and emission features in the energy spectra of the bursts. The absorption features have been observed[62,63] in a number of spectra, generally in the energy range from about 30 to 60 keV, as can be seen in the spectrum of the 25 March 1978 burst[63] shown in FIGURE 5. In X-ray binaries, such absorption features[12] appear to be the result of cyclotron absorption in intense magnetic fields of a few times 10^{12} gauss, which strongly suggests that magnetic neutron stars are the sources of most gamma ray bursts. Moreover, the narrowness of the observed absorption features, implying a small range of effective magnetic field strengths, further suggests that the soft burst emission (<0.1 MeV) comes from a relatively small region close to the polar cap of a neutron star and is observed at large angles to the axis of the field. The soft continuum spectra are in fact quite consistent[64] with gyrosynchrotron emission in such fields.

As can be seen in the spectrum of the 25 March 1978 burst, however, this soft component accounts for only a fraction ($\sim 20\%$) of the observed burst emission. Most of the emission in this burst is seen in a spectrally distinct hard component between ~ 0.25 and 6 MeV. Similar hard components, with energies extending as high as 20 MeV, have been observed[65] in many other bursts. The $e^{\pm}$ pair production opacity of these hard photons imposes a strong constraint on the minimum size of the emission region. This size greatly exceeds that of a neutron star polar cap, unless the star is uncomfortably close or the emission is highly beamed.

To reconcile these features, it has been suggested[66] that the bulk of the observed burst energy was initially ejected from the polar cap of a neutron star in a highly collimated jet of $e^\pm$ pairs that disrupted and isotropized far above the star to form a fireball[68] that expanded until it became transparent to photon-photon pair production and the observed photons escaped. In such a model, the emission timescale is determined by the size at which the fireball becomes transparent. Thus, the observed duration can give a measure of the total energy and hence the distance of the burst.[67]

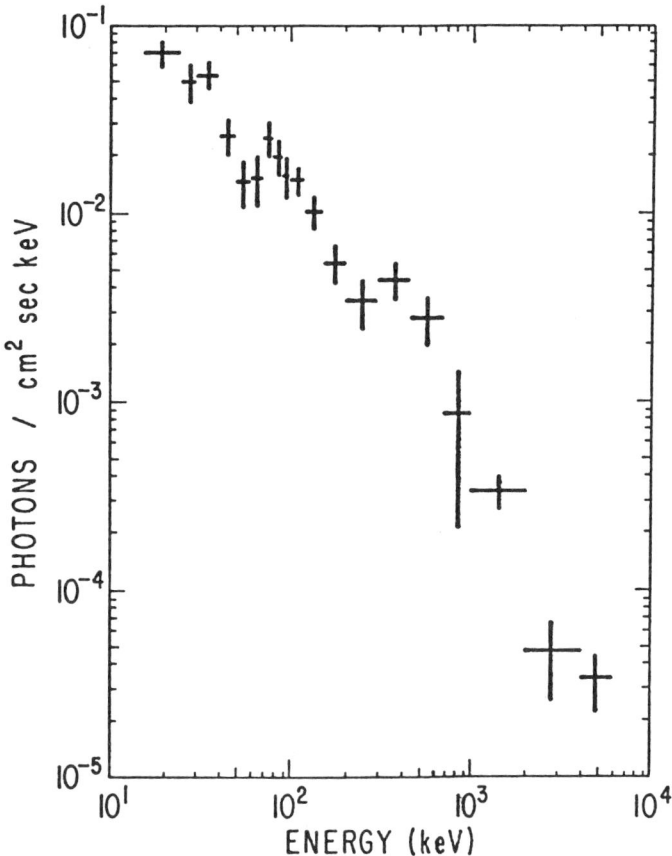

FIGURE 5. Observed[63] gamma ray spectrum of the 25 March 1978 burst.

There is also evidence for possible redshifted $e^\pm$ annihilation line emission in the spectra of some gamma ray bursts. The most commonly observed emission line in burst spectra falls in the energy range from 0.40 to 0.46 MeV, as seen[62] by low-resolution NaI detectors in the spectra of a third of the most intense gamma ray bursts. Such line emission is generally thought to be optically thin $e^\pm$ annihilation radiation redshifted by the strong gravitational field of a neutron star. However, in an optically thick region, stimulated annihilation radiation[69] could also produce a line at about 0.43 MeV without a gravitational redshift. A well-defined line at ~0.43 MeV (FIGURE 6) was also seen[70,71] in the spectrum of the 5 March 1979 burst, thus suggesting that the source of this burst was also a neutron star. However, as mentioned above, other characteristics of this burst seem to place it in a different class from that of the typical galactic burst.

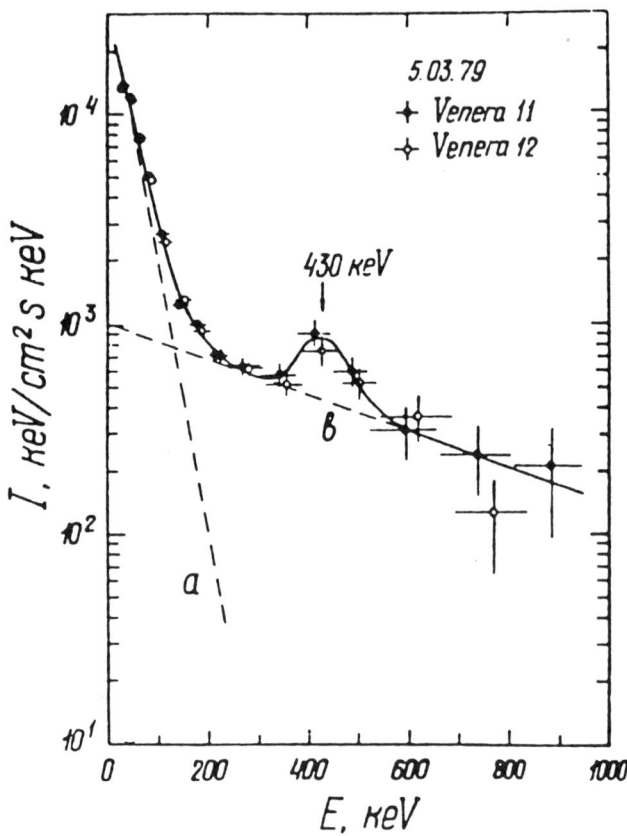

FIGURE 6. The spectrum[71] of the impulsive emission spike of the 5 March 1979 gamma ray burst.

Current theoretical ideas on gamma ray bursts generally involve strongly magnetized neutron stars. These ideas have developed, in part, as a result of the detailed observations and modeling[72,73] of the 5 March 1979 burst, even though it is quite likely that the underlying energy source of this burst is not typical of all gamma ray bursts.

The most probable energy source of gamma ray bursts is either gravitational or nuclear. Magnetic field annihilation, responsible for rapid energy generation in solar flares, is insufficient energetically. Gravitational energy can be released in a burst from a neutron star when a large amount of matter is impulsively accreted onto its surface, either in an asteroid or comet impact[74,75] or a sporadic dumping of an accretion disk by magnetospheric instabilities.[76] Such accretion releases about 100 MeV/nucleon, which is the potential energy at the neutron star surface. Gravitational energy could also be released in a corequake of a neutron star.[72,77] Such quakes could result[78] from a collapse following a phase transition from ordinary nuclear matter to a new state containing a Bose-Einstein condensate of pions.[79] Pion condensates are believed to exist above a critical density, about twice the nuclear density, and are believed to have lower energies per baryon and a significantly softer equation of state than ordinary nuclear matter. As a result of accretion or reduced centrifugal forces due to a slowing rate of rotation, the core density of a neutron star may increase beyond the critical density, thus resulting in a supercompressed metastable state that could eventually collapse to the pion condensed state. Such a collapse could release[80] about 10^{48} erg in a time no longer than the free-fall time (10^{-4} sec). As much as 10% of this energy could go into neutron star vibrations if the oscillation amplitude is on the order of the radius change (~ 10 m). Neutron star quakes can set up neutron star vibrations that dissipate mainly by gravitational radiation (e.g., reference 81). A fraction of the vibrational energy, however, can be converted[72,78] into magnetoacoustic waves that dissipate by accelerating particles in the magnetosphere. Radiation from these particles would then be responsible for the observed gamma ray emission.

Alternatively, impulsive energy release from neutron stars could result from a nuclear detonation of degenerate matter accumulated over a relatively long period of time by slow accretion of gas.[82,83] Such detonations release several MeV per nucleon from the burning of helium to the iron peak nuclei. All three of these processes, impulsive accretion, corequakes, or nuclear detonations, appear to be quite capable of providing the 10^{37} to 10^{39} ergs required for typical galactic gamma ray bursts. However to account for the $\sim 10^{44}$ ergs of the 5 March 1979 burst, such large amounts of accreted matter are required that the accretion and nuclear detonation appear to be ruled out. The energy released in corequakes, however, appears to be adequate for this burst.

GALACTIC CENTER

Intense positron annihilation radiation at 0.511 MeV has been observed from the direction of the Galactic Center for over a decade. This emission was first reported in a series of balloon observations with low-resolution NaI detectors, starting in 1970.[84-86] However, it was not until 1977 that the annihilation line energy of 0.511 MeV was clearly identified with high-resolution Ge detectors.[87] The latter observation also revealed that the line is very narrow (FWHM < 3.2 keV) and that it shows evidence for three-photon positronium continuum emission below 0.511 MeV, thus implying that $\sim 90\%$ of the positrons annihilate via positronium. Therefore, the observed intensity of $\sim 10^{-3}$ photons/cm^2 sec implies an annihilation rate of $\sim 2 \times 10^{43}$ positrons/sec or an annihilation radiation luminosity of $\sim 3 \times 10^{37}$ ergs/sec at the 10-kpc distance of the Galactic Center.

Subsequent Ge detector observations[88,89] on HEAO-3 have confirmed the narrowness (FWHM < 2.5 keV) of the line and have provided more precise information on the line-center energy (510.90 ± 0.25 keV, see FIGURE 7). These measurements also showed that the direction of the source is coincident with that of the Galactic Center (within the ±4° observational uncertainty). Most important, the HEAO-3 observations revealed that the line intensity varies with time, decreasing by a factor of three in six months from $(1.85 \pm 0.21) \times 10^{-3}$ photons/cm^2 sec in the fall of 1979 to $(0.65 \pm 0.27) \times 10^{-3}$ photons/cm^2 sec in the spring of 1980. This decrease, confirmed by later observations,[90-92] implies that the sizes of both the annihilation region and the positron source are less than the light-travel distance of 10^{18} cm. The reported annihilation line fluxes from the Galactic Center as a function of time during the last 15 years are shown in FIGURE 8.

The nature of the positron annihilation region is further constrained by the observed line width and intensity variations. The line width (FWHM < 2.5 keV) requires[93] a gas temperature in the annihilation region less than 5×10^4 K and the intensity variation requires that the density of gas at this site be high enough ($>10^5$ cm^{-3}) so that the positrons can slow down and annihilate in less than half a year. Such regions appear to exist in both the peculiar warm clouds[94] and the compact nonthermal

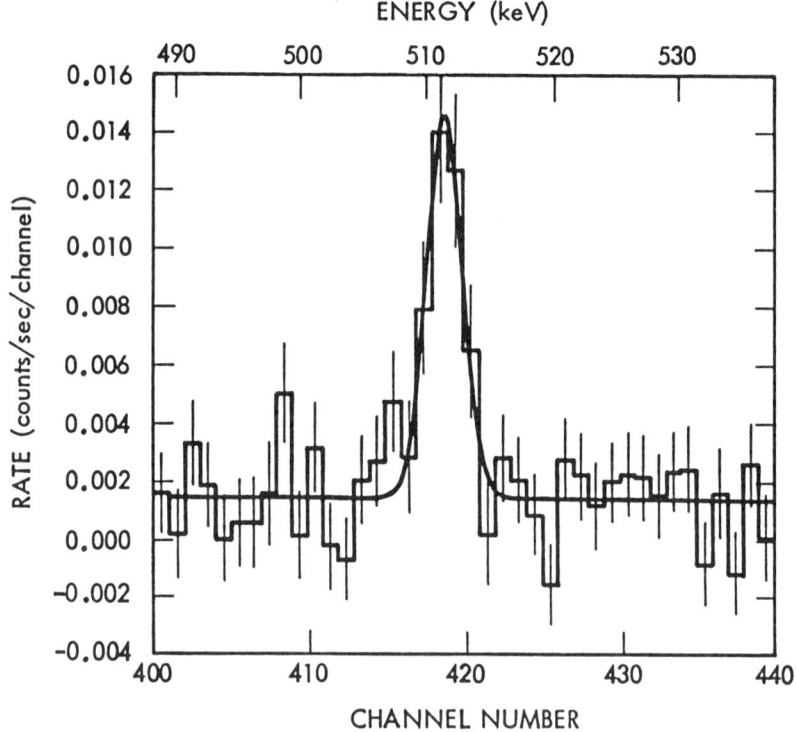

FIGURE 7. Gamma ray spectrum near 0.511 MeV observed[88] from the direction of the Galactic Center.

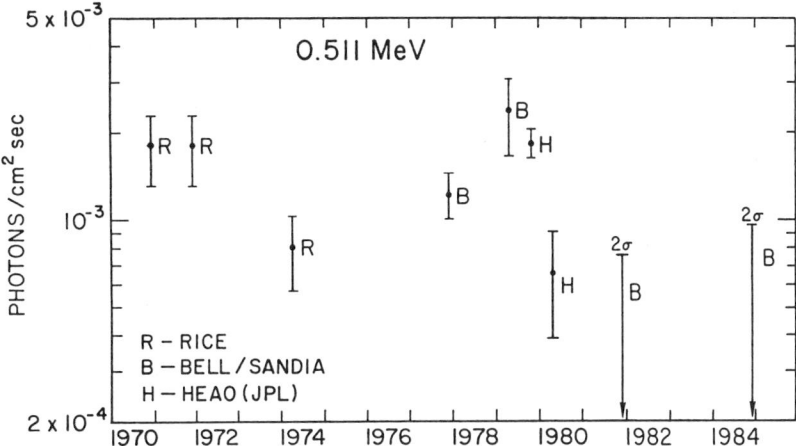

FIGURE 8. Observed 0.511-MeV fluxes and upper limits from the direction of the Galactic Center.

source[95] within the central parsec of the Galaxy. While previous theoretical studies[93] suggested that the line width also constrains the ionization fraction of the ambient gas to values greater than ~10%, it has recently been pointed out[96] that when the results of new laboratory measurements[97] of positron annihilation in neutral H are taken into account, this constraint is no longer valid.

The nature of the positron source is strongly constrained[98] by the observed variation of the 0.511-MeV intensity and by observations at other wavelengths. The decrease of a factor of three in the line intensity in six months clearly excludes any of the multiple extended sources, such as cosmic rays, pulsars,[99] supernovae,[100] or primordial black holes,[101] previously proposed. Instead, it essentially requires[102] a single compact ($<10^{18}$ cm) source, which is apparently located either at or close to the Galactic Center and which is inherently variable on timescales of six months or less. However, because the observed line-center energy shows no evidence for any gravitational redshift, the annihilation site must be removed by at least 10^3 Schwartzschild radii from this compact object.

The strongest constraints on the positron production processes are set[98] by observations of the accompanying continuum emission at energies $>m_e c^2$.[89,103] These require a high positron production efficiency, such that more than 10% of the total radiated energy $>m_e c^2$ goes into electron-positron pairs. Under the conditions of positron production on timescales comparable to that of the observed variation and in an optically thin, isotropically emiting region, only photon-photon pair production among ~MeV photons can provide the required high efficiency. Moreover, the absolute luminosity of the annihilation line requires that the photon-photon collisions take place in a very compact source ($d < 5 \times 10^8$ cm). Pair production in an intense radiation field around an accreting black hole of $<10^3$ M_0 appears to be a possible source.[98,104] However, if the gamma ray continuum is beamed, the observed continuum cannot be used to determine the photon density at the source. In this case, a photon density high enough to produce pairs at the observed rate may be present in a much larger source region than that estimated for isotropic gamma ray emission. Such pair sources may be associated with jets in massive million-solar mass black holes.[98,102,105-108] However, the total gamma ray luminosity in these models is much higher ($\sim 10^{40}$ erg/sec) than that of the isotropic model ($\sim 10^{38}$ erg/sec). Another important difference between the $<10^3 \, M_0$ and the $\sim 10^6 \, M_0$ black hole models is that while dynamical considerations imply that the more massive hole should reside at the nucleus of the Galaxy, the currently determined positional uncertainty of the line source ($\pm 4°$) would allow a variety of locations for the less massive object. Future imaging experiments with much better angular resolution could therefore differentiate between the models.

GALACTIC NUCLEOSYNTHESIS

The search for gamma ray lines from nucleosynthetic radionuclei in our galaxy has been carried on for over a decade to test current theories of the explosive nucleosynthetic origin of most nuclei heavier than helium. This search has at last resulted in the first observation of such a line from ^{26}Al, made with the high-resolution Ge spectrometer on HEAO-3.[3,109] That this line should be detectable was pointed out earlier,[110,111] but the observed intensity turned out to be nearly an order of magnitude greater than was predicted.

A rich variety of explosive nucleosynthetic lines have been proposed from both supernovae and novae. The most abundant radionucleus expected[112] from explosive nucleosynthesis in supernovae is ^{56}Ni, which decays with an 8.8-day mean-life to ^{56}Co, which, in turn, decays with a mean-life of 114 days to ^{56}Fe; 20% of the ^{56}Co decays are via positron emission. Nucleosynthesis of ^{56}Ni in supernovae is thought[113] to be the primary source of galactic ^{56}Fe.

The bulk of the gamma rays[114] and positrons[115] from the ^{56}Ni decay chain, however, are absorbed in the expanding nebula and their energies emerge only as lower energy radiation. The characteristic light curves of Type I supernovae, in fact, appear to follow the ^{56}Ni and ^{56}Co decay[114] and optical lines from both ^{56}Co and the resulting ^{56}Fe have been detected[116] in the spectrum of an extragalactic supernova, SN 1972e. Any such direct gamma ray line emission escaping from the nebula would be detectable for only a few years after the supernova explosion.

Gamma ray lines from other longer-lived radionuclei, such as 1.1-yr ^{57}Co, 3.8-yr ^{22}Na, and 68-yr ^{44}Ti from supernovae, have also been suggested.[112,117,118] However, these too could only be detectable for, at most, about 100 years after the explosion.

There are, however, three much longer-lived ($>10^5$ yr) sources of nucleosynthetic gamma ray lines, namely, β^+-decay positrons, ^{26}Al, and ^{60}Fe, which could give a direct measure of the overall galactic average rate of explosive nucleosynthesis. Since a fraction of the positrons from ^{56}Co decay are expected[114,115] to escape into the interstellar medium and since in the tenuous interstellar gas the positron lifetime against annihilation is quite long ($\sim 10^5$ yr in a density of 1 H cm^{-3}), positrons should accumulate from several thousand supernovae, assuming that galactic supernovae occur about once every 30 years. Their annihilation should thus produce[100,119] diffuse galactic gamma ray line emission at 0.511 MeV. Furthermore, estimates (e.g., reference 102) of the rate of positron production by other types of sources suggest that the principal source of galactic positrons should in fact be those escaping from ^{56}Co decay produced in Type I supernovae.

Recent observations[120,121] of galactic 0.511-MeV emission with wide ($\gtrsim 50°$) field-of-view detectors reveal considerably higher line intensities than would be expected from the Galactic Center source alone, which suggests that there may be a spatially diffuse source of 0.511-MeV line emission in the Galaxy. Conclusive measurements of such diffuse line emission can thus provide information on the average rate of galactic nucleosynthesis of ^{56}Fe during the last 10^5 years.

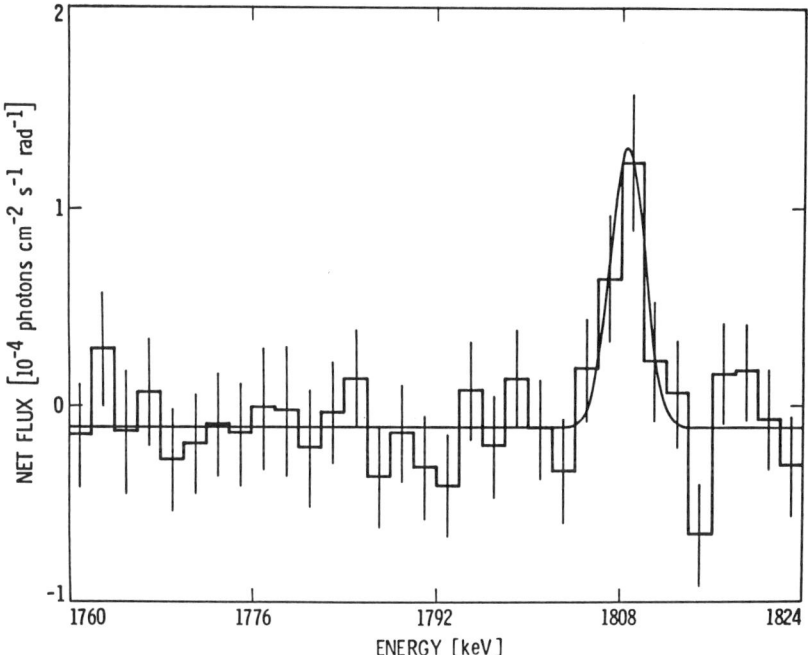

FIGURE 9. Observed[109] gamma ray spectrum near 1.809 MeV from the galactic plane in the direction of the Galactic Center.

Similarly, the long-lived radionuclei, ^{60}Fe (mean-life $\sim 4 \times 10^5$ yr) and ^{26}Al (mean-life $\sim 1 \times 10^6$ yr), which are also expected from explosive nucleosynthesis, should accumulate from $\sim 10^4$ or more supernovae and be well distributed through the interstellar medium before they decay. Diffuse galactic line emission is thus expected at 1.809 MeV from ^{26}Al decay to ^{26}Mg[110,111] and at 1.332 MeV, 1.173 MeV, and 0.059 MeV from ^{60}Fe to ^{60}Co and its subsequent decay to ^{60}Ni.[122]

Diffuse galactic line emission at 1.809 MeV from ^{26}Al has now been measured[3,109] and confirmed.[123] The measured line, shown in FIGURE 9, has a width (FWHM) $\lesssim 3.0$ keV, which is quite consistent with that expected solely from galactic rotation. The intensity varies with galactic longitude from 4.8 (± 1.0) × 10^{-4} photons/cm² sec rad in the direction of the galactic center[109] to less than 40% of that in the direction of the anticenter.[123] This intensity is roughly an order of magnitude greater than that predicted[110,111] from supernova production.

The observed flux corresponds to a total mass of about 3 M_0 of ^{26}Al in the interstellar medium. Assuming steady state, this implies a present galactic production of ~3 × 10^{-6} M_0/yr of ^{26}Al. By comparison, the estimated present production rate of ^{27}Al is of the order of $10^{-4} M_0$/yr. Thus, the production ratio of ^{26}Al/^{27}Al in the ^{26}Al source must be $>3 \times 10^{-2}$ because otherwise too much ^{27}Al would be produced. The calculated[124] yields of Type II supernovae, however, give a ratio of only (1 to 2) × 10^{-3}, which, like the predicted intensity, is an order of magnitude too low.

There are, however, other possible sources of ^{26}Al: Novae,[125,126] red giants,[127] and O and Wolf-Rayet stars.[128] For novae, the calculated[125,126] production ratio of ^{26}Al/^{27}Al is of the order of unity, which is more than sufficient. Moreover, estimates[3,109,129] of the current galactic production rate of ^{26}Al by novae come quite close to the required rate inferred from the observations. Calculations of the ^{26}Al/^{27}Al ratio from pulsating red giants[127] is also of the order of unity and that in the winds of O and Wolf-Rayet stars[128] is about 4 × 10^{-2}, which would be just sufficient. However, the estimated total galactic production rate from these sources appears to be less than that of novae. Thus, it seems at the present that the bulk of the ^{26}Al in the interstellar medium is most likely produced by novae, while the bulk of the ^{27}Al may come from Type II supernovae with only about 10% of it coming from novae. The recent discovery[130] of a new low-lying resonance for ^{26}Al production in the ^{25}Mg(p, γ) reaction, however, suggests that new theoretical calculations of the yields for the various sources are needed.

SS433

Intense, time-variable, and very narrow gamma ray line emission has recently been observed from SS433 with the high-resolution Ge spectrometer flown on HEAO-3. This instrument is particularly sensitive to very narrow lines (widths less than a few keV). The line with the strongest intensity and highest statistical significance was seen[4] at 1.497 MeV (see FIGURE 10). In addition, spectral features at ~1.2 MeV[4] and ~6.695 MeV[131] were also reported. All of these lines have very narrow widths (FWHM < 10 keV). Searches for these very narrow lines have subsequently been carried out with a Ge spectrometer flown on a balloon[132] and the NaI spectrometer on SMM,[133] whose energy resolution is much lower than that of the Ge spectrometer. Although no lines were detected in either of these searches, these negative results could be due to the time variability of the SS433 gamma ray source.

Two different identifications of the 1.497-MeV line have been proposed, both of which assume that this line is blueshifted emission from the approaching jet. The first suggestion[4] identifies the line with the 1.369-MeV line from ^{24}Mg* excited by inelastic collisions, while the other[134] associates it with a line at 1.380 MeV from the fusion reaction ^{14}N(p, γ) ^{15}O in a very narrow resonance at a proton energy of 0.278 keV. The optically determined[135] Doppler shifts of the approaching jet of SS433 at the epoch of the gamma ray observations are consistent with both of these identifications, as is the possible association of the 1.2-MeV feature with the redshifted counterpart of the

1.497-MeV line from the receding jet. Moreover, the inelastic excitations and fusion models, based on these identifications, each predict another line at either 6.129 MeV from ^{16}O* de-excitations[136] or 6.175 MeV from ^{15}O* de-excitations.[134] The observed feature at ~6.695 MeV could be identified with either of these lines. The two models also predict other lines that have not yet been observed.

If the observed 1.497-MeV line is due to ^{24}Mg de-excitations, then the fact that the gamma ray and optical Doppler shifts are similar implies that the Mg nuclei are moving essentially at the flow speed of the jets. This corresponds to a kinetic energy of ~33 MeV/nucleon. At this energy, the 1.369-MeV line can be produced in nuclear reactions with either ambient protons or moving protons, provided that the proton

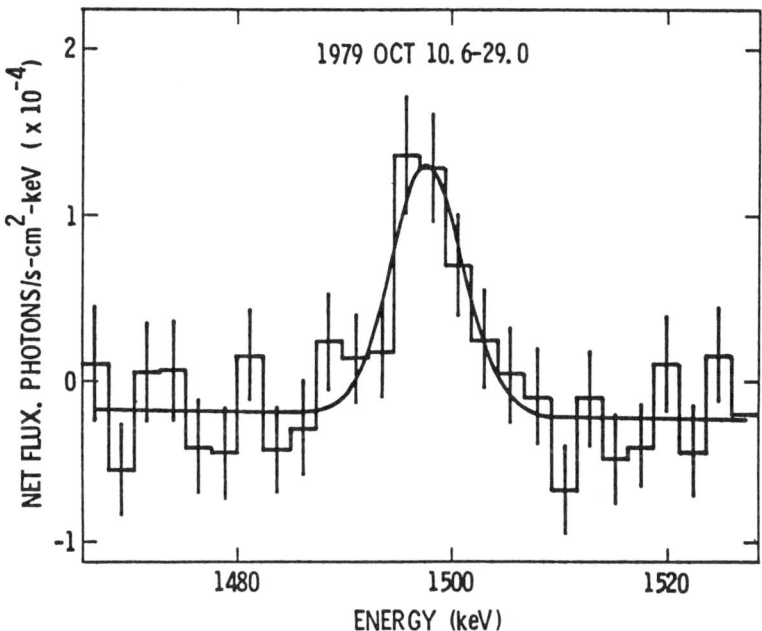

FIGURE 10. Observed[4] gamma ray spectrum within ±30 keV of the 1.5-MeV line from the direction of SS433.

velocity in the Mg rest-frame exceeds ~$0.07c$, which corresponds to the effective threshold energy (~2 MeV) for exciting the 1.369-MeV level. However, unless the relative proton velocity is less than ~$0.09c$, corresponding to a rest-frame energy less than ~4 MeV, the recoil of the excited Mg nuclei in a gas would broaden the line to a width that is larger than that observed.[4] Therefore, for inelastic excitations in a gas,[137] the velocity differential between the protons and the Mg nuclei must lie in a very narrow range so that the protons have sufficient energy to excite the line, but not too much energy to broaden it excessively. Moreover, if the 6.695-MeV line is confirmed with a very narrow width, excitations in a gas can be ruled out because at proton velocities $<0.09c$ ^{16}O cannot be excited.

These constraints can be eliminated[136] by a line-narrowing effect[138,139] involving de-excitations of nuclei embedded in dust grains. The grains also offer a simple explanation[136] to the fact that the strongest very narrow line is at 1.369 MeV from ^{24}Mg. For local galactic abundances[43] and de-excitations in a gas, the strongest lines are generally at other energies, depending on the proton energy in the Mg rest frame. Since at ~4 MeV the strongest line is at 1.634 MeV from ^{20}Ne de-excitations, a very strong depletion of Ne relative to Mg is required if the 1.497-MeV line is due to Mg de-excitations in a gas. In grains, on the other hand, Ne and other volatiles are naturally depleted.

Very narrow gamma ray lines can be produced from the de-excitation of nuclei embedded in dust grains if the sizes of the grains are large enough ($\gtrsim 10^{-4}$ cm) and the lifetimes of the nuclear levels are long enough ($\gtrsim 10^{-12}$ sec). If these two conditions are met, an excited nucleus produced in a grain loses its recoil energy by Coulomb collisions and stops in the grain before it de-excites. Thus, the line is not broadened by the recoil following de-excitation. A variety of very narrow grain lines are expected[138,139] with relative intensities depending on the elemental abundances in the grains, as well on the details of the interaction model.

In the jet-grain interaction model,[136] refractory grains were assumed in which the abundances of Mg, Si, and Fe were assumed to be the same as the local galactic abundances,[43] while the more volatile elements were depleted. Thus, the C, N, and O abundances were reduced relative to the local galactic abundances by a factor, f, and the H, He, Ne, and S abundances were set to zero, It was also assumed that the grains, moving with the jet velocity, interact with a stationary ambient medium. This corresponds to a thin-target interaction model in which the bombarding proton energy in the grain rest frame has the fixed value of 33 MeV. Alternatively, the gamma ray lines may be produced while the grains, moving at the speed of the jet flow, sweep up the ambient protons. This would occur if the bulk of the heavy elements were in the grains and the radiation pressure that accelerates the jets couples primarily to these elements and not to the hydrogen. This corresponds to thick-target interactions where the bombarding protons in the jet rest frame have initially 33 MeV, but later produce the gamma rays as they slow down and eventually stop in this frame.

The relative intensities of very narrow lines for these abundances in the thin- and thick-target cases are shown in TABLE 3. The line at 4.438 MeV from ^{12}C is not shown because even in grains this line is broad, owing to the very short (0.06 psec) lifetime of the 4.439-MeV level. Also shown are relative intensities for MgO, a very refractory compound with a very high melting temperature, which is a feature that is important for the survival of the grains.[136]

As can be seen, in all cases the strongest very narrow line can be at 1.369 MeV, provided that the depletion factor, f, is small enough. As already pointed out, the 6.129-MeV line can be associated with the reported feature at ~6.7 MeV. The confirmation of this feature and the measurement of its relative intensity would determine the depletion factor. An upper limit on the 1.634-MeV line, reported recently,[140] appears to be in conflict with the thin-target ratio given in TABLE 1, but not with the thick-target ratios. The thin-target ratio in TABLE 1 is lower than that given in reference 141, where the contribution of Si spallation to the 1.369-MeV line was ignored. There is as yet no data on the other lines shown in TABLE 1. As can be seen, such data would provide important information on the composition of the grains.

In the absence of grains, the 1.497-MeV line could still be identified[137] with the 1.369-MeV line from inelastically excited ^{24}Mg, provided that the excitations were due to protons with velocities relative to the ^{24}Mg nuclei less than 0.09c. At higher relative velocities, the line width would be larger than observed. However, the composition of the gas in which these interactions take place must be quite different from the local galactic composition.[43] For such a composition, the intensity of the 1.634-MeV line produced by protons of a few MeV is larger by about an order of magnitude than that of the 1.369-MeV line, in conflict with the fact that the upper limit on the ^{20}Ne line intensity is considerably lower than the observed intensity of the 1.369-MeV line.

In the fusion model[134] for gamma ray production of SS433, the line at 1.380 MeV results from the de-excitation of the 7.556-MeV level of ^{15}O to the ground state via a state at 6.176 MeV. The 7.556-MeV level is populated by p-^{14}N reactions through a narrow resonance at a proton energy of 0.278 MeV.[142,143] The low energy and narrow width of this resonance lead to a very narrow width for the 1.380-MeV line, provided that the temperature of the ^{14}N nuclei in the jets is sufficiently low ($<10^8$ K). This implies that the protons and the ^{14}N nuclei must have different temperatures. This has

TABLE 3. Relative Very Narrow Line Intensities

Photon Energy (MeV)	Excitation Process	(O:Mg:Si:Fe) (22f:1:1.1:1) Thin Target (ref. 138)	Thick Target	(O:Mg:Si:Fe) (1:1:0:0) Thick Target
0.847	^{56}Fe(p,p')^{56}Fe*	0.5	0.70	0.0
0.931	^{56}Fe(p,pn)^{55}Fe*	0.6	0.41	0.0
1.317	^{56}Fe(p,pn)^{55}Fe*	0.5	0.31	0.0
1.369	^{24}Mg(p,p')^{24}Mg* ^{28}Si(p,x)^{24}Mg*	1.0	1.00	1.00
1.634	^{24}Mg(p,x)^{20}Ne*	0.5	0.27	0.33
1.779	^{28}Si(p,p')^{28}Si*	0.4	0.63	0.0
6.129	^{16}O (p,p')^{16}O*	4.0f	4.4f	0.24

profound implications on the energetics of the system, as discussed below. Regarding relative line intensities, the de-excitation of the 7.556-MeV level implies additional lines at 6.176, 0.764, 6.793, 2.374, and 5.183 MeV, with intensities relative to that of the 1.380-MeV line of 1, 0.40, 0.40, 0.28, and 0.28, respectively. The 6.176-MeV line could be identified with the ~6.7-MeV feature, but the fact that this feature is observed[131] to be much weaker than the 1.497-MeV line argues against the fusion model. Searches for the other predicted lines have not yet been carried out.

Gamma ray line production by inelastic excitations is accompanied by energy loss to Coulomb collisions. If the gamma ray lines were due to fusion, the line production would also be accompanied by Coulomb losses because of the multitemperature nature of the particles implied by the observed line widths. However, the rate of Coulomb energy loss for a given rate of gamma ray line production is much larger for fusion than for inelastic excitation because the line production cross section for fusion in the resonance (~0.1 mb) is much smaller than that for inelastic excitation (~200 mb). The observed gamma ray line luminosity of SS433 of ~10^{37} erg/sec implies a Coulomb energy loss $>10^{47}$ erg/sec for the fusion model. The Coulomb energy loss for inelastic excitation[136,137] is model dependent, with the lowest value, $\sim 4 \times 10^{40}$ erg/sec, obtained in the thick-target jet-grain model. Since even this value is highly super-Eddingtonian for a stellar size object, the bulk of the Coulomb energy loss should go into mass motion in the jets. This Coulomb energy loss will also heat the grains, but the estimated temperature, <3000 K, is below the melting point of MgO. The survival of grains in the environment of the jets of SS433 has not yet been studied in detail, except for the suggestion[144] that the presence of clumps of dense matter (e.g., grains) may be a prerequisite for the acceleration of the jets by line locking. Crucial tests of the proposed models for gamma ray line production in SS433 will come from the confirmation of the already reported lines and from further observations of the relative intensities and widths of the predicted lines.

CYGNUS X-3

Intense, ultrahigh-energy gamma ray emission has been observed from the compact galactic object Cygnus X-3. The first observations,[145,146] at photon energies $>10^{12}$ eV, were carried out with ground-based detectors at the Crimean Astrophysical Observatory, which measure the Cherenkov light pulses that accompany the air showers produced by the gamma rays. These very high energy observations were confirmed by several other Cherenkov observations (e.g., references 147 and 148). Cygnus X-3 was also observed[149] at ~ 100 MeV by the spark chamber flown on SAS-2 in Earth orbit, confirming earlier observations[150] at ~ 40 MeV with a balloon-borne detector. However, the spark chamber on COS B only set[151] upper limits at 70 to 3000 MeV, which are lower than the SAS-2 observations, probably indicating time variability. Cygnus X-3 has recently been observed in the 10^{15} to 10^{16} eV range by ground-based air shower arrays at Kiel[5] and Haverah Park.[152] The gamma ray spectrum of Cygnus X-3 (from reference 5) is shown in FIGURE 11.

All of the gamma ray observations are characterized by a 4.8-hour period, most probably the period of the binary system in which one of the components is a compact object. The 4.8-hour modulation of Cygnus X-3 is also observed in X rays[153] and at infrared wavelengths,[154] but not at radio wavelengths.[155] The underground detection of muons produced in the Earth's atmosphere or crust by primary quanta originating at Cygnus X-3 has been reported recently.[156] This muonic signal also shows the 4.8-hour modulation, and the direction of the primaries are consistent with an origin at Cygnus X-3.

The gamma ray observations of Cygnus X-3 imply the acceleration of charged particles to very high energies. These particles are most likely protons producing gamma rays via π^0 meson decay, probably in the atmosphere of the companion star (e.g., reference 157). Gamma ray production by directly accelerated electrons is unlikely in view of the very rapid energy loss rates expected for electrons. Thus, the observed 10^{16}-eV photons require the acceleration of protons to at least $\sim 10^{17}$ eV.

The mechanism capable of such acceleration is not understood. It has been proposed that the acceleration could be driven by the rotational energy of the compact object[158] or the accretion disk[159] or that the acceleration could be due to shocks,[160] but the properties of all of these mechanisms remain to be worked out in detail. However, protons from just a few sources similar to Cygnus X-3 could power the galactic cosmic rays at energies of $\sim 10^{17}$ eV[157] (or even at higher energies), thus suggesting that the acceleration mechanism that operates in compact galactic sources might be capable of accelerating particles to the highest energies observed in the cosmic rays ($>10^{19}$ eV).

The production of π^0 mesons should be accompanied by other secondary products. Of particular interest are neutrons and neutrinos that could, in principle, be detected. Neutrons of energy $\sim 10^{18}$ eV can reach Earth from Cygnus X-3 in one lifetime and produce detectable air showers, provided, of course, that the primary protons are

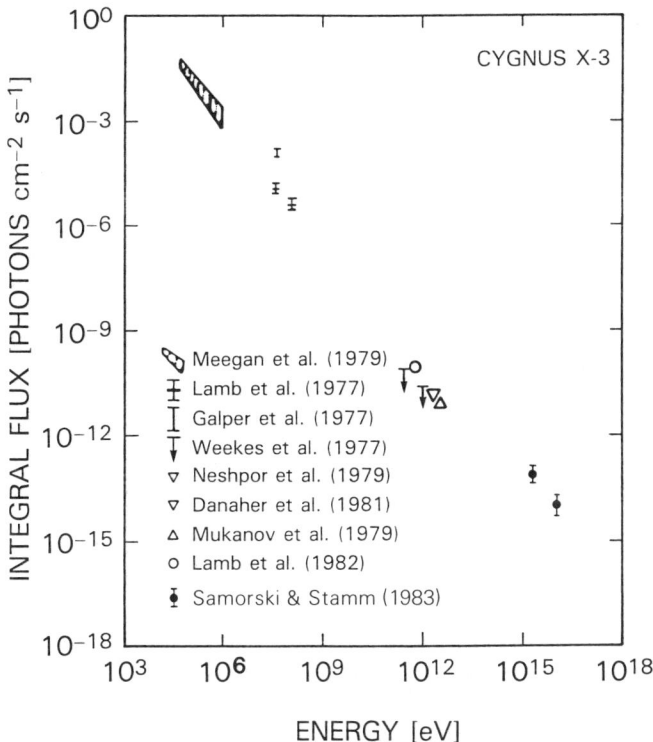

FIGURE 11. The integral gamma ray spectrum of Cygnus X-3.[5]

accelerated to such high energies. The neutrinos could be observed with future detection systems such as DUMAND.[161] However, neutrons or neutrinos, as well as gamma rays, cannot be responsible for the reported underground observations.[156] The upper limits on the neutron flux set by hadronic air shower observations of cosmic rays imply a much smaller muon production than the observed muon signal. Likewise, the observed gamma ray fluxes are also too small to produce sufficient muons.[156] Neutrinos appear also to be ruled out as a possible explanation[156] because the underground muonic signal depends on the zenith angle of Cygnus X-3, contrary to the expectation for neutrino-produced muons. As suggested in reference 162, the recent observations of Cygnus X-3 may indicate the existence of a new elementary particle or the unexpected interactions of existing ones.

SUMMARY

Important advances have taken place recently in gamma ray astronomy and we have attempted to highlight some of these in the present review. The solar flare data, including a remarkably detailed gamma ray line spectrum, provide insights into problems of particle acceleration and confinement and allow the determination of elemental abundances by a powerful new technique. A gamma ray line from recently synthesized radioactive aluminum has been observed and confirmed by independent observations. These detections provide evidence for ongoing nucleosynthesis in the galaxy. Ultrahigh-energy gamma rays have been observed and confirmed from several compact galactic objects, indicating that particle acceleration to very high energies occurs at these objects and constitutes an important ingredient in their overall energetics.

We have also provided detailed discussions of gamma bursts, of the gamma ray observations of the Galactic Center, and of SS433. The recent gamma ray bursts studies have provided much new insight into the nature of their sources, with magnetized neutron stars emerging as the best candidates. Recent observations of the Galactic Center provided only upper limits on the 0.511-MeV line flux, but a variety of theoretical and laboratory studies have elaborated considerably the physical processes that govern the production of pairs and the annihilation of the positrons. Gamma ray lines have been reported from the compact galactic object SS433. The theoretical implications of these observations are very exciting, but before further progress is possible, new data are needed to confirm the observations and differentiate between the theories.

We have not discussed several other aspects of gamma ray astronomy, particularly the diffuse galactic and extragalactic emissions, several other galactic sources, and the active galactic nuclei, which are very powerful gamma ray emitters. Significant progress in these and other areas is expected when data from the Gamma Ray Observatory, to be launched in 1988, will become available.

REFERENCES

1. RAMATY, R. & R. E. LINGENFELTER. 1982. Annu. Rev. Nucl. Part. Sci. **32**: 235.
2. FORREST, D. J. 1983. *In* Positron-Electron Pairs in Astrophysics. M. L. Burns *et al.*, Eds.: 3. American Institute of Physics. New York.
3. MAHONEY W. A., J. C. LING, A. S. JACOBSON & R. E. LINGENFELTER. 1982. Astrophys. J. **262**: 742.
4. LAMB, R. C., J. C. LING, W. A. MAHONEY, G. R. RIEGLER, W. A. WHEATON & A. S. JACOBSON. 1983. Nature **305**: 37.
5. SAMORSKI, M. & W. STAMM. 1983. Astrophys. J. **286**: L17.
6. DOWTHWAITE, J. C. *et al.* 1984. Astrophys. J. **286**: L35.
7. BIGNAMI, G. F. & W. HERMSEN. 1983. Annu. Rev. Astron. Astrophys. **21**: 67.
8. ZYSKIN, YU. L. & D. B. MUKANOV. 1983. Sov. Astron. Lett. **9**: 117.
9. CAWLEY, M. F. *et al.* 1985. 19th International Cosmic Ray Conference Papers **1**: 1973.

10. DOWTHWAITE, J. C. et al. 1984. Nature **309**: 691.
11. BALTRUSAITIS, R. M. et al. 1985. Astrophys. J. **293**: L69.
12. VOGES, W. et al. 1982. Astrophys. J. **263**: 803.
13. PROTHEROE, R. J. et al. 1984. Astrophys. J. **280**: L47.
14. HARDING, A. K. & F. W. STECKER. 1985. Astrophys. J. **291**: 471.
15. PROTHEROE, R. J. & R. W. CLAY. 1985. Nature **315**: 205.
16. BAKER, R. E. et al. 1981. 17th International Cosmic Ray Conference Papers **1**: 222.
17. STECKER, F. W. 1975. In Origin of Cosmic Rays. J. L. Osborne & A. W. Wolfendale, Eds.: 267. Reidel. Dordrecht.
18. RAMATY, R., R. J. MURPHY, B. KOZLOVSKY & R. E. LINGENFELTER. 1983. Sol. Phys. **86**: 395.
19. CHUPP, E. L. 1984. Annu. Rev. Astron. Astrophys. **22**: 359.
20. MURPHY, R. J. & R. RAMATY. 1985. Adv. Space Res. (COSPAR) **4**(no.7): 127.
21. MURPHY, R. J., R. RAMATY, D. J. FORREST & B. KOZLOVSKY. 1985. 19th International Cosmic Ray Conference Papers **4**: 249.
22. LINGENFELTER, R. E. & R. RAMATY. 1967. In High Energy Nuclear Reactions in Astrophysics. B. S. P. Shen, Ed.: 99. Benjamin. New York.
23. CHUPP, E. L., D. J. FORREST, P. R. HIGBIE, A. N. SURI, C. TSAI & P. P. DUNPHY. 1973. Nature **241**: 333.
24. HUDSON, H. E. et al. 1980. Astrophys. J. **236**: L91.
25. PRINCE, T., J. C. LING, W. A. MAHONEY, G. R. RIEGLER & A. S. JACOBSON. 1982. Astrophys. J. **255**: L81.
26. YOSHIMORI, M. et al. 1985. J. Phys. Soc. Japan **54**: 487.
27. CHUPP, E. L. et al. 1981. Astrophys. J. **244**: L171.
28. LINGENFELTER, R. E., E. J. FLAMM, E. H. CANFIELD & S. KELLMAN. 1965. J. Geophys. Res. **70**: 4077 and 4087.
29. CHUPP, E. L. et al. 1982. Astrophys. J. **263**: L95.
30. CHUPP, E. L. et al. 1983. 18th International Cosmic Ray Conference Papers **10**: 334.
31. DEBRUNNER, H., E. FLUCKIGER, E. L. CHUPP & D. J. FORREST. 1983. 18th International Cosmic Ray Conference Papers **4**: 75.
32. EFIMOV, YU. E. & G. E. KOCHAROV. 1983. 18th International Cosmic Ray Conference Papers **10**: 276.
33. EVENSON, P., P. MEYER & K. R. PYLE. 1983. Astrophys. J. **274**: 875.
34. RAMATY, R. 1986. In Physics of the Sun, vol. II. P. A. Sturrock, Ed.: 291. Reidel. Dordrecht.
35. MCGUIRE, R. E. & T. T. VON ROSENVINGE. 1985. Adv. Space Res. (COSPAR) **4**(no. 7): 127.
36. RAMATY, R., B. KOZLOVSKY & A. N. SURI. 1977. Astrophys. J. **214**: 617.
37. IBRAGIMOV, I. A. & G. E. KOCHAROV. 1977. Sov. Astron. Lett. **3**: 221.
38. ELLISON, D. C. & R. RAMATY. 1985. Astrophys. J. **298**: 400.
39. RAMATY, R. 1979. In Particle Acceleration in Astrophysics. J. Arons et al., Eds.: 135. American Institute of Physics. New York.
40. SHARE, G. H., E. L. CHUPP, D. J. FORREST & E. RIEGER. 1983. In Positron-Electron Pairs in Astrophysics. M. L. Burns et al., Eds.: 15. American Institute of Physics. New York.
41. CLIVER, E. W., D. J. FORREST, R. E. MCGUIRE & T. T. VON ROSENVINGE. 1983. 18th International Cosmic Ray Conference Papers **10**: 342.
42. MURPHY, R. J., D. J. FORREST, R. RAMATY & B. KOZLOVSKY. 1985. 19th International Cosmic Ray Conference Papers **4**: 253.
43. MEYER, J. P. 1985. Astrophys. J. Suppl. Ser. **57**: 173.
44. KLEBESADEL, R. W., I. B. STRONG & R. A. OLSON. 1973. Astrophys. J. **182**: 85
45. LINGENFELTER, R. E., H. S. HUDSON & D. M. WORRALL, Eds. 1982. Gamma-Ray Transients and Related Astrophysical Phenomena. American Institute of Physics. New York.
46. WOOSLEY, S. E., Ed. 1984. High Energy Transients in Astrophysics. American Institute of Physics. New York.
47. MAZETS, E. P. et al. 1981. Astrophys. Space Sci. **80**: 1.

48. MEEGAN, C. A., G. J. FISHMAN & R. B. WILSON. 1984. In High Energy Transients in Astrophysics. S. E. Woosley, Ed.: 422. American Institute of Physics. New York.
49. HIGDON, J. C. & R. E. LINGENFELTER. 1984. In High Energy Transients in Astrophysics. S. E. Woosley, Ed.: 568. American Institute of Physics. New York.
50. HIGDON, J. C. & R. E. LINGENFELTER. 1985. 19th International Cosmic Ray Conference Papers **1**: 37.
51. CLINE, T. L. 1981. Ann. N.Y. Acad. Sci. **375**: 314.
52. EVANS, W. D. et al. 1980. Astrophys. J. **237**: L7.
53. CLINE, T. L. 1980. Comments Astrophys. **9**: 13.
54. CLINE, T. L. 1982. In Gamma-Ray Transients and Related Astrophysical Phenomena. R. E. Lingenfelter et al., Eds.: 17. American Institute of Physics. New York.
55. GOLENETSKII, S. V., V. N. ILYINSKII & E. P. MAZETS. 1984. Nature **307**: 41.
56. ROTHSCHILD, R. E. & R. E. LINGENFELTER. 1984. Nature **312**: 737.
57. KLEBESADEL, R. W., E. E. FENIMORE, J. G. LAROS & J. TERRELL. 1982. In Gamma-Ray Transients and Related Astrophysical Phenomena. R. E. Lingenfelter et al., Eds.: 1. American Institute of Physics. New York.
58. HJELLMING, R. M. & S. P. EWALD. 1981. Astrophys. J. **246**: L137.
59. SCHAEFER, B. E. 1981. Nature **294**: 722.
60. SCHAEFER, B. E. et al. 1984. Astrophys. J. **286**: **L1**.
61. PEDERSEN, H. et al. 1984. Nature **312**: 46.
62. MAZETS, E. P., S. V. GOLENETSKII, R. L. APTEKAR, YU. A. GURYAN & V. N. ILYINSKII. 1981. Nature **290**: 378.
63. HUETER, G. J. 1984. In High Energy Transients in Astrophysics. S. E. Woosley, Ed.: 373. American Institute of Physics. New York.
64. LIANG, E. P. 1982. Nature **299**: 321.
65. NOLAN, P. L. et al. 1984. In High Energy Transients in Astrophysics. S. E. Woosley, Ed.: 399. American Institute of Physics. New York.
66. HUETER, G. J. & R. E. LINGENFELTER. 1983. In Positron-Electron Pairs in Astrophysics. M. L. Burns et al., Eds.: 89. American Institute of Physics. New York.
67. LINGENFELTER, R. E. & G. J. HUETER. 1984. In High Energy Transients in Astrophysics. S. E. Woosley, Ed.: 558. American Institute of Physics. New York.
68. CAVALLO, G. & M. J. REES. 1978. Mon. Not. R. Astron. Soc. **183**: 359.
69. RAMATY, R., J. M. MCKINLEY & F. C. JONES. 1982. Astrophys. J. **256**: 238.
70. MAZETS, E. P., S. V. GOLENETSKII, V. N. ILYINSKII, R. L. APTEKAR & YU. A. GURYAN. 1979. Nature **282**: 587.
71. MAZETS, E. P., S. V. GOLENETSKII, YU. A. GURYAN & V. N. ILYINSKII. 1982. Astrophys. Space Sci. **84**: 173.
72. RAMATY, R. et al. 1980. Nature **287**: 122.
73. RAMATY, R., R. E. LINGENFELTER & R. W. BUSSARD. 1981. Astrophys. Space Sci. **75**: 193.
74. HARWIT, M. & E. E. SALPETER. 1973. Astrophys. J. **187**: L97.
75. COLGATE, S. A. & A. G. PETCHEK. 1981. Astrophys. J. **248**: 771.
76. LAMB, F. K. 1984. In High Energy Transients in Astrophysics. S. E. Woosley, Ed.: 179. American Institute of Physics. New York.
77. TSYGAN, A. I. 1975. Astron. Astrophys. **44**: 21; **49**: 159.
78. ELLISON, D. C. & D. KAZANAS. 1983. Astron. Astrophys. **128**: 102.
79. HAENSEL, P. & R. SCHAEFFER. 1982. Nucl. Phys. **A381**: 519.
80. HAENSEL, P. & M. PROSZYNSKI. 1982. Astrophys. J. **258**: 306.
81. WANG, Q. D. & T. LU. 1984. Phys. Lett. **148B**: 211.
82. WOOSLEY, S. E. & R. E. TAAM. 1976. Nature **263**: 101.
83. WOOSLEY, S. E. 1982. In Gamma-Ray Transients and Related Astrophysical Phenomena. R. E. Lingenfelter et al., Eds.: 273. American Institute of Physics. New York.
84. JOHNSON, W. N., F. R. HARNDEN & R. C. HAYMES. 1972. Astrophys. J. **172**: L1.
85. JOHNSON, W. N. & R. C. HAYMES. 1973. Astrophys. J. **184**: 103.
86. HAYMES, R. C. et al. 1975. Astrophys. J. **201**: 593.
87. LEVENTHAL, M., C. J. MACCALLUM & P. D. STANG. 1978. Astrophys. J. **225**: L11.
88. RIEGLER, G. R. et al. 1981. Astrophys. J. **248**: L13.

89. RIEGLER, G. R. et al. In Positron-Electron Pairs in Astrophysics. M. L. Burns et al., Eds.: 230. American Institute of Physics. New York.
90. LEVENTHAL, M., C. J. MACCALLUM, A. F. HUTERS & P. D. STANG. 1982. Astrophys. J. **260**: L1.
91. PACIESAS, W. S. et al. 1982. Astrophys. J. **260**: L7.
92. LEVENTHAL, M. & C. J. MACCALLUM. 1985. 19th International Cosmic Ray Conference Papers **1**: 213.
93. BUSSARD, R. W., R. RAMATY & R. J. DRACHMANN. 1979. Astrophys. J. **228**: 928.
94. LACY, J. H., C. H. TOWNES, T. R. GEBALLE & D. J. HOLLENBACH. 1980. Astrophys. J. **241**: 132.
95. KELLERMANN, K. I., D. B. SHAFFER, B. G. CLARK & B. J. GELDZAHLER. 1977. Astrophys. J. **214**: L61.
96. BROWN, B. L. 1985. Astrophys. J. **292**: L67.
97. BROWN, B. L., M. LEVENTHAL, A. P. MILLS, JR. & D. W. GIDLEY. 1984. Phys. Rev. Lett. **53**: 2347.
98. LINGENFELTER, R. E. & R. RAMATY. 1982. In Galactic Center. G. Riegler & R. Blandford, Eds.: 148. American Institute of Physics. New York.
99. STURROCK, P. A. & K. B. BAKER. 1979. Astrophys. J. **234**: 612.
100. RAMATY, R. & R. E. LINGENFELTER. 1979. Nature **278**: 127.
101. OKEKE, P. N. & M. J. REES. 1980. Astron. Astrophys. **81**: 263.
102. RAMATY, R. & R. E. LINGENFELTER. 1981. Philos. Trans. R. Soc. London **A301**: 671.
103. RIEGLER, G. R., J. C. LING, W. A. MAHONEY, W. A. WHEATON & A. S. JACOBSON. 1985. Astrophys. J. **294**: L13.
104. MCKINLEY, J. M. 1986. In Proc. Third International Workshop on Positron-Gas Scattering. Wayne State University. Michigan. In press.
105. BLANDFORD, R. D. 1982. In Galactic Center. G. Riegler & R. Blandford, Eds.: 177. American Institute of Physics. New York.
106. LINGENFELTER, R. E. & R. RAMATY. 1983. In Positron-Electron Pairs in Astrophysics. M. L. Burns et al., Eds.: 267. American Institute of Physics. New York.
107. KARDASHEV, N. S., I. D. NOVIKOV, A. G. POLNAREV & B. E. STERN. 1983. In Positron-Electron Pairs in Astrophysics. M. L. Burns et al., Eds.: 253. American Institute of Physics. New York.
108. BURNS, M. L. 1983. In Positron-Electron Pairs in Astrophysics. M. L. Burns et al., Eds.: 281. American Institute of Physics. New York.
109. MAHONEY, W. A., J. C. LING. W. A. WHEATON & A. S. JACOBSON. 1984. Astrophys. J. **286**: 578.
110. RAMATY, R. & R. E. LINGENFELTER. 1977. Astrophys. J. **213**: L5.
111. ARNETT, W. D. 1977. Ann. N.Y. Acad. Sci. **302**: 90.
112. CLAYTON, D. D., S. A. COLGATE & G. J. FISHMAN. 1969. Astrophys. J. **155**: 75.
113. WOOSLEY, S. E., T. S. AXELROD & T. A. WEAVER. 1981. Comments Nucl. Part. Phys. **9**: 185.
114. COLGATE, S. A. & C. MCKEE. 1969. Astrophys. J. **157**: 623.
115. ARNETT, W. D. 1979. Astrophys. J. **230**: L32.
116. AXELROD, T. S. 1980. Ph.D. thesis. University of California, Santa Cruz.
117. CLAYTON, D. D. 1974. Astrophys. J. **188**: 155.
118. CLAYTON, D. D. 1975. Astrophys. J. **198**: 151.
119. CLAYTON, D. D. 1973. Nature (London) Phys. Sci. **244**: 137.
120. ALBERNHE, F. et al. 1981. Astron. Astrophys. **94**: 214.
121. DUNPHY, P. P., E. L. CHUPP, D. L. FORREST, 1983. In Positron-Electron Pairs in Astrophysics. M. L. Burns et al., Eds.: 237. American Institute of Physics. New York.
122. CLAYTON, D. D. 1971. Nature **234**: 291.
123. SHARE, G. H. et al. 1985. Astrophys. J. **292**: L61.
124. WOOSLEY, S. E. & T. A. WEAVER. 1980. Astrophys. J. **238**: 1017.
125. WALLACE, R. K. & S. E. WOOSLEY. 1981. Astrophys. J. Suppl. Ser. **45**: 389.
126. HILLEBRANDT, H. & F. K. THIELEMANN. 1982. Astrophys. J. **255**: 617.
127. NORGAARD, H. 1980. Astrophys. J. **236**: 895.
128. DEARBORN, D. S. P. & J. B. BLAKE. 1985. Astrophys. J. **288**: L21.

129. CLAYTON, D. D. 1984. Astrophys. J. **280**: 144.
130. CHAMPAGNE, A. E., A. J. HOWARD & J. D. PARKER. 1983. Astrophys. J. **269**: 686.
131. WHEATON, W. A., J. C. LING, W. A. MAHONEY & A. S. JACOBSON. 1984. Bull. Am. Astron. Soc. **16**: 472.
132. MACCALLUM, C. J., A. F. HUTERS, P. D. STANG & M. LEVENTHAL. 1985. Astrophys. J. **291**: 486.
133. GELDZAHLER, B. J. *et al.* 19th International Cosmic Ray Conference Papers **1**: 187.
134. BOYD, R. N., M. WIESCHER, G. H. NEWSON & G. W. COLLINS II. 1984. Astrophys. J. **276**: L9.
135. MARGON, B. 1984. Annu. Rev. Astron. Astrophys. **22**: 507.
136. RAMATY, R., B. KOZLOVSKY & R. E. LINGENFELTER. 1984. Astrophys. J. **283**: L13.
137. HELFER, H. L. & M. P. SAVEDOFF. 1984. Astrophys. J. **283**: L49.
138. LINGENFELTER, R. E. & R. RAMATY. 1977. Astrophys. J. **211**: L19.
139. RAMATY, R., B. KOZLOVSKY & R. E. LINGENFELTER. 1979. Astrophys. J. Suppl. Ser. **40**: 487.
140. WHEATON, W. A., J. C. LING, W. A. MAHONEY & A. S. JACOBSON. 1985. 19th International Cosmic Ray Conference Papers **1**: 183.
141. NORMAN, E. B. & D. BODANSKY. 1984. Nature **308**: 212.
142. AJZENBERG-SELOVE, F. 1981. Nucl. Phys. **A360**: 143.
143. FOWLER, W. A., G. R. CAUGHLAN & B. A. ZIMMERMAN. 1967. Annu. Rev. Astron. Astrophys. **5**: 525.
144. PEKAREVICH, M., T. PIRAN & J. SHAHAM. 1984. Astrophys. J. **283**: 295.
145. VLADIMIRSKY, B. M., A. A. STEPANIAN & V. P. FOMIN. 1973. 13th International Cosmic Ray Conference Papers **1**: 456.
146. STEPHANIAN, A. A., V. P. FOMIN, YU. I. NESHPOR, B. M. VLADIMIRSKY & YU. L. ZYSKIN. 1982. *In* Very High Energy Gamma Ray Astronomy. P. V. Ramana Murthy & T. C. Weekes, Eds.: 43. Tata Institute. Bombay, India.
147. DENAHER, S., D. J. FEGAN, N. A. PORTER & T. C. WEEKES. 1981. Philos. Trans. R. Soc. London **A301**: 637.
148. DOWTHWAITE, J. C. *et al.* 1983. Astron. Astrophys. **126**: 1.
149. LAMB, R. C., C. E. FICHTEL, R. C. HARTMAN, D. A. KNIFFEN & D. J. THOMPSON. 1977. Astrophys. J. **212**: L63.
150. GALPER, A. M. *et al.* 1975. 14th International Cosmic Ray Conference Papers **1**: 95.
151. HERMSEN, W. *et al.* 1985. 19th International Cosmic Ray Conference Papers **1**: 95.
152. LLOYD-EVANS, J. *et al.* 1983. Nature **305**: 784.
153. PARSIGNAULT, D. R., E. SCHREIER, J. GRINDLAY & H. GURSKY. 1976. Astrophys. J. **209**: L73.
154. BECKLIN, E. E. *et al.* 1973. Nature (London) Phys. Sci. **245**: 302.
155. GREGORY, P. C. *et al.* 1973. Nature (London) Phys. Sci. **239**: 114.
156. MARSHAK, M. L. *et al.* 1985. Phys. Rev. Lett. **54**: 2079.
157. HILLAS, A. M. 1984. Nature **312**: 50.
158. EICHLER, D. & W. T. VESTRAND. 1984. Nature **307**: 613.
159. CHANMUGAM, G. & K. BRECHER. 1985. Nature **313**: 767.
160. KAZANAS, D. & D. C. ELLISON. 1986. Nature **319**: 380.
161. STENGER, V. J. 1980. DUMAND-80, Hawaii Dumand Center, Honolulu, vols. 1 and 2.
162. WATSON, A. 1985. Nature **315**: 454.

PART VII. NUMERICAL ASTROPHYSICS

Simulations of the Formation of Large-Scale Structure

SIMON D. M. WHITE

Steward Observatory
University of Arizona
Tucson, Arizona 85721

INTRODUCTION

Recent developments in several fields have led to a surge of activity in studies of the nature and origin of galaxies and larger structures in the universe. The ability to measure redshifts for large numbers of galaxies has permitted the completion of several extensive surveys designed to map out the three-dimensional distribution of galaxies in large regions of space. Such studies suggest a richer structure than was apparent from the distribution of galaxies on the sky. Groups and clusters of galaxies frequently appear to be organized in loose sheets and chains that extend for many tens of megaparsecs, while very large volumes of space contain no bright galaxies at all. These observational data are discussed in recent reviews by Oort (1983)[1] and Davis (1985).[2] At the same time, a fruitful interaction between particle physics and cosmology has led to a much more definite model for the early universe than has heretofore been available. The inflationary scenario offers solutions to a number of fundamental cosmological puzzles and provides a specific mechanism for producing the fluctuations from which galaxies and all other structures must develop. In addition, it requires the present universe to have critical density. This requirement is consistent with cosmic nucleosynthesis of the light elements only if most of the matter in the universe is in a nonbaryonic and therefore "exotic" form. This work is reviewed in a number of papers in the proceedings of the Inner Space/Outer Space meeting held at Fermilab in May 1984. A final development that has made possible a confrontation between these two very different lines of research has been the increasing sophistication of numerical studies of the nonlinear phases of structure formation. These permit the predictions of the inflationary scenario to be extrapolated forward for comparison with observed large-scale structure.

Recently, I reviewed studies of the growth of structure from the initial conditions predicted by the inflationary model for the Inner Space/Outer Space proceedings.[3] This article discussed both the linear development of fluctuations in the early universe and the numerical studies of their subsequent evolution. I will not repeat a full review here, but will instead summarize the results obtained over the last five years from simulations of the growth of structure.

NUMERICAL STUDIES OF STRUCTURE FORMATION

Aarseth, Gott, and Turner (1979[4] and subsequent papers) were the first workers to make extensive use of numerical simulations to produce dynamically consistent models of the galaxy distribution for detailed comparison with observation. Their models followed the hierarchical clustering of 1000–4000 points from an initial distribution that was purely random or was made up of randomly oriented lines of particles. These initial conditions were chosen for convenience and were not based on any detailed theory for the prior evolution of structure. The results showed that a number of the statistical properties of the galaxy distribution could be matched by the models. However, interpretation of this work was rendered difficult by the arbitrariness of the initial conditions. Subsequent studies have shown that evolution from these initial conditions leads to some significant quantitative disagreements with the observed correlation properties of galaxies (Efstathiou and Eastwood, 1981),[5] as well as to qualitative disagreement with the large-scale morphology of the galaxy distribution (Davis et al., 1982).[6]

In parallel with this work, Russian cosmologists were carrying out a systematic numerical survey of the growth of structure from initial conditions with a large coherence length. They were motivated by the "pancake" theory for the evolution from adiabatic initial fluctuations (Zel'dovich, 1970).[7] One-, two-, and three-dimensional simulations were carried out (Doroshkevich et al., 1980;[8] Klypin and Shandarin, 1983[9]) and were later repeated with greater numerical resolution by Melott (1982, 1983),[10,11] by Centrella and Melott (1983),[12] and, using a different numerical technique, by Frenk, White, and Davis (1983).[13] In all these studies, the initial conditions were chosen to be similar, highly idealized, qualitative representations of the predictions of the adiabatic theory and in most of them, comparison with observation was limited to remarks about the morphology of the nonlinear structure and its characteristic scales. Quantitative comparisons were hampered by the unrealistic initial conditions and by the limited small-scale resolution of many of the three-dimensional models.

The work of Aarseth, Gott, and Turner (1979)[4] was based on N-body integrators that obtained particle accelerations by direct summation of the forces due to other objects. Such schemes allow arbitarily high spatial resolution, but have a limited dynamic range in mass because of the relatively small number of particles they employ. All the early work on the adiabatic theory used Fast Fourier Poisson (FFT) solvers to find forces on particles. This procedure has limited spatial resolution because of the grid imposed during the force calculation, but it has much better mass resolution because its speed allows the use of many more particles (up to 10^6). A hybrid scheme that overcomes the limited resolution of the FFT methods at the expense of a substantial increase in computer time was introduced by Efstathiou and Eastwood (1981).[5] Detailed tests of the performance of currently available integration methods suggest that this hybrid scheme is to be preferred in most cosmological situations (Efstathiou et al., 1985).[14]

The first simulations that used initial conditions modeled explicitly on detailed calculations of the earlier evolution of structure were the neutrino-dominated universe models of White, Frenk, and Davis (1983).[15] These authors showed that the sheet and filament structures that form in such models are not sensitive to the numerical techniques employed. Their amplitude on large-scales is much too great to be consistent with observation if galaxies are assumed to form in pancakes in proportion to the amount of shocked baryonic material available. Further analysis of these simulations showed them to predict neutrino clusters with such large masses ($10^{16}\ M_\odot$) that their presence would be very difficult to disguise, whatever the history of galaxy formation (White, Davis, and Frenk, 1984).[16] Detailed hydrodynamical studies of the shocking and cooling of gas in neutrino "pancakes" suggest that it might be difficult to get galaxies to form at all in a neutrino-dominated universe (Shapiro, Struck-Marcell, and Melott, 1983;[17] Bond et al., 1984[18]).

If massive neutrinos no longer appear to be an attractive candidate for the mass required to close the universe, a plethora of more exotic candidate particles have been suggested to take their place. Almost all of these act in the same way insofar as they affect the evolution of structure, and they pass under the generic name of cold dark matter (CDM). The density fluctuations at late times in a CDM-dominated universe are not predicted to have a coherence length. As a result, clustering is expected to grow hierarchically from small scales to large in such a universe. An extensive series of simulations by Davis et al. (1985)[19] show that the morphology of the predicted CDM distribution is qualitatively similar to the observed hierarchical distribution of galaxies. Reasonable quantitative agreement is also obtained for open models with $\Omega = 0.2$. However, the CDM distribution in Einstein de Sitter models is in serious disagreement with observation. This is a manifestation of the well-known fact that the observed M/L ratios of groups and clusters of galaxies are much lower than that required to close the universe. Thus, if $\Omega = 1$, then the mass distribution must differ from the observed galaxy distribution. In a CDM universe, a difference in distribution of the right kind is inferred if galaxies are assumed to form only at rare high peaks of CDM density distribution (Bardeen, this conference). Davis et al. (1985)[19] use their simulations to show that a recipe of this kind may be able to reconcile a closed CDM universe with most aspects of the observed galaxy distribution. A major remaining difficulty may be that, like the original simulations of Aarseth, Gott, and Turner, such models will turn out to have too little large-scale coherence to explain the long filamentary structure and the large voids found in the real universe.

REFERENCES

1. OORT, J. H. 1983. Annu. Rev. Astron. Astrophys. **21:** 273.
2. DAVIS, M. 1985. Proceedings of the Inner Space/Outer Space meeting. University of Chicago Press. Chicago.
3. WHITE, S. D. M. 1985. Proceedings of the Inner Space/Outer Space meeting. University of Chicago Press. Chicago.
4. AARSETH, S. J., J. R. GOTT & F. J. TURNER. 1979. Astrophys. J. **239:** 13.
5. EFSTATHIOU, G. & J. W. EASTWOOD. 1981. Mon. Not. R. Astron. Soc. **194:** 503.
6. DAVIS, M., J. HUCHRA, D. W. LATHAM & J. TONRY. 1982. Astrophys. J. **253:** 423.
7. ZEL'DOVICH, Y. B. 1970. Astron. Astrophys. **5:** 84.
8. DOROSHKEVICH, A. G., E. V. KOTOK, I. D. NOVIKOV, A. N. POLYUDOV, S. F. SHANDARIN & YU. S. SIGOV. 1980. Mon. Not. R. Astron. Soc. **192:** 321.
9. KLYPIN, A. A. & S. F. SHANDARIN. 1983. Mon. Not. R. Astron. Soc. **204:** 891.
10. MELOTT, A. 1982. Phys. Rev. Lett. **48:** 894.
11. MELOTT, A. 1983. Mon. Not. R. Astron. Soc. **202:** 593.
12. CENTRELLA, J. & A. MELOTT. 1983. Nature **305:** 196.

13. FRENK, C. S., S. D. M. WHITE & M. DAVIS. 1983. Astrophys. J. **271:** 417.
14. EFSTATHIOU, G., M. DAVIS, C. S. FRENK & S. D. M. WHITE. 1985. Astrophys. J. Suppl. Ser. **57:** 241.
15. WHITE, S. D. M., C. S. FRENK & M. DAVIS. 1983. Astrophys. J. Lett. **274:** L1.
16. WHITE, S. D. M., M. DAVIS & C. S. FRENK. 1984. Mon. Not. R. Astron. Soc. **209:** 27P.
17. SHAPIRO, P. R., C. STRUCK-MARCELL & A. MELOTT. 1983. Astrophys. J. **275:** 413.
18. BOND, J. R., J. CENTRELLA, A. S. SZALAY & J. R. WILSON. 1984. Mon. Not. R. Astron. Soc. **210:** 515.
19. DAVIS, M., G. EFSTATHIOU, C. S. FRENK & S. D. M. WHITE. 1985. Astrophys. J. **292:** 371.

Gravitational Radiation, Gravitational Collapse, and Numerical Relativity[a]

TSVI PIRAN[b] AND RICHARD F. STARK[c]

[b]*Racah Institute for Physics*
Hebrew University
Jerusalem, Israel
and
Institute for Advanced Study
Princeton, New Jersey 08540

[c]*Center for Radiophysics and Space Research*
Cornell University
Ithaca, New York 14853

Numerical relativity is the numerical solution of Einstein's equations, or as put vividly by L. Smarr[1] at the eighth Texas symposium, "Generation of Space-Time by a Computer." The idea was conceived by B. DeWitt in the late fifties shortly after the application of computers to hydrodynamics. May and White[2] and Hahn and Lindquist[3] obtained the first solutions in the mid-sixties, but only in the last decade have we obtained enough computer power and enough physical insight into Einstein's equations[4] to apply numerical relativity to interesting problems in relativistic astrophysics.

Several review papers discussing various aspects of numerical relativity have been published recently.[5-13] In this article, we discuss the problem of generation of gravitational radiation by a rotating, axisymmetric, stellar collapse and demonstrate via its solution the different aspects of numerical relativity.

Rotating stellar collapse is probably the most frequent potential strong source of gravitational radiation. (Black hole and neutron star collisions might be stronger sources, but are much less frequent.) As such, there is a great interest in the generation of gravitational radiation in a rotating collapse event. We would like, therefore, to calculate the total energy, as well as the waveform of the emitted radiation. There have been a few attempts to calculate this using perturbation calculations. However, the collapsing configuration involves dynamical strong fields and in order to obtain a general result we must turn to numerical relativity. The rotation induces deviations from spherical symmetry. The code must be at least two-dimensional (counting only spatial dimensions). Since rotation induces both modes of gravitational radiation, the code must be general enough to allow for a nondiagonal metric.

The first step in numerical relativity is the reformulation of the familiar Einstein equations in such a way that they can be solved numerically. In other words, we have to

---

[a]We gratefully acknowledge the support of a Royal Society (London) Israel Academy Programs Award for 1981–83 at the Hebrew University (RFS), NSF grant no. Phy83-05288 to Cornell University (RFS), and NSF grant no. Phy84-07219 to the IAS (TP) and BSF foundation.

choose first what are we solving for (which variable, how do we specify coordinate conditions, etc.) and then what is given and where (initial data, boundary conditions, etc.). We know three ways to do so.

The first route, advocated recently by Stewart[9,14] and his collaborators,[15,16] by Isaacson, Welling, and Winicour,[17] and by d'Iverno and Smallwood,[18] is to solve a characteristic problem. Here one specifies the initial data on two intersecting null surfaces and evolves this data along null characteristics. The main advantage of this formalism is that it reduces the partial differential equations of general relativity to ordinary differential equations along the characteristics. (For recent developments with this approach, we refer the reader to the review paper of Stewart.[9])

Regge calculus is a second formalism.[19] This is the gravitational analogue of finite element methods. Space-time is approximated by a piecewise linear space and the resulting Einstein equations are solved for the edge lengths. Recent developments may make this approach useful for numerical relativity.[20] However, a serious drawback is the fact that we do not know, yet, how to formulate hydrodynamics on a Regge lattice.

The $3 + 1$[21] approach has so far been the classical route for numerical relativity.[4] It also seems most appropriate for our problem. We specify an initial stellar configuration on an initial spacelike hypersurface and follow its evolution into the future on a chosen foliation of the space-time into three-dimensional spacelike hypersurfaces. The variables are the three metric of the spacelike hypersurface, $g_{ij}(i, j = 1, 2, 3)$, its extrinsic curvature, K_{ij}, and the hydrodynamic variables (baryon number, n, density, ρ, energy density, e, pressure, p, and four velocity, u^μ) describing the hydrodynamical state of a perfect fluid. The application of the $3 + 1$ approach to numerical relativity has been discussed extensively elsewhere[4,7,8] and we will not review it again here.

The second step, and probably the crucial one, is the coordinate choice. Smarr,[11] Bardeen,[22] Bardeen and Piran,[23] and Piran[8,24] have discussed the importance of the gauge conditions for numerical solutions. It seems that this point cannot be overstressed. In fact, one of us had constructed already in 1979 an axisymmetric code for rotating collapse. This code was based on cylindrical area coordinates, using $\sqrt{(x^2 + y^2)}$ and z as variables. However, it was discarded when the advantage of the radial gauge, which we will discuss shortly, became apparent. Furthermore, Nakamura[13] in work reported at the eleventh Texas symposium has solved numerically the general relativistic rotating collapse problem. He found that configurations with small amounts of angular momentum form black holes. With a large amount of angular momentum, the collapsing object forms a bouncing thick disk. However, because of the coordinate choice, these calculations were not accurate enough to estimate the amount of gravitational radiation emitted. As we shall see later, the energy of the emitted radiation is less than 10^{-3} of the total energy involved. It is not surprising that it is difficult to calculate it numerically, unless the scheme is specifically designed for this.

To calculate the energy and waveform of the gravitational radiation, Bardeen and Piran[23] constructed a coordinate system—the radial gauge with polar slicing—with which we could both follow the collapsing matter and the emitted radiation. The radial gauge is a special case of coordinates suggested by Geroch[25] at the sixth Texas symposium in an attempt to prove the positive energy theorem. It is a natural generalization of the spherical Schwarzschild coordinates. In this gauge, one can show that the ADM mass is positive definite. However, as we shall see later, not every spacelike hypersurface can be expressed in these coordinates. We overcome this problem by using the polar slicing[23] in conjunction with the radial gauge.

To see how the construction of the radial coordinate system arises naturally from the interest in gravitational radiation, consider asymptotic propagation of a gravitational plane wave along the x direction:

$$ds^2 \simeq -dt^2 + dx^2 + (1 - h_+)dy^2 + 2h_x dy dz + (1 + h_+)dz^2 \qquad (1)$$

with $h_+, h_x \ll 1$. A radially propagating wave will appear as a perturbation on a flat metric expressed in polar coordinates:

$$ds^2 \simeq -dt^2 + dr^2 + r^2[(1 - h_+)d\theta^2 + 2\sin\theta h_x d\theta d\phi + (1 + h_+)\sin^2\theta d\phi^2]. \qquad (2)$$

Adding the effect of the mass, we convert this to a radially propagating perturbation over a Schwarzschild metric:

$$ds^2 \simeq -\left(1 - \frac{2M(r)}{r}\right)dt^2 + \frac{dr^2}{\left(1 - \frac{2M(r)}{r}\right)} + r^2[(1 - h_+)d\theta^2$$

$$+ 2\sin\theta h_x d\theta d\phi + (1 + h_+)\sin^2\theta d\phi^2] \simeq -\left(1 - \frac{2M(r)}{r}\right)dt^2$$

$$+ \frac{dr^2}{\left(1 - \frac{2M(r)}{r}\right)} + r^2[(1 - h_+)d\theta^2 + (1 + h_+)(\sin\theta d\phi + h_x d\theta)^2]. \qquad (3)$$

The approximate expression for the integrated Schwarzschild mass, $M(r)$, includes a contribution from the energy density of the asymptotic gravitational radiation:

$$M(r) = M_0 + \frac{1}{32\pi} \int_{r_0}^{r} r'^2 dr' \sin\theta d\theta d\phi (\dot{h}_+^2 + \dot{h}_x^2), \qquad (4)$$

where M_0 is the mass at r_0 and the dot ($\cdot$) denotes a time derivative. To obtain a coordinate system that holds everywhere, we replace $(1 - h_+)$ by $1/(1 + h_+)$ and generalize g_{tt} and g_{ti} by introducing the lapse function, N, and the shift vector, N^i:

$$ds^2 = (-N^2 + N_i N^i)dt^2 - 2N_i dx^i dt + \frac{dr^2}{\left(1 - \frac{2m(r,\theta)}{r}\right)}$$

$$+ r^2\left[\frac{d\theta^2}{1 + h_+} + (1 + h_+)(\sin\theta d\phi + h_x d\theta)^2\right]. \qquad (5)$$

These are radial "area" coordinates since the area of a $d\theta d\phi$ element is $r^2 \sin^2\theta d\theta d\phi$, just as in flat space.

The mass aspect, $m(r, \theta)$, is now a generalization of the integrand in equation (4):

$$m(r, \theta) = \int \left\{ \frac{1}{2} \sqrt{\left(1 - \frac{2m}{r}\right)} \left(\sin\theta \frac{\partial}{\partial\theta} \frac{1}{\sqrt{\left(1 - \frac{2m}{r}\right)}} \right)_{,r,\theta} \right.$$

$$+ \frac{1}{4\sin\theta} \frac{\partial}{\partial\theta} \cdot \left(\frac{1}{\sin\theta} \frac{\partial}{\partial\theta} \sin^2\theta\, h_+ \right) \Bigg\} dr$$

$$+ 4\pi \int r^2 dr \left\{ E + K_+^2 + K_x^2 + K_{12}^2 + K_{13}^2 + \frac{3}{4} K_T^2 - KK_T \right.$$

$$\left. + \frac{1}{4}\left(1 - \frac{2m}{r}\right) \frac{h_{+,r}^2}{(1+h_+)^4} + (1+h_+)^4 h_{x,r}^2 \right\}. \quad (6)$$

We define the average mass at r, $M(r)$, as:

$$M(r) \equiv \frac{1}{4\pi} \int \sin\theta\, d\theta\, d\phi\, m(r, \theta). \quad (7)$$

With our gauge condition, the integrand is positive definite and $M(r)$ tends asymptotically to the ADM mass. Asymptotically, the integrand is just the energy density of a radially propagating pulse of gravitational radiation. We use the integrand therefore for a definition of the gravitational energy density everywhere. In addition, as will be shown later, it is valid even at surprisingly small radii.

The shift vector, N^i, is chosen so that the spatial coordinate conditions,

$$g_{r\theta} = g_{r\phi} = 0, \quad (8)$$

and so that the area condition is preserved along the $\partial/\partial t$ trajectory.

These coordinates have one obvious potential difficulty. The line element becomes singular for $r = 2m(r, \theta)$. This is just the regular Schwarzschild coordinate singularity on a horizon of a black hole. To avoid this singularity, we chose a time coordinate, via N, such that the spacelike slices never intersect the black hole horizon.[d] This is a stronger singularity avoidance feature than the singularity avoidance of the commonly used maximal slicing condition.[26] This time coordinate is determined by a condition on the trace of K, trK:

$$trK = K_r^r. \quad (9)$$

This condition has the nice geometric feature that it keeps the area condition along the normals to the three-dimensional hypersurfaces, as well as along the time direction.[e] It

---

[d] In the case of spherical symmetry, it can be shown that the spacelike hypersurfaces do not intersect the horizon, but only approach it infinitely closely. We expect that this also holds true in the general nonspherical case, but we have not been able to prove this. In the following discussion, the boundary of the region where N is exponentially small is denoted the horizon.

[e] In fact, this condition is singular at the origin where it must be replaced by a different condition.

is, though, compatible with the spatial gauge conditions. Before a black hole forms, this lapse function vanishes and stops the evolution in that region. The spacelike hypersurfaces warp around the event horizon, but do not cross it (see FIGURE 1). At the same time, the metric function increases at the boundary of the collapsed region. The vanishing of the lapse prevents g_{rr} from diverging.

To see how this happens, consider the quantity, $Ng_{rr}^{1/2}$ (see FIGURE 2). For the Schwarzschild metric, $N(1 - 2m/r)^{-1/2} = 1$. Under spherical symmetry, the general equation for $N(1 - 2m/r)^{-1/2}$ reduces to

$$\frac{\partial(Ng_{rr}^{1/2})}{\partial r} = \frac{4\pi re}{\left(1 - \frac{2M}{r}\right)} (Ng_{rr}^{1/2})$$

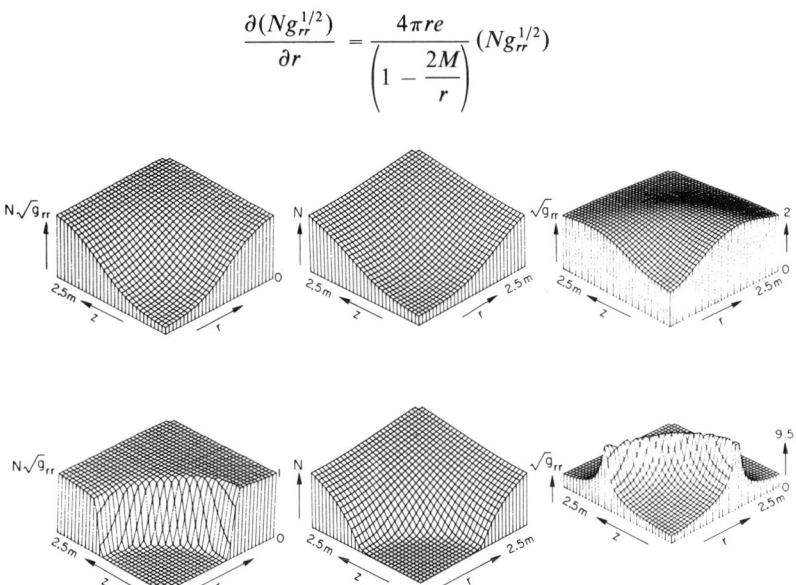

FIGURE 1. The metric function, $g_{rr}^{1/2} = (1 - 2m/r)^{-1/2}$ (right), the lapse function, N(middle), and $Ng_{rr}^{1/2}$ (left) just before a black hole tries to form (top) and then later (bottom). At the location of the horizon, $g_{rr}^{1/2}$ increases, but both N and $Ng_{rr}^{1/2}$ vanish inside the horizon. The scale of N and $N\sqrt{g_{rr}}$ graphs is from 0 to 1. The scale of $\sqrt{g_{rr}}$ is 0–2 (top) and 0–9.5 (bottom). These results are for $a/M = 0.9$ at $t = 22\,M$ (top) and $t = 46\,M$ (bottom).

and using the boundary condition of $Ng_{rr}^{1/2} = 1$ at $r \to \infty$, we obtain

$$N = \left(1 - \frac{2M}{r}\right)^{1/2} \exp\left\{-\int_r^\infty \frac{4\pi re}{\left(1 - \frac{2M}{r}\right)} dr\right\}. \tag{10}$$

Before a black hole forms, $(1 - 2m/r)$ becomes very small at some region near $r \simeq r_h$. The integral in the exponent becomes large and therefore N is exponentially small in the region of $r < r_h$. The evolution freezes in the region of $r < r_h$, but continues for $r > r_h$. We expect a similar behavior also in the nonspherical case.

Bardeen and Piran[23] and Piran[8] describe a numerical scheme based on this coordinate system. This is a partially conserved scheme. (The Hamiltonian constraint,

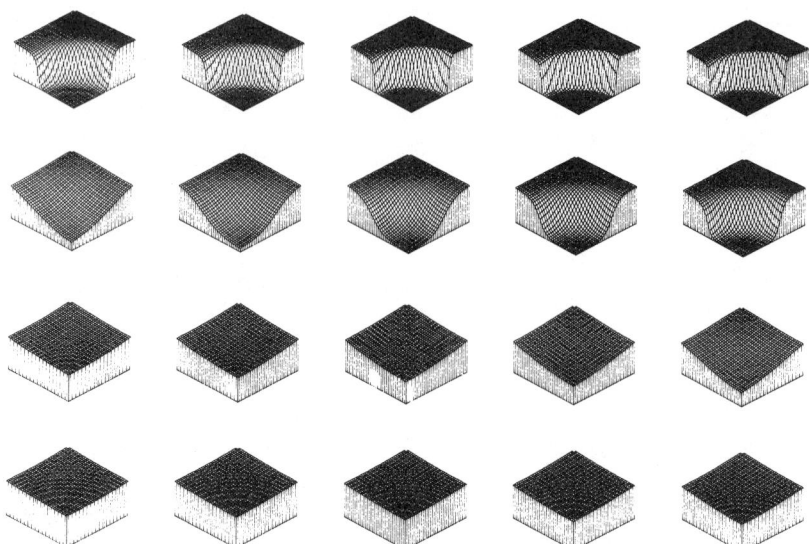

FIGURE 2. Formation of a black hole is best seen by $Ng_{rr}^{1/2}$. $Ng_{rr}^{1/2}$ shown is for collapse with $a/M = 0.9$. The initial data is at the lower left corner and subsequent time steps are to the right and upwards.

but not the momentum constraints, is imposed.) Stark and Piran[27] constructed a numerical code based on this scheme. The numerical methods used were partially tested on a cylindrical code[28] and they follow generally those described by Wilson.[10] The numerical details of this code will be discussed elsewhere.[29] Here we demonstrate first the accuracy of this code and the validity of its results and then discuss the actual rotating collapse problem.

A numerical solution is always an approximation to the exact solution. To obtain a numerical solution, we replace a continuum equation by a finite difference equation. For example, the equation,

$$\frac{\partial f}{\partial t} = \frac{\partial g}{\partial x}, \tag{11}$$

is replaced by

$$\frac{f_i^{n+1} - f_i^n}{dt} = \frac{g_{i+1/2}^{n+1/2} - g_{i-1/2}^{n+1/2}}{dx}, \tag{12}$$

where the functions, f and g, are defined on a numerical grid:

$$t_n = t_0 + n\, dt, \quad x_i = x_0 + i\, dx. \tag{13}$$

The solution of the finite difference equation can be viewed as the first terms in an expansion of the exact solution in the numerical parameter: $(dx/\ell) \sim (dt/\tau) \ll 1$. ℓ and τ here are typical length and time scales, respectively, over which the functions, f

and g, vary. The advantage of this expansion over a perturbation expansion with a physical parameter is that (dx/ℓ) is an artificial parameter. Therefore, we can obtain solutions in any physical region, including where there is no natural small physical expansion parameter.

In a typical problem, there will be more than one numerical expansion parameter. In particular, a typical second expansion parameter arises in numerical relativity when we deal with infinite systems. The outer edge of the numerical grid is finite—L—and we must have $(\ell/L) \ll 1$. Both conditions mean that the number of grid points in any direction, N, must satisfy

$$N \simeq \left(\frac{L}{dx}\right) = \left(\frac{L}{\ell}\right)\left(\frac{\ell}{dx}\right) \gg 1. \qquad (14)$$

In a D-dimensional system, the total number of points, N^D, will be large. This imposes demands on the speed, as well as on the storage of the computer that we use. Currently the largest available computers are just sufficient to solve three-dimensional general relativistic problems. (By three-dimensional, we mean either dynamic configurations that depend on two spatial dimensions or three-dimensional static configurations.) A comparison of the performance of different computers for the calculations we describe below is given in TABLE 1.

Since a numerical code provides only an approximate solution, one has to demonstrate the accuracy of the code. One way to do so is to show that the solution, $f(dx/\ell)$, converges as $dx/\ell \rightarrow 0$ and that the series, f_L, converges as $L \rightarrow \infty$. In FIGURE 3, we present comparisons between collapse solutions with grid sizes of dx, $2dx$, and $8dx$. This demonstrates convergence.

These are not the only tests that a code should pass. In addition to all numerical tests of convergence, we must compare the numerical solution with some known solutions and demonstrate the ability to reproduce them. An agreement between the numerical solution and an analytic solution in some specific cases gives us some assurance on the validity of the numerical solution, as well as a measure of the errors in it. In other branches of physics, one can compare the numerical solutions also with experiments. In general relativity, such comparison is impossible directly and hence the utmost importance of the available exact or perturbative solutions.

A long series of tests will be discussed elsewhere in detail.[29] Here we describe three tests. The first is a study of a relativistic spherical polytrope. The polytrope was stable

TABLE 1. Comparison of Running Times on Different Computers

Computer	CPU Time per Time Step with 80 × 8 Grid
Vax 780/11	20 sec
AP FPS 164	3.5 sec
Cyber 170/855	2.3 sec
IBM 3081	2.1 sec
Cray 1[a]	0.18 sec

[a]Vectorization reduced the running time by a factor of 50%; another 25% may be obtained by further vectorization.

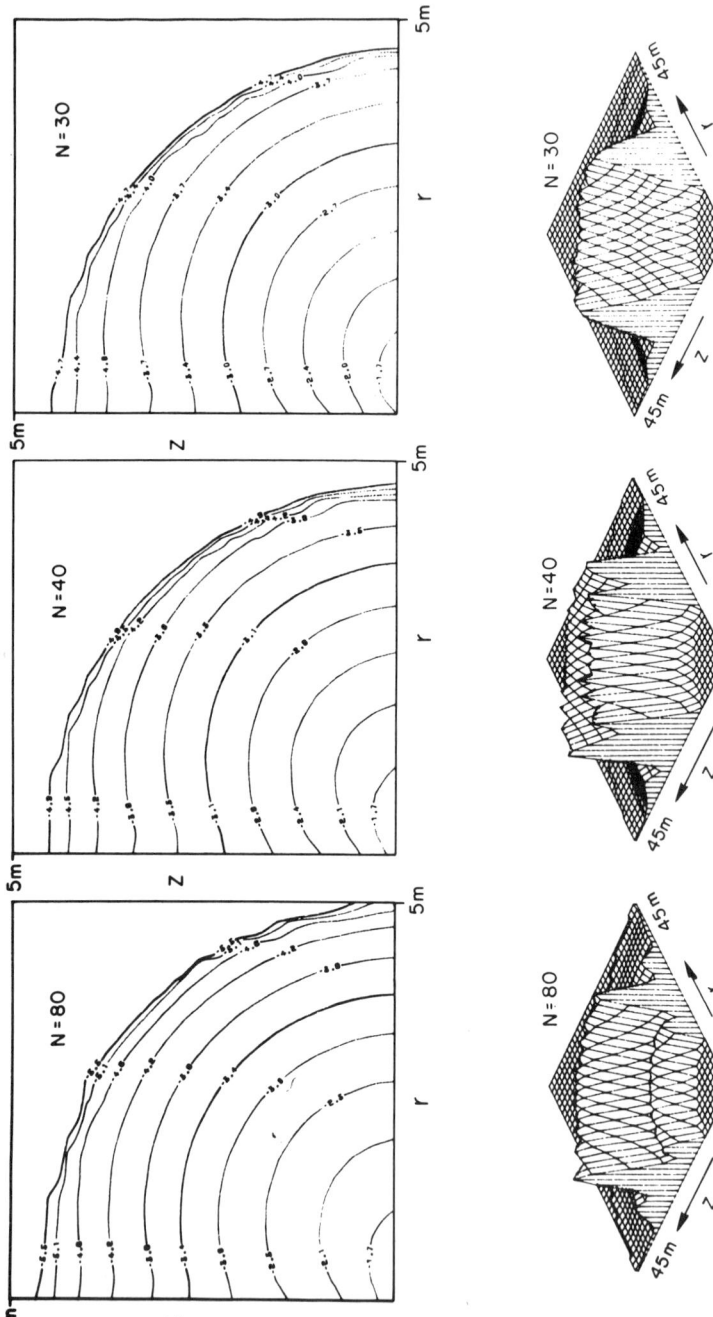

FIGURE 3. Comparison of calculations with three different grid sizes. Upper row: Logarithmic coordinate density contours at $t = 15\,M$ for $a/M = 0.9$, with three grid sizes—80 (left), 40 (middle), and 30 (right). Lower row: The energy density of the $+$ mode (divided by $\sin^4\theta$) at $t = 50\,M$ for $a/M = 0.9$, with the same grid structures.

for $\Gamma > \Gamma_{crit}$ and when perturbed, it oscillated with a period equal to the lowest normal mode calculated from perturbation expansion. With $\Gamma < \Gamma_{crit}$, the polytrope was unstable. Γ_{crit} was equal within three digits to the one calculated with perturbation expansion (including general relativistic corrections).

A free weak pulse of gravitational radiation (Brill wave) propagates as expected for small amplitude propagating waves (see FIGURE 4). One should observe that our outgoing boundary conditions allow the pulse to propagate out of the numerical grid without any reflection. When the amplitude of the initial pulse increased, the ingoing part achieved high enough energy density (when it bounced around the origin) to bound itself and form a black hole.

ADM ENERGY DENSITY. LOW AMPLITUDE + WAVE

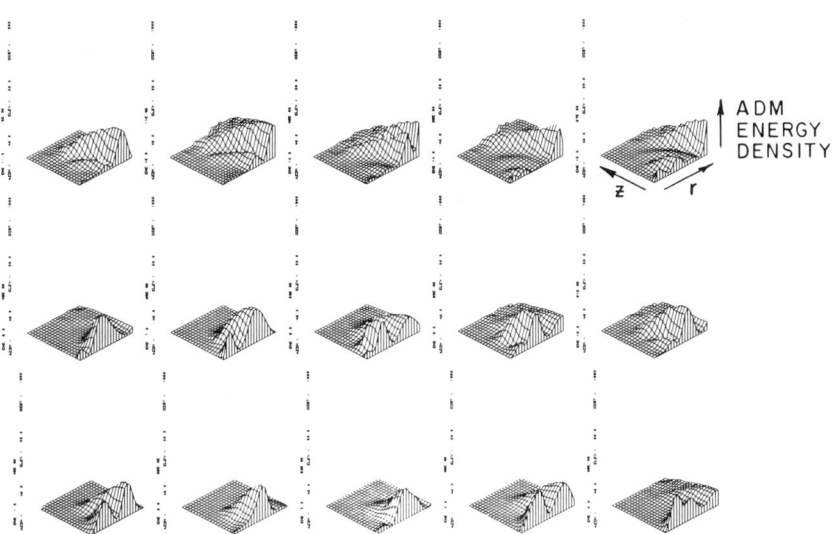

FIGURE 4a. Energy density of propagating Brill waves. The initial configuration is at the lower left corner. The vertical scale decreases by a factor of 10^{-2} as the evolution proceeds. Note that there are not reflected waves from the outer boundary.

Petrich, Shapiro, and Wasserman[30] have calculated recently the waveform of the gravitational radiation emitted by a collapsing spheroidal dust shell onto a Schwarzschild black hole. This solution is obtained by a careful superposition of gravitational radiation emitted by test particles falling into a black hole. It is valid in the limit where the mass of the shell, μ, is negligible compared with the mass of the black hole, M. A comparison of the perturbation calculations and our calculation is shown in FIGURE 5. After demonstrating that our code can calculate properly the propagation of gravitational radiation and that it can follow, correctly, the motion of the oscillating polytrope,

ADM ENERGY DENSITY. HIGH AMPLITUDE + WAVE

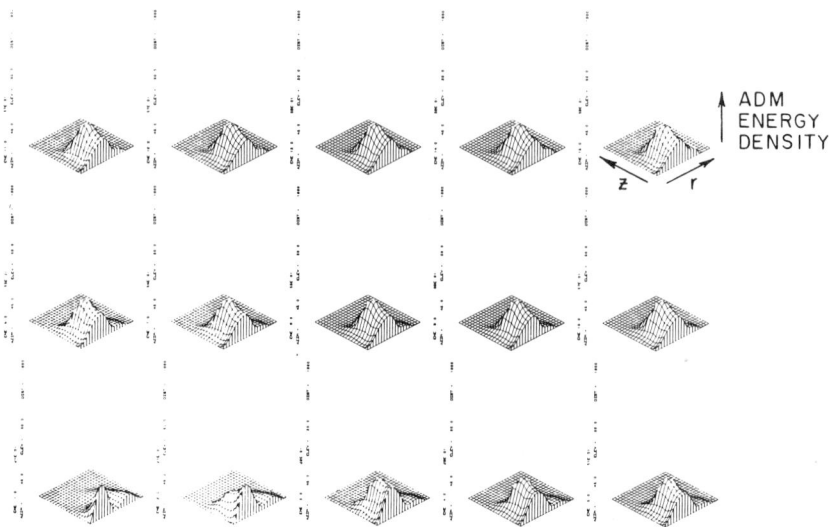

FIGURE 4b. The same calculation as in FIGURE 4a when the initial density is increased. The evolution freezes since a black hole has formed.

LAPSE. HIGH AMPLITUDE + WAVE

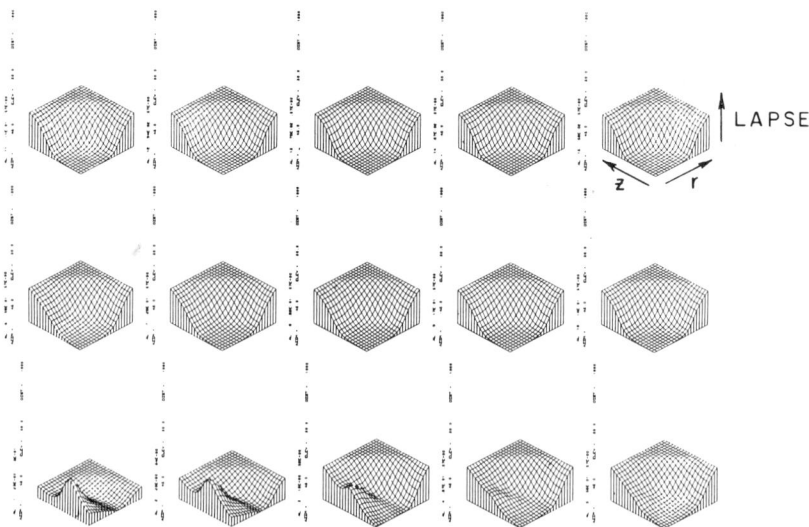

FIGURE 4c. The lapse function corresponding to FIGURE 4b.

this calculation demonstrates the missing link: the code can calculate correctly the generation of gravitational radiation.

We turn now to the rotating collapsing configurations. We consider perfect fluids with an adiabatic equation of state:

$$p = (\Gamma - 1)e. \qquad (15)$$

(We use $\Gamma = 2$ in all the calculations reported here.) We do not allow for complicated equations of state, neutrino transport, nuclear reactions, changing compositions, and all the detailed features, which we heard about in Wilson's lecture,[31] that take place in

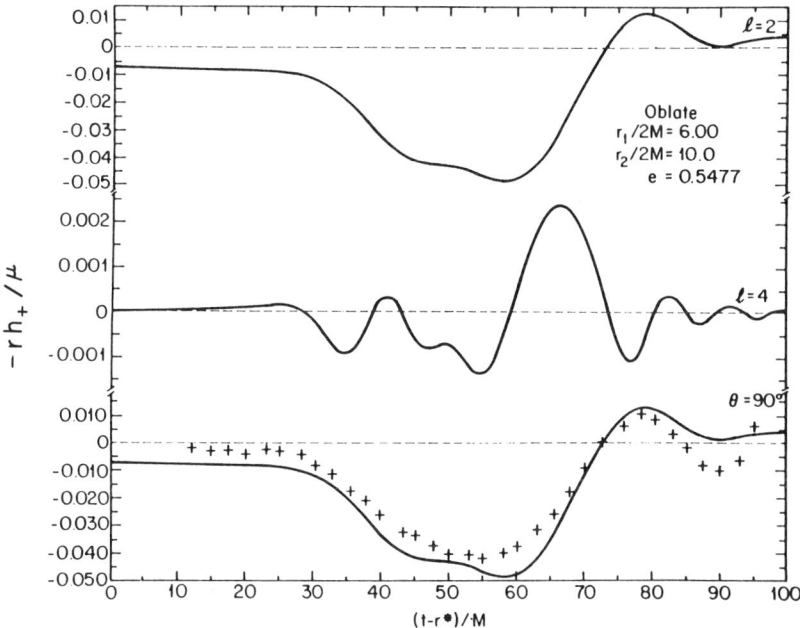

FIGURE 5. Calculation of a collapsing dust shell on a Schwarzschild black hole (from Petrich, Shapiro, and Wasserman[30]). The equatorial waveform is calculated analytically (solid line) and numerically (crosses), while the upper curve is the $\ell = 2$ mode, the middle curve is the $\ell = 4$ mode, and the lower curve is the total emission. The + marks are the numerical results.

supernovae collapse. However, the production of gravitational radiation is determined by the overall hydrodynamic behavior of the matter and this can be reproduced by such an equation of state. Note that there is no scale in the problem (with this equation of state) and therefore these results do not depend explicitly on the total mass, M.

Our idealized initial conditions do not attempt to reproduce a rotating star in a late stage of its evolution. We construct a static (spherical and nonrotating) polytrope. The initial radius of all configurations discussed here is $6\,M$ (distances are measured in gravitational units) and the central density is $3.15 \times 10^{-3}\,M^{-2}$. In cgs units, this amounts to a radius of $8.5 \times 10^5\,(M/M_\odot)$ cm and a central density of 1.8×10^{15}

$(M/M_\odot)^{-2}$ (gm/cm^3); a quite compact, but not unbelievable, unrealistic neutron star or neutron starlike core. The motivation of choosing such a compact configuration is that it reduces significantly the amount of computer time needed to follow the collapse. To induce the collapse, we reduce the pressure by a constant factor (0.01). Angular momentum is added at the initial moment of the collapse as a rigid rotation. In principle, one could solve for a stationary rotating polytrope[32] and then reduce the pressure. However, at least with low angular momentum, we do not expect that such an initial configuration will lead to a drastically different result.

To imitate better a realistic collapse, we also add the option of an initial radial infall velocity. This initial radial velocity has the form,

$$V(r) = V_0 \frac{\sqrt{2r}}{R_*^{3/2}} \quad \text{for } r < R_*, \tag{16}$$

where R_* is the outer radius of the star and V_0 is a specified constant. When $V_0 = 1$, $V(R_*)$ is the Newtonian free-fall velocity from infinity. (Unless specified otherwise, $V_0 = 0.1$.)

For a particular pressure reduction factor, adiabatic index, and radial velocity, the configuration is characterized by the single dimensionless parameter, $(a/M) = Jc/GM^2$ (where J is the total angular momentum of the star). The cosmic censorship conjecture and Nakamura's calculations suggest that black holes form only for small a/M. When a/M is large, the centrifugal forces will prevent the collapse in the radial direction.

Without angular momentum ($a/M = 0$), the configuration remains spherical. (Sphericity is maintained numerically to better than 1%.) A black hole begins to form after about 24 M (time is measured in gravitational units), after the central density has increased by a factor of 50 relative to the initial central density. Once a black hole forms, the evolution freezes and this density does not increase further. This is expressed by the fact that the lapse function, N, vanishes (see FIGURE 1). Strictly speaking, as explained earlier, a black hole never forms in our calculations since our spacelike hypersurfaces never intersect the horizon; rather they warp at an infinitesimal distance around it. The metric function, g_{rr}, increases on the black hole's horizon. Numerical problems at this place will prevent us from continuing the calculations indefinitely. The black hole formation is almost complete by $t = 30\ M$, but small amounts of matter continue to fall as long as $t \simeq 60\ M$. The effect of the initial radial infall velocity is to speed up slightly the formation of the black hole. Without it ($V_0 = 0$), the black hole begins to form at about $t \simeq 30\ M$ and it swallows most of the matter at $t \simeq 34\ M$. Examination of the flow pattern of the material reveals that the matter falls in without any shocks. At latter stages, the highest velocity is just outside the black hole's horizon where there is still a small amount of infalling matter.

Some amount of gravitational radiation is produced, which is solely due to numerical errors. The energy of this radiation, a few $\times\ 10^{-8}\ M$, is an indication of the accuracy of the calculations.

The qualitative picture of the matter flow does not change with a/M values of up to about 0.75. This is demonstrated, for example, by the radial flow pattern for $a/M = 0.4$, which is shown in FIGURE 6. Some deviation from spherical symmetry develops eventually. However, this happens when the star is so compact that a black hole forms and hides the nonspherical matter inside it. The most noticeable effect of the rotation is the generation of some amount of gravitational radiation, which we will discuss later.

Even though the matter distribution is eventually flattened once a black hole forms, the lapse function and g_{rr} contour lines become practically spherical. This is not surprising in view of the fact that the horizon of a Kerr black hole is located at $R_{+} = M + \sqrt{(M^2 - a^2)}$, where R is the Kerr radial coordinate. Now the relation between the

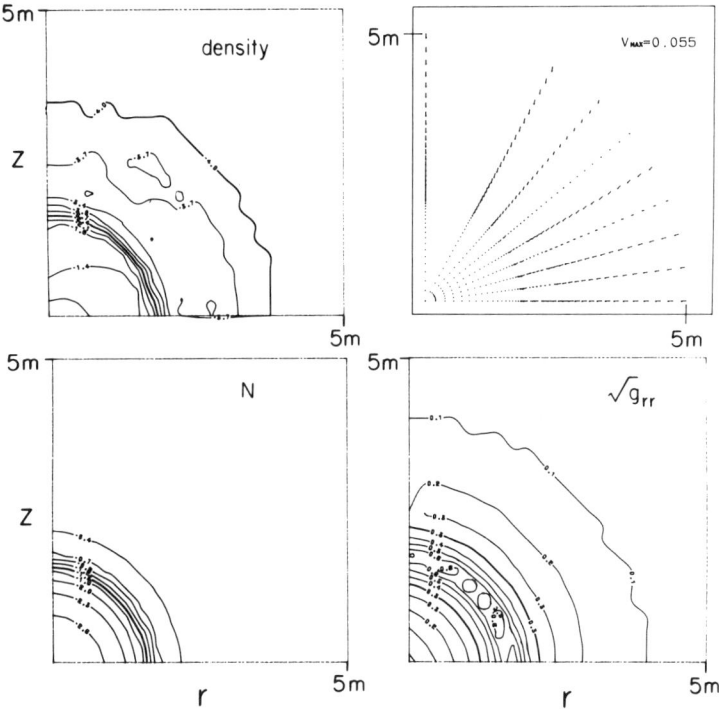

FIGURE 6. Collapse with $a/M = 0.475$ at $t = 27\,M$, just before formation of a black hole. The velocity field is at the top right and the maximal velocity is $0.055c$. There is only very small deviations from sphericity. The logarithmic coordinate density contour lines are at the top left, the logarithmic lapse contour lines are at the bottom left, and the logarithmic $\sqrt{g_{rr}}$ contour lines are at the bottom right.

radial coordinates, r and θ, and the Kerr coordinates, R and Θ, are:

$$r^2 \simeq R^2 + a^2; \quad \theta \simeq \Theta. \tag{17}$$

We anticipate that the horizon will be located approximately on the sphere, $r_{+}^2 \simeq 2M[M + \sqrt{(M^2 - a^2)}]$. For a Kerr black hole,

$$g_{rr} = \frac{r^2(r^2 - a^2 \sin^2 \theta)}{[(r^2 - a^2)(r^2 - 2M[r^2 - a^2]^{1/2})]}, \tag{18}$$

and for $r \gg r_{+}$, where this approximation is valid, g_{rr} is a weak function of θ.

The effect of the angular momentum becomes noticeable for $a/M > 0.7$. The flow pattern changes and while matter on the equator still falls in radially, matter up to 30° away from the pole falls parallel to the rotation axis, towards the equator (see FIGURE 7). The collapse evolves slower when angular momentum is added. For $a/M = 0.9$, a black hole begins to form only at $t \simeq 34\ M$ and the black hole formation ends at $t \simeq 50\ M$. It is the inner part that forms a black hole. The material collapses towards the equator and bounces off it, but a black hole forms at about that time and freezes further motion of the fluid at that stage. In the outer region, the matter bounces off the equator and a shock wave forms. This shock is relatively weak since only a negligible fraction of the matter is in this region.

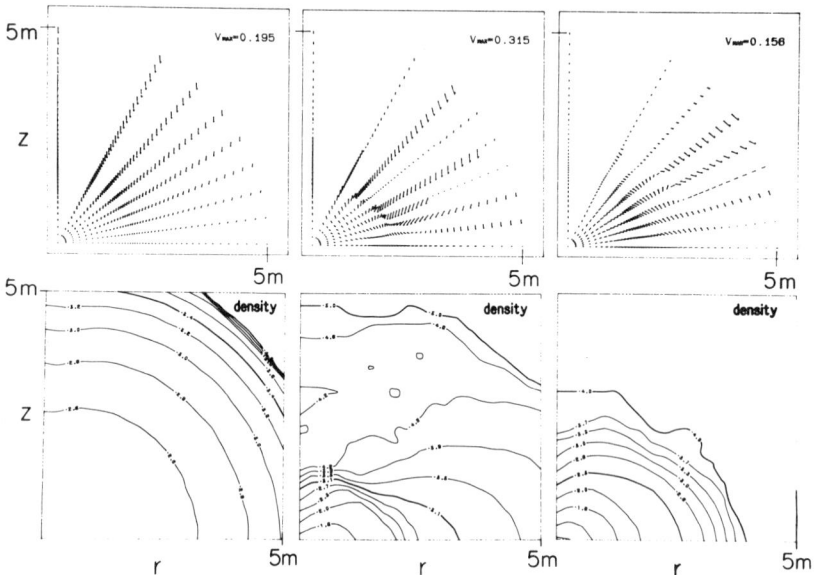

FIGURE 7. Collapse with $a/M = 1.2$. The velocity field (top) is at $t = 7\ M$, $t = 22\ M$, and $t = 30\ M$. Notice the bounce from the equator at large radii. Logarithmic coordinate density contours (bottom) are also at the same time.

The flow pattern changes drastically when a/M increases further. The centrifugal force becomes so strong that the initial centrifugal force dominates over the horizontal component of the gravitational field and the combined forces point outwards away from the axis of symmetry. The pressure is not sufficient to balance the vertical component of the gravitation force and the resulting velocity field points towards the equator and away from the rotation axis (see FIGURE 8). The matter collapses towards the equator and moves away from the center. The result is a flattened disk whose maximal density is on an equatorial ring slightly away from the origin. Eventually the inner region recollapses, now in the horizontal direction along the equator. At this stage, a black hole forms. This happens much later than when a black hole formed

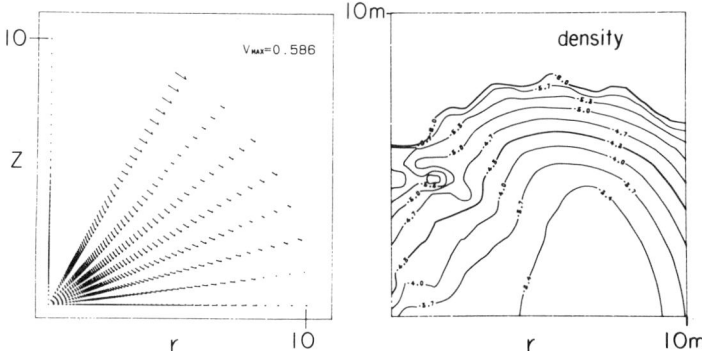

FIGURE 8. Velocity field for $a/M = 1.8$ at $t = 8.5\,M$. The centrifugal force is so large that the velocity (left) is directed away from the rotation axis. The logarithmic coordinate density contours for $a/M = 1.8$ are at $t = 20\,M$ (right). The resulting configuration is a thick disk.

without angular momentum. This recollapse is typical for a small transition region. With higher a/M values, the final configuration is a thick disk.

The high angular momentum configurations are not a realistic approximation to anything that will happen in nature. High amounts of angular momentum are not unlikely. However, in this case, the initial configuration produced by the addition of rigid rotation to both a spherical and compact configuration becomes unrealistic since the initial centrifugal force is just too large.

A typical waveform of the emitted gravitational radiation from collapse to a black hole resembles remarkably the negative of the waveform emitted by a single test particle falling into a black hole (see FIGURE 9). Both waveforms are related to the

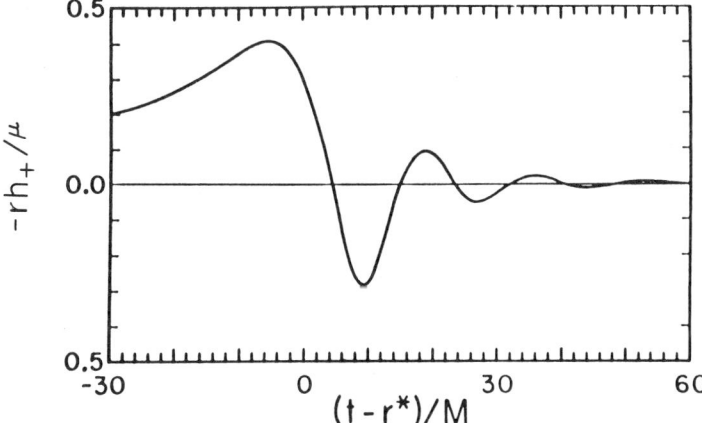

FIGURE 9a. Equatorial waveform emitted by a test particle falling into a black hole (from Detweiler[33]).

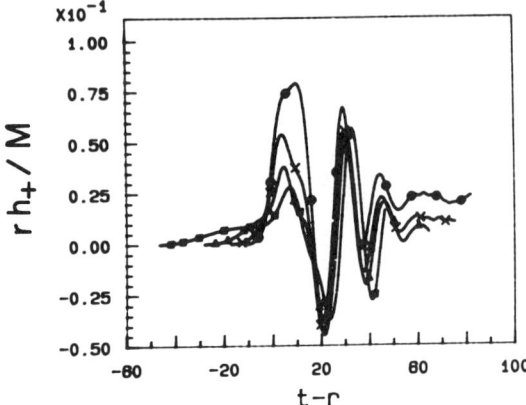

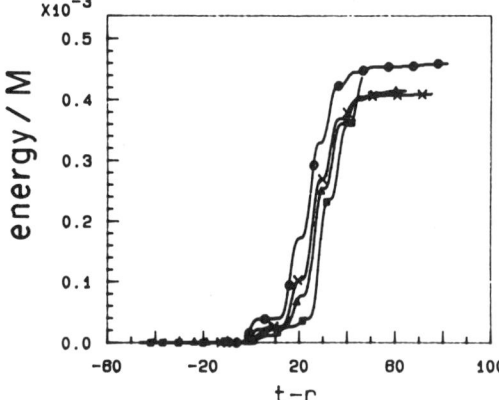

FIGURE 9b. The equatorial + mode waveform for $a/M = 0.9$ (top). The waveform is measured at four different radii ($22\,M$: circles; $28\,M$: crosses; $36\,M$: triangles; $47\,M$: squares). Time integrated energy emitted versus retarded time is in the middle and the x mode waveform for the same collapse is at the bottom.

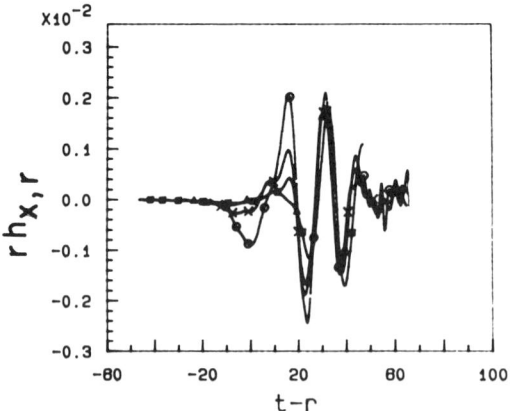

normal modes of oscillation of a black hole. However, the calculated wave form from collapse has a reduced amplitude. The shape of the waveform does not change significantly when the angular momentum is increased, even though its amplitude is increased by a large factor. The quadrupole nature of the + mode is shown by the $\sin^4 \theta$ dependence of the energy density. This dependence is factored out in FIGURE 10 and therefore the plotted amplitude does not vary in the angular direction.

Most of the energy is emitted in the + mode, with

$$\epsilon_+ \simeq 10\epsilon_x, \tag{19}$$

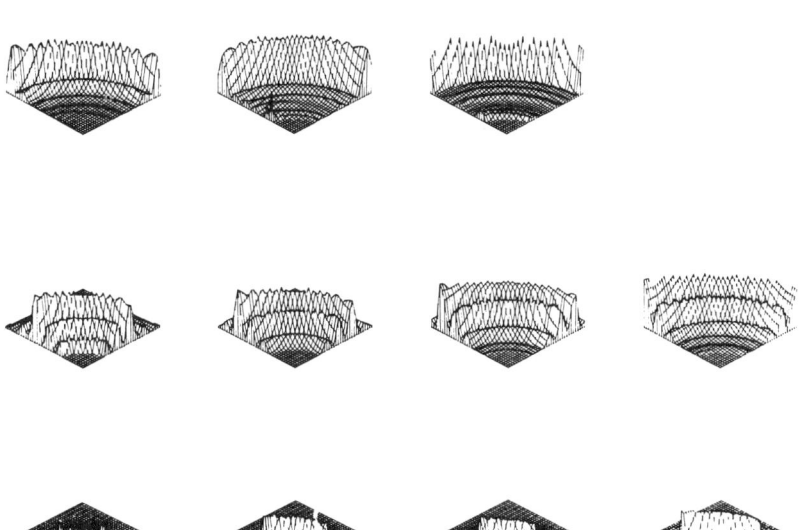

FIGURE 10. Energy density (divided by $\sin^4 \theta$) of the + mode gravitational wave for a collapse with $a/M = 0.9$. Lower left is at $t = 30\, M$, while the evolution continues to the right and then upwards (upper right is $t = 80\, M$). The vertical scale varies between the different graphs. Note the propagation of waves out of the boundary without reflection.

where ϵ_+ and ϵ_x are the efficiencies of conversion of mass to gravitational radiation in the + and x modes, respectively.

A power spectrum, $\nu^2 f_\nu^2$ (f_ν is the Fourier transform of the waveform), of the + mode signal for $a/M = 0.9$ is shown in FIGURE 11. The pulse is impulsive with a bandwidth comparable to the frequency. The spectrum peaks at $0.06\, M^{-1}$ and ranges from $0.03\, M^{-1}$ to $0.075\, M^{-1}$. In cgs units, these are $12\, (M/M_0)^{-1}$ kHz and $(6-14)\, (M/M_0)^{-1}$ kHz, respectively. The x mode is emitted preferably at 45° to the equator. The typical power spectrum of the x mode peaks at slightly higher frequencies than those of the + mode (see FIGURE 11).

The efficiency, $\epsilon \equiv E_{GW}/M$, is shown in FIGURE 12 as a function of a/M. We estimate, by comparing various solutions and by comparing results obtained at

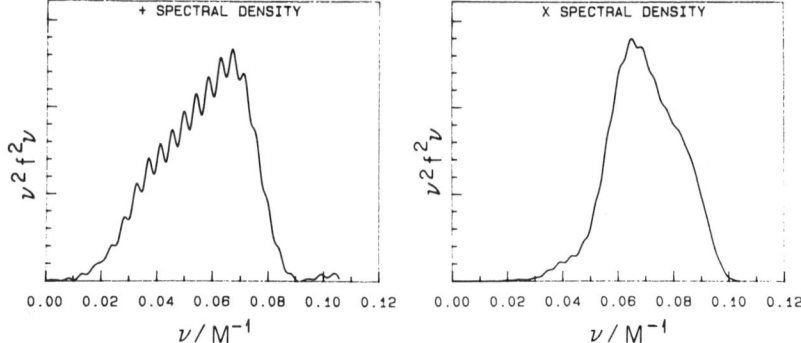

FIGURE 11. Power spectrum (in arbitrary units), $\nu^2 f_\nu^2$, of the + mode (left) and the x mode (right) for collapse with $a/M = 0.9$. (The small-scale oscillation is a numerical artifact.)

different radial distances, that the error is about 30%. The calculated efficiency can be fitted with an $(a/M)^4$ dependence:

$$\epsilon \simeq 1 \times 10^{-3} (a/M)^4 \qquad (20)$$

for $a/M < 0.85$. For higher a/m, ϵ levels off at about 6×10^{-4}.

For higher angular momentum configurations, a lower frequency component appears in the emitted gravitational radiation pulses. This is due to the matter motion

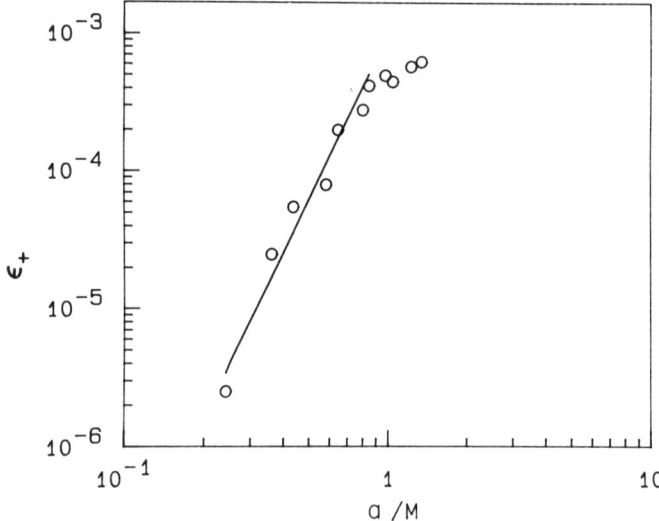

FIGURE 12. Efficiency of generation of gravitational radiation versus the angular momentum parameter, a/M. Circles correspond to calculated points; the solid line is $10^{-3} (a/M)^4$.

during the initial flattening phase and not due to the formation of a deformed black hole. For low angular momentum, this contributes little to the total emitted energy. It becomes important for higher a/M, when a black hole does not form.

These calculations represent an idealized model with an adiabatic equation of state, somewhat artificial initial data, and axial symmetry. Clearly, this does not model realistically a stellar collapse. Can this serve as an indication to what will happen? The answer is probably yes if the axial symmetry is indeed maintained. This is supported by supernova calculations and by the fact that the emitted gravitational radiation corresponds to the normal modes of the black hole that forms. However, it is possible that a bar instability develops during the collapse. This might lead to a fission of the star and to subsequent generation of a lot of angular momentum as the debris spiral in and coalesce. The axial symmetry suppresses this instability and does not allow us to follow this scenario.

A three-dimensional code could determine whether a bar instability develops in rotating collapsing objects. However, a hybrid method may be simpler. The finite number of points in any given direction in a numerical solution is equivalent, in essence, to suppression of the high modes of the exact solution. The cutoff is at $\lambda \simeq 2dx$. The higher frequency modes vary too rapidly to be described on the finite numerical grid. In specific cases, it might be advantageous to replace finite differencing in the ϕ direction by an expansion in $e^{im\phi}$,

$$f(r, \theta, \phi) = \sum_m a_m(r, \theta) e^{im\phi}, \tag{21}$$

with coefficients that depend on r and θ. To observe a bar instability, it is sufficient to use $m \leq 2$.

We have presented preliminary results of the emission of gravitational radiation by a rotating gravitational collapse. We have demonstrated that numerical relativity can solve with high accuracy problems in relativistic astrophysics that cannot be solved otherwise. The efficiency of emission of gravitational radiation increases with angular momentum up to about $a/M \simeq 0.85$, where it levels off at about 6×10^{-4}. Most of the radiation is emitted in the + mode and the waveform resembles the one emitted by a test particle falling into a black hole, but with a reduced amplitude and an opposite sign. These results represent only one series of configurations and they have many limitations as far as being a realistic approximation for an astrophysical collapse. Still, we expect that, unless a bar instability develops and the system deviates strongly from axial symmetry, these results can serve as a good indication for both the efficiency and the waveform of gravitational radiation emitted by potential stellar rotating collapse events.

ACKNOWLEDGMENTS

We would like to thank many colleagues for their helpful discussions: T.P. thanks J.M. Bardeen for many conversations, remarks, and suggestions; R.F.S. thanks S.L. Shapiro, S.A. Teukolsky, and I. Wasserman for valuable discussions and encouragement.

REFERENCES

1. SMARR, L. 1977. Ann. N.Y. Acad. Sci. **302:** 569–604.
2. MAY, M. M. & R. H. WHITE. 1966. Phys. Rev. **141:** 1232–1240.
3. HAHN, S. G. & R. W. LINDQUIST. 1964. Ann. Phys. **29:** 304–331.
4. See, for example, YORK, J. W., JR. 1979. *In* Sources of Gravitational Radiation. L. L. Smarr, Ed. Cambridge University Press. Cambridge, England; YORK, J. W., JR. 1983. *In* Gravitational Radiation. N. Deruelle & T. Piran, Eds. North-Holland. Amsterdam.
5. CENTRELLA, J. *In* The Origin and Evolution of Galaxies. B. Jones & J. Jones, Eds.
6. PIRAN, T. 1980. Ann. N.Y. Acad. Sci. **375:** 1.
7. PIRAN, T. 1982. *In* Proceedings of the Second Marcel Grossmann Meeting. R. Ruffini, Ed. North-Holland. Amsterdam.
8. PIRAN, T. 1983. *In* Gravitational Radiation. N. Deruelle & T. Piran, Eds. North-Holland. Amsterdam.
9. STEWART, J. 1984. Numerical relativity. Preprint.
10. WILSON, J. 1979. *In* Sources of Gravitational Radiation. L. Smarr, Ed. Cambridge University Press. Cambridge, England.
11. SMARR, L. 1979. *In* Sources of Gravitational Radiation. L. Smarr, Ed. Cambridge University Press. Cambridge, England.
12. NAKAMURA, T. 1981. Numerical relativity. Prog. Theor. Phys. Suppl. **70:** 202–214.
13. NAKAMURA, T. 1984. Ann. N.Y. Acad. Sci. **442:** 56.
14. STEWART, J. M. 1984. *In* Radiation Hydrodynamics. M. L. Norman & K. H. Winkler, Eds. Reidel. Amsterdam.
15. STEWART, J. M. & H. FRIEDRICH. 1982. Proc. R. Soc. London **A384:** 427.
16. CORKILL, R. W. & J. M. STEWART. 1983. Proc. R. Soc. London **A386:** 373.
17. ISAACSON, R. A., J. S. WELLING & J. WINICOUR. 1983. J. Math. Phys. **24:** 1824.
18. D'IVERNO, R. A. & J. SMALLWOOD. 1980. Phys. Rev. **D22:** 1233.
19. REGGE, T. 1961. Nuovo Cimento **19:** 558.
20. PIRAN, T. & R. WILLIAMS. 1985. In preparation.
21. ARNOWITT, R., S. DESER. & C. W. MISNER. 1962. *In* Gravitation, Introduction to Current Research. L. Witten, Ed. Wiley. New York.
22. BARDEEN, J. M. 1982. *In* Gravitational Radiation. N. Deruelle & T. Piran, Eds. North-Holland. Amsterdam.
23. BARDEEN, J. M. & T. PIRAN. 1983. Phys. Rep. **96:** 205.
24. PIRAN, T. *In* Proceedings of the Third Marcel Grossmann Meeting. Hunin, Ed. Science Press. Peking.
25. GEROCH, R. 1973. Ann. N.Y. Acad. Sci. **224:** 108.
26. EARDLEY, D. & L. SMARR. 1978. Phys. Rev. **D17:** 2529.
27. STARK, R. F. & T. PIRAN. 1985. Phys. Rev. Lett. **55:** 891; SRARK, R. F. & T. PIRAN. 1986. *In* Proceedings of the Fourth Marcel Grossmann Meeting. R. Ruffini, Ed. North-Holland. Amsterdam (In press); Paper in preparation.
28. PIRAN, T. 1980. J. Comput. Phys. **35:** 254.
29. STARK, R. F. & T. PIRAN. In preparation.
30. PETRICH, L. I., S. L. SHAPIRO & I. WASSERMAN. 1984. Preprint.
31. WILSON, J. *et al.* 1986. Ann. N.Y. Acad. Sci. **470:** 267–293.
32. BUTTERWORTH, E. M. & J. R. IPSER. 1976. Astrophys. J. **204:** 200.
33. DETWEILER, S. L. 1979. *In* Sources of Gravitational Radiation. L. Smarr, Ed. Cambridge University Press. Cambridge, England.

Stellar Core Collapse and Supernova[a,b]

J. R. WILSON,[c] R. MAYLE,[c,d] S. E. WOOSLEY,[c,e]
AND T. WEAVER[c]

[c]*Lawrence Livermore National Laboratory*
Livermore, California 94550

[d]*Physics Department*
University of California
Berkeley, California 94720

[e]*Board of Studies in Astronomy and Astrophysics*
Lick Observatory
University of California
Santa Cruz, California 95064

INTRODUCTION

Massive stars that end their stable evolution as their iron cores collapse to a neutron star or black hole have long been considered good candidates for producing Type II supernovae. For many years, the outward propagation of the shock wave produced by the bounce of these iron cores has been studied as a possible mechanism for the explosion (cf., Bowers and Wilson, 1982;[1] Arnett, 1980,[2] 1983;[3] Brown, Bethe, and Baym, 1982;[4] Hillebrandt, 1984;[5] Bruenn, 1985;[6] Cooperstein, 1982;[7] Kahana, Baron, and Cooperstein, 1984[8]). For the most part, the results of these studies have not been particularly encouraging, except perhaps in the case of very low mass iron cores (Hillebrandt, Nomoto, and Wolff, 1984;[9] Cooperstein, 1982[7]) or very soft nuclear equations of state (Baron, Cooperstein, and Kahana, 1985[10]). The shock stalls, overwhelmed by photodisintegration and neutrino losses, and the star does not explode. More recently, slow late time heating of the envelope of the incipient neutron star has been found to be capable of rejuvenating the stalled shock and producing an explosion after all (Wilson, 1985;[11] Bethe and Wilson, 1985[12]). The present paper discusses this late time heating and presents results from numerical calculations of the evolution,

---

[a]Disclaimer: This document was prepared as an account of work sponsored by an agency of the United States Government. Neither the United States Government nor the University of California nor any of their employees, makes any warranty, express or implied, or assumes any legal liability or responsibility for the accuracy, completeness, or usefulness of any information, apparatus, product, or process disclosed, or represents that its use would not infringe privately owned rights. Reference herein to any specific commercial products, process, or service by trade name, trademark, manufacturer, or otherwise, does not necessarily constitute or imply its endorsement, recommendation, or favoring by the United States Government or the University of California. The views and opinions of the authors expressed herein do not necessarily state or reflect those of the United States Government thereof, and shall not be used for advertising or product endorsement purposes.

[b]This research has been supported by the National Science Foundation (AST-81-08509 and AST-84-18185) and, at Livermore, by the U. S. Department of Energy through contract no. W-7405-ENG-48 and by the Institute for Geophysics and Planetary Physics (IGPP).

core collapse, and subsequent explosion of a number of recent stellar models. For the first time, they all, except perhaps the most massive, explode with reasonable choices of input physics.

PREEXPLOSIVE MODELS

A number of presupernova stellar models were considered that varied in mass and the input physics used to study their stable evolution. Because of these variations, the iron cores at the time of collapse spanned a range of masses from 1.31 $M_\odot$ to 2.05 $M_\odot$ (TABLE 1). Several of these models have been published previously and the detailed evolution of the remainder is the subject of papers now in preparation.

TABLE 1. Summary of Runs

Initial Stellar Mass	Model	Initial Iron Core Mass	Explosion Energy[a]	Residual Mass in Baryons	Final Neutron Star Mass
10	A old	1.40	1.8×10^{50}	1.47	1.33
11	A	0.20	3.0	1.42	1.31
12	C	1.31	3.8	1.35	1.26
15	A old	1.56	3.4	1.75	1.53
15	C	1.35	3.5	1.42	1.31
20	C	1.70	Not run through collapse		
25	A old	1.37	4.5	1.66	1.46
25	B	1.63	11.0	1.83	1.58
25	C	2.05	10.0	2.44	1.96 B.H.?
50	A	1.79	7.5	1.86	1.60
100	A	1.85	0	>2.95	B.H.?
			A Rate $C(\alpha, \gamma)0 = 1$		
	Model =		B Rate $C(\alpha, \gamma)0 = 2.5$		
			C Rate $C(\alpha, \gamma)0 = 3.0$		

[a]As calculated in the core bounce code for the inner mantle only. The final kinetic energy at infinity will be smaller.

The star of lowest main sequence mass considered here, Model 10A, was calculated by Woosley, Weaver, and Taam (1980)[13] using the older version of the stellar evolution program KEPLER described by Weaver, Zimmerman, and Woosley (1978).[14] In particular, the Fowler, Caughlan, and Zimmerman (1975)[15] reaction rate for $^{12}C(\alpha, \gamma)^{16}O$ was employed, as well as earlier vintages of electron capture rates, screening corrections, initial composition, and prescription for silicon burning. A similar core evolution up through the onset of degenerate neon shell flashes has recently been determined by Nomoto (1984)[16,17] for a 2.6 $M_\odot$ helium core. In both cases, the evolution is characterized by temperature inversion in the core and off-center burning. In particular, the neon, oxygen, and silicon burning all ignite off-center (in the Woosley, Weaver, and Taam study) and the hydrogen and helium shells are ejected by a particularly violent neon flash roughly five years before the star explodes. Previous studies of core bounce in this star have either shown no prompt hydrodynamical explosion (Bowers and Wilson, 1982;[1] Bruenn, 1985[6]) or else a very weak one (Hillebrandt, 1982).[18] Similarities between this star and the observed properties of the Crab Nebula, especially the lack of substantial heavy element nucleosynthesis, have been noted (Arnett, 1975;[19] Woosley, Weaver, and Taam, 1980;[13] Nomoto, 1982,[20] 1984;[16,17] Nomoto et al., 1982;[21] Hillebrandt, 1982[18]) and imply that the progenitor of the Crab was a star of roughly this mass. A particularly bright light curve could be a consequence of the shock interacting with the tenuous envelope ejected just before the explosion (Weaver and Woosley, 1979).[22]

Only one other model employed in this paper was calculated using the old version of KEPLER. Model 15A is identically the 15 $M_\odot$ presupernova star published by Weaver, Zimmermann, and Woosley (1978).[14] All other stars have been evolved using a revised version of the program (Weaver, Woosley, and Fuller, 1985)[23] that incorporates the new weak interaction rates of Fuller, Fowler, and Newman (1982, 1985),[24-26] an improved implementation of weak processes at an earlier stage of the star's life (near oxygen core depletion), revisions to nuclear screening corrections, finer time step and zoning criteria, and Cameron (1982)[27] initial abundances. Because of the smaller helium abundance employed in the new studies, helium core masses are smaller than in the earlier studies. For example, the new 11 $M_\odot$ star has a helium core at the end of its life of 2.4 $M_\odot$, which is slightly smaller than the helium core in the old 10 $M_\odot$ star (see above). Helium core masses in the larger stars are also reduced by about 10%. More importantly, the revised weak rates and prescription for their implementation lead to iron core masses that are considerably smaller than in Weaver, Zimmerman, and Woosley. Model 25A is an example with an iron core mass reduced from its 1978 value, 1.61 $M_\odot$, to 1.37 $M_\odot$ (Weaver, Woosley, and Fuller, 1985).[23]

The same revised code and 1975 version of $^{12}C(\alpha, \gamma)^{16}O$ were also used for Models 11A, 50A, and 100A. The 11 $M_\odot$ star had an evolution that differed markedly from all the other stars examined here, but was quite similar to that reported for a 9.6 $M_\odot$ star by Nomoto[17,18] (see also, Miyaji et al., 1980[28]). The helium core mass in the Nomoto study was 2.4 $M_\odot$, roughly the same as here at the end of the evolution. However, just after depletion of carbon in the center of the 11 $M_\odot$ star, the helium core mass was 2.6 $M_\odot$ (hydrogen dredge-up of the helium core during off-center carbon burning accounts for the difference). In a fashion similar to Model 10A discussed above, the evolution of Model 11A is characterized by temperature inversion (owing especially to plasma neutrino losses) and off-center burning. A neon core develops (60% neon and 20% each of magnesium and oxygen) containing very nearly a Chandrasekhar mass (1.45 $M_\odot$). That this mass exceeds 1.37 $M_\odot$ marks a critical departure between the present work and that of Nomoto.[16,17] For the smaller mass, neon and subsequent burning stages never ignite (under stable conditions), whereas in Model 11A, both neon and oxygen burning do occur in a shell from about 0.6 $M_\odot$ to 1.3 $M_\odot$ (FIGURE 1) and, in a second convective burning stage, from 1.3 $M_\odot$ to 1.4 $M_\odot$. Thus at the end of its life, this star consists of a cold ($T \sim 5 \times 10^8$ K) central core of about 0.5 $M_\odot$ of neon surrounded by silicon and sulfur with a total mass slightly exceeding the (zero entropy) Chandrasekhar mass (FIGURE 2). Further cooling then leads to the collapse of the core containing combustible nuclear fuel.

It must be acknowledged at this point that the onset of collapse in this star has not been calculated in an entirely self-consistent fashion. Electron capture reactions were

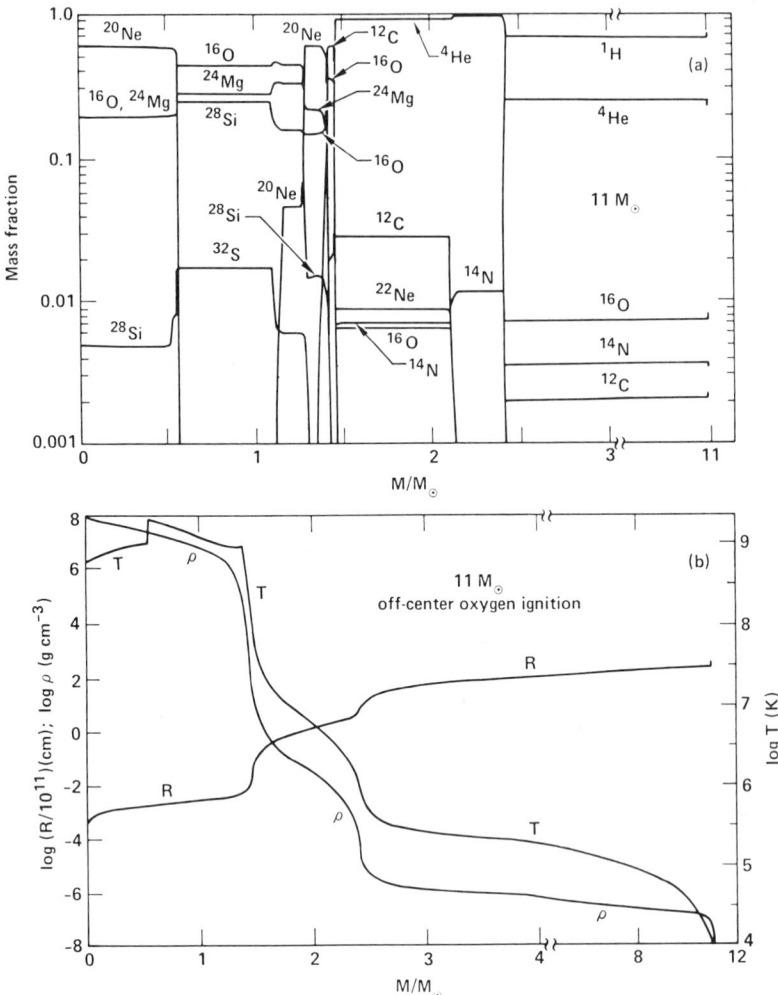

FIGURE 1. Composition (a) and structure (b) of an 11 $M_\odot$ star (Model 11A) just after the off-center ignition of oxygen burning. The temperature at the base of the convective oxygen shell at this point is 1.87×10^9 K. The central density and temperature are 8.19×10^7 g cm^{-3} and 5.63×10^8 K, respectively. The surface luminosity, effective temperature, and radius, all of which should not change prior to the stellar explosion, are 1.47×10^{38} erg s^{-1}, 4290 K, and 2.47×10^{13} cm, respectively. The neutrino luminosity is 1.21×10^{42} erg s^{-1}. Note scale breaks at 3.0 $M_\odot$ (a) and 4.0 $M_\odot$ (b).

not included in the calculation of neon and oxygen burning although oxygen burns at such a high density off-center here that such captures would certainly have been important, probably triggering core collapse at an earlier stage. At a density of ~10^8 g cm^{-3}, for example, capture on the products of oxygen burning [e.g., ^{33}S$(e^-, \nu)^{33}$P; see Woosley, Arnett, and Clayton, 1972[29]] would lead to $Y_e \sim 0.49$, reducing the zero

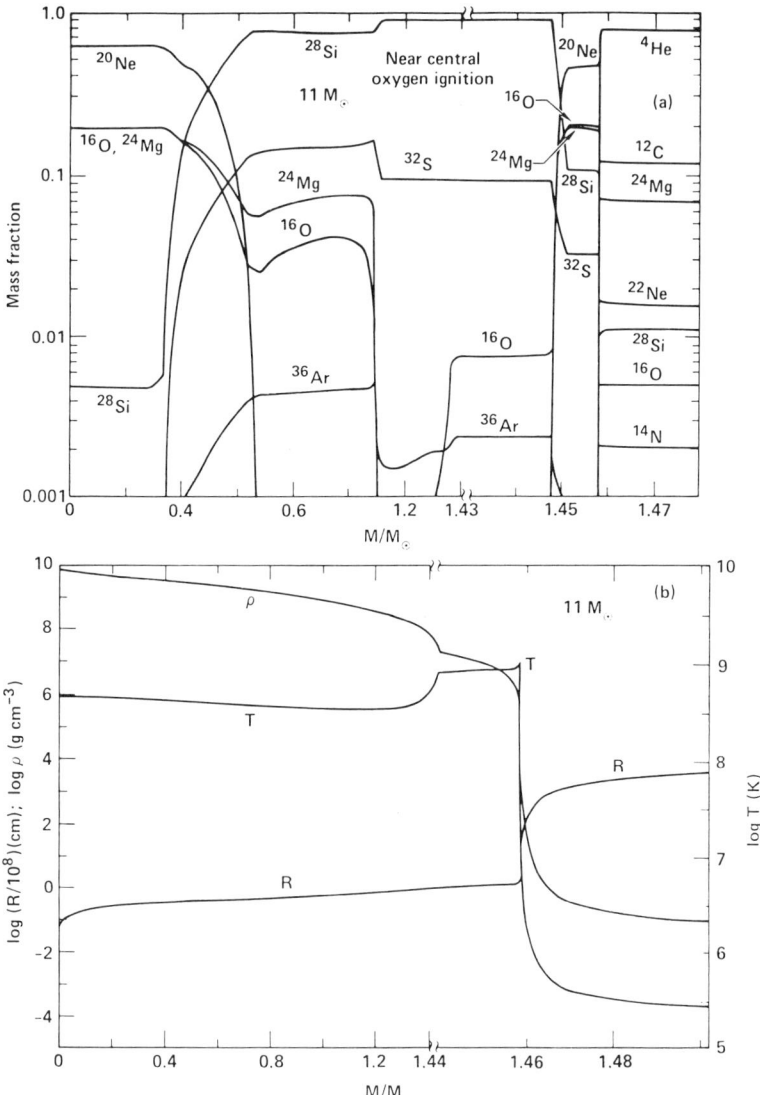

FIGURE 2. Composition (a) and structure (b) of the central regions of an 11 $M_\odot$ star (Model 11A) just before the onset of collapse triggered by oxygen ignition and electron capture. The central density at this point is 7.1×10^9 g cm^{-3} (runaway will occur at 2.5×10^{10} g cm^{-3}) and the central temperature is 4.5×10^8 K. The total entropy in the inner 1.4 $M_\odot$, in dimensionless units, S/Nk, is ~0.65 and the electron entropy is ~0.1. Partial mixing between the neon core and silicon shell is apparent in the region of 0.4 to 0.5 $M_\odot$, brought about by the inversion of mean molecular weight. The temperature at the base of the convective helium shell, which by now has penetrated into the hydrogenic envelope, is greater than 4×10^8 K. The actual combustion region was not well resolved in the present calculation even though mesh sizes of $\leq 10^{-4}$ $M_\odot$ were employed in the helium burning region. Note scale breaks at 1.43 $M_\odot$ (a) and 1.44 $M_\odot$ (b).

entropy Chandrasekhar mass to 1.38 $M_\odot$ and causing the core to collapse. We do not think that the continued evolution is especially sensitive to this detail, but a self-consistent calculation would obviously be desirable. It is interesting to note that the contraction of the core is not without nucleosynthetic import in the overlying layers. We find that the heating of the helium burning shell leads to extension of the helium convective zone into the hydrogen envelope. The mixing of protons into a superheated helium shell has obvious and far-reaching implications for r- and s-process nucleosyntheses, which will be discussed elsewhere.

As the core contracts to 2.5×10^{10} g cm^{-3}, oxygen burning, greatly enhanced by electron screening, ignites at a temperature of $\sim 5 \times 10^8$ K. We note that at this point the core is also hovering on the verge of general relativistic instability (Shapiro and Teukolsky, 1983)[30] because the mean adiabatic index, Γ, is so nearly equal to $4/3$. Future calculations should also include post-Newtonian gravity. Owing to the extremely degenerate nature of the core, the nuclear runaway that ensues will lead to complete combustion to iron group nuclei. Here the evolution becomes quite similar to that described by Nomoto[16,17] for the 2.4 $M_\odot$ helium core. Iron group products capture electrons efficiently, thus leading to a sudden drop in Y_e at the center of the star to about 0.40. The thermal increment to the pressure from burning is inconsequential compared to the great loss in electron degeneracy pressure and the core begins to implode dynamically. When first, neon and magnesium and, later, silicon and sulfur, fall down, they are heated by compression and burn explosively so that a standing combustion front exists at ~ 110 to 170 km. Once the collapse is well under way, energy transport by convection is negligible.

Models 12C, 15C, and 25C also used the improved version of KEPLER, but incorporated additional revisions to the rate for $^{12}C(\alpha, \gamma)^{16}O$. Recent measurements by Kettner et al. (1982)[31] and reanalysis of new and old data by Langanke and Koonin (1984)[32] have led to a revised rate, as tabulated by Caughlan et al. (1984),[33] that under typical conditions in massive stars is about three times the old value used in all other calculations (Fowler, Caughlan, and Zimmerman, 1975).[15] Such a large revision has, as we shall see, major implications not only for nucleosynthesis, but also for the structure of the presupernova star. The iron core mass in Models 12C and 15C are not greatly altered from, say, Model 25A or from the 20 $M_\odot$ star studied by Weaver, Woosley, and Fuller[23] (although, as it turns out, the extent of the oxygen burning shell is much larger with the new rate). However, Model 25C has a much larger iron core. To examine the dependence on $^{12}C(\alpha, \gamma)^{16}O$, a third 25 $M_\odot$ star was also studied in which the old FCZ75 rate was simply multiplied by a constant factor of 2.5. The iron core mass was found to have an intermediate value (TABLE 1).

Why should there be such a great variation in the iron core mass for stars of 15 and 25 $M_\odot$ and for two stars, both of 25 $M_\odot$, differing only in the magnitude of the reaction rate for $^{12}C(\alpha, \gamma)^{16}O$? Models 15C and 25C, for example, which utilize otherwise identical physics including electron capture rates, have iron core masses at collapse that differ by about 0.6 $M_\odot$ (TABLE 1). A comparable difference characterizes Models 25A and 25C. Why is the evolution so radically altered? The answer apparently resides in the nature of carbon and neon burning and how they affect the entropy structure of the stellar core. In Model 15C and Model 25A (the 25 $M_\odot$ star with low α rate), carbon burning ignites as a well-developed, exoergic, convective burning stage. In fact, the 15 $M_\odot$ star goes through three distinct stages of carbon convective burning before igniting neon burning at its center. The first stage depletes carbon in a region out to about 0.4

$M_\odot$, the second convective shell goes from ~0.4 out to 1.0 $M_\odot$, and finally a third stage burns carbon out to about 1.5 $M_\odot$. The initial carbon abundance following helium burning in this star is 0.14. While carbon burning goes on, neutrino losses cool the core. During the first carbon convective stage, the central conditions are $T_c = 8.2 \times 10^8$ K and $\rho_c = 3.8 \times 10^5$ g cm^{-3}, while at the end of the third stage (at neon ignition), $T_c = 1.6 \times 10^9$ K and $\rho_c = 9.2 \times 10^6$ g cm^{-3}. This loss of entropy from the core allows it to become sufficiently degenerate so that the core size can become approximately equal to the Chandrasekhar mass. Thus, a strong carbon burning shell is established at the edge of a (semi-)degenerate core. From this point onwards, there exists a sharp increase in the entropy at $M \sim 1.5\ M_\odot$. This restrains the outward extension of convective shells during oxygen and silicon burning so that when the star finally collapses, it does so with an iron core of 1.33 $M_\odot$, which is close to the traditional Chandrasekhar mass for material with equal numbers of neutrons and protons. At least a portion of the decrease from ~1.5 to 1.33 $M_\odot$ is a consequence of electron capture during and shortly after oxygen burning.

The 25 $M_\odot$ star with the most recent α-capture rate is quite another story. Because the carbon abundance following helium depletion is so low, ~9% by mass, carbon burning and, as it turns out, neon burning as well, never ignite in the center of the star as exoergic, convective burning stages. The trace abundances of carbon and neon burn away radiatively, without the nuclear energy generation ever exceeding neutrino losses. Because there is no cooling stage and because the 25 $M_\odot$ star had a larger entropy in its core to start with, the core does not become especially degenerate and so is not sensitive to the Chandrasekhar mass. Carbon is depleted radiatively out to a mass of about 2.5 $M_\odot$. Neon burns radiatively out to about 1.5 $M_\odot$. Thus the next fully developed burning stage after helium burning is oxygen burning. Prior to silicon core ignition, oxygen burns in a convective core first out to about 1.35 $M_\odot$ and then in a convective shell out to 2.4 $M_\odot$. When silicon does ignite, it is within a 2.4 $M_\odot$ core comprised of almost pure silicon and sulfur. The large entropy increase associated with the oxygen shell is then at 2.4, not 1.4 $M_\odot$ as it was in the 15 $M_\odot$ model. Silicon burns out to 1.3 $M_\odot$, but even with $Y_e \sim 0.46$, the core is too small to collapse given its thermal content. Thus, a silicon convective shell burns out to ~2.1 $M_\odot$ before the core collapses.

It is interesting that the property of a massive star that most sensitively determines its final evolutionary state (neutron star or black hole), namely, the iron core mass at collapse, is so sensitive to occurrences in the relatively early life of the star. The 50 and 100 $M_\odot$ models used here both had old (smaller) α-capture rates and the iron core masses are expected to be even larger, with revised rates, than in Model 25C. Studies since the meeting confirm this. A new 50 $M_\odot$ model with revised alpha-rate collapses with an iron core mass of 2.5 $M_\odot$.

Since the baryon mass remaining after the delayed explosion always exceeds the initial iron core mass, it appears that the compact remnants of stars that have main sequence mass $\gtrsim 20\ M_\odot$ [with exact value sensitive to $^{12}C(\alpha, \gamma)^{16}O$] will have quite different properties from those of lower mass. In particular, since nuclear equations of state suggest an upper bound to the mass of a stable neutron star of ~2.0 $M_\odot$, stars heavier than ~25 $M_\odot$ may leave black holes and lighter stars (but heavier than ~8 $M_\odot$) will leave neutron stars. It is also important to note that an object that eventually, after cooling and deleptonization, becomes a black hole, may be the residual of an explosion that ejects matter, produces explosive nucleosynthesis, and exhibits a light curve not markedly discrepant with observations of Type II supernovae.

CORE COLLAPSE AND SLOW EXPLOSION

Core collapse is initiated in all models, except 11A, by a combination of electron capture and photodisintegration (FIGURES 3 and 4). Generally speaking, photodisintegration is the dominant mechanism leading to instability in stars heavier than about 15 $M_\odot$. As the collapse commences, most of the iron core falls in as a unit, that is, homologously, but as the density increases, the pressure does not rise as rapidly as density to the four-thirds power. Hence, a decreasing fraction of the core collapses as a unit. Taking Model 25C as prototypical, just before nuclear density is achieved at the center, the sonic point is located at a mass point of 0.55 $M_\odot$ so that only 0.5 to 0.6 $M_\odot$ is collapsing homologously. After the central part of the core rises above nuclear density, the pressure rapidly increases and halts the infall of the inner 0.5 $M_\odot$ of the core. As the pressure wave moves out through the more slowly falling material, a shock wave first arises at about 0.7 to 0.8 $M_\odot$. The outward propagation of this shock wave through the star has been investigated intensively over the last few years. Only in the case of very low core masses has the shock proceeded outwards with adequate energy to be considered the basis of the supernova phenomenon (Hillebrandt, Nomoto, and Wolff, 1984).[9] The success or failure of the shock in escaping the iron core depends upon subtle effects in the equation of state and neutrino cooling.

In this paper, we consider the subsequent behavior of the stellar cores in models in which the shock initially fails to produce a supernova. A few hundredths of a second following bounce, matter from the outer parts of the core and the surrounding stellar mantle are falling nearly freely onto an almost stationary accretion shock. Below the shock, matter settles inward relatively slowly and accumulates on the dense core (FIGURE 5). Consider the energy budget of the slowly settling matter at a radius of $r \sim 10^7$ cm. Since the region of interest is outside the neutrinosphere, the heating by the hot radiating core is approximately

$$\dot{E}_+ = \kappa_a(T_p)(L_\nu Y_n + L_{\bar{\nu}} Y_p)/4\pi r^2 \qquad (1)$$
$$\approx \kappa_a(T_p) L_\nu/4\pi r^2,$$

where $L_\nu \approx L_{\bar{\nu}}$ at late times and matter is presumed to be dissociated into free nucleons in the region of interest. Here κ_a is the absorption opacity, T_p, the neutrinosphere temperature, L_ν, the electron neutrino luminosity, and r, the radius. Heating due to electron scattering is initially small and is ignored here. The cooling rate of the matter on the other hand is

$$\dot{E}_- = \kappa_e(T_m) a' c T_m^4, \qquad (2)$$

where κ_e is the emission opacity, T_m, the local matter temperature, and a' is the radiation constant for neutrinos (7/16ths of the photon radiation constant).

The matter in the region of interest is only moderately degenerate. For this case,

$$\kappa_a = 1.33 \, \sigma_0 \left(\frac{\epsilon_\nu}{m_e c^2}\right)^2 / m_H$$
$$\approx 11.0 \times 10^{-19} \, T_p^2 \, \text{cm}^2 \, \text{g}^{-1}, \qquad (3)$$

where $\kappa_e = 11.0 \times 10^{-19} \, T_m^2$ and where the temperatures are measured in MeV. One

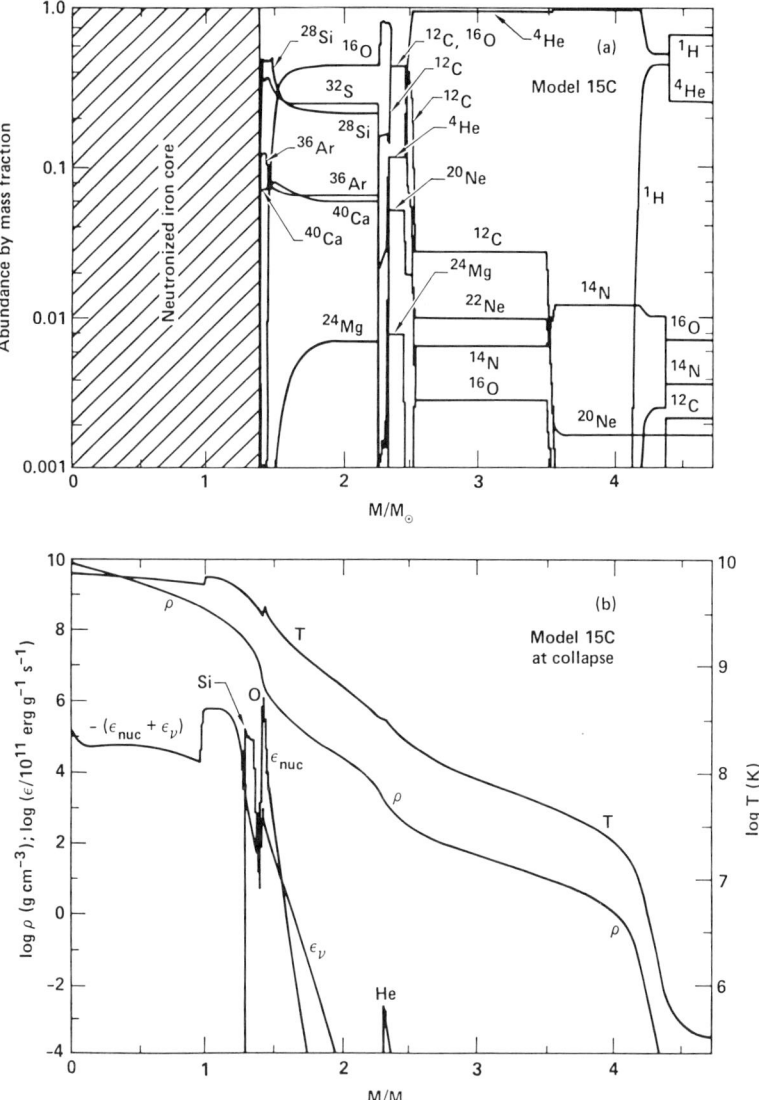

FIGURE 3. Composition (a) and structure (b) of a 15 $M_\odot$ star (Model 15C) at the onset of core collapse (as defined by collapse velocity equal to 1000 km s^{-1}). Interior to the iron core (1.33 $M_\odot$), energy is being lost due to a combination of neutrino emission from electron capture and photodisintegration. Both nuclear energy generation and neutrino losses are inherently negative there. Peaks in the nuclear energy generation, ϵ_{nuc}, are apparent at the silicon, oxygen, and helium burning shells. Also plotted are the neutrino losses due to plasma processes (ϵ_ν in the region outside the iron core). Note the rapid falloff in density just outside the iron core and the substantial abundances of the elements silicon through calcium in the oxygen convective shell. For the most part, the oxygen shell is ejected (in this model only) without much explosive modification.

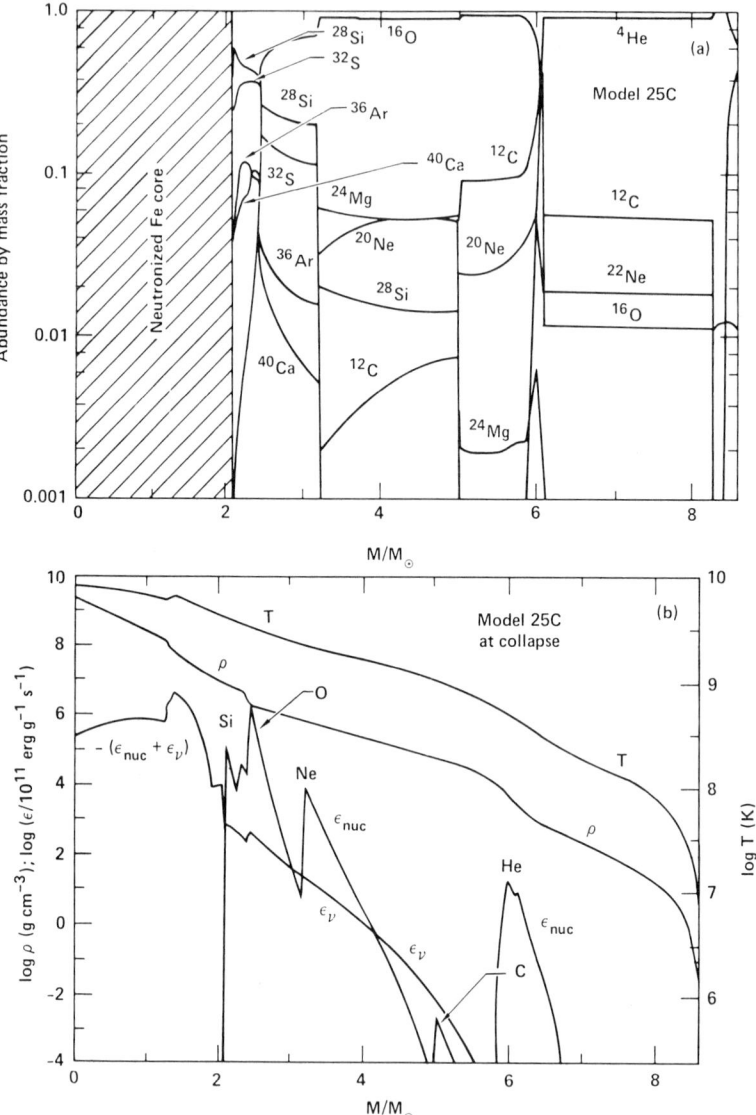

FIGURE 4. Composition (a) and structure (b) of a 25 $M_\odot$ star (Model 25C) at the onset of core collapse. Note the large iron and the less rapid falloff in density (compared to FIGURE 3) outside that iron core. Notation and sampling time are as in FIGURE 3.

can also write

$$L_\nu = 4\pi r_p^2 a' c T_p^4/4,$$

$$\dot{E}_{net} = 2.0 \times 10^{18} T_p^6 \left[\left(\frac{r_p}{2r_m}\right)^2 - \left(\frac{T_m}{T_p}\right)^6\right]. \quad (4)$$

Thus if $T_m < T_p (r_p/2r_m)^{1/3}$, heating of the matter will occur. For a core mass of 1.5 $M_\odot$, the gravitation energy is $GE = 2 \times 10^{19}/r_7$ ergs g^{-1} (r_7 is the radius in units of 10^7 cm), which is also approximately the internal energy of the material in the region of interest. The low density of matter there implies that the nuclei will disintegrate at a low temperature, about 1 to 2 MeV. This requires an energy of about 8×10^{18} ergs g^{-1}. If the entropy in this region is low, less than 10 k, then the degeneracy of the electrons may be an important sink of energy, and if the entropy is high, pairs and radiation are big sinks of energy. Thus, the heat capacity of matter in this region is large and the

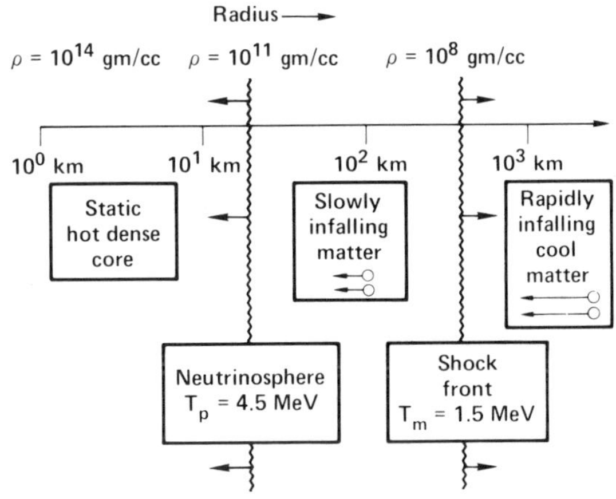

FIGURE 5. Schematic representation of typical conditions in a star a few seconds after core bounce.

temperatures stay moderately low. A little outside of a radius of 10^7 cm, the matter will be in the dissociation range and may have a temperature low enough that the core radiation can heat it.

As a specific example, consider the 25 $M_\odot$ model at a relevant time, when $T_p = 4.5$ MeV, $L_\nu = 1.3 \times 10^{53}$ erg s^{-1}, and $E_+ = 1.2 \times 10^{21}/r^2$ erg g^{-1} s^{-1}. For a core mass of 1.5 $M_\odot$, the local gravitational energy is of the order of $2 \times 10^{19}/r_7$ erg g^{-1} s^{-1} and a characteristic heating time scale is 0.02 sec. In practice, since losses are always important, the time scale for heating is usually a few tenths of a second. Early on, the scattering on electrons of neutrinos of all types contributes about 20% to heating compared to nuclear absorption. After the heating has progressed and the matter has expanded, pairs form that help keep the temperature down and also provide more electrons for heating by scattering. The heating by scattering on pairs becomes important at later times. For a more detailed discussion of the heating process, see Bethe and Wilson (1985).[12]

After material moves below a radius of about 10^7 cm, it loses pressure support and falls rapidly down onto the dense core. This accretion onto the core adds considerably to the neutrino luminosity. The luminosity associated with accretion, which is taken as the mass flux multiplied by the gravitational potential at the core, is sometimes as large as the neutrino energy flux emerging from the core itself. Heating of the matter by the neutrinos slows down the infall of matter, which in turn leads to a decrease of luminosity and heating. Thus, the accretion process may (and in some cases does) become unstable and oscillatory.

In FIGURE 6, we give a radius versus time plot for selected mass points. On this graph are indicated the time intervals over which particular physical processes are important, along with the approximate composition of the several regions. In FIGURE 7, the density radius profile at the end of a model calculation is presented. Note the extreme disparity between the core density and the bubble density, while the bubble and envelope densities are quite comparable.

SUMMARY OF RUNS

TABLE 1 gave a summary of the results for the 11 cases studied (the 20 $M_\odot$ model remains under study as of the time of this writing). FIGURES 8a–j are the corresponding radius versus time plots. For the lower range of masses (10 to 12 $M_\odot$), the shock wave moves out continuously. Note, however, that the shock wave is moving out much more slowly than matter is falling through the shock except at late times. The mantle heating and shock propagation are a continuous process. Beginning with the 15 $M_\odot$ model and

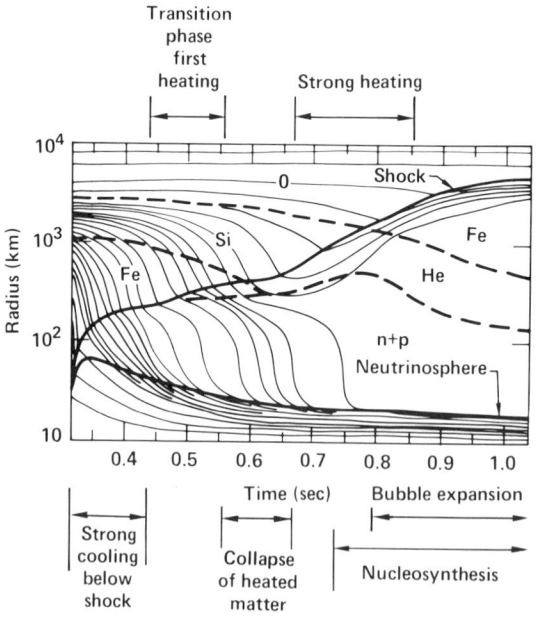

FIGURE 6. Radius as a function of time for selected mass points for Model 25B. Indicated compositions are the major composition between the dashed lines. The approximate time intervals of the important processes are indicated above and below the graph.

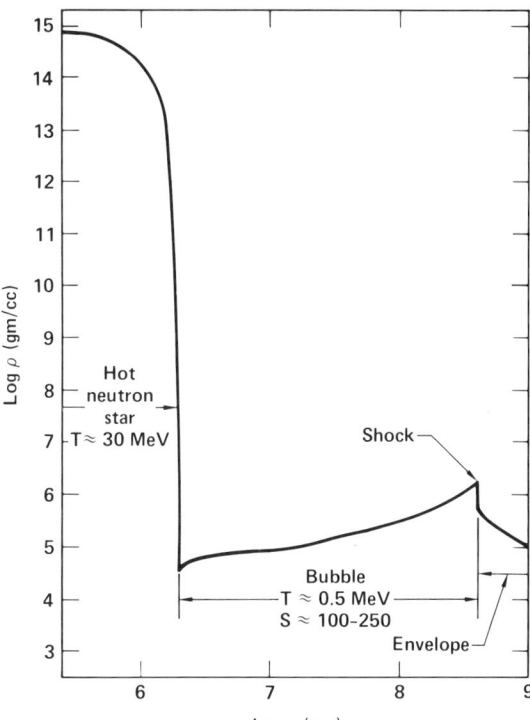

FIGURE 7. Density as a function of radius for Model 25B at a late time, 1.05 s.

all heavier models, the shock does not progress monotonically in radius with time. The retreat of the shock wave leads to more matter falling onto the central core, which enhances the luminosity and leads to further heating. This interplay between heating and infall is best seen in FIGURE 8j and FIGURE 9.

The 11 $M_\odot$ model is different from all the others. As stated earlier, this star first evolves a 2.4 $M_\odot$ He core that later becomes unstable within its inner 1.45 $M_\odot$. At the time the stellar evolution calculation was linked to the supernova code, the composition consisted of a sequence of layers, 0.20 $M_\odot$ "Fe" interior to a roughly stationary deflagration front, surrounded by 0.30 $M_\odot$ of O + Ne + Mg, 0.90 $M_\odot$ of Si, and 0.05 $M_\odot$ of O + Ne + Mg. The central density at the time the two calculations were linked was 8×10^{10} g cm^{-3} and the central temperature was 0.9 MeV. The iron behind the deflagration front (located at 100 to 170 km) is highly neutronized, owing to its high density. The unburned layers of oxygen, neon, magnesium, and silicon have very low entropy, about 0.3 k. Hillebrant, Nomoto, and Wolff (1984)[9] found that a somewhat similar model exploded by direct propagation of the bounce shock outward. FIGURE 10 shows how the density profiles compare between our model and theirs. The much steeper density profile and lower core mass of Nomoto's model should increase the likelihood of a prompt explosion, but our model explodes only after a long time and only as a consequence of neutrino energy transport. FIGURE 11 shows the region of net neutrino heating superimposed on the radius time graph. In order to explore the dependence of the results on the calculated composition, the Si region in Model 11A was artificially switched to O + Ne + Mg in the hope that the energy released by nuclear burning would enhance the possibility of a prompt explosion. These changes made only a small difference in the overall behavior.

Our calculations have used the equation of state described in Bowers and Wilson[1,34] modified to agree with the Bethe, Brown, Cooperstein, and Wilson (1983)[35] equation of state near nuclear density. With this equation of state, the bounce has a very low

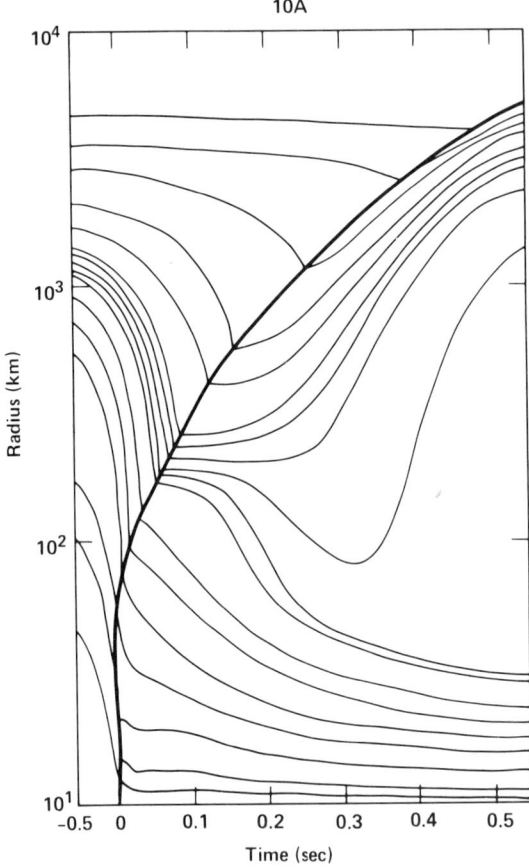

FIGURE 8a. Radius as a function of time for selected mass points for all the models of TABLE 1. Time on the graphs starts shortly before bounce.

amplitude, as can be seen from FIGURES 8. We find that by stiffening the equation of state below nuclear density to increase the mass of the homologously collapsing core and by softening the equation of state at and above nuclear density in order to increase the amplitude of the bounce, a prompt explosion follows the bounce. The changes in the equation of state we found necessary to produce immediate explosions were large, but not ridiculous. The low mass stars of 10–12 $M_\odot$ will require very careful modeling since

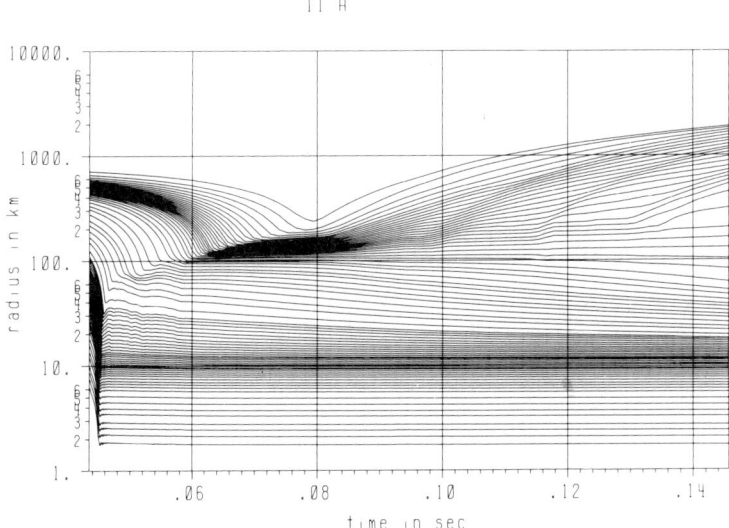

FIGURE 8b. Radius as a function of time for selected mass points for all the models of TABLE 1. Time on the graphs starts shortly before bounce.

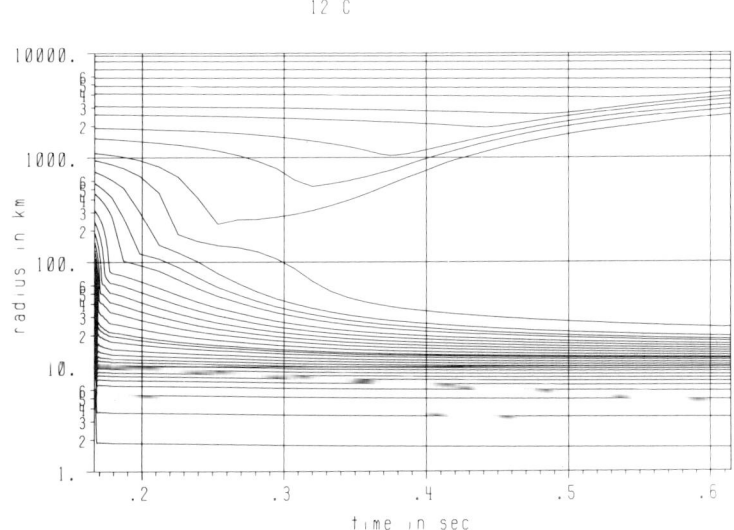

FIGURE 8c. Radius as a function of time for selected mass points for all the models of TABLE 1. Time on the graphs starts shortly before bounce.

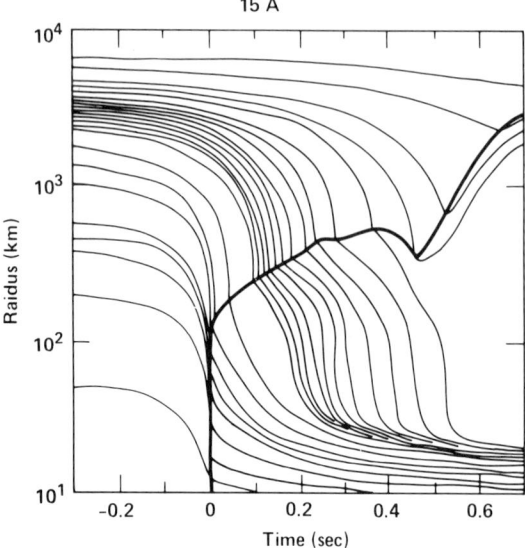

FIGURE 8d. Radius as a function of time for selected mass points for all the models of TABLE 1. Time on the graphs starts shortly before bounce.

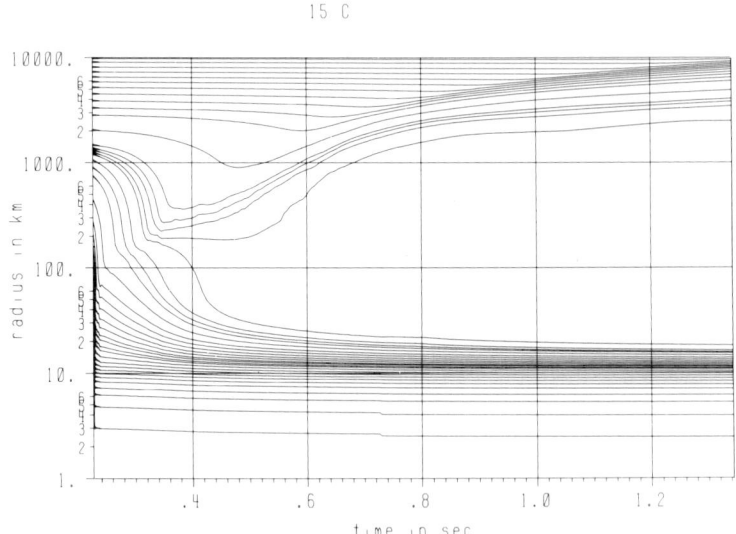

FIGURE 8e. Radius as a function of time for selected mass points for all the models of TABLE 1. Time on the graphs starts shortly before bounce.

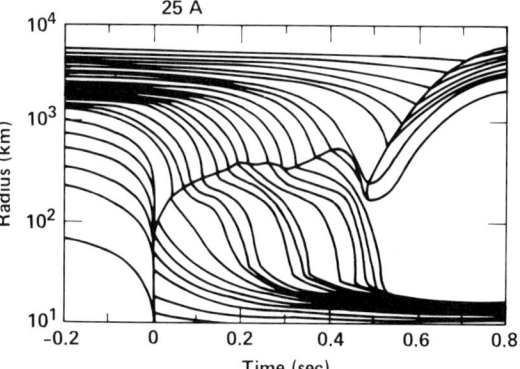

FIGURE 8f. Radius as a function of time for selected mass points for all the models of TABLE 1. Time on the graphs starts shortly before bounce.

they are on the edge between exploding by prompt shock propagation and slow neutrino heating. This is quite important because the prompt explosions are more energetic, 10^{51} ergs or more, than the slow heating explosions.

Models 25 A, B, and C provide an interesting sequence since they have widely varying "Fe" core masses inside the same total mass of star. At the time of the start of collapse, the entropies of the inner 0.5 $M_\odot$ of the three models are 1.02, 1.30, and 1.50, respectively, in units of Boltzman's constant. In FIGURE 12, the density profiles for the three models are shown at times when the central zone has reached a density of 10^{14} g cm^{-3}. The inner cores are seen to have converged to very similar density distributions.

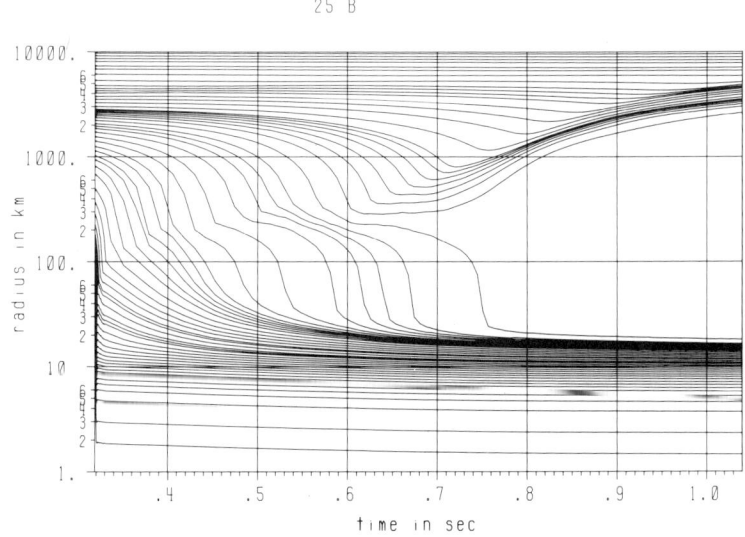

FIGURE 8g. Radius as a function of time for selected mass points for all the models of TABLE 1. Time on the graphs starts shortly before bounce.

284 ANNALS NEW YORK ACADEMY OF SCIENCES

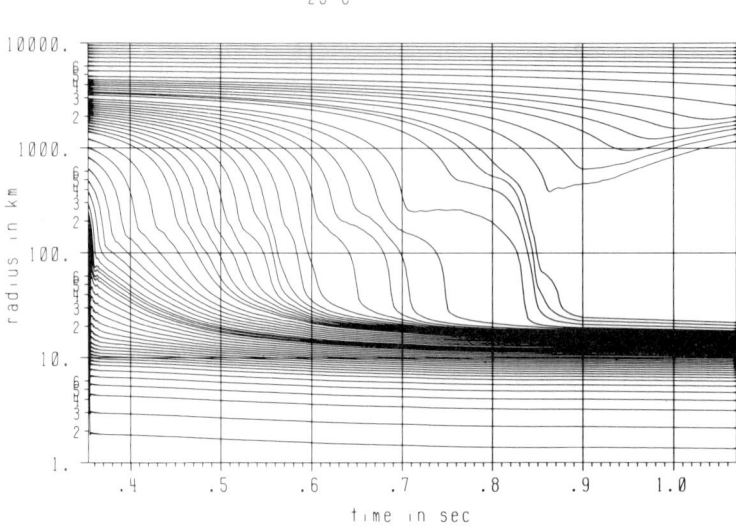

FIGURE 8h. Radius as a function of time for selected mass points for all the models of TABLE 1. Time on the graphs starts shortly before bounce.

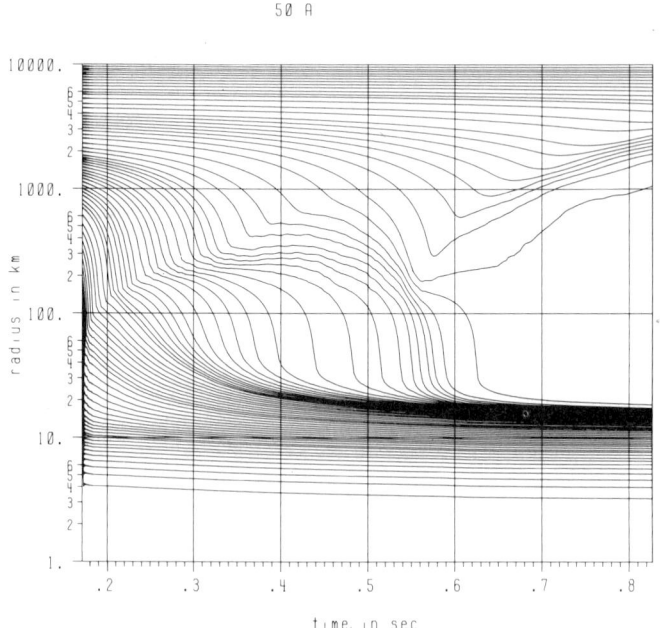

FIGURE 8i. Radius as a function of time for selected mass points for all the models of TABLE 1. Time on the graphs starts shortly before bounce.

100 A

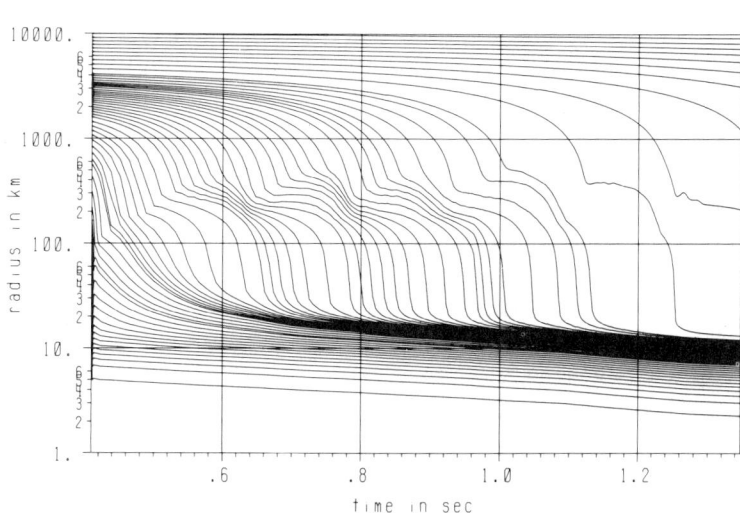

FIGURE 8j. Radius as a function of time for selected mass points for all the models of TABLE 1. Time on the graphs starts shortly before bounce.

The entropies of the inner 0.5 $M_\odot$ at this central density are 1.32, 1.62, and 1.80. The entropies have all risen about 0.30 units, but this is not enough to affect the structure of the inner core. The average lepton numbers in the central core at the beginning of collapse were 0.431, 0.438, and 0.443, but at a central density of 10^{14} g cm^{-3}, they had come close to being the same at a value of 0.383. The central core is about the same in the three models. The principal difference after collapse is the higher densities of matter immediately outside the core. From FIGURES 8f, 8g, and 8h, we see the time from collapse to explosion is about the same, 0.45 s. The energy of explosions of Models 25B and 25C are about the same, but Model 25A's energy is low (see TABLE 1). The

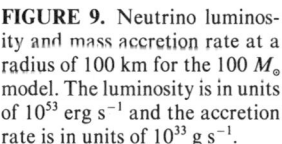

FIGURE 9. Neutrino luminosity and mass accretion rate at a radius of 100 km for the 100 $M_\odot$ model. The luminosity is in units of 10^{53} erg s^{-1} and the accretion rate is in units of 10^{33} g s^{-1}.

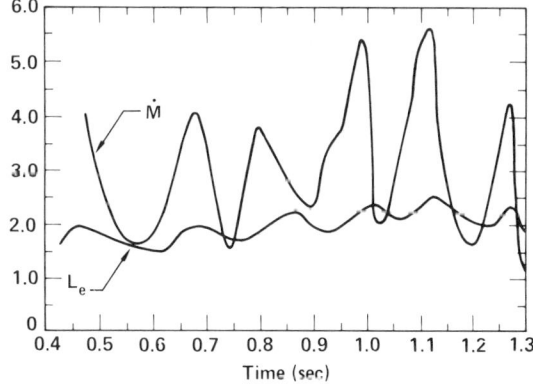

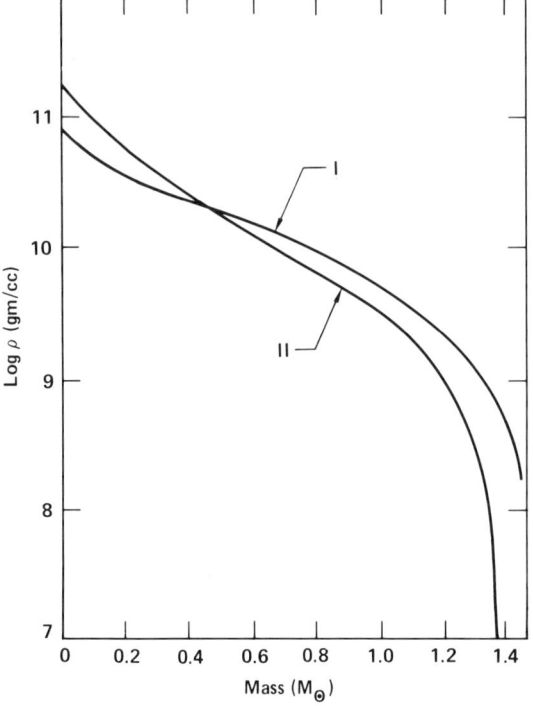

FIGURE 10. Initial density distributions for Model 11A of the present study, I, and for Nomoto's 2.2 $M_\odot$ He core model, II, at a time near the onset of core collapse. The density profile for Nomoto's model is adopted from Hillebrandt, Nomoto, and Wolff (1984).[9]

instability of the accretion process is seen in all three 25 $M_\odot$ models (FIGURES 8f, g, h), while at the lower mass, 15 $M_\odot$, only one model showed the instability (see FIGURE 8d). The biggest difference in the models would be in the explosive nucleosynthesis of the ejected matter. The density of the ejected matter just outside the collapsing core is much lower in Model 25A than in Model 25C and very little of the ejecta in Model 25A

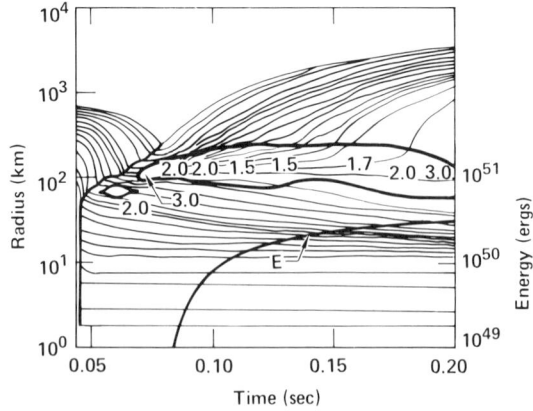

FIGURE 11. Radius as a function of time for select mass points in Model 11A. The region of net heating is interior to the heavy lines. The numbers inside are the heating rates in 10^{20} erg g^{-1} s^{-1} at the positions of the numbers. The curve labeled E is the energy of matter above the escape energy.

is heated to a sufficiently high temperature for explosive reprocessing. Nucleosynthesis in Model 25C will be discussed in more detail in the next section.

At late times, the stellar cores may become unstable to convection. In FIGURE 13, the entropy and electron number are shown for Model 25C at the time of neutrino heating. Between masses of 1.2 $M_\odot$ and 2.1 $M_\odot$, the core has a small decrease of entropy and so is weakly unstable in the Rayleigh-Taylor sense. All the core from 0 to 2.2 $M_\odot$ is unstable by the salt-finger effect due to the negative slope of the lepton number. If the core has sufficiently large initial nonradial perturbations for growth to occur in the available time, they could increase the heat flow appreciably and lead to more energetic explosions. A calculation made using a mixing length model for salt-finger

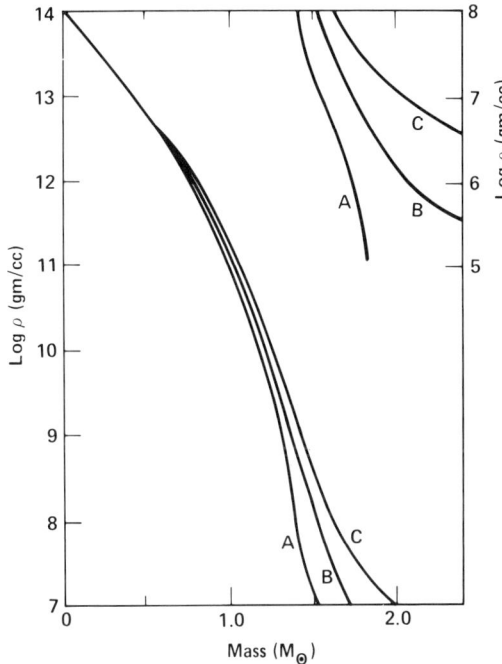

FIGURE 12. Density as a function of enclosed mass for Models 25A, 25B, and 25C.

convection in the 25 $M_\odot$ Model C resulted in the explosion energy increasing from 1×10^{51} ergs to 2×10^{51} ergs.

A weakness of the numerical model at present is that it does not include the redshift effect due to gravity. In FIGURE 14, we show the magnitude of the redshifts for the several models. The luminosity at the neutrinosphere is primarily a local effect and should not depend on the redshift, but the luminosity should fall off with radius outside the neutrinosphere due to gravity. In addition, the cross sections will be decreased at large distances by the redshift reduction of the neutrino energy. The low mass calculations are probably realistic in this regard, but the 50 and 100 $M_\odot$ models are rather suspect.

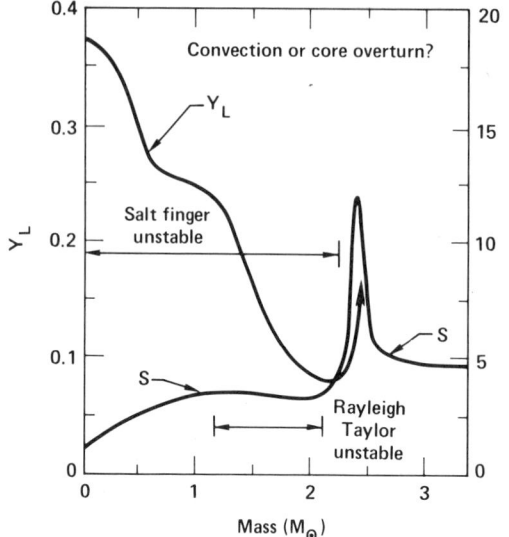

FIGURE 13. Dimensionless entropy, S, and lepton number, Y_L, as a function of mass for Model 25C at a time, 0.86 s. This is the beginning of the explosion.

FIGURE 14. Gravitational redshifts at the neutrino photosphere as a function of the mass of several models. The top curve is the redshift at the end of the calculation. The bottom curve is the redshift at the time when the star first has some matter with energy in excess of the escape energy. (The point for the 100 $M_\odot$ model was arbitrarily taken at 1 s since mass ejection was never achieved by that model.)

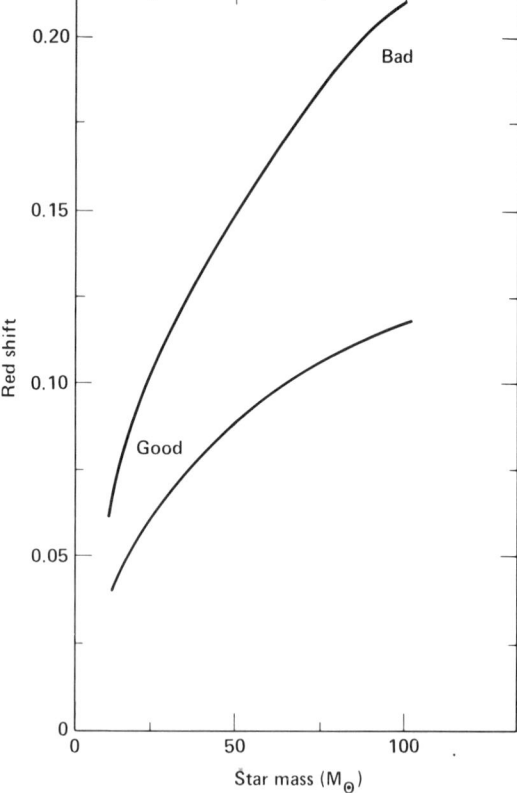

MANTLE AND ENVELOPE EJECTION

Explosive Nucleosynthesis

The passage of the shock wave through the overlying mantle of the star, in addition to providing the impulse for its ejection, leads to high temperatures and nuclear reactions that were followed in detail for Models 15C and 25C (Weaver and Woosley, 1985).[36] FIGURE 15 shows the explosive nucleosynthesis in Model 25C, and nucleosynthetic results of Models 15C and 25C are both shown compared to solar abundances in

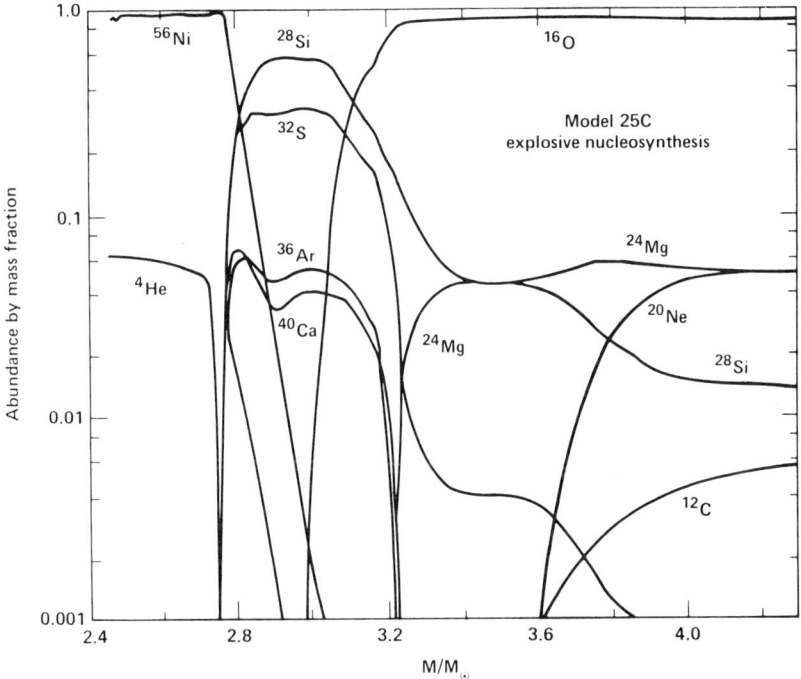

FIGURE 15. Composition of the inner regions of a 25 $M_\odot$ supernova following shock passage and ejection. The preexplosive composition (see FIGURE 4) is substantially altered out to about 3.6 $M_\odot$. Composition interior to the mass cut, 2.44 $M_\odot$, is not plotted.

FIGURE 16. In the case of Model 15C, the abundances ejected are almost entirely, with the obvious exception of the iron group, produced in the preexplosive stages of evolution and merely shoved off the star by the explosion. The diminished importance of explosive nucleosynthesis, as compared, for example, to past studies (Weaver and Woosley, 1980)[37] is a consequence of both the low explosion energy and the rapid fall off of density around the core of the preexplosive star. The large abundances of silicon through calcium are especially a result of an extensive oxygen burning convective shell just outside the core (FIGURE 3).

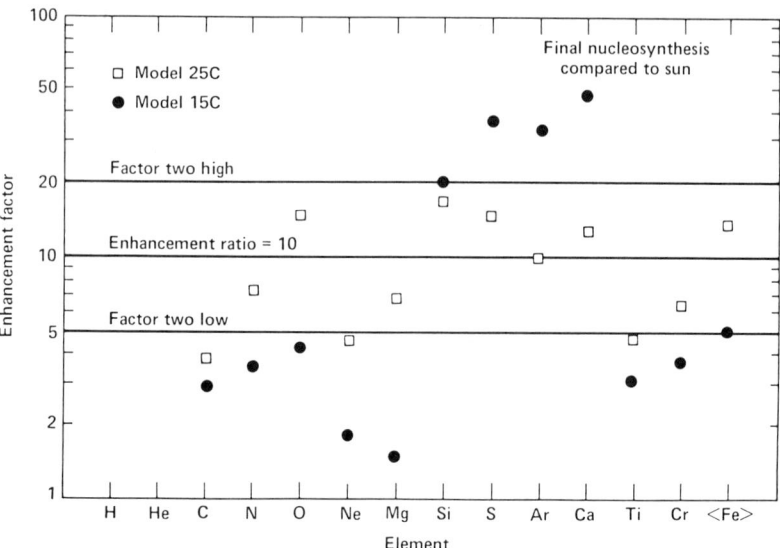

FIGURE 16. Comparison of bulk elemental nucleosynthesis in 15 and 25 $M_\odot$ supernovae (Models 15C and 25C) to solar abundances (Cameron, 1982).[27] The enhancement factor is the mass fraction of a given element in the ejecta compared to its mass fraction in the sun (which was also its initial abundance in the calculation). The deficient production of carbon, neon, and magnesium, especially in the 15 $M_\odot$ study, is a consequence of the revised rate for $^{12}C(\alpha, \gamma)^{16}O$. The large productions of silicon through calcium in the 15 $M_\odot$ model do not exist if this material falls back onto the neutron star following hydrodynamical interaction with the stellar envelope. Such was the result in the present study, but this occurrence is quite model dependent.

An interesting phenomenon observed in the 15 $M_\odot$ delayed explosion was mass reimplosion. The explosion energy is so low that, after hydrodynamical exchange of energy occurs throughout the mantle and the mantle interacts with the envelope, a portion of the inner mantle finds itself endowed with less than the escape velocity. About 0.7 $M_\odot$ of heavy elements, including all the iron group, and most of the silicon through calcium fell back onto the core. Aside from its nucleosynthetic consequences, this occurrence obviously has many important implications for future study, e.g., delayed black hole formation. However, the details of this fallback are very sensitive to the energy of the explosion, which we regard, at least for the time being, as highly uncertain. Such finely tuned effects may also be sensitive to the manner in which the shock wave generated in the core bounce code was coupled back into the stellar evolution code.

It is interesting to note that the energy liberated by the nuclear reactions was not an inconsequential fraction of the final explosion energy, even in Model 15C where the final kinetic energy at infinity was calculated to be 2×10^{50} erg (30% of which was generated by nuclear burning during shock wave passage). Because of the larger explosion energy, there was no reimplosion in the 25 $M_\odot$. This larger explosion energy and especially the flatter density gradient near the core (FIGURES 4 and 12) also imply a considerable amount of explosive nucleosynthesis. Nuclear energy generation formed an even greater fraction of the final kinetic energy at infinity. The initial shock contained roughly 10^{51} erg as calculated in the core bounce program, but the binding energy of the mantle was also about 10^{51} erg. Nuclear burning gave 4×10^{50} erg and the final kinetic energy was also 4×10^{50} erg. Thus without the energy from nuclear burning following shock wave passage, this star may well have reimploded.

The final abundances for Model 25C are shown compared to the sun in FIGURE 16 and are in remarkable agreement. Isotopic nucleosynthesis is currently under study in the same model and preliminary results indicate that this good agreement extends to the odd-Z elements and the less abundant, neutron-rich isotopes of most of the even-Z elements. Species in the silicon through calcium group, which used to be underproduced (Woosley and Weaver, 1982),[38] and in the iron group are now produced in large quantities by explosive oxygen and silicon burning. The improvement results chiefly from a more gradual density gradient surrounding the exploding core. The production of neon and magnesium, which used to be major overproductions, has now declined to where their underproduction is actually becoming troublesome. The changes in abundances of these two elements, both produced in the preexplosive star, is a direct

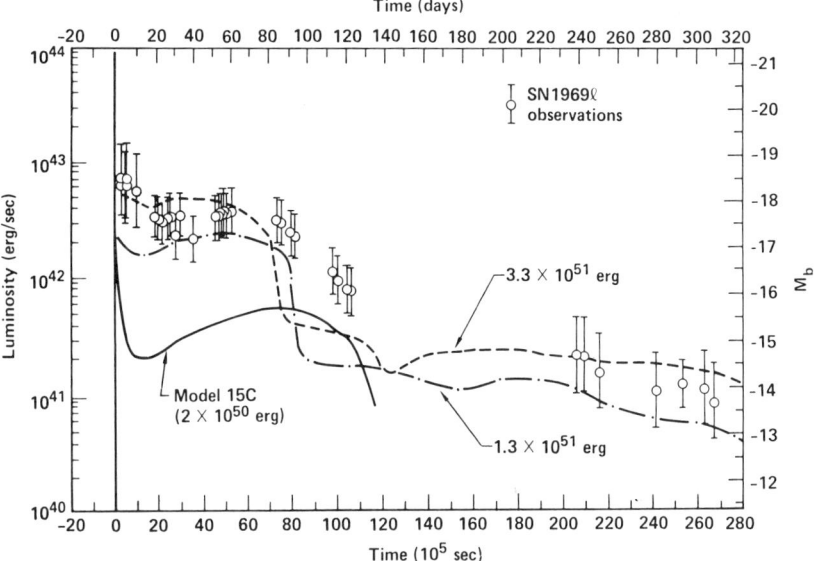

FIGURE 17. Light curve of a 15 $M_\odot$ supernova (Model 15C) that explodes by the delayed mechanism (solid line). The qualitative nature of the curve, including the duration of the event, are in agreement with observations of supernova 1969l, but only if the distance to NGC 1058 is reduced by a factor of about three compared to currently accepted values. A more energetic explosion would give a brighter light curve as evidenced by the two models previously calculated by Weaver and Woosley (1980).[37] The dashed line was an explosion characterized by total kinetic energy at infinity of 3.3×10^{51} erg. The dot-dashed line was a similar model with 1.3×10^{51} erg of kinetic energy. The expansion velocity of Model 15C is much too slow to agree with observations of 19691 (Kirshner et al., 1973;[39] Weaver and Woosley, 1980[37]). The experimental points are UBV data of Ciatti et al. (1971),[40] as transformed by the method of Schurmann et al. (1978),[41] and integrated multifrequency scans of Kirshner et al. (1973).[39]

consequence of the larger $^{12}C(\alpha, \gamma)^{16}O$ reaction rate and the smaller abundance of carbon after helium burning (neon and magnesium are the principal products of carbon burning). Interestingly, the new reaction rate also implies that carbon does not owe its origin to massive stars.

Light Curve for Model 15C

The propagation of the shock through Model 15C was also followed using KEPLER and the optical light curve was calculated. In FIGURE 17, this light curve is compared both to previous calculations by Weaver and Woosley (1980)[37] of a much more energetic explosion in a similar 15 $M_\odot$ star and to observations of Type II supernova 1969l. The general shape of the light curve from the delayed explosion calculated here agrees well with the observations and, if the distance to the galaxy NGC 1058 were readjusted, it might agree in absolute magnitude as well. However, measurements of the photospheric velocity in this same supernova are about three to five times greater during the plateau stage than in the calculation. Thus Model 15C in particular, and perhaps all of the delayed explosions in general, are too low in energy to have been supernova 1969l. Perhaps more energetic (prompt) explosions do occur in a limited range of lower mass stars (but with an adequate frequency to explain observations like 1969l), but it seems highly unlikely that prompt hydrodynamical explosions will ever be calculated for iron core masses as great as the 2.1 $M_\odot$, which characterizes Model 25C. Then the supernova that we most commonly see may be hydrodynamical in origin, but the elements in our own bodies would attest to the success of delayed explosions.

REFERENCES

1. BOWERS, R. L. & J. R. WILSON. 1982. Astrophys. J. **263**: 366.
2. ARNETT, W. D. 1980. Ann. N.Y. Acad. Sci. **336**: 366.
3. ARNETT, W. D. 1983. Astrophys. J. Lett. **263**: L55.
4. BROWN, G. E., H. A. BETHE & G. BAYM. 1982. Nucl. Phys. **A375**: 481.
5. HILLEBRANDT, W. 1984. Ann. N.Y. Acad. Sci. **422**: 197.
6. BRUENN, S. 1985. Astrophys. J. Suppl. Ser. **58**: 771.
7. COOPERSTEIN, J. 1982. Ph.D. thesis, SUNY–Stony Brook.
8. KAHANA, S., E. BARON, & J. COOPERSTEIN. 1984. *In* Problems of Collapse and Numerical Relativity. D. Bancel & M. Signore, Eds.: 163. Reidel. Dordrecht.
9. HILLEBRANDT, W., K. NOMOTO & R. G. WOLFF. 1984. Astron. Astrophys. **133**: 175.
10. BARON, E., J. COOPERSTEIN & S. KAHANA. 1985. Phys. Rev. Lett. **55**: 126.
11. WILSON, J. R. 1985. *In* Numerical Astrophysics. J. Centrella, J. LeBlanc & R. Bowers, Eds.: 422. Jones and Bartlett. Boston.
12. BETHE, H. A. & J. R. WILSON. 1985. Astrophys. J. **295**: 14.
13. WOOSLEY, S. E., T. A. WEAVER & R. E. TAAM. 1980. *In* Type I Supernovae. J. C. Wheeler, Ed.: 96. University of Texas. Austin.
14. WEAVER, T. A., G. B. ZIMMERMAN & S. E. WOOSLEY. 1978. Astrophys. J. **225**: 1021.
15. FOWLER, W. A., G. R. CAUGLAN & B. A. ZIMMERMAN. 1975. Ann. Rev. Astron. Astrophys. **13**: 69.
16. NOMOTO, K. 1984. *In* Stellar Nucleosynthesis. C. Chiosi & A. Renzini, Eds.: 239. Reidel. Dordrecht.
17. NOMOTO, K. 1984. Astrophys. J. **277**: 791.
18. HILLEBRANDT, W. 1982. Astron. Astrophys. Lett. **110**: L3.

19. ARNETT, W. D. 1975. Astrophys. J. **195**: 727.
20. NOMOTO, K. 1982. *In* Supernovae: A Survey of Current Research. M. J. Rees & R. J. Stoneham, Eds.: 205. Reidel. Dordrecht.
21. NOMOTO, K., W. M. SPARKS, R. A. FESEN, T. R. GULL, S. MIYAJI & D. SUGIMOTO. 1982. Nature **299**: 803.
22. WEAVER, T. A. & S. E. WOOSLEY. 1979. Bull. Am. Astron. Soc. **11**: 724.
23. WEAVER, T. A., S. E. WOOSLEY & G. M. FULLER. 1985. *In* Numerical Astrophysics. J. Centrella, J. LeBlanc & R. Bowers, Eds.: 374. Jones and Bartlett. Boston.
24. FULLER, G. M., W. A. FOWLER & M. J. NEWMAN. 1982. Astrophys. J. Suppl. Ser. **48**: 279.
25. FULLER, G. M., W. A. FOWLER & M. J. NEWMAN. 1982. Astrophys. J. **252**: 715.
26. FULLER, G. M., W. A. FOWLER & M. J. NEWMAN. 1985. Astrophys. J. **293**: 1, 57–A6.
27. CAMERON, A. G. W. 1982. *In* Essays in Nuclear Astrophysics. C. A. Barnes, D. D. Clayton & D. N. SCHRAMM, Eds.: 23. Cambridge University Press. Cambridge.
28. MIYAJI, S., K. NOMOTO, K. YOKOI & D. SUGIMOTO. 1980. Publ. Astron. Soc. Japan **32**: 303.
29. WOOSLEY, S. E., W. D. ARNETT & D. D. CLAYTON. 1972. Astrophys. J. **175**: 731.
30. SHAPIRO & TEULKOSKY. 1983. Black Holes, White Dwarfs, and Neutron Stars, p. 160. John Wiley & Sons. New York
31. KETTNER, K. U., H. W. BECKER, L. BUCHMANN, J. GORRES, H. KRAWINKEL, C. ROLFS, P. SCHMALBROCK, H. P. TRAUTVETTER & A. VLIEKS. 1982. Z. Phys. **A308**: 73.
32. LANGANGKE, K. & S. KOONIN. 1984. Preprint MAP-56, Kellogg Radiation Lab., Cal-Tech. Submitted to Nucl. Phys. A.
33. CAUGHLAN, G. R., W. A. FOWLER, M. J. HARRIS & B. A. ZIMMERMAN. 1984. Orange Aid preprint no. 400. Kellogg Lab., Cal-Tech, California.
34. BOWERS, R. L. & J. R. WILSON. 1982. Astrophys. J. Suppl. Ser. **50**: 115.
35. BETHE, H. A., G. E. BROWN, J. COOPERSTEIN, JR. & J. R. WILSON. 1983. Nucl. Phys. **A403**: 625.
36. WEAVER, T. A. & S. E. WOOSLEY. 1985. Bull. Am. Astron. Soc. **16**: 971.
37. WEAVER, T. A. & S. E. WOOSLEY. 1980. Ann. N.Y. Acad. Sci. **336**: 335.
38. WOOSLEY, S. E. & T. A. WEAVER. 1982. Essays in Nuclear Astrophysics. C. A. Barnes, D. D. Clayton & D. N. Schramm, Eds. Cambridge University Press. Cambridge.
39. KIRSHNER, R. P., J. B. OKE, M. V. PENSTON & L. SEARLE. 1973. Astrophys. J. **185**: 303.
40. CIATTI, F. L., L. ROSINO & F. BERTOLA. 1971. Mem. Soc. Astron. Ital. **42**: 163.
41. SCHURMANN, S. R., W. D. ARNETT & S. W. FALK. 1979. Astrophys. J. **230**: 11.

PART VIII. LATE STAGES OF STELLAR EVOLUTION

Evolution of Supernova Progenitors and Supernova Models[a]

KEN'ICHI NOMOTO

Department of Earth Science and Astronomy
College of Arts and Sciences
University of Tokyo
Meguro-ku, Tokyo 153, Japan
and
Department of Physics
Brookhaven National Laboratory
Upton, New York 11973

SUPERNOVAE OF TYPE I AND II

Supernovae are the stellar explosions releasing energies of about 10^{51} ergs and shining as bright as a whole galaxy. Because of their enormous energy impulse, supernovae have strong influences on the interstellar medium. Supernovae are important sights of nucleosynthesis and eject a lot of heavy elements into space. Some supernovae leave neutron stars or black holes behind. Thus supernovae are one of the key events in the evolution of stars, galaxies, and the universe.

The understanding of supernovae and their remnants is being greatly accelerated by the recent progresses in observations at all wave bands and in theoretical modeling.

From the observational side, supernovae are classified as Type I and Type II. This classification is based on spectroscopy: Type I supernovae (SN-I) are hydrogen-deficient, while Type II supernovae (SN-II) show hydrogen lines in their spectra near maximum light (e.g., see reference 1). SN-I and SN-II have several other distinct features (Doggett and Branch, 1985;[2] Trimble, 1982[3] and references therein):

Light Curves: SN-I light curves are quite uniform and characterized by an exponential tail. SN-II are classified into two classes, namely, a plateau type and a linear decline type. Light curves are successfully reproduced by the ^{56}Ni decay model for SN-I and by the shock heating model for SN-II with plateau. However, SN-II of linear decline have not been modeled although ^{56}Ni decay could also power this light curve.

Population: SN-II appear only in spiral galaxies and are concentrated in spiral arms. SN-I have been observed in all types of galaxies, including ellipticals, and are not concentrated in the spiral arms.

---

[a]This work was supported in part by the Japanese Ministry of Education, Science, and Culture through research grant nos. 58340023, 59380001, and 60540152, and by the U.S. Department of Energy under contract no. DE-AC02-76CH00016.

Guided by these facts, the origin of supernovae has been extensively explored. The currently most-favored candidates of supernova progenitors are massive red-supergiants and white dwarfs.

Massive Star Progenitors: Stars more massive than 8 $M_\odot$ evolve into a catastrophic stage when the core starts to collapse. The ultimate outcome of collapse depends on the stellar mass. For very massive stars, the collapse results in a black hole formation, while in less massive stars, the collapse is transformed into an explosion that leaves a neutron star behind.

In the explosion, a shock wave generated at the core bounce travels outwards to reach the surface. Whether the resulting heating-up of the photosphere produces a supernova light curve depends on the radius of the presupernova star. Very massive stars tend to lose their hydrogen-rich envelope to become helium stars with a relatively small radius at the presupernova stage; the resultant optical light is rather dim unless a substantial amount of ^{56}Ni is produced. On the other hand, less massive stars retain their hydrogen-rich envelope and evolve to become red-supergiants; then the shock heating of the photosphere gives rise to a bright optical outburst that reproduces well the SN-II light curves. Therefore, we have two kinds of compact remnants (black holes or neutron stars) and two types of optical appearance (dim supernovae or SN-II) from massive progenitors, but their relations are not clarified yet.

White Dwarf Progenitors: White dwarfs are cold degenerate stars, but they can be rejuvenated if they are in close binaries and accrete matter from companion stars. The outcome of such a rejuvenation depends on the accretion rate and composition of the white dwarf. When the white dwarf mass grows close to the Chandrasekhar limit, thermonuclear explosion of the white dwarf occurs. A large amount of ^{56}Ni is produced and the radioactive decays of ^{56}Ni to ^{56}Co and ^{56}Fe power the SN-I-like light curve. The white dwarf is disrupted completely and no neutron star is left behind. In some cases, the accretion induces a collapse of a white dwarf to form a neutron star.

Helium Star Progenitors: Massive helium stars in close binaries could give rise to another class of supernovae. In these stars, collapse and bounce of an iron core generate a shock wave, but the shock heating mechanism that produces supernova light curves does not work because of tidal loss of the hydrogen-rich envelope. However, in relatively massive helium stars, say more massive than 4 $M_\odot$, some amount of ^{56}Ni could be produced.[4] Then the radioactive decays could power an optical outburst that might resemble SN-I. This model might account for a peculiar class of SN-I.[5] Moderate mass helium stars (1–2.5 $M_\odot$) could also become peculiar SN-I that leave neutron stars behind, but whether such stars are actually formed is not known yet.

Here I will summarize these currently most-favored models for supernovae and discuss to what extent these models are established by theoretical modeling and observations.

In the next section, theoretical models of SN-I are discussed and compared with observations. In the third section, it is shown that a merging pair of CO white dwarfs (which has been proposed to be a SN-I progenitor) forms a single neutron star. In the fourth section, presupernova evolution of massive stars, in particular of the 10–13 $M_\odot$ range, are discussed. In the final section, modes of neutron star formation are summarized and the observations of young supernova remnants are compared with the neutron star cooling model.

WHITE DWARF MODELS FOR TYPE I SUPERNOVAE: CARBON DEFLAGRATION SUPERNOVAE

The origin of SN-I has long been difficult to identify because presupernova evolution involves complicated processes of mass loss. Based on recent developments, however, SN-I are now objects for which the detailed comparison between models and observations is possible. Since SN-I do not have a thick hydrogen-rich envelope, newly synthesized elements during explosion could be observed in the spectra. Thus, SN-I provide us with direct evidence for the thermonuclear explosion and associated nucleosynthesis (e.g., see Wheeler, 1982[6] for a review and references therein).

Accreting white dwarfs have been considered to be promising candidates of SN-I progenitors. The mechanism of explosion was proposed to be a thermonuclear explosion in the electron-degenerate cores.[7] These ideas have been confirmed by recent extensive numerical calculations of the evolution of white dwarfs.

The outcome of such a white dwarf rejuvenation depends primarily on the mass accretion rate and thus several types of explosion could occur. Among them the most plausible model for SN-I is the carbon deflagration of a carbon-oxygen (CO) white dwarf.[8] The currently most-favored evolutionary model from the beginning of accretion through the thermonuclear explosion is described next (see Nomoto, 1984[9] and references therein).

Progenitor System

The evolutionary scenario starts from a close binary system consisting of intermediate mass stars. As a result of Roche lobe overflow, the primary star of this close binary becomes a white dwarf composed of carbon-oxygen. When the secondary star evolves, it begins to transfer matter over to the white dwarf either as a stellar wind from a red giant or as Roche lobe overflow. In other words, the white dwarf accretes hydrogen-rich matter.

The outcome of the accretion of hydrogen-rich matter depends on the accretion rate. If $\dot{M}$ is lower than $10^{-8}\ M_\odot\ \mathrm{yr}^{-1}$, the resultant hydrogen flash is strong enough to eject the accreted material in a nova-like manner. If $\dot{M}$ is higher, on the contrary, the flash is weaker so that a significant fraction of the accreted materials is processed into helium and a white dwarf mass grows.

When a sufficient amount of helium layer is built up, a helium flash is ignited. The strength of the flash again depends on the accretion rate of helium. The accretion at a rate higher than $1 \times 10^{-8}\ M_\odot\ \mathrm{yr}^{-1}$ leads to a relatively weak helium flash; the recurrence of such a flash increases the CO core mass until a carbon deflagration is initiated at the center. The slower accretion induces a helium detonation that in most cases disrupts the white dwarf completely.

The above new critical rate of $1 \times 10^{-8}\ M_\odot\ \mathrm{yr}^{-1}$ is lower than the previous value of $4 \times 10^{-8}\ M_\odot\ \mathrm{yr}^{-1}$ (FIGURE 1) because the NCO reaction [^{14}N (e^-, ν) ^{14}C (α, γ) ^{18}O] was found to trigger a helium flash at a significantly lower density.[10] Such dependences of outcome on $\dot{M}$ and the initial mass of CO white dwarfs are summarized in FIGURE 1, where the region for the double detonation should be narrower if the NCO reaction is taken into account.

Therefore, the accretion in the range of 10^{-8}–$10^{-6}\ M_\odot\ \mathrm{yr}^{-1}$ leads to a carbon deflagration. The candidates of a companion star that transfer matter over to the white dwarf at such a relatively high rate are subgiants undergoing Roche lobe overflow and red giants undergoing wind-type mass loss. These rather rapidly accreting systems could be observed as symbiotic stars. The cataclysmic variables may not be progenitors of SN-I because a relatively slow accretion may cause too strong a flash to increase the white drawf mass.

Carbon Deflagration Models and Nucleosynthesis

The carbon deflagration is initiated at the center of a CO white dwarf when its mass gets close to the Chandrasekhar limit. The explosive carbon burning front then propagates outwards on the time scale for convective heat transport. [The nuclear energy release is only 20% of the Fermi energy of degenerate electrons, which is too small to initiate a shock-driven detonation wave (e.g., see reference 8).]

Hereafter, hydrodynamical behavior of the carbon deflagration and the associated nucleosynthesis are shown based on the most detailed modeling (model W7) by Nomoto, Thielemann, and Yokoi (NTY).[11] The propagation of the deflagration wave is shown in FIGURE 2. It takes 1.2 s for the front to reach the surface region. This is significantly slower than the detonation wave that takes only 0.2 s.[12] Hence, the white dwarf expands to weaken the deflagration and to finally quench it.

At the deflagration wave, explosive nucleosynthesis is going on to synthesize various elements depending primarily on the peak temperature, T_p. In the inner layer of 0.7 $M_\odot$, T_p is higher than 5×10^9 K so that nuclear reactions are rapid enough to incinerate materials into nuclear statistical equilibrium (NSE) composition. At freezing-out phase, NSE composition becomes mostly ^{56}Ni. In the intermediate region of 0.7–1.0 $M_\odot$, $T_p = (4.5$–$5) \times 10^9$ K so that the materials undergo incomplete Si burning to yield ^{40}Ca, ^{36}Ar, ^{32}S, and ^{28}Si. In still outer layers of $M_r < 1.28\ M_\odot$, T_p is

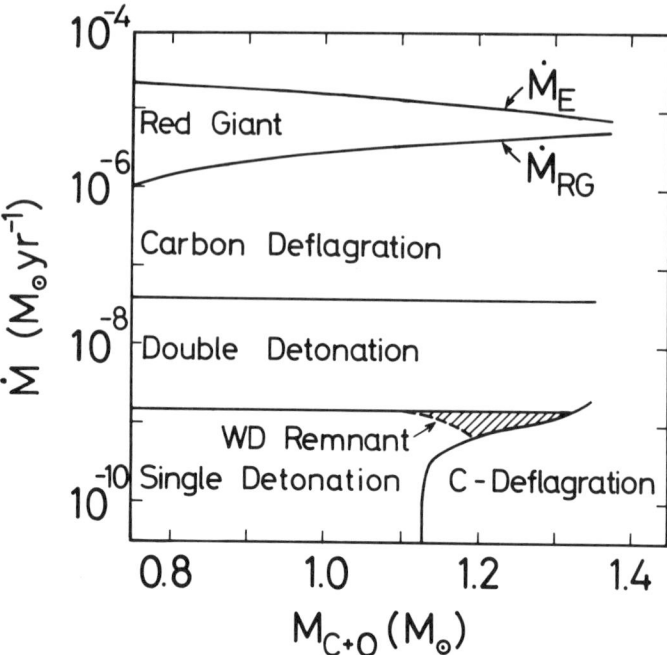

FIGURE 1. The triggering mechanisms of Type I supernovae (SN-I) in accreting C + O white dwarfs, which depend on the accretion rate of helium, $\dot{M}$, and the initial mass, M_{CO}. $\dot{M}_E$ is the Eddington's critical rate for helium. The rapid accretion ($\dot{M} \gtrsim \dot{M}_{RG}$) forms a red-giant-like envelope. The carbon deflagration models can account for many of the observed features of SN-I. The detonation models are not consistent with the early time spectra of SN-I. For $\dot{M} < 10^{-9} M_\odot$ yr^{-1}, both single detonation and carbon deflagration are possible depending on M_{CO}. The single detonation results in either a total disruption or an explosion that leaves a white dwarf remnant behind.

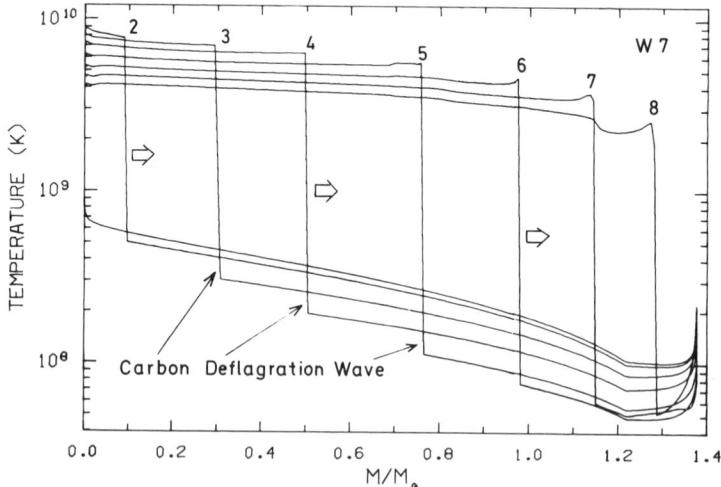

FIGURE 2. Propagation of the carbon deflagration wave and the associated change in the temperature distribution. Stage numbers 1–9 correspond to $t(s)$ = 0.0 (#1), 0.60 (#2), 0.79 (#3), 0.91 (#4), 1.03 (#5), 1.12 (#6), 1.18 (#7), 1.24 (#8), and 3.22 (#9), respectively. Time is measured from the initiation of the deflagration.

lower and explosive burning of oxygen and neon synthesize ^{32}S, ^{28}Si, ^{24}Mg, etc. In the outermost layers, C + O remain unburnt. The resultant composition structure after freeze-out is shown in FIGURE 3 (Thielemann, Nomoto, and Yokoi, 1986; TNY).[13]

Comparison with Observations

The carbon deflagration model (W7) can account for many of the observed features of SN-I:

(1) The explosion energy of this model (1.3×10^{51} erg) and the expansion velocity near the surface ($v_{exp} \simeq 10^4$ km s^{-1}) are consistent with the observed velocity of SN-I near maximum light.

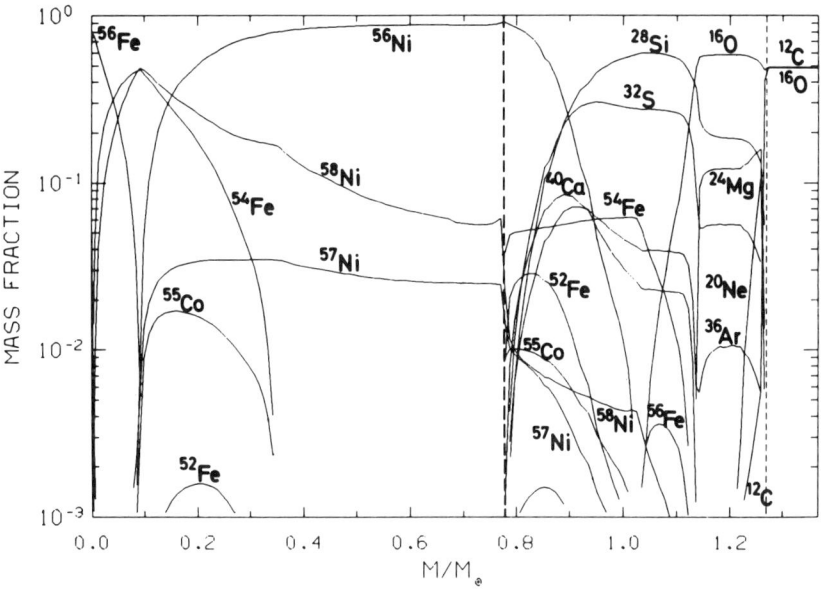

FIGURE 3. Composition structure after explosive nucleosynthesis in the carbon deflagration wave (TNY).[13] At $M_r < 0.7\ M_\odot$, the white dwarf undergoes incineration into almost NSE composition. In the intermediate region at $0.7 \lesssim M_r/M_\odot < 1.3$, the white dwarf undergoes partial explosive burning. In the outer layer ($M_r \gtrsim 1.3\ M_\odot$), C + O remains unburnt.

(2) The ejected amount of ^{56}Ni (0.6 $M_\odot$) is large enough to reproduce a characteristic SN-I light curve by the radioactive decays of ^{56}Ni → ^{56}Co → ^{56}Fe.
(3) Provided that the composition stratification in the outer layers are removed by mixing, the synthetic spectra at early times are in good agreement with the observed spectra of typical SN-I 1982b in NGC 4536 (FIGURE 4 taken from Branch et al., 1985[14]). The spectral features are identified as Ca, S, Si, Mg, and O; moreover, the feature at 3200 Å could be due to Co II.[15]

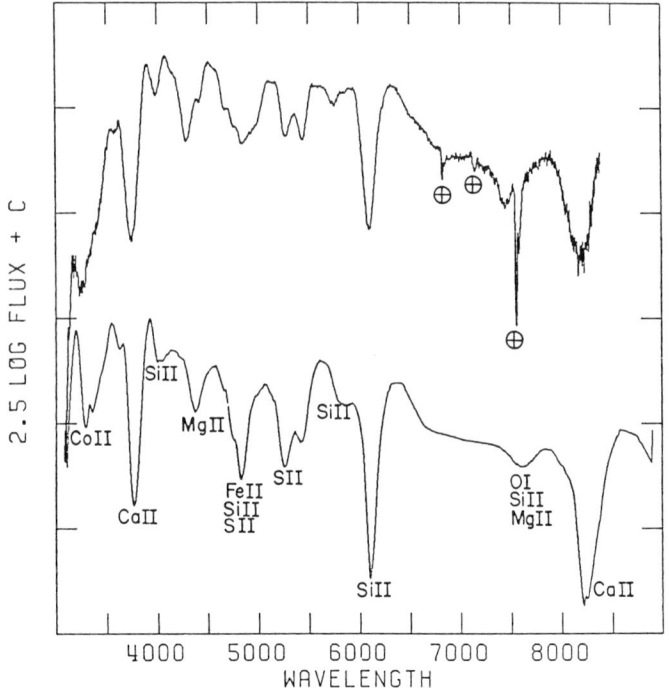

FIGURE 4. The maximum-light spectrum of SN 1981b (top), from Branch et al. (1983),[15] is compared to a synthetic spectrum for the carbon deflagration model (W7 of NTY) 15 days after the explosion.[14] In this model, the outer layer is assumed to be mixed.

(4) At later times, the observed features are identified as those of Fe and Ca. The Ca absorption line shows that there may be a strong gradient in the calcium abundance near $v_{exp} \simeq 8000$ km s^{-1}.[15] The carbon deflagration model shows that the transition in the composition structure from the NSE to incomplete Si burning products occurs at $v_{exp} \simeq 8000$ km s^{-1}. Thus, the model and observations are quite consistent.

(5) In late time optical spectra, Fe and Co emission features observed in SN 1972e[16] fit better to the theoretical spectra of the deflagration model than those of the detonation models.[17] Recent IR observations of SN-I 1983n in M83 detected the [Fe II] emission line at 1.644 μm.[18] The inferred mass of Fe is 0.1–0.8 $M_\odot$. This confirms the prediction of the radioactive decay model by Axelrod (1980).[19] (However, SN 1983n belongs to a peculiar class of SN-I, so it may not be a carbon deflagration supernovae.)

(6) The "missing iron" in the remnant of SN-I has been a problem. However, X-ray spectra of Tycho obtained with the satellite Tenma showed that the Fe line emissions originate from low ionization state; this suggests that iron in Tycho is largely overabundant.[20,21] They also reported the overabundances of Si, S, Ar, and Ca relative to solar abundances.

Remaining Issues

The above agreements are good enough to support the carbon deflagration model for SN-I. However, there remain several problems:

(1) Propagation speed of the carbon deflagration wave: Certainly 2-D/3-D calculations are needed.[22]

(2) Mechanism of mixing in the outer layers, which is required by the early time spectra: In the carbon deflagration model, the outer layers are convectively unstable before the free expansion phase (NTY),[11] but the actual mixing process and some other turbulent processes have not been studied yet.

(3) Overproduction of ^{58}Ni (and slightly ^{54}Fe): In the inner layer of the exploding white dwarf, electron-captures on NSE elements produce some neutron-rich iron peak elements. Among them, ^{58}Ni is overabundant relative to ^{56}Fe with respect to the solar values as seen in FIGURE 5.[13] [The overproduced ^{54}Fe as reported in NTY[11] and Woosley *et al.* (1984)[17] were found to be processed into ^{58}Ni via ^{54}Fe (α,γ) ^{58}Ni reaction.[13]] This might be ascribed to the uncertainty in the treatment of the carbon deflagration. Also contribution from white dwarfs formed from population II stars may be important because the smaller neutron excess due to the smaller ^{22}Ne abundance leads to a smaller production of ^{58}Ni and ^{54}Fe.[13,23] Another possibility is that presupernova accretion would be more rapid to ignite carbon burning at a lower central density than the present models.

(4) Progenitor binary system: Recently, double white dwarf systems have been

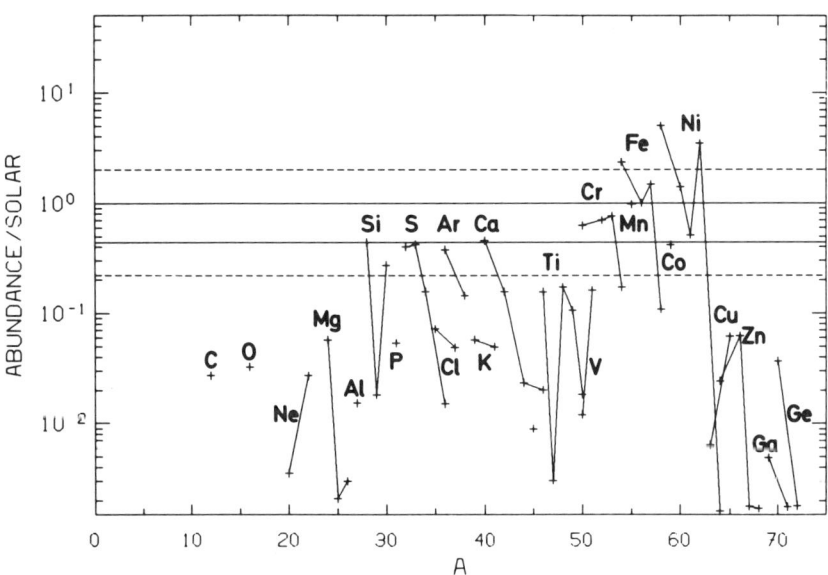

FIGURE 5. Nucleosynthesis in carbon deflagration supernovae (NTY).[11] The abundances of stable isotopes relative to the solar values are shown. The ratio is normalized to ^{56}Fe.

proposed as SN-I progenitors.[24,25] However, the outcome of this kind of system would not be SN-I, as will be discussed in the next section.

(5) The existence of a spectral subclass of Type I supernovae (SN 1983n in M83 and SN 1984ℓ in NGC 991; see Wheeler, 1985[26]): These SN-I are characterized by the absence of the Si feature. They probably have lower maximum luminosity than typical SN-I. Possible models include: (a) Single He-detonation supernovae that produce mostly ^{56}Ni with some unburnt C + O material; (b) a carbon deflagration initiated at a much higher central density; it produces less ^{56}Ni, but more ^{54}Fe, ^{56}Fe, and ^{58}Ni than the NTY models (this should be a rare event); and (c) massive helium stars ($M > 4\ M_\odot$) in close binaries where explosion energy from core bounce is available to compensate

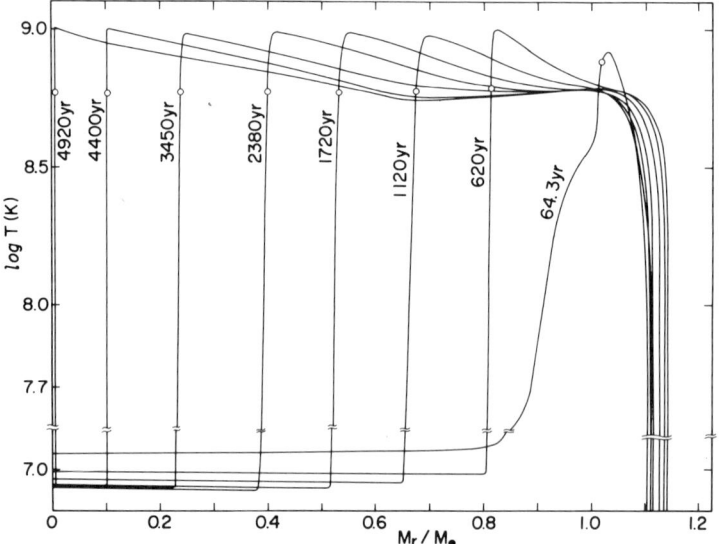

FIGURE 6. Propagation of the carbon burning front from the outer layer of the accreting CO white dwarf through the center ($\dot{M} = 1 \times 10^{-5}\ M_\odot\ \mathrm{yr}^{-1}$). Change in the temperature distribution against M_r is shown for several stages. Open circles indicate the carbon flashing layer. Time is measured from the carbon ignition near the surface.[29]

smaller ^{56}Ni production.[5] Less massive helium stars have too thin of a Si layer to synthesize significant amounts of ^{56}Ni. Therefore, this progenitor model predicts that the subclass of SN-I belongs to a population even younger than typical SN-II. Certainly, detailed numerical calculations of hydrodynamical models and synthetic spectra are required.

Hubble Constant and Type I Supernovae

The agreement between the carbon deflagration model and the SN-I observations is so good that the maximum bolometric luminosity, L_{max}, of SN-I is determined from the model. Since SN-I are quite uniform both in maximum luminosities and spectra (except for the above subclass), SN-I can be used as a standard candle to determine the distance to the parent galaxies and the Hubble constant (Arnett, Branch, and Wheeler, 1984, ABW).[27] The radioactive decay model for the SN-I light curve gives $L_{max} = 2.2 \times 10^{43} M_{Ni}/M_\odot$ erg s^{-1} (ABW). From the well-observed six SN-I, ABW derived $L_{max} = 1.9 \times 10^{43} (H_0/50)^{-2}$, where H_0 is the Hubble constant in units of km s^{-1} Mpc^{-1}. These two relations give $H_0 = 46 (M_{Ni}/M_\odot)^{-1/2}$. ABW estimated that H_0 is in the range of 39-73, with a best estimate of 59.

Using the above carbon deflagration models (NTY), a more restricted range of H_0 can be obtained. The best estimate relies on the model W7 because of good agreement between theoretical spectra and the observed early time spectra (FIGURE 4); then $M_{Ni} = 0.58 M_\odot$ gives $H_0 = 60$. Another model with slightly slower deflagration (model C6 of NTY) produces $M_{Ni} = 0.48 M_\odot$, which corresponds to $H_0 = 66$. The velocity profile of the model C6 is marginally consistent with the observations; in other words, for the models with smaller M_{Ni}, the expansion velocity, v_{exp}, at the outer layer would be lower than 10^4 km s^{-1} and v_{exp} at a layer containing Ca would be significantly lower than 8000 km s^{-1}. If $M_{Ni} > 1 M_\odot$, on the other hand, the amount of carbon burning products, O and Mg, would probably be too small to be compatible with the early time spectra. Therefore, the favored range of ^{56}Ni mass is $0.48 M_\odot \lesssim M_{Ni} \lesssim 1 M_\odot$, with the most plausible value being $0.58 M_\odot$. This corresponds to $46 \lesssim H_0 \lesssim 66$, with the favored value of 60.

MERGING OF DOUBLE WHITE DWARFS TO FORM A SINGLE NEUTRON STAR

Formation of a Double White Dwarf System

Some close binary systems consisting of initially intermediate mass stars form double white dwarfs at their final stages of evolution. In this evolutionary process, the primary star first becomes a white dwarf. When the secondary star starts to transfer matter, its rate is too high for matter to be swallowed up by the white dwarf; the materials then escape from the system and double white dwarfs are left.

Merging of Double White Dwarfs

Two white dwarfs are then brought closer together because the angular momentum is carried away from the system by gravitational radiation. If the white dwarfs are initially close enough (orbital separation $< 3 R_\odot$), they will become sufficiently close in less than a Hubble time so that the less massive of the two fills its Roche lobe and begins to transfer matter over to the more massive white dwarf.

Since the white dwarf radius gets larger as it loses mass, the mass transfer to the more massive white dwarf should be very rapid; the accretion rate, $\dot{M}$, may well be close to the Eddington limit, $\dot{M}_{Edd}$. The subsequent merging process is so complicated that only an approximate investigation has been done. Nomoto and Iben (1985),[28] Saio and Nomoto (1985),[29] and Woosley and Weaver (1986)[30] calculated such a rapid accretion onto a white dwarf for the case of a CO-CO white dwarf pair. Their results have shown that the two merging CO white dwarfs produce a single neutron star in the following way: The rapid accretion heats up the surface layer of the accreting white dwarf very efficiently. At the same time, heat is conducted into the inner cold layer. If $\dot{M} \geq \dot{M}_{crit} \simeq 3 \times 10^{-6} M_\odot \text{ yr}^{-1} \simeq 0.15 \dot{M}_{Edd}$, carbon ignition occurs in the outer layer before the white dwarf mass reaches the Chandrasekhar limit. For $\dot{M} < \dot{M}_{crit}$, on the contrary, white dwarf interior is more thermally relaxed so that carbon is ignited at the center. The central carbon ignition would lead to a Type I supernova induced by carbon deflagration.

The off-center carbon burning for $\dot{M} \geq \dot{M}_{crit}$ is not explosive, but only a weak flash because the ignition density is as low as $\sim 10^6$ g cm^{-3}. The maximum temperature attained during the flash is 1×10^9 K, which is too low to form a shock wave. The next inner layer is heated up by heat inflow from the carbon burning layer and eventually a

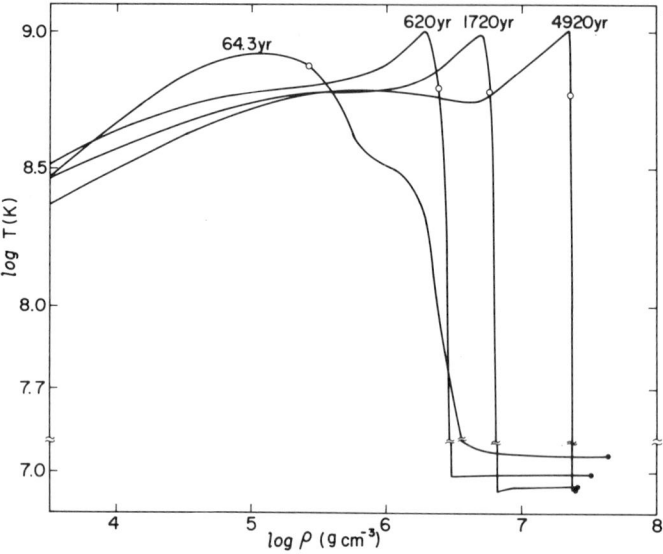

FIGURE 7. Same as FIGURE 6, but for the temperature distribution against the density.

weak carbon flash is ignited. In this way, the carbon burning front propagates inward in a self-sustained manner. This is essentially similar to a conductive carbon deflagration although burning is not explosive.

FIGURES 6 and 7 show such a propagation of the burning front and an associated change in temperature distribution in the white dwarf;[29] this model assumes that a CO white dwarf accretes CO from a companion at a rate of $\dot{M} = 1 \times 10^{-5} M_\odot \text{ yr}^{-1}$. The burning front propagates all the way through the central region in ~ 5000 yr and C + O is converted into O + Ne + Mg. Most of the released nuclear energy is lost by neutrinos so that the white dwarf is still bound; in other words, the CO white dwarf is completely changed into the ONeMg white dwarf in a quiescent manner. If the total mass of the system is larger than 1.4 $M_\odot$, the ONeMg white dwarf eventually undergoes electron-captures on ^{24}Mg and ^{20}Ne. The resultant decrease in the Chandrasekhar limit triggers a collapse of the white dwarf to form a neutron star.[31,32] If the system consists of a CO-ONeMg white dwarf pair, the merging would also lead to a neutron star formation.

Another type of pair, that is, He-He double white dwarfs, would evolve analogously (Nomoto and Sugimoto, 1977,[33] for the case of rapid accretion) to form a single CO white dwarf due to propagating helium shell burning. Because of too small of a total mass, however, it will not become a supernova.

Formation of a Single Millisecond Pulsar?

The single neutron star originating from the merging of two white dwarfs has the following properties:

(1) The mass ejection at the neutron star formation must be so small that a gaseous remnant could easily disappear. Therefore, the neutron star would become an isolated star.

(2) The parent white dwarf could gain angular momentum during the merging process. Therefore, the resultant neutron star could rotate very fast.

(3) The progenitors of CO or ONeMg white dwarfs are 5–10 $M_\odot$ stars. Thus, a location of the resultant neutron star may not be far from the galactic plane. Also its space velocity must be quite low because of the small mass ejection at its formation.

What this kind of isolated fast-spinning neutron star looks like depends on the surface magnetic field. Suppose that the magnetic field of a neutron star originates from a fossil field of the white dwarf and no enhancement of the field occurs after the neutron star formation. Then the neutron star could be very weakly magnetized if the magnetic field of the white dwarf had already decayed during its long-term cooling. The neutron star could keep rotating rapidly for a relatively long time. These features could account for many of the properties of the single millisecond pulsar.[34] In this connection, it is an interesting suggestion that white dwarfs in cataclysmic variables can be classified into two classes, namely, with and without strong magnetic field.[35]

However, if the magnetic field of the white dwarf is strong, or its strength is enhanced during collapse[36] or in the hot envelope of a neutron star,[37] the neutron star originating from the merging white dwarfs could be observed as a normal pulsar. This is also an interesting origin of isolated pulsars.

Helium Stars with Extended Envelope

A single helium star with moderate mass has been proposed as a Type I supernova progenitor.[38] The evolutionary origin of such a helium star is not clear, but it might be mixing during main-sequence[38] or extensive mass loss. Merging of a IIc CO white dwarf pair could also yield a similar object.[24,25] If the mass of the helium star is in the range of 1.5–2.5 $M_\odot$, it forms a degenerate core and a red-giant size helium envelope.

Afterwards, the core mass, M_{CO}, grows as the helium shell burning converts helium into C + O; this is equivalent to the accretion of CO at a rate of $\dot{M}_{CO} = 7 \times 10^{-6}$ $(M_{CO}/M_\odot - 0.6)$ $M_\odot$ yr^{-1}.[39] This rate exceeds $\dot{M}_{crit} \simeq 3 \times 10^{-6}$ $M_\odot$ yr^{-1} for $M_{CO} >$

1.0 $M_\odot$. Therefore, carbon is ignited off-center exactly in the same way as in rapidly accreting white dwarfs. Inward propagation of the carbon burning front then changes the CO core into the ONeMg core.

Subsequent evolution should lead to a collapse of the ONeMg core, which induces a supernova explosion, leaving a neutron star behind. Since these helium stars have a red-giant size envelope, shock heating could give rise to a Type I supernova-like light curve near the maximum.[40] However, these stars have a negligible Si layer so that ^{56}Ni decays cannot work. Unless pulsar activity produces a SN-I-like light curve tail, the optical outburst would be quite different from typical SN-I. (The quench of the inwardly propagating carbon burning found in the previous calculation was just due to too coarse of a zoning.[41])

PRESUPERNOVA EVOLUTION OF MASSIVE STARS AND TYPE II SUPERNOVA MODELS

The currently most-favored model for Type II supernovae (SN-II) is based on core collapse and bounce in massive stars. The long-standing central question is what is the mechanism that transforms collapse into explosion. In these days, combined efforts of extensive numerical calculation and theoretical analysis have been paid to answer this. Through these studies, it has been recognized that the hydrodynamical behavior of collapse-bounce is quite sensitive to the presupernova structure, e.g., mass and composition of the collasping core, its central entropy, and the density distribution. Therefore, presupernova stellar evolution needs to be carefully reexamined. In view of this, I would like to discuss several characteristics of massive star evolution.

Evolution as a Function of Stellar Mass

FIGURE 8 shows evolutionary paths of the central temperature and density for stars of masses of 30 $M_\odot$,[42] 12 $M_\odot$,[43] and 9 $M_\odot$.[44] These curves show three typical evolutionary routes through core collaspe; the stars of 30 $M_\odot$ and 9 $M_\odot$ represent the evolution of a nondegenerate and strongly degenerate star, respectively, and the 12 $M_\odot$ star is an intermediate case.

Their evolutionary behavior is governed by the gravothermal specific heat, c_g, of the core. The central temperature and density of a core of mass, M_1, obeys a relation, $T_c^3/\rho_c \propto M_1^2$, for ideal gas (e.g., see reference 45, p. 170) and entropy is given as $s_c \propto \ell n$ $(T_c^{3/2}/\rho_c)$ + const $\propto \ell n$ $(M_1^2/T_c^{3/2})$ + const. These relations show that as central entropy is lowered by radiative and neutrino losses, the star contracts to increase ρ_c and T_c. In other words, c_g ($\propto -1.5$) is negative. As the star evolves, central entropy decreases further and eventually electrons become degenerate if M_1 is smaller than a certain critical mass. (The above relation shows that entropy is lower for smaller M_1 with the same ρ_c.) Then c_g changes its sign from negative to positive and thus T_c decreases as entropy is lost from the core.

In 8–10 $M_\odot$ stars, this occurs during the gravitational contraction of an ONeMg core whose mass is smaller than 1.37 $M_\odot$. No neon burning is ignited under nondegenerate conditions and further evolution is promoted by the increase in mass of the degenerate ONeMg core up to the Chandrasekhar limit, M_{Ch}. Eventually, the central density reaches the threshold for electron-captures on ^{24}Mg and ^{20}Ne that triggers a collapse of the core.

For more massive stars (larger than about 13 $M_\odot$), entropy in the core is sufficiently high that electrons are always nondegenerate and thus $c_g < 0$. Therefore, these stars are gravothermally unstable and evolve rather straightforwardly through alternating phases of gravitational contraction and nuclear burning up to the iron core formation.

Stars in the mass range of 10–13 $M_\odot$ are intermediate between the above two cases and develop semidegenerate cores. They show interesting evolutionary behavior, which currently attracts much attention. I will discuss this in some detail in the next section.

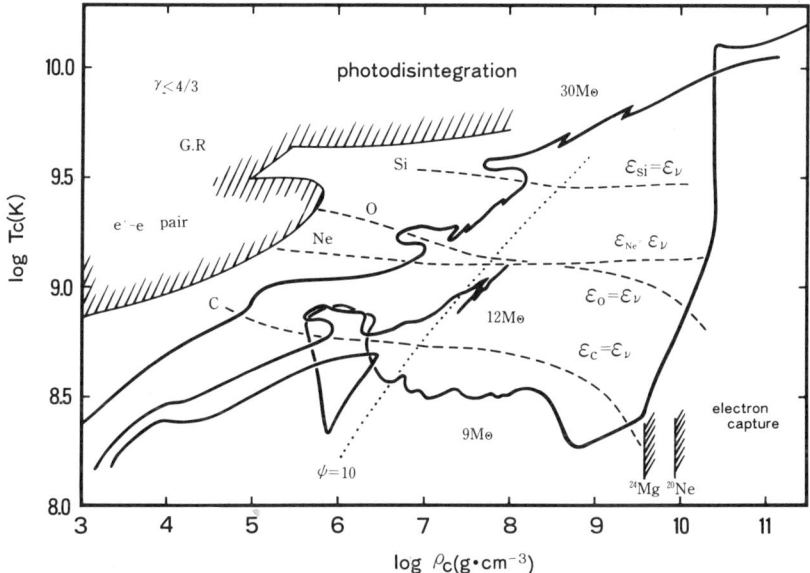

FIGURE 8. Evolution of the stars with masses of 30 $M_\odot$, 12 $M_\odot$, and 9 $M_\odot$ in the central temperature and density plane. Dashed lines show the ignition loci for carbon, neon, oxygen, and silicon burning. The dotted line with $\psi = 10$ (where ψ denotes the chemical potential of an electron in units of kT) indicates a region where electron degeneracy is significant. In the shaded region (upper left), the star is dynamically unstable. The shades in the lower right indicate the region where electron-captures are significant.

It should be noted, though, that the exact values of the above critical stellar masses may depend on the treatment of overshooting at the convective core edge (i.e., these values could shift to somewhat lower values).[46–49]

Evolution of 10–13 $M_\odot$ Stars

The evolution of stars in the mass range of 10–13 $M_\odot$ is complicated and is sensitive to stellar masses because neon shell flashes are affected by electron degeneracy. Stars in this range undergo nondegenerate carbon burning and form an ONeMg core whose mass, M_{ONM}, is in the range of 1.37–1.5 $M_\odot$. The core mass, M_{ONM}, is large enough to ignite neon, yet the core is semidegenerate and the degree of degeneracy depends sensitively on the ratio of M_{ONM}/M_{Ch}.

In such a core, combined effects of electron degeneracy and neutrino cooling produce a temperature inversion. This leads to an off-center neon ignition. Because of electron degeneracy, neon shell burning is unstable to a flash. FIGURE 9 shows a chemical evolution of a 10.4 $M_\odot$ star (corresponding to a helium star of a mass of M_{He} = 2.6 $M_\odot$) through the stage where neon is ignited at M_r = 0.88 $M_\odot$.[44] (See also Habets[46,47] for a detailed discussion of the evolution up to neon ignition and its mass dependence for helium stars corresponding to this mass range.)

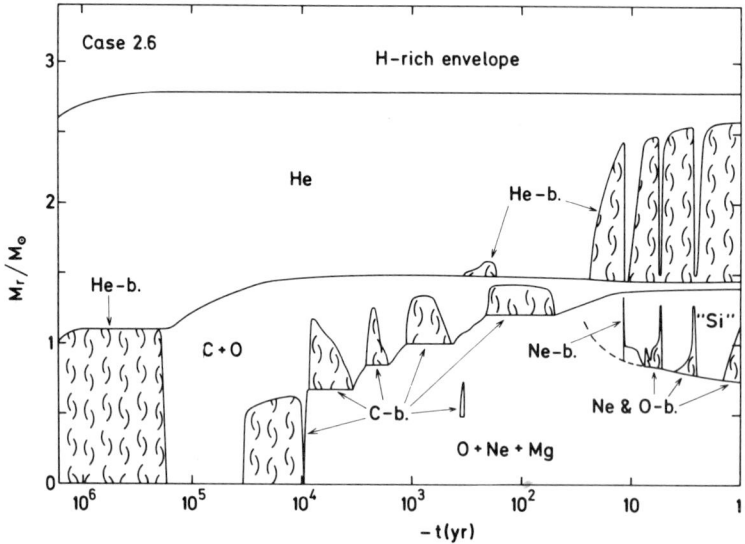

FIGURE 9. Chemical evolution of a 10.4-$M_\odot$ star from helium burning through the ignition of an off-center neon/oxygen flash at M_r = 0.88 $M_\odot$. Curled regions are convective, owing to nuclear burning.

The subsequent behavior of neon shell burning is sensitive to the stellar mass and is crucial for the ultimate fate of the star.[50] FIGURES 10 and 11 show new results for the 12 $M_\odot$ star (M_{He} = 3 $M_\odot$).[43] For this star, a neon flash is first ignited at M_r = 0.30 $M_\odot$ and subsequent neon and oxygen burning form a layer composed of ^{28}Si, ^{30}Si, and ^{34}S. The neon burning layer with a temperature as high as 2×10^9 K propagates inward all the way through the center to form a Si-S core. The composition profile during the propagation is shown in FIGURE 11. During the inward propagation, the density at the burning shell is relatively low ($\rho < 10^8$ g cm^{-3}). Neon burning is ignited layer by layer so that the released energy in one flash is too small to induce major dynamical effects.

For a slightly smaller mass star of 11.2 $M_\odot$ (M_{He} = 2.8 $M_\odot$), on the other hand, the

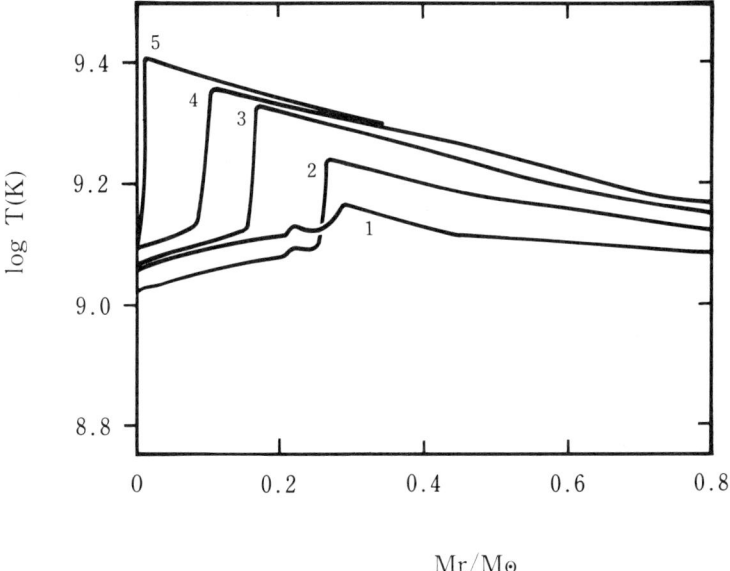

FIGURE 10. Propagation of a neon flashing front for the 12-$M_\odot$ star. The associated change in the temperature profile from stages 1 to 5 is shown.[43]

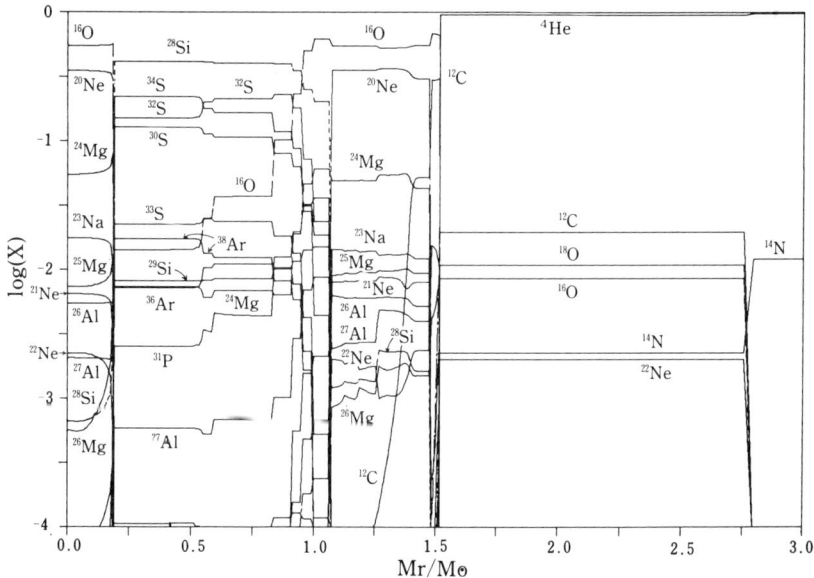

FIGURE 11. Composition structure of the 12-$M_\odot$ star when the neon/oxygen flashing front reaches a shell at $M_r = 0.19\ M_\odot$.[43]

inward neon burning shell reaches such high density layers as $\rho > 10^8$ g cm^{-3}. Then the neon shell flashes become so explosive as to cause some dynamical effects.[43,50]

The mechanism of the propagation of the neon burning shell is different from the carbon shell burning described in the third section of this paper. The carbon burning shell propagates inward due to heat conduction; the time scale of neutrino cooling (~ 300 yr) is too long to quench carbon burning as compared with the conduction time

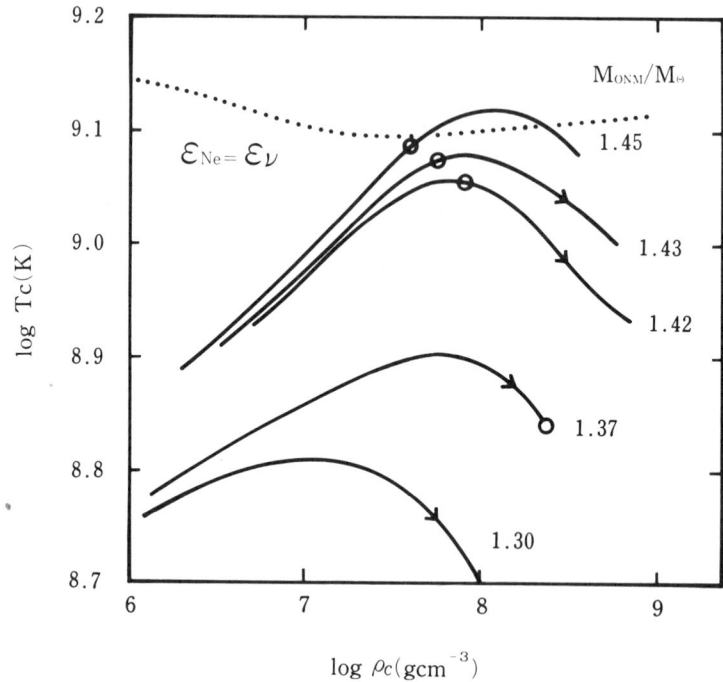

FIGURE 12. The gravitational contraction of semidegenerate O + Ne + Mg stars in the mass range of $M_{ONM} = 1.30$–1.45 $M_\odot$. The change in the sign of the gravothermal specific heat (i.e., the decrease in T_c) is shown in the central temperature and density plane. The dotted line is the neon ignition line, although neon burning is artificially suppressed in the contraction models. For 1.46 $M_\odot > M_{ONM} \gtrsim 1.37$ $M_\odot$, off-center neon ignition occurs at the stage marked by open circles. Afterwards, the central temperature reaches the neon ignition for $M > 1.44$ $M_\odot$, while smaller mass stars are cooled down without igniting neon.

scale. On the other hand, inward propagation of neon burning is induced by compressional heating due to gravitational contraction of the ONeMg core rather than heat conduction. In the 12 $M_\odot$ model, it takes only two years for the neon burning front to propagate through the center; this is much faster than conduction (see also references 50 and 51).

However, this may not be the case for smaller mass cores. Whether the neon burning shell reaches the center depends on the competition between neutrino cooling and compressional heating and thus depends on the core mass. In other words, the central temperature could start to decrease before reaching the neon ignition line if the gravothermal specific heat of a degenerate core changes its sign (see references 52 and 53).

In order to clarify the mass dependence of neon shell burning, simple O + Ne + Mg star models are calculated for several masses of M_{ONM} as shown in FIGURE 12. Here the stars are undergoing gravitational contraction (with neon burning artificially suppressed). The result shows that for 1.46 $M_\odot \gtrsim M_{ONM} > 1.37\ M_\odot$, the neon ignition temperature is reached first in the outer shell at the stage marked by the open circles in FIGURE 12. Afterwards, the central temperature also reaches the neon ignition line for $M_{ONM} > 1.44\ M_\odot$, while T_c starts to decrease without igniting neon for the smaller mass stars. In the actual stellar models, this critical mass may be significantly smaller and could even be as low as 1.35 $M_\odot$ because neon-oxygen shell burning produces a layer with smaller Y_e ($\simeq 0.48$ in FIGURE 11) and reduces the effective value of M_{Ch}. For smaller M_{Ch}, a core can contract further. Therefore, it is crucially important to use such an extensive reaction network for neon-oxygen burning as in FIGURE 11.

Based on these models, we can classify the evolution as follows:

(a) For $M \gtrsim 13\ M_\odot$, temperature inversion is negligibly small so that neon burning is ignited at the center.
(b) For 13 $M_\odot \gtrsim M > 11\ M_\odot$, neon is ignited off-center and the burning layer propagates inward all the way through the center. For less massive and denser cores, the neon shell flashes are so strong as to induce some dynamical effects.
(c) For 11 $M_\odot > M \gtrsim 10\ M_\odot$, the core is more strongly degenerate when neon ignition takes place at the outer shell of larger M_r. Hence, the time scale of compressional heating is longer because of slower contraction. On the other hand, the neutrino cooling time scale is as short as 10 yr. Therefore, neon and oxygen burning could be quenched by neutrino cooling; inward heat conduction is too slow. However, this mass range could be significantly narrower than the present prediction or even disappear because of small Y_e in the neon-oxygen burning layer.
(d) For $M < 10\ M_\odot$, neon is not ignited; in other words, the maximum temperature attained throughout the core during contraction is lower than the neon ignition temperature of 1.3×10^9 K.[44]

The evolution for cases (a) and (b) after central neon-oxygen burning also depends on the mass of a Si-S core relative to the Chandrasekhar limit. It is important to note that Y_e is reduced to as low as 0.48 during oxygen burning (even before appreciable electron-captures start) because oxygen burning produces a significant amount of neutron-rich species such as ^{30}Si and ^{34}S (FIGURE 11; reference 43). During the contraction of Si-S core, Y_e will still decrease due to electron-captures.[54,55] Therefore, electron degeneracy may not become strong and Si ignition is expected to take place at the center for most cases. Thus, stars more massive than $\sim 11\ M_\odot$ will evolve through the iron core formation although the behavior of Si-burning at high densities has not been fully investigated.

Presupernova Structure and Collapse-Bounce Mechanism for Massive Stars

The presupernova structure and the hydrodynamic behavior of core collapse depend on the stellar mass and may tentatively be classified into four groups: (1) $M > \sim 20\ M_\odot$, (2) $11-\sim 20\ M_\odot$, (3) $10-11\ M_\odot$, and (4) $8-10\ M_\odot$.

Stars More Massive than 11 $M_\odot$

These stars develop an iron core. The core collapse is induced by photodisintegration of iron nuclei and electron-captures. The collapsing core breaks into two parts: an inner part that collapses almost homologously and an outer quasi-free-fall layer. The inner core bounces around $\rho_c \simeq 4 \times 10^{14}$ g cm^{-3} because the equation of state becomes very stiff due to the transition into homogenous nuclear matter. At bounce, a shock wave forms near the outer edge of the homologous core and propagates outward.

The problem, though, is that the shock gets weaker mainly because the dissociation of the nuclei absorbs energy from the shock. The shock wave then stalls and does not give rise to explosion (e.g., see reference 56).

In order for the prompt bounce-shock mechanism to work, the initial shock wave should be more energetic and/or the amount of heavy nuclei in the outer layer should be smaller to lower the energy dissipation in the shock. In these respects, a larger homologous core is favorable (e.g., see reference 57). The energy transferred from the rebound core to the newly formed shock depends also on the high density equation of state of nuclear matter.

One of the recent efforts is concerned with the uncertainty of the stiffness of nuclear matter. Baron et al. (1985)[58] demonstrated that the shock is successful for the 12–15 $M_\odot$ stars if the incompressibility of the nuclear matter is significantly smaller than the standard value.

Another effort is to calculate neutrino transport with more accurately taking into account neutrino-electron scattering because the neutronization process during the collapse determines the mass of a homologous core, M_{HC}. It turns out that the core deleptonization proceeds appreciably even after the neutrino trapping density is reached because of enhanced neutrino transport.[59] This effect lowers Y_e and thus M_{HC}. In other words, the shock propagation would be more difficult than in the models calculated with crude neutrino transport. Whether the prompt shock ejection succeeds still seems to be a delicate problem.

For stars more massive than 20 $M_\odot$, even a powerful shock by Baron et al.[58] seems to fail. However, such a stalled shock can be revived ~ 1 s later by neutrino interaction with matter behind the shock (see reference 53 for details).

8–11 $M_\odot$ Stars

This mass range is divided into two groups:

(a) Stars of 10–11 $M_\odot$ have a small degenerate ONeMg core (typically 0.3 $M_\odot$) that is surrounded with a Si-rich layer up to $M_r = 1.4$–1.5 $M_\odot$, though this mass range might not exist at all.

(b) Stars of 8–10 $M_\odot$ have a strongly degenerate ONeMg core of 1.28–1.34 $M_\odot$ that is surrounded with a C + O layer up to the helium burning shell at M_r = 1.38 $M_\odot$.

For both cases (a) and (b), core collapse is triggered by electron-captures on ^{24}Mg and ^{20}Ne that lower the Chandrasekhar limit.[32] One of the important differences from the collapsing iron core is that the ONeMg core (especially for 8–10 $M_\odot$ stars) contains a substantial amount of nuclear fuel. Therefore, oxygen combustion is ignited when the central density reaches ~2.5 × 10^{10} g cm^{-3}. Of course, the nuclear energy release is

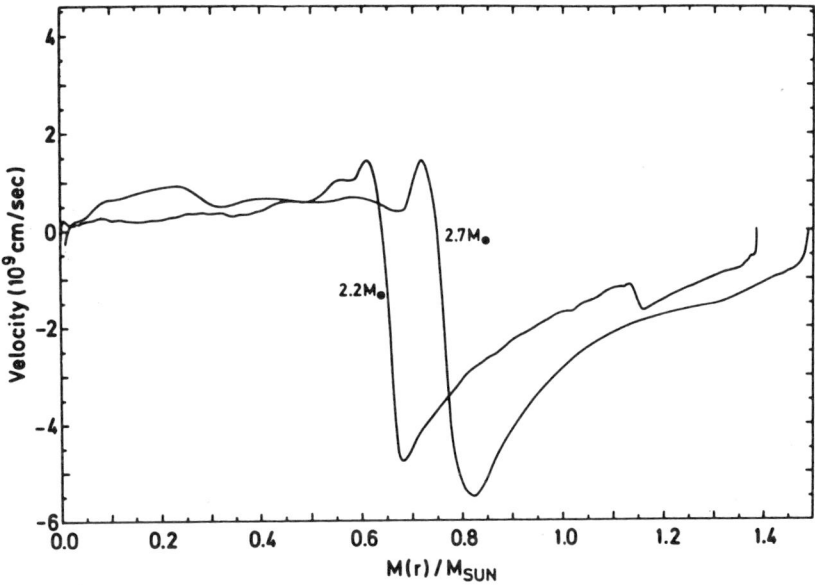

FIGURE 13. Velocity profiles just after core-bounce for two helium core models with masses of 2.7 $M_\odot$ and 2.2 $M_\odot$, respectively.[60] Because of the slightly smaller Y_e at neutrino trapping, the mass of a homologous core is smaller and thus the shock forms at a smaller M_r for the 2.2-$M_\odot$ core. On the other hand, the velocity of the still infalling matter is significantly reduced in the 2.2-$M_\odot$ case due to the deceleration of matter in O-burning shell.

much smaller than the gravitational potential energy of the core and thus cannot directly cause mass ejection. However, oxygen combustion has a significant effect on the hydrodynamics of collapse by decelerating the infall for 8–10 $M_\odot$ stars (reference 60; FIGURE 13). As a result of such a deceleration, the density in the infalling region is significantly lower than in the iron core (FIGURE 14). Therefore, the ram pressure exerted by the infalling matter is much smaller. This favors the successful shock. On the other hand, the mass of the homologous core is smaller in the ONeMg core than in the iron core because of a smaller Y_e (~0.34 according to Hillebrandt et al., 1984[60]); this is due to a higher entropy produced by oxygen combustion and also to a slower collapse. The smaller M_{HC} yields smaller shock energy. The net outcome is still the successful shock.[60]

The 10–11 $M_\odot$ stars have an important difference from the 8–10 $M_\odot$ stars, though collapse is triggered by electron-captures on ^{24}Mg and ^{20}Ne: In the collapse of 10–11 $M_\odot$ stars, the original ONeMg layer is so small that it is contained in the homologous part, while the outer free-falling layer has undergone neon-oxygen burning and is composed of Si-rich elements. Therefore, nuclear energy release must be much smaller than oxygen combustion and could even be negative. Accordingly, the deceleration effect on the infalling material would be negligible. I think this is why prompt explosion does not occur for the Wilson's (1985)[53] 11-$M_\odot$ star model. These stars have another important feature, namely, the density drops sharply over ten orders of

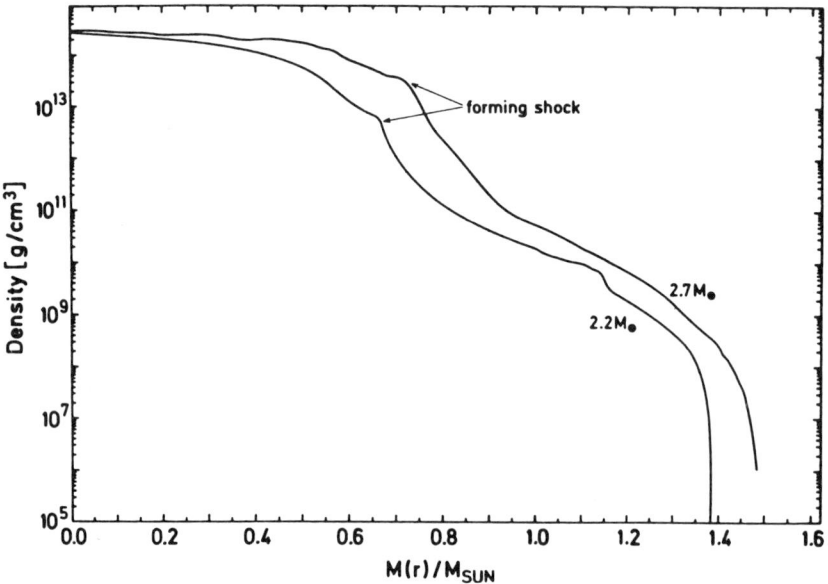

FIGURE 14. Density profiles corresponding to FIGURE 13. The presence of the O-burning shell reduces significantly the density of the infalling material.

magnitude near the helium burning shell ($M_r = 1.4\ M_\odot$). This would reduce the effects of late time neutrino heating.[53] However, the shock wave can be strengthened by the very steep density gradient once it manages to reach there.[61] Every effect seems to be marginal for this mass range, though, so careful investigation is needed.

NEUTRON STAR FORMATION IN SUPERNOVA EXPLOSIONS AND COOLING OF NEUTRON STARS

The identification of a neutron star in supernova remnants is one of the possible observational tests of the theoretical supernova models. As discussed in the previous sections, there exist several modes of neutron star formation:

(1) Type II supernova explosions that are triggered by collapse of the iron (or ONeMg) core of massive stars ($M_B > M \gtrsim 8\ M_\odot$).
(2) Iron core collapse in massive helium stars ($M_B/3 > M_{He} \gtrsim 2.5\ M_\odot$) in close binary systems; some of these might form a subclass of SN-I.
(3) Collapse of an ONeMg core of moderate mass helium stars ($2.5\ M_\odot > M_{He} \gtrsim 1\ M_\odot$); these are single stars with an extended helium envelope.
(4) Merging of double C + O white dwarfs; this could leave a quite isolated pulsar (possibly a millisecond pulsar).
(5) Collapse of accreting ONeMg white dwarfs in close binary systems; this is a possible origin of low mass X-ray binaries.

On the other hand, certain types of supernova explosions do not leave neutron stars behind:

(6) Pair instability supernovae of very massive objects ($M \gtrsim 100\ M_\odot$), which result in either complete disruption or black hole formation.
(7) Explosion of very massive stars induced by iron core collapse ($100\ M_\odot > M \gtrsim M_B$), which leaves a black hole behind.
(8) Type I supernova explosion of accreting C + O white dwarfs, which disrupts the star completely. (Some exploding C + O white dwarfs leave a white dwarf remnant so that a search for a white dwarf in the central region of supernova remnants might be interesting.)

The Einstein X-ray Observatory had an ability to detect thermal radiation from hot neutron stars. The detection and nondetection (upper limit) of compact X-ray sources in young supernova remnants provide important constraints on the models for the neutron star formation summarized above. The observed X-ray flux and upper limit should be compared with the theoretical cooling curve of neutron stars.

FIGURE 15 shows such a comparison for a neutron star model with a high density equation of state by Friedman and Pandharipande.[62] (FP model with $M = 1.3\ M_\odot$ and $R = 11$ km.[63]) In this "standard" cooling curve, cooling due to such exotic particles as pions and quarks are not included. The upper and lower limits of the theoretical curve reflect the uncertainties involved in the nucleon superfluidity. FIGURE 16 shows that it takes more than 300 years for the neutron star core to become isothermal.[63] For the upper limit curve in FIGURE 15, the time scale to reach the isothermal state is longer than 10^4 yr, even for the FP model. Some results are:

(a) The detection points of RCW103 and the Crab are consistent with the cooling curve. Tuohy et al. (1983)[64] suggested that the point source in RCW103 is due likely to blackbody radiation from the surface of the neutron star, though future spectral observation is needed to confirm it
(b) The upper limit to Cas A, Tycho, and SN1006 are well below the lower limit of the cooling curve. This implies that either a neutron star was not formed by a supernova explosion or a neutron star exists, but cooled very rapidly owing to exotic particles. The former interpretation is consistent with the failure of detection of synchrotron nebulosity in these remnants.[65] Since Tycho and SN 1006 are the likely remnants of SN-I, the interpretation in terms of the absence of a neutron star is consistent with the complete disruption model for SN-I.

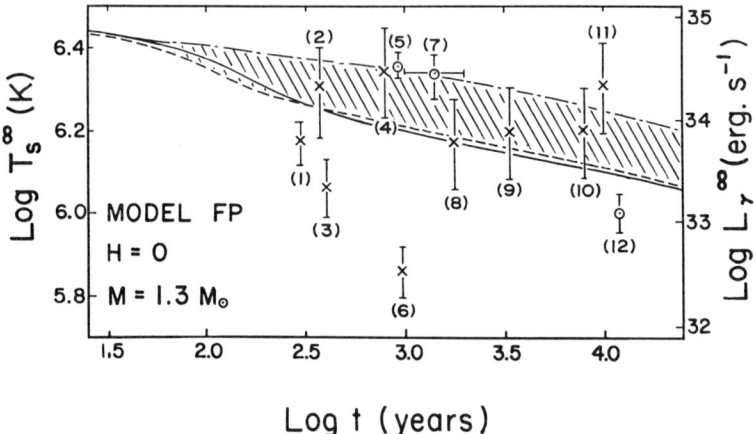

FIGURE 15. Comparison between the data from the Einstein Observatory for young supernova remnants and the theoretical cooling curves of neutron star. The numbers refer to: (1) Cas A, (2) Kepler, (3) Tycho, (4) 3C58, (5) Crab, (6) SN1006, (7) RCW103, (8) RCW86, (9) W28, (10) G350.0-18, (11) G22.7-0.2, and (12) Vela.

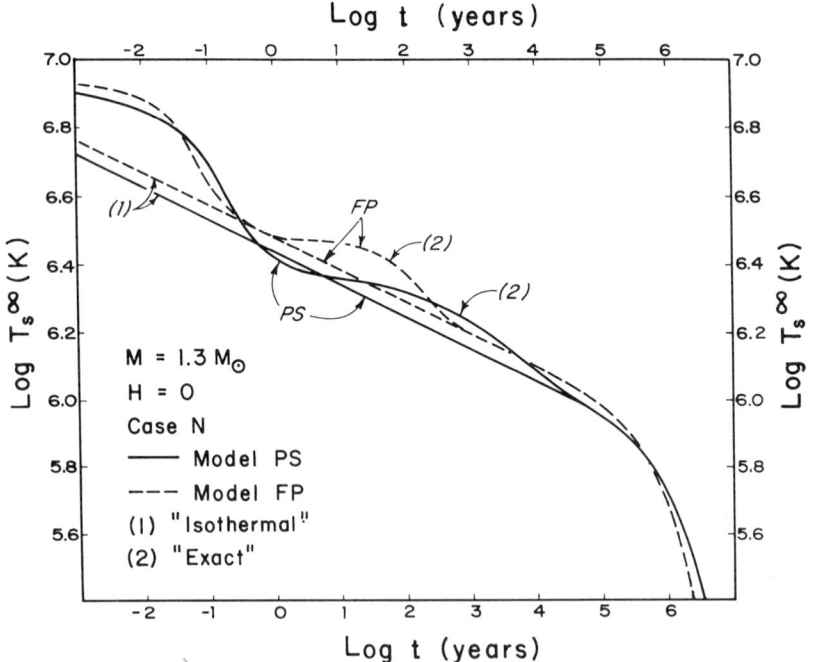

FIGURE 16. Cooling curves obtained with (1) "isothermal" and (2) "exact" methods are compared for the PS (stiff) and FP models. It takes 500 yr (FP) and 30,000 yr (PS) for the core of these models to become isothermal.

(c) The surface temperature of Vela could lead to a new important interpretation because it is below the "standard cooling" curve. (Neutron star models employing other equations of state lead to the same conclusion.) This result may suggest the existence of "nonstandard cooling" due to pions or quarks.[66] If so, the mass of the Vela neutron star is large enough to allow exotic particles to form. Before reaching such an important conclusion, however, more careful investigation is required concerning the effects of magnetic field, etc.[66]

It is certainly important to obtain spectral information by future X-ray (and UV) satellite for these point sources so as to see if the X ray is thermal radiation from a hot neutron star or of nonthermal origin. Chemical abundances of the supernova remnants provide another important test. [See Kafatos and Henry (1985)[67] for the Crab nebula and Danziger and Gorenstein (1983)[68] for other supernova remnants.]

ACKNOWLEDGMENTS

I have greatly benefited from discussion with G. Brown, S. Kahana, J. Lattimer, A. Burrows, S. Woosley, and D. Branch during my visit to Stony Brook and Santa Cruz (October '84) and with W. Hillebrandt, F-K. Thielemann, and E. Müller at the Max Planck Institute (December '84). I would like to thank D. Sugimoto, S. Tsuruta, I. Iben, H. Saio, and M. Hashimoto for collaboration, discussion, and comments. It is a pleasure to thank M. Livio, I. Lichtenstadt, and Z. Barkat for discussion and the hospitality in Jerusalem.

REFERENCES

1. OKE, J. B. & L. SEARLE. 1974. Annu. Rev. Astron. Astrophys. **12:** 315.
2. DOGGETT, J. B. & D. BRANCH. 1985. Astron. J. **90:** 2303.
3. TRIMBLE, V. 1982. Rev. Mod. Phys. **54:** 1183.
4. WEAVER, T. A. & S. E. WOOSLEY. 1980. Ann. N.Y. Acad. Sci. **336:** 335.
5. WHEELER, J. C. & R. LEVREAULT. 1985. Astrophys. J. Lett. **294:** L17.
6. WHEELER, J. C. 1982. *In* Supernovae: A Survey of Current Research. M. J. Rees & R. J. Stoneham, Ed.: 167. Reidel. Dordrecht.
7. HOYLE, F. & W. A. FOWLER. 1960. Astrophys. J. **132:** 565.
8. NOMOTO, K., D. SUGIMOTO & S. NEO. 1976. Astrophys. Space Sci. **39:** L37.
9. NOMOTO, K. 1984. *In* Stellar Nucleosynthesis. C. Chiosi & A. Renzini, Eds.: 205, 238. Reidel. Dordrecht.
10. HASHIMOTO, M., K. NOMOTO, K. ARAI & K. KAMINISI. 1984. Phys. Rep. Kumamoto Univ. **6:** 75; 1986. Astrophys. J. **307**. In press.
11. NOMOTO, K., F-K. THIELEMANN & K. YOKOI. 1984. Astrophys. J. **286:** 644 (NTY).
12. ARNETT, W. D. 1969. Astrophys. Space Sci. **5:** 180.
13. THIELEMANN, F-K., K. NOMOTO & K. YOKOI. 1986. Astron. Astrophys. In press (TNY).
14. BRANCH, D., J. B. DOGGETT, K. NOMOTO & F-K. THIELEMANN. 1985. Astrophys. J. **294:** 619.
15. BRANCH, D., C. H. LACY, M. L. MCCALL, P. G. SUTHERLAND, A. UOMOTO, J. C. WHEELER & B. J. WILLS. 1983. Astrophys. J. **270:** 123.
16. KIRSHNER, R. P. & J. B. OKE. 1975. Astrophys. J. **200:** 574.
17. WOOSLEY, S. E., T. S. AXELROD & T. A. WEAVER. 1984. *In* Stellar Nucleosynthesis. C. Chiosi & A. Renzini, Eds.: 263. Reidel. Dordrecht.
18. GRAHAM, J. R., W. P. S. MEIKLE, D. A. ALLEN, A. J. LONGMORE & P. M. WILLIAMS. 1986. Mon. Not. R. Astron. Soc. In press.

19. AXELROD, T. S. 1980. Ph.D. thesis, University of California at Santa Cruz.
20. TSUNEMI, H., K. YAMASHITA, K. MASAI & S. HAYAKAWA. 1986. Astrophys. J. In press.
21. HAMILTON, A. J. S., C. L. SARAZIN, A. E. SZYMKOWIAK & M. H. VARTANIAN. 1985. Astrophys. J. Lett. **297:** L5.
22. MÜLLER, E. & W. D. ARNETT. 1982. Astrophys. J. Lett. **261:** L107; 1986. Astrophys. J. Submitted.
23. TRURAN, J. W. & W. D. ARNETT. 1971. Astrophys. Space Sci. **11:** 430.
24. IBEN, I., JR. & A. V. TUTUKOV. 1984. Astrophys. J. Suppl. Ser. **55:** 335.
25. WEBBINK, R. F. 1984. Astrophys. J. **277:** 355.
26. WHEELER, J. C. 1985. *In* Supernovae as Distance Indicators. N. Bartel, Ed.: 34. Springer-Verlag. New York.
27. ARNETT, W. D., D. BRANCH, & J. C. WHEELER. 1985. Nature **314:** 337.
28. NOMOTO, K. & I. IBEN, JR. 1985. Astrophys. J. **297:** 531.
29. SAIO, H. & K. NOMOTO. 1985. Astron. Astrophys. **150:** L21.
30. WOOSLEY, S. E. & T. A. WEAVER. 1986. *In* Nucleosynthesis and Its Implication on Nuclear and Particle Physics. J. Audouze & T. van Thuan, Eds. Reidel. Dordrecht. In press.
31. NOMOTO, K., S. MIYAJI, D. SUGIMOTO, & K. YOKOI. 1979. *In* IAU Colloq. 53, White Dwarfs and Variable Degenerate Stars. H. M. Van Horn & V. Weidemann, Eds.: 56. University of Rochester. Rochester, New York.
32. MIYAJI, S., K. NOMOTO, K. YOKOI & D. SUGIMOTO. 1980. Publ. Astron. Soc. Japan **32:** 303.
33. NOMOTO, K., & D. SUGIMOTO. 1977. Publ. Astron. Soc. Japan **29:** 765.
34. BACKER, D. C., S. R. KULKARNI, C. HEILES, M. M. DAVIS & W. M. GOSS. 1982. Nature **300:** 615.
35. KING, A. R., J. FRANK & H. RITTER. 1985. Mon. Not. R. Astron. Soc. **213:** 181.
36. FLOWERS, E. & M. A. RUDERMAN. 1976. Astrophys. J. **215:** 302.
37. BLANDFORD, R. D., J. H. APPLEGATE & L. HERNQUIST. 1983. Mon. Not. R. Astron. Soc. **204:** 1025.
38. WHEELER, J. C. 1978. Astrophys. J. **225:** 212.
39. UUS, U. 1970. Nauchn. Inf. Acad. Nauk. USSR **17:** 25.
40. LASHER, G. 1975. Astrophys. J. **201:** 194.
41. NOMOTO, K. 1982. *In* Supernovae: A Survey of Current Research. M. J. Rees & R. J. Stoneham, Eds.: 205. Reidel. Dordrecht.
42. SUGIMOTO, D. & K. NOMOTO. 1974. *In* IAU Symposium 66, Late Stages of Stellar Evolution. R. J. Taylor, Ed.: 105. Reidel. Dordrecht.
43. HASHIMOTO, M. & K. NOMOTO. 1985. In preparation.
44. NOMOTO, K. 1984. Astrophys. J. **277:** 791.
45. SUGIMOTO, D. & K. NOMOTO. 1980. Space Sci. Rev. **25:** 155.
46. HABETS, G. M. H. J. 1985. Ph.D. thesis, University of Amsterdam.
47. HABETS, G. M. H. J. 1985. Astron. Astrophys. Submitted.
48. CASTELLANI, V., A. CHIEFFI, L. PULONE & A. TORNAMBE. 1985. Astrophys. J. Lett. **294:** L31.
49. BERTELLI, G., A. BRESSAN & C. CHIOSI. 1985. Astron. Astrophys. **150:** 33.
50. WOOSLEY, S. E., T. A. WEAVER & R. E. TAAM. 1980. *In* Type I Supernovae. J. C. Wheeler, Ed.: 96. University of Texas. Austin.
51. IKEUCHI, S., K. NAKAZAWA, T. MURAI, R. HOSHI & C. HAYASHI. 1972. Prog. Theor. Phys. **48:** 1890.
52. BARKAT, Z., Y. REISS & G. RAKAVY. 1974. Astrophys. J. Lett. **193:** L21.
53. WILSON, J. R., R. MAYLE, S. E. WOOSLEY & T. A. WEAVER. 1986. This volume.
54. WEAVER, T. A., S. E. WOOSLEY & G. M. FULLER. 1985. *In* Numerical Astrophysics. J. Centrella, J. Leblanc & R. Bowers, Eds.: 374. Jones and Bartlett. Boston.
55. THIELEMANN, F-K. & W. D. ARNETT. 1985. Astrophys. J. **295:** 604.
56. HILLEBRANDT, W. 1982. *In* Supernovae: A Survey of Current Research. M. J. Rees & R. J. Stoneham, Eds.: 123. Reidel. Dordrecht.
57. BURROWS, A. & J. LATTIMER. 1983. Astrophys. J. **270:** 735.
58. BARON, E., J. COOPERSTEIN & S. KAHANA. 1985. Nucl. Phys. **A440:** 744; 1985. Phys. Rev. Lett. **55:** 126.

59. BRUENN, S. W. 1985. Astrophys. J. Suppl. Ser. **58:** 771.
60. HILLEBRANDT, W., K. NOMOTO & R. G. WOLFF. 1984. Astron. Astrophys. **133:** 175.
61. HILLEBRANDT, W. 1983. Astron. Astrophys. **110:** L3.
62. FRIEDMAN, B. & V. R. PANDHARIPANDE. 1981. Nucl. Phys. **A361:** 502.
63. NOMOTO, K. & S. TSURUTA. 1986. Astrophys. J. Lett. In press.
64. TUOHY, I., G. P. GARMIRE, R. N. MANCHESTER & M. A. DOPITA. 1983. Astrophys. J. **268:** 778.
65. HELFAND D. J. & R. H. BECKER. 1984. Nature **307:** 215.
66. TSURUTA, S. 1986. Comments Astrophys. In press.
67. KAFATOS, M. & R. B. C. HENRY, Eds. 1985. The Crab Nebula and Related Supernovae Remnants. Cambridge University Press. Cambridge.
68. DANZIGER, J. & P. GORENSTEIN, Eds. 1983. IAU Symposium 101, Supernova Remnants and Their X-Ray Emission. Reidel. Dordrecht.

Accretion onto the White Dwarf and X-Ray Production in Nonmagnetic Cataclysmic Variables

A. R. KING

Astronomy Department
University of Leicester
Leicester LE1 7RH, England

INTRODUCTION

Cataclysmic variables (CVs) are short-period binary systems (periods ~ a few hours) in which a white dwarf accretes from a close, low-mass companion star. For most of their lifetimes, they derive their luminosity from the gravitational energy released by the accreting matter. Nuclear burning, although rather more efficient ($\sim 6 \times 10^{18}$ erg g^{-1} compared with $\sim 2 \times 10^{17}$ erg g^{-1}), is confined to bright ($\sim 10^{38}$ erg s^{-1}), but short (~ 10 yr) and infrequent ($\sim 10^5$ yr) nova explosions. Unlike in many neutron-star and black-hole binaries, the companion star is faint, so that accretion dominates the output of the system at all wavelengths from the visible/near IR to X rays. More than one hundred CV systems are known, with typical distances of 50–500 pc. For these reasons, they are ideal systems for the study of accretion processes (e.g., see references 1 and 2 for comparisons with other close binaries). In particular, many of them show convincing evidence of possessing accretion discs; indeed, the present confidence that the theory of such discs[1,3] may not be too far from describing reality is largely based on a confrontation with observations of CVs.

A central problem in accretion theory is how the matter actually lands on the accreting object. For neutron stars and white dwarfs, the problem is subdivided according to whether the star possesses a magnetic field strong enough to disrupt the accretion flow. The magnetic cataclysmic variables are a distinct subclass of CVs, readily recognizable because of their coherently pulsed X-ray emission, and progress in understanding them has been quite rapid of late (see references 4–6 for reviews and 7 and 8 for discussions of their likely evolutionary status and relation to nonmagnetic CVs). In the nonmagnetic systems under discussion here, we must investigate the boundary layer region where the inner edge of the accretion disc interacts with the white dwarf surface. As is usual in studying accretion processes, the most fruitful approach has been to work back and forth between theory and observation: in particular, X-ray observations have provided powerful insights.

In the remainder of this review, I shall first discuss X-ray observations of nonmagnetic CVs and relate them to current theories of the boundary layer. I will argue that hard (>2 keV) emission probably arises in a hot corona surrounding the white dwarf (formed from the accretion flow because of thermal instabilities in the boundary layer region) and then show that there is considerable observational support for this idea. Because of conductive losses from the corona into the white dwarf body, hard X-ray production is likely to be quite inefficient (~ 1–10%), with most of the boundary layer accretion energy ultimately emerging as soft X-ray emission from the heated white dwarf surface. Finally, this picture offers a possible explanation of the quasi-periodic soft X-ray pulsations seen during the dwarf nova outbursts of some systems, especially in terms of the channeling of the conduction losses by transient, dynamo-generated surface magnetic fields.

X-RAY EMISSION FROM CVs

At least 50 CVs are known X-ray sources.[9] Although all are rather weak (10^{-12}–10^{-10} erg cm^{-2} s^{-1}), magnetic systems are noticeably stronger than nonmagnetic ones. Since there is little reason to suppose that all magnetic systems have systematically higher accretion rates than nonmagnetic ones,[8] this suggests that hard X-ray production is rather less efficient in the latter; we shall see that there are probably quite good reasons for this. Only two nonmagnetic CVs, SS Cyg[10] and U Gem,[11] are bright enough ($\lesssim 10^{32}$ erg s^{-1}) to have so far allowed much observational study beyond simple detection. Both show, in varying degrees at different epochs, two-component X-ray spectra: a hard component, probably thermal bremsstrahlung at temperatures of $\sim 10^8$ K ($kT \sim 10$ keV), and a soft, roughly blackbody component, with (very uncertain) characteristic temperatures of $(1–3) \times 10^5$ K ($kT \sim 10$–30 eV). The latter component is heavily attenuated by interstellar photoelectric absorption, which renders accurate estimates of the total luminosity very difficult. It is likely that all nonmagnetic CVs have (with varying proportions) two-component spectra of this type,[12] although the soft X-ray component will be unobservable if its characteristic temperature falls below $\sim 10^5$ K.

Both SS Cyg and U Gem are dwarf novae: at irregular intervals (weeks, months), they brighten optically by factors of ~ 10, with these outbursts normally lasting $\lesssim 10$ days. There is little doubt that this behavior reflects an increased rate of accretion onto the white dwarf, although the reasons for this are still unclear and thus a subject of active research (e.g., see references 13–15). The outbursts offer an excellent opportunity to study the dependence of X-ray production on accretion rate: interestingly, the behaviors of U Gem and SS Cyg are strikingly different. Both hard and soft X-ray components in U Gem increase strongly at outburst,[9,11] while the brighter system, SS Cyg, shows a sharp decrease in hard X-ray emission, with brightening in soft X rays.[10,11] This behavior reverses on the decline from outburst. FIGURE 1 shows optical and hard and soft X-ray light curves from a recent EXOSAT observation of most of an outburst of SS Cyg.

The obvious interpretation is that at sufficiently high accretion rates, $\dot{M}$, the production of hard X rays is suppressed in favor of soft X-ray emission. While U Gem never reaches this regime, in SS Cygni, $\dot{M}$ attains these values at outburst. However, it is important to note that the suppression of hard X rays does not occur because of increased absorption within a dense accretion flow: if this were the case, hard X-ray spectra of SS Cyg should show marked low-energy cutoffs due to varying intrinsic absorption (which are never seen). This was first demonstrated by Swank,[16] and a series of EXOSAT observations confirm this (M.G. Watson, personal communication). Accordingly, the hard X-ray production mechanism must become inefficient at high $\dot{M}$ and "turn off" the emission.

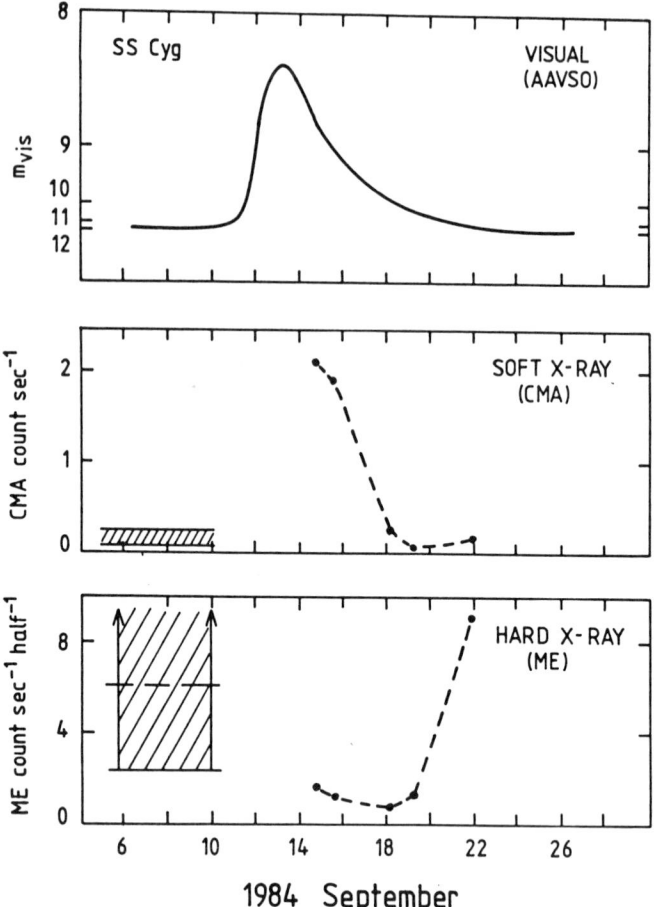

FIGURE 1. An outburst of SS Cygni in September 1984. Shown are the optical light curve, kindly supplied by the American Association of Variable Star Observers, and the hard (2–10 keV) and soft (0.04–2 keV) X-ray light curves observed with EXOSAT. There is no evidence of varying intrinsic absorption in the hard X-ray spectra. The hatched areas in the two lower panels show the typical ranges of soft and hard X-ray emission in the quiescent state.

MODELS

I shall discuss further aspects of the X-ray data below; however, those outlined above already present a challenge to theory. Temperatures in the accretion discs in CVs are unlikely to exceed 5×10^4 K (e.g., see reference 3), so the X-ray production must involve the boundary layer.[17] A slowly rotating white dwarf of mass, M, and radius, R_*, accreting steadily at a rate, $\dot{M}$, has a total accretion luminosity,

$$L_{acc} = \frac{GM\dot{M}}{R_*}. \tag{1}$$

Since a Kepler orbit at R_* has a binding energy of $GM/2R_*$ per unit mass, the accretion disc only releases $1/2 \, L_{acc}$; the remaining fraction of L_{acc} is emitted as the accreting matter lands on the white dwarf surface. This "boundary layer" luminosity, L_{BL}, is only reduced significantly below $1/2 \, L_{acc}$ for white dwarfs rotating at angular velocities very close to breakup.[17] For any plausible boundary layer structure, a fraction, η, of the order of one-half (or more) of the luminosity must ultimately be radiated as optically thick emission from the white dwarf surface, either because of heating from optically thin regions above or, as I shall suggest later, because of direct transport of the accretion energy under the white dwarf photosphere. If the emitting area is a fraction, f, of the surface of a 1 $M_\odot$ white dwarf ($R_* = 5 \times 10^8$ cm), we find an effective blackbody temperature,

$$T_b = 5 \times 10^4 \, (\eta L_{33}/f)^{1/4} \text{ K}, \qquad (2)$$

where L_{33} is L_{BL} in units of 10^{33} erg s^{-1}. This is in reasonable agreement with the detection of soft X-ray components during outbursts of U Gem and SS Cyg, where L_{33} is probably $\gtrsim 1$ (especially as Pringle[17] has argued that much of the emission from an optically thick boundary layer may come from an equatorial ring occupying a surface fraction, $f \lesssim 0.1$).

The estimate [equation (2)] means that mechanisms for heating optically thin gas to temperatures of $\sim 10^8$ K must be sought to explain the hard X-ray emission. Pringle and Savonije[18] noted that the Kepler velocity of material circulating in the boundary layer near R_* would, if strongly shocked, produce temperatures of

$$T_s = \frac{3GMm_H}{16kR_*} \sim 3 \times 10^8 \text{ K}, \qquad (3)$$

although there are problems in producing strong shocks in the highly sheared boundary layer flow. In both this model and that of Tylenda,[19] the suppression of hard X-ray emission at high accretion rates is attributed to absorption, which as we have seen above is now ruled out by observation. King and Shaviv[20] assume that the turbulent dissipation (no doubt involving shocks, etc.) in the boundary layer can be represented by an eddy viscosity and they compare the timescale t_{heat} for viscous heating of the boundary layer gas with the (optically thin) radiative cooling time, t_{rad}, and the adiabatic expansion timescale, t_{exp}. It turns out that the ratios, t_{heat}/t_{rad}, t_{heat}/t_{exp}, and t_{rad}/t_{exp}, depend solely on the gas temperature, T, and the product, ρb, of the gas density, ρ, and the size, b, of a boundary layer element. For low values of ρb, such as might be expected to prevail at low accretion rates, radiative cooling is always ineffective and viscosity heats the gas to temperatures of the order of T_s [equation (3)], where t_{exp} becomes comparable to t_{heat} and prevents further heating. Gas at such temperatures cannot be confined to the equatorial region and will expand out to form an X-ray emitting corona around the white dwarf of scale height, $H \sim 0.3 \, R_*$. For high values of ρb, which are expected to characterize most of the boundary layer gas at high accretion rates, t_{rad} is always the shortest timescale so that the bulk of the gas can cool and settle on the white dwarf without ever reaching hard X-ray temperatures.

Clearly, this mechanism does satisfy the requirement that high accretion rates can turn off the hard X-ray emission without invoking reabsorption: the thermal instability is largely suppressed at some critical accretion rate, $\dot{M}_{crit}$, corresponding to the value, $\rho b \sim 1$ g cm^{-2}, at which t_{rad} becomes the shortest timescale. Actually to work out the value of $\dot{M}_{crit}$ requires a self-consistent solution of the boundary layer flow, which is a formidable undertaking even if a steady state is assumed. One might hope, however, that the parameters calculated by Pringle and Savonije[18] for the region of the accretion disc just outside the boundary layer might be used to estimate ρb for accretion rates near $\dot{M}_{crit}$. In this case, one gets $\dot{M}_{crit} \sim 10^{16}$ g s^{-1}, which is in rough agreement with what is suggested by observation.

THE WHITE DWARF CORONA

The considerations given above show that the existence of an X-ray emitting corona around the white dwarf is quite plausible. EXOSAT observations of SS Cyg[21] in quiescence (i.e., between outbursts) offer direct evidence in support of this. From FIGURE 2, one sees that the hard (2–10 keV) X rays vary by large factors on timescales down to a few tens of seconds. Assuming that this behavior results from radiative cooling, we must demand that the radiative (bremsstrahlung) cooling time of the X-ray gas, given in general by

$$t_{rad} \sim 30\, \rho_{-10}^{-1}\, T_8^{1/2} \text{ s} \qquad (4)$$

(where ρ_{-10} is the gas density in units of 10^{-10} g cm^{-3} and T_8 is its temperature in units of 10^8 K), is of the order of a few $\times$ 10 s. Since T_8 is actually measured by the X-ray spectrum to be ~ 1 (see below), this implies gas densities of $\rho_{-10} \sim 1$ in the emitting region. The distance (~ 100 pc) and the measured 2–10 keV flux imply an emission measure, $V\rho^2 T^{1/2}$, for the gas, where V is the volume of the emitting region, which must lie close to the white dwarf surface. Putting in the values of $\rho \sim 10^{-10}$ and $T \sim 10^8$, as derived above, gives $V \sim 10^{27}$ cm^3. This is a volume comparable to that occupied by the white dwarf itself ($R_* \lesssim 10^9$ cm) and it clearly implies a corona of scale height of $H \leq 0.5\, R_*$ surrounding a large fraction of its surface.

In fact, one can show[20] that a corona with just these parameters is the likely end result of the thermal instability discussed in the previous section. In addition to t_{rad} [equation (4)], two important timescales for the coronal gas are the hydrodynamic time, $t_{hyd} = H/c_s$ (c_s = sound speed), and the thermal conduction time, $t_{cond} = 5\rho c_s^2 H^2/2K_0 T^{7/2}$, where K_0 is a constant of the order of 10^{-6} in cgs units. t_{hyd} gives the timescale on which ρ and H can adjust to changes in, say, T, while t_{cond} measures the rate at which electron thermal conduction into the white dwarf surface will try to cool the gas. Any such conductive losses must be promptly reradiated, presumably as soft X rays, from the heated surface.

Now let us ask what an element of coronal gas will do if it is allowed to cool from a temperature of $\sim$few $\times 10^8$ K. Constructing the ratios of t_{rad}/t_{hyd}, t_{rad}/t_{cond}, and t_{hyd}/t_{cond}, one finds that they depend only on T and ρ (FIGURE 3). Unless ρ is quite high (5×10^{-10} g cm^{-3}), t_{cond} will initially be the shortest timescale. H and ρ, varying on timescales of $t_{hyd} > t_{cond}$, will remain effectively constant as the gas cools by conduction (A $\rightarrow$ B in FIGURE 3). This removes a large fraction of the thermal energy, causing it to be radiated as soft X rays rather than hard, and it suggests a reason for the low efficiency of hard X-ray production in nonmagnetic CVs noted earlier. (Note the broken diagonal lines of $L_x = constant$ on the figure.) At B, t_{hyd} decreases below t_{cond}, which is still shorter than t_{rad}, so the gas can now adjust ρ and H quasi-hydrostatically in response to the cooling. This is the regime extensively investigated in the study of solar flares: the precise track followed by the gas requires detailed calculation, but it is probably something like B → C as indicated. In any case, the gas must ultimately reach the region where $t_{hyd} < t_{rad} < t_{cond}$, where it is quasi-hydrostatic and cooling radiatively. Most of the observed hard X-ray emission must come from gas in this region since the gas spends the longest time there. The gas is both cooler ($T \sim 10^5$ K) and denser near the base at the white dwarf surface and so cools faster there than in the

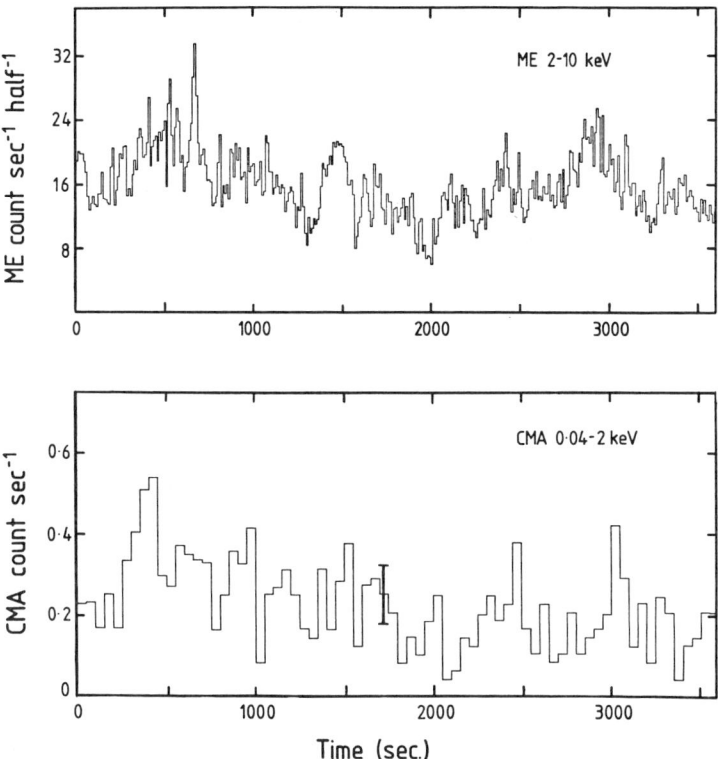

FIGURE 2. Hard and soft X-ray light curves of SS Cyg in quiescence obtained with EXOSAT. (Adapted from reference 21.)

hotter, more tenuous regions above. Thus, matter condenses at the base (ρ increasing) at the expense of the hotter zones, where ρ decreases slowly. This in turn means that the hard X-ray emission ($T \sim 10^8$ K) will decrease, while the soft X rays ($T \ll 10^8$ K) brighten. If a fraction, f, of the white dwarf surface is involved, the hard X-ray luminosity, L_x, is proportional to $4\pi R^2 f H \rho^2 T^{1/2}$. The reasoning above shows that ρ, H ($\propto T$), and T all decrease together in the X-ray emitting gas so that

$$L_x \sim T^{3/2+2\epsilon}, \quad \epsilon \text{ small.} \tag{5}$$

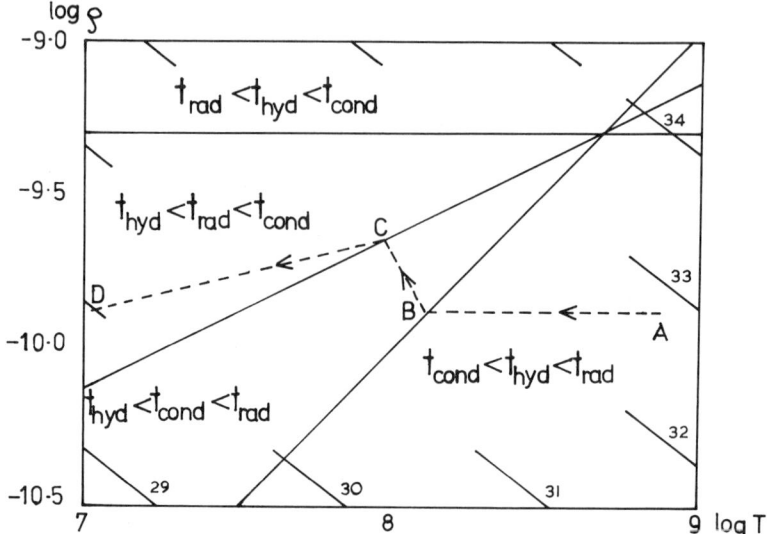

FIGURE 3. Cooling plane for coronal gas in a nonmagnetic CV. Moderate density gas follows tracks A-B-C-D. Shown are the lines, $L_x = constant$ (broken, for clarity), labeled by the value of log L_x. (Adapted from reference 20.)

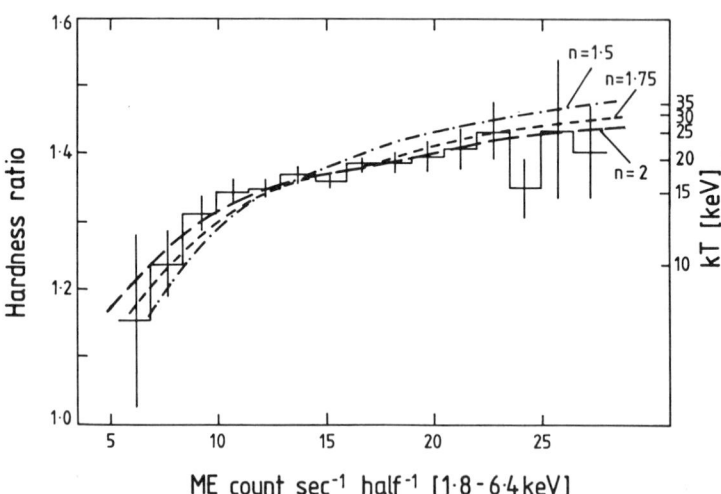

FIGURE 4. Mean "hardness ratio" = (3.3–6.4 keV flux/1.8–3.3 keV flux) for SS Cygni plotted against the 1.8–6.4 keV count rate. The equivalent bremsstrahlung temperatures are shown on the right-hand axis. The dashed curves show the relations $L_x \propto T^n$ folded through the detector response. (Adapted from reference 21.)

The derivation of this result given in reference 20 neglected velocity gradients in the quasi-hydrostatic gas, which is not in general justified. However, a full hydrodynamical calculation[22] confirms the result.

COMPARISON WITH OBSERVATIONS OF SS CYG

Let us see how the model suggested in the previous section compares with observation. First, a tendency for the X-ray spectrum of SS Cyg to harden with increasing luminosity was already suspected[23] before this model was developed; indeed, it was one of the main motivations for it. The detailed form of the correlation was unknown, however. FIGURE 4 shows the results from a recent EXOSAT observation.[21] Clearly, a correlation, $L_x \propto T^n$, with $n \simeq 2$, is in reasonable agreement with the data and

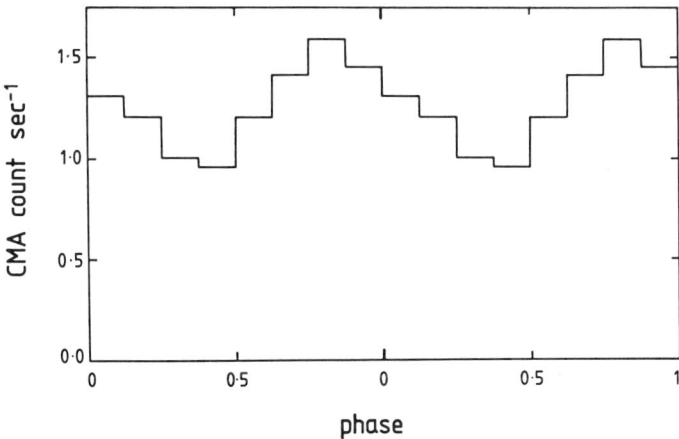

FIGURE 5. A quasi-periodic oscillation (period ~ 9.6 s) seen by EXOSAT in the soft X-ray emission of SS Cyg during the outburst of September 1984. A short string of data has been folded to produce a mean light curve.

it is just what is expected from equation (5). Note further from FIGURE 3 that the behavior of equation (5) is only expected in regimes characterized by $\rho \sim 10^{-10}$ g cm^{-3} and $T \sim 10^8$ K; thus, the X-ray variability timescale (few $\times$ 10 s) is correctly given by t_{rad} [equation (4)]. We can now reverse the argument given in the first paragraph of the previous section: the normalization required of the $L_x \propto T^2$ curve of FIGURE 4, together with the distance of ~ 100 pc to SS Cyg, now imply that the corona must cover a substantial fraction, $0.1 < f \lesssim 1$, of the white dwarf.

QUASI-PERIODIC SOFT X-RAY PULSATIONS

During their outbursts, both SS Cyg and U Gem show quite large amplitude ($\sim 30\%$) pulsations in soft X rays.[24,25] (See FIGURE 5.) These remain coherent (with periods of ~10 s in SS Cyg and ~27 s in U Gem) only for a few cycles, before the phase of the pulsation suffers an apparently random shift. Accordingly, the pulsations can show up in power spectra of short strings of data, but disappear if long strings are analyzed. Strikingly, there is no accompanying pulsation observed in hard X rays.[26]

It seems possible to understand this behavior in terms of the kind of coronal model suggested above.[27] We recall that the soft X rays largely result from conduction into the white dwarf from the corona. In the sun, a precisely analogous process gives rise to the so-called Ca^+ emission network. Because even very weak magnetic fields completely control the directionality of transport processes like conduction, this network is expected to map the surface magnetic field of the sun, a suggestion very strongly supported by observation. These magnetic fields themselves arise because of dynamo action, driven ultimately by differential rotation in the sun's outer layers. The white dwarf surface in a CV is subject to extremely strong torques because of the constant accretion, especially at outburst, of high angular momentum material at its equator. Thus, it seems quite reasonable to suppose that dynamo action will occur and that the resulting field structure will make the soft X-ray emission from the surface appear patchy. In the sun, large-scale bright spots appear in the Ca^+ network in solar active regions, where the new magnetic flux is thought to be generated. On an accreting white dwarf, such a spot, carried around by the star's rotation, will cause a periodic modulation in soft X rays. As the surface layers are subject to torques and not rigidly fixed to the core, this rotation period will be slightly variable. A quantitative analysis[27] shows that the soft X-ray spots are likely to last only a few rotation periods before disappearing, with a new spot erupting at some other, essentially random, location on the surface. This will naturally result in a random phase shift in the observed pulsation so that the main features of the observations are reproduced. In particular, no hard X-ray pulsation is expected since conduction and magnetic fields are quite irrelevant for the hard X-ray emission.

DISCUSSION

The last two sections show that the ideas of the sections entitled MODELS and THE WHITE DWARF CORONA do meet with some success in explaining a number of observational results, namely:

(i) the low efficiency of hard X-ray production in nonmagnetic CVs;
(ii) the switch from hard to soft emission at high accretion rates, not accompanied by absorption;
(iii) the temperatures, luminosities, and variability timescales of the emission;
(iv) the $L_x \propto T^2$ relation for SS Cyg;
(v) the quasi-periodic soft X-ray pulsations seen in SS Cyg and U Gem at outburst and possibly in other systems (see reference 27 for discussion).

However, it is important to bear in mind that detailed hydrodynamical calculations are still required, particularly of the expansion of the boundary layer flow and of some parts of the "cooling plane" (FIGURE 3). As in all accretion disc phenomena, the use of a turbulent viscosity is no doubt a vast oversimplification of what must be occurring in reality. These are difficult problems and again one must hope to be guided to some extent by observations. Here too, much remains to be done; in particular, only a small region of the cooling plane (FIGURE 3) is as yet sampled by observation, so repeated monitoring of bright sources will be fruitful. A further urgent necessity is to try to widen the observational base to include other systems: almost all of the data so far used have come from SS Cyg, with a little from U Gem; one would like to know how typical (if at all) these systems are.

A natural development will be to try to apply those ideas to accretion onto nonmagnetic neutron stars. Clearly, the very different physical processes (cooling, radiation pressure, two-fluid behavior, etc.) prevent a simple scaling. It is sobering, however, to realize how complicated even the comparatively simple case of accretion onto a white dwarf appears to be.

ACKNOWLEDGMENTS

As is obvious from the references, all of this work has involved collaborations with a number of people to whom I am greatly indebted. I would particularly like to thank M. Watson for help in preparing this review. The work with G. Shaviv was supported by a U.K. Science and Engineering Council Visiting Fellowship. Janet Mattei of the AAVSO alerted us to an outburst of SS Cyg and kindly supplied the optical light curve shown in FIGURE 1. I thank the Royal Society and the conference organizers for travel support.

REFERENCES

1. FRANK, J., A. R. KING & D. J. RAINE. 1985. Accretion Power in Astrophysics. Cambridge University Press. London/New York.
2. LEWIN, W. H. G. & E. P. J. VAN DEN HEUVEL. 1983. Accretion-Driven Stellar X-Ray Sources. Cambridge University Press. London/New York.
3. PRINGLE, J. E. 1981. Annu. Rev. Astron. Astrophys. **19**: 137.
4. LIEBERT, J. & H. S. STOCKMAN. 1985. In Cataclysmic Variables and Low-Mass X-ray Binaries. J. Patterson & D. Q. Lamb, Eds. Reidel. Dordrecht.
5. WARNER, B. 1983. Cataclysmic Variables and Related Objects. M. Livio & G. Shaviv, Eds.: 155. Reidel. Dordrecht.
6. KING, A. R. 1983. See reference 5, p. 181; KING, A. R. 1985. In Recent Results on Cataclysmic Variables. ESA SP-236, p. 133.
7. CHANMUGAM, G. & A. RAY. 1984. Astrophys. J. **285**: 252.
8. KING, A. R., J. FRANK & H. RITTER. 1985. Mon. Not. R. Astron. Soc. **213**: 181.
9. CORDOVA, F. A. & K. O. MASON. 1983. See chapter 4 of reference 2.
10. RICKETTS, M. J., A. R. KING, & D. J. RAINE. 1979. Mon. Not. R. Astron. Soc. **186**: 233.
11. SWANK, J. H., E. A. BOLDT, S. S. HOLT, R. E. ROTHSCHILD & P. J. SERLEMITSOS. 1978. Astrophys. J. **226**: 133; MASON, K. O., M. LAMPTON, P. A. CHARLES & S. BOWYER. 1978. Astrophys. J. **226**: L129.
12. CORDOVA, F. A. & K. O. MASON. 1984. Mon. Not. R. Astron. Soc. **206**: 879.
13. BATH, G. T. 1973. Nature Phys. Sci. **246**: 84.
14. MEYER, F. & E. MEYER-HOFMEISTER. 1981. Astron. Astrophys. **104**: L10.
15. FAULKNER, J., D. N. C. LIN & J. PAPALOIZOU. 1983. Mon. Not. R. Astron. Soc. **205**: 355.
16. SWANK, J. Quoted in reference 9.
17. PRINGLE, J. E. 1977. Mon. Not. R. Astron. Soc. **178**: 195.
18. PRINGLE, J. E. & G. J. SAVONIJE. 1979. Mon. Not. R. Astron. Soc. **187**: 777.
19. TYLENDA, R. 1981. Acta Astron. **31**: 127.

20. KING, A. R. & G. SHAVIV. 1984. Nature **308:** 519.
21. KING, A. R., M. G. WATSON & J. HEISE. 1985. Nature **313:** 220.
22. KING, A. R. & M. D. SMITH. 1986. Mon. Not. R. Astron. Soc. Submitted.
23. RICKETTS, M. J. & A. BENNETTS. 1982. Personal communication.
24. CORDOVA, F. A., T. J. CHESTER, I. R. TUOHY & G. P. GARMIRE. 1980. Astrophys. J. **235:** 163.
25. CORDOVA, F. A., T. J. CHESTER, K. O. MASON, S. M. KAHN & G. P. GARMIRE. 1984. Astrophys. J. **278:** 739.
26. SWANK, J. 1979. IAU Colloq. **53:** 135.
27. KING, A. R. 1985. Nature **313:** 221; HAMEURY, J. M., A. R. KING & J. P. LASOTA. 1985. Nature **317:** 597.

PART IX. STATUS REPORTS

The Hubble Space Telescope

NETA A. BAHCALL

Space Telescope Science Institute
Baltimore, Maryland 21218

INTRODUCTION

The Hubble Space Telescope (HST) will be the first long-lived international optical observatory in space. Its location above the obscuring and distorting effects of the earth's atmosphere will provide HST with unique capabilities that will yield a major improvement in observational optical astronomy. The main unique capabilities of HST are:

(i) High angular resolution: $\sim 0.1''$;
(ii) Faint stellar limiting magnitude: $\sim 28^m$;
(iii) UV observations: ≥ 1150 Å.

The telescope will be equipped with a complement of six scientific instruments, including cameras and spectrographs that will take advantage of HST's unique capabilities. (See the following section and TABLE 1). It is expected that some of the instruments will be changed after several years in orbit. HST will be placed in low earth orbit (500 km) by NASA's Space Transportation System (FIGURE 1) in 1986. The telescope can be maintained and repaired in orbit, including change of scientific instruments, as necessary, over at least fifteen years. Communications to and from HST will occur through the Tracking and Data Relay Satellite System (TDRSS), a geostationary communication satellite system. Because of its low earth orbit, most of HST observations will be preplanned.

The HST data received at Goddard Space Flight Center will be transmitted to the Space Telescope Science Institute (ST ScI) in Baltimore, Maryland, which has the responsibility for science operations with HST. The ST ScI will select scientific observations, prepare observing plans and commands to the spacecraft, receive, reduce, and archive the data, and distribute it to the observers. The HST and ST ScI will serve the international astronomical community, who will submit to the ST ScI proposals for specific scientific programs. All proposals will be peer-reviewed. A high oversubscription of the available observing time is expected.

THE TELESCOPE AND ITS SCIENTIFIC INSTRUMENTS

The HST consists of a 2.4-meter (94-inch) mirror, of Ritchey-Chrétien optical design, with an $f/24$ Cassegrain configuration (FIGURE 2). The image formed by this mirror at the focal plane is of extremely high quality. Seventy percent of a star's energy falls within a radius at the focal plane of less than 0.1 arcsec at 6328 Å. (For more details, see references 1 and 2.) The telescope, using appropriate scientific instruments, can cover the wavelength range from the far ultraviolet to the far infrared.

TABLE 1. Scientific Instruments[a]

Instrument	Field	Resolution	Band (Å)	Limit
Wide Field/Planetary Camera	2.7'; 1.2' □	0.1"; 0.04"	1150–11,000Å	28^m
Faint Object Camera	11"; 22" □	0.02"; 0.04"	1200–6000	28^m
Faint Object Spectrograph	0.1" to 4.3"	3Å; 30Å	1150–7000	$22; 26^m$
High Resolution Spectrograph	0.25"; 2"	0.03; 0.15; 1.5Å	1100–3200	$11; 14; 17^m$
High Speed Photometer	0.4"; 1"; 10"	16 μsec	1200–8000	24^m
Fine Guidance System	69' □	0.003"	4670–7000	17^m

[a] IDTs/PIs: Westphal; Macchetto; Harms; Brandt; Bless; Jeffreys.

The field of view is approximately 14 arcmin in radius. Its central part, with highest resolution, is viewed by five focal plane instruments. Three Fine Guidance Sensors (FGS) view three annular segments near the edge of the field. The central 2.7 × 2.7 arcmin are reflected off to the Wide-Field and Planetary Camera (WF/PC); four quadrants that are approximately 6 arcmin wide around this central area are used by the four axial instruments.

The total of six scientific instruments and their main characteristics are summarized in TABLE 1.

The Wide-Field and Planetary Camera (WF/PC) will have the largest field of view available on HST. Its field of view is covered by a mosaic of four charge-coupled devices (CCDs), totaling 1600 × 1600 pixels and covering the wavelength range from 1150 to 11,000 Å. The WF/PC can be operated in two modes corresponding to focal

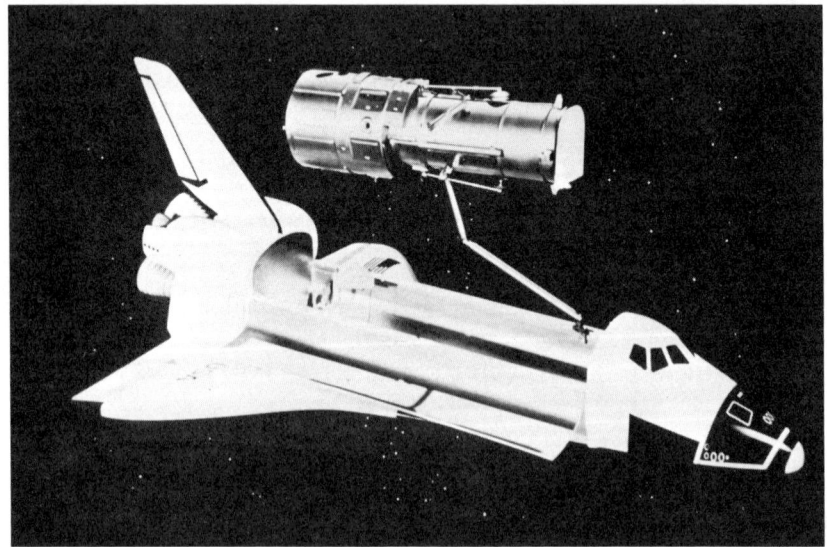

FIGURE 1. The space telescope being deployed by shuttle.

ratios of $f/12.9$ or $f/30$. The first yields a 2.7×2.7 arcmin² field of view with a pixel size of $0.1''$; the second has a 68.7×68.7 arcsec² field with a $0.043''$ pixel. The WF/PC is expected to detect very faint objects ($m_v \simeq 28$ in a one-hour exposure). It is provided with a large set of filters, gratings, and polarizers. It will be used effectively for deep survey observations and any imaging projects requiring a wide field of view.

The Faint Object Camera (FOC) will provide very high resolution images of small fields in order to fully exploit the optical capabilities of HST. Three scales are available with the FOC: $f/48$, with a field of view of 22.5×22.5 arcsec² and a pixel size of 0.044 arcsec; $f/96$, with a field of 11.2×11.2 arcsec² and a pixel of 0.022 arcsec; and $f/288$, with a field of 3.8×3.8 arcsec with a pixel of 0.0075 arcsec. A long-slit 0.1×20 arcsec may also be used with the $f/48$ mode to provide spectroscopy at resolution of 2000.

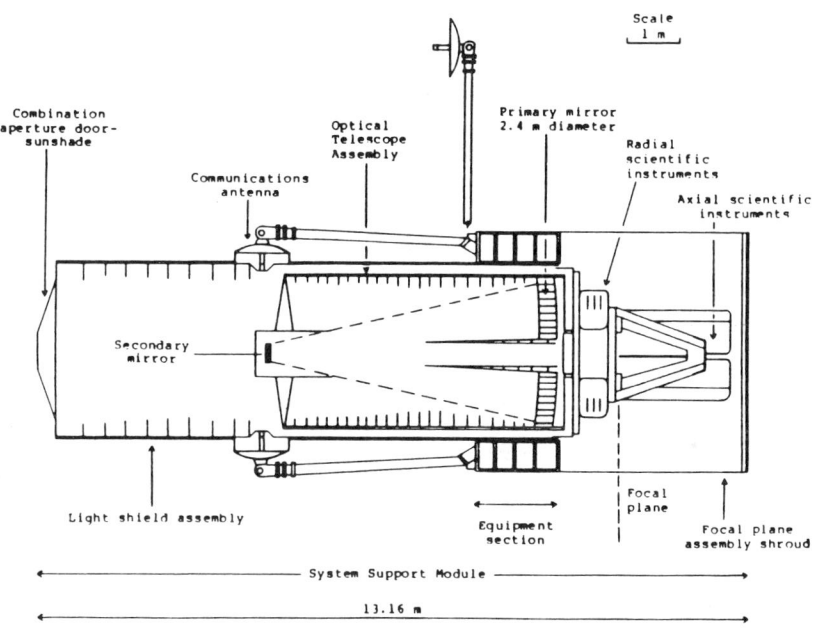

FIGURE 2. A space telescope conceptual drawing.

The Faint Object Spectrograph (FOS) will provide high efficiency spectroscopy (resolution $\simeq 250$ and 1300) of faint sources between 1150 and 7000 Å. An assortment of entrance apertures is available for isolating a desired portion of an extended object. A polarization analyzer is provided for linear and circular spectropolarimetry. The detectors can be used in a time-resolved mode, with integration times from 18 ms to ~ 100 s. The detectors are two 512-diode Digicons. It can reach $\sim 26^m$ in the low resolution mode and $\sim 22^m$ at high resolution.

The High Resolution Spectrograph (HRS) may be used for low, medium, and high resolution spectroscopy (resolution = 2000, 20,000, and 100,000, respectively). The

HRS is intended to be used strictly in the UV (1100 to 3200 Å). The detectors are 512-diode Digicons.

The High Speed Photometer (HSP) is designed to take advantage of the lack of atmospheric scintillation for a telescope in orbit, as well as to provide good UV performance. Integrations as short as 10 μs are possible over a broad wavelength range (1200–8000 Å) and polarimetry is also possible. A variety of filters, polarizers, and apertures are available.

The three Fine Guidance Sensors (FGS) are capable of measuring the relative positions of stars to 0.0016 arcsec accuracy over a 20 arcmin^2 field of view. In normal operations, two of the sensors are used for spacecraft attitude control, allowing the third to be used for astrometric measurements within its annular 4 × 16 arcmin field. Observing modes are provided for resolving close binaries (in the separation range of 0.01 to 0.1 arcsec with magnitude difference ≤2.5^m) and tracking moving targets, in addition to measuring parallaxes and proper motions.

Because of its low earth orbit and the limited connection time with TDRSS, only a small amount (≤20%) of real-time observations could be carried out on HST. Most observations will need to be preplanned months in advance and HST will be operated in a far more automated mode than most astronomical observatories. Most observations will be strung together into an observing sequence well in advance (six months) of the observations. This is done in order to optimize the efficiency of use of HST by minimizing slewing-time, instrument changes, and calibration times, and to take into account all viewing and orbital constraints.

SPACE TELESCOPE SCIENCE INSTITUTE

The ST ScI is an independent research institute operated by the Association of University for Research in Astronomy, Inc. (AURA) on behalf of NASA. The main responsibility of the ST ScI is the scientific planning and operations of the HST.

ST ScI will be responsible for:

(i) Managing the solicitation and selection of the observational program for HST;
(ii) Preparing observing plans and schedules;
(iii) Monitoring and calibrating the scientific instruments;
(iv) Providing technical support to observers;
(v) Receiving, calibrating, and archiving the scientific data;
(vi) Distributing the data to the observers (both general observers and archival researchers);
(vii) Providing facilities for the reduction and analysis of HST data;
(viii) Providing funding for U.S. observers.

SCIENCE WITH THE SPACE TELESCOPE

The high angular resolution and sensitivity of HST and its broad wavelength coverage will allow astronomers to perform scientific observations currently unfeasible with ground-based telescopes. Several articles and scientific symposia[1,3,4] have addressed the potential scientific returns to be expected from HST in the various fields of astronomy. Recently (1985), several scientific working-groups set up by the ST ScI and its Advisory Committee have investigated and recommended a dozen major scientific projects of fundamental importance to astronomy that could be addressed with HST.

The reports of the working-groups have been made public and their suggested projects are included in the larger, more general list provided below.

In this section, I list some projects that are of high scientific importance and that require the unique capabilities of HST (i.e., cannot be done from the ground). It should be emphasized that only projects that require high spatial resolution, faint stellar images, or UV observations will be accepted for HST observing. [For example, HST is not unique in reaching faint limiting magnitudes of extended objects (e.g., galaxies) unless high resolution is required.] The projects are divided by scientific subdisciplines.

Cosmology

(i) Determination of accurate distances to nearby galaxies and the Hubble constant. HST will provide an extension of a factor of ten in the distance over which properties of the standard candles can be measured.

(ii) Determination of the cosmological deceleration parameter, q_0. This fundamental determination may be attempted by observing the dependence of standard candles (such as the size of galaxies) with redshift to $z \sim 1$ or by using supernovae as standard candles (if they prove to be standard).

(iii) Confirmation of the expansion of the universe by studying the surface brightness dependence on the inverse of $(1 + z)^4$ to large redshifts.

Evolution

(i) The evolution of galaxies and clusters of galaxies with redshift to $z \sim 1$. Galaxy morphology determination at high redshifts will be a most exciting study with HST and will provide new insight into the nature of galactic evolution.

(ii) Evolution in QSO absorption lines. This will allow an understanding of the evolution of matter in the universe from nearby to very large redshifts.

QSOs and AGNs

(i) A study of the physics of the nuclear regions of QSOs and AGNs, including the central source and the line regions. High resolution imaging and spectroscopy will be uniquely suited to HST capabilities and they are expected to reveal exciting discoveries.

(ii) A study of the host galaxies around QSOs, which can be seen with high resolution only with HST.

(iii) QSO absorption line survey and high resolution observations in the UV.

(iv) Gravitational lenses. The high resolution structure and properties of these systems will be investigated.

Galaxies and Clusters

(i) The evolution of galaxies and clusters (see above).
(ii) Detailed properties of nearby galaxies using the unique high resolution and sensitivity of HST.

Stars and Interstellar Medium

(i) Stellar populations in our and other galaxies.
(ii) Detailed studies of globular clusters.
(iii) Detailed, high resolution studies of supernovae and supernovae remnants.

Planetary Astronomy

(i) Atmospheres, features, satellites, and rings of planets.
(ii) Primitive bodies in the solar system.
(iii) Companions of nearby stars.

Surveys

(i) Deep surveys of the sky will reveal faint stars, galaxies, supernovae, variable objects, and, most importantly, unknown types of objects.

ALLOCATION OF OBSERVING TIME

Observing time on HST will be allocated to openly solicited, peer-reviewed, and competitively selected proposals from the international astronomical community. During the first two and a half years of HST operations, the scientists associated with the development of HST and its instruments are guaranteed an average of 30% of observing time. Scientists from ESA member states will receive, on the average, at least 15% of the available time.

The total observing time on HST, assuming a 35% efficiency (due to earth occultation, SAA, and slewing) will be of the order of 3000^h per year. Since a high oversubscription is expected in the requested observing time, several guidelines aimed at optimizing the scientific program on HST have been adopted by the ST ScI following the recommendations of the Space Telescope Advisory Committee (STAC). In order to ensure a broad distribution of scientific possibilities and scopes of projects and to allow the conduct of some large projects if they are of fundamental importance to astronomy, approximately equal amounts of time will be allocated to small, medium, and large ($>100^h$ observing time) projects. Among the large projects will be several Key Projects identified by the community through scientific working-groups and the STAC. These Key Projects will be specifically identified in HST's Call for Proposals, and proposals to carry out these projects will be submitted by groups in the community. All proposals will be peer-reviewed for selection. The main selection criteria will be the scientific merit of the proposal and its need of HST's unique capabilities. Technical feasibility and demands made on HST and its resources will also be considered.

The review of proposals will be carried out by half a dozen scientific discipline panels, each reviewing and ranking proposals within their own discipline. A cross-discipline Time Allocation Committee will make the final recommendation of the combined science program to the ST ScI director. No a priori quota of observing time will be given to any discipline.

All data will be kept proprietary to the observers for a period of one year after the observations. At that time, the data will be placed in a public archive and will be available to interested scientists. It is expected that the HST public archive, with its enormous volume of astronomical data, will be a highly useful source of scientific investigations for many years to come.

REFERENCES

1. BAHCALL, J. & L. SPITZER. 1982. The Space Telescope. Sci. Am. **247**(1): 40–51.
2. O'DELL, C. R. 1980. The Space Telescope. *In* Annual Reviews Monograph: Telescope for the 1980s. G. Burbidge & A. Hewitt, Eds.: 129–193. Annual Reviews, Inc. Palo Alto, California.
3. HALL, D. N. B., Ed. 1982. The Space Telescope Observatory. IAU Commission 44 (NASA CP-2244), 18th General Assembly, Patras, Greece.
4. LONGAIR, M. S. & J. W. WARNER. 1979. Scientific Research with the Space Telescope. IAU Colloquium #54 (NASA CP-2111), Princeton, New Jersey.

The Advanced X-Ray Astrophysics Facility[a]

HARVEY TANANBAUM

Harvard-Smithsonian Center for Astrophysics
Cambridge, Massachusetts 02138

INTRODUCTION

The Advanced X-ray Astrophysics Facility (AXAF) is a sophisticated observatory designed to carry out X-ray observations from space, above the X-ray absorbing atmosphere of the Earth. Much of the design for AXAF has been predicated on the highly successful HEAO-2/Einstein X-ray observatory that NASA operated from 1978 until 1981. AXAF has been under study for several years and is now in the design/definition stage, Phase B. Two teams of industrial contractors have undertaken the study of the overall observatory to understand the requirements, to develop design concepts, and to formulate an overall technical basis for building AXAF. Proposals for scientific instruments have been received and reviewed, and the announcement of the payload selection is expected in early 1985. At this time (December 1984), AXAF is one of the leading candidates for a formal new start to begin construction in FY87, with a planned launch date sometime in 1991.

AXAF will be designed for a 15-year lifetime. AXAF, in orbit, along with the Hubble Space Telescope, the Gamma-ray Observatory, and the Space Infrared Telescope Facility, plus large, ground-based optical and radio telescopes and arrays, will constitute the primary facilities needed to carry out major astronomical programs for the rest of this century.

We now recognize the importance of X-ray observations in the study of our violent universe where explosions frequently accelerate particles to high energies and heat gases to high temperatures. Such explosions produce large amounts of X-ray radiation. Some objects are so hot they can be seen only in X rays, while others have radically different appearances at X-ray wavelengths.

AXAF will probe the nature of cosmic violence on all distance scales, from stars to superclusters of galaxies. It will address fundamental questions, such as the existence of stellar black holes; the existence of supermassive black holes in the centers of galaxies; the contribution of hot gas to the mass of the universe; the presence of dark matter in galaxies, clusters of galaxies, and superclusters; the age and size of the universe; the mechanisms by which particles are accelerated to high energies in stars; the validity of basic physical theory (strong interactions and general relativity) in neutron stars; and the details of stellar evolution and supernova explosions.

The capability of AXAF to carry out such broad objectives was key in obtaining the recommendation of the Astronomy Survey Committee (Field Committee) for

---

[a] This work was supported under NASA contract no. NAS8-32667.

AXAF as the number one priority, major new program for all of ground-based and space astronomy in the U.S. for the next decade.

In the next section, a brief overview is presented of the AXAF as an observatory. Then several specific examples of science projects are discussed to indicate some of the reasons for the widespread support of this program.

THE AXAF OBSERVATORY

A free-flying AXAF has been under study for several years now. FIGURE 1 shows the reference configuration developed by NASA's Marshall Space Flight Center and

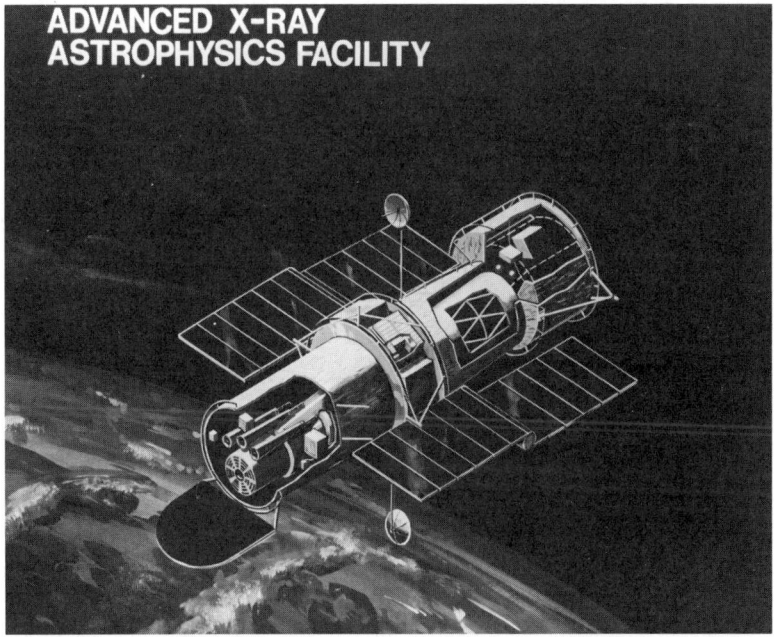

FRONTISPIECE. Artist's concept of the Advanced X-ray Astrophysics Facility (AXAF).

our group at the Smithsonian Astrophysical Observatory as a starting point for more detailed industry system studies. The X-ray telescope is indicated at one end with an array of scientific instruments at the other end, which are connected by a long metering structure or optical bench. Since only one focal plane instrument at a time can use the telescope, the observatory will require either a movable mirror or a mechanism to interchange instruments.

The overall length is 14 meters, the diameter is about 4 meters, and the weight is approximately 8000 kg. The figure shows articulated solar arrays, which are part of the 1200-watt electrical power system. There are antennae for communicating with the

Tracking and Data Relay Satellites (and ultimately the ground) and a spacecraft module that contains standard spacecraft subsystems hardware.

Much of the spacecraft hardware can be based on Space Telescope subsystems and/or multimission modular spacecraft (MMS) subsystems, as are appropriate. Many of the spacecraft requirements are relatively modest: for example, ≤ 64 kb/s telemetry, ~ 30 arc sec pointing accuracy (the telescope has a $1°$ field of view), and $2°$ C temperature control for the X-ray mirror. Other specifications such as the 0.5" X-ray telescope resolution and the 0.5" ex post facto aspect or pointing determination are somewhat more demanding, but still readily achievable.

AXAF is planned to have a 15-year lifetime; therefore, there will be a need for on-orbit servicing, maintenance, and repair. This same capability will also allow us to upgrade the AXAF scientific instrumentation to take advantage of technological

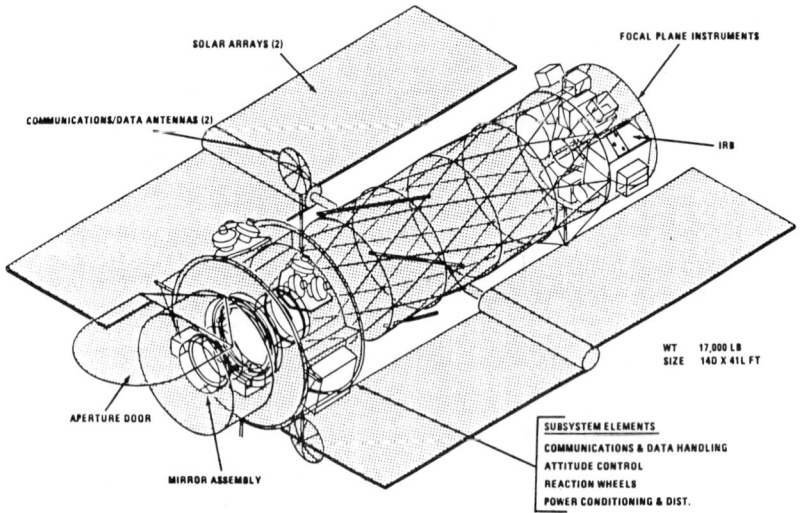

FIGURE 1. The AXAF reference configuration.

advances over its 15-year lifetime. Concepts and life-cycle costs are being analyzed with long-lifetime incorporated into the design from the beginning; this relies heavily on Space Telescope work already accomplished in this area. AXAF will be serviceable by the space shuttle and/or the space station (when it is available). FIGURE 2 shows one servicing approach using the shuttle, with AXAF shown docked on a special adapter fixture. It shows an orbital replacement unit being transferred by astronauts using the shuttle remote manipulator system (robot arm). Other specialized servicing equipment and tools are also indicated in the figure.

Perhaps the most technically challenging side to AXAF is the high resolution X-ray mirror. The AXAF design consists of six nested sets of grazing incidence X-ray mirrors, with the largest having a 1.2 m diameter (twice that of HEAO-2). FIGURE 3 shows the effective mirror area in cm^2 as a function of energy from 0.1 to 10 keV for

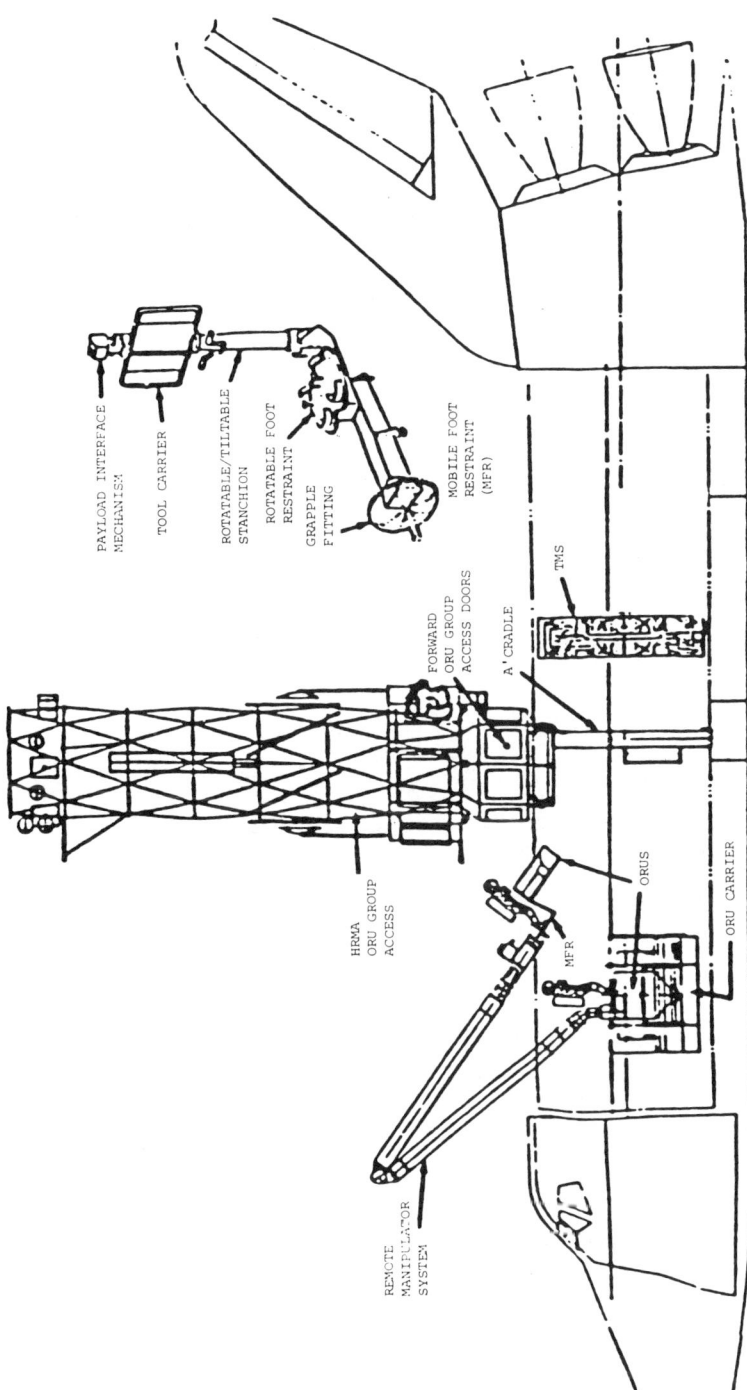

FIGURE 2. The AXAF Service Mission Concept using space shuttle. The space station will also be used when available.

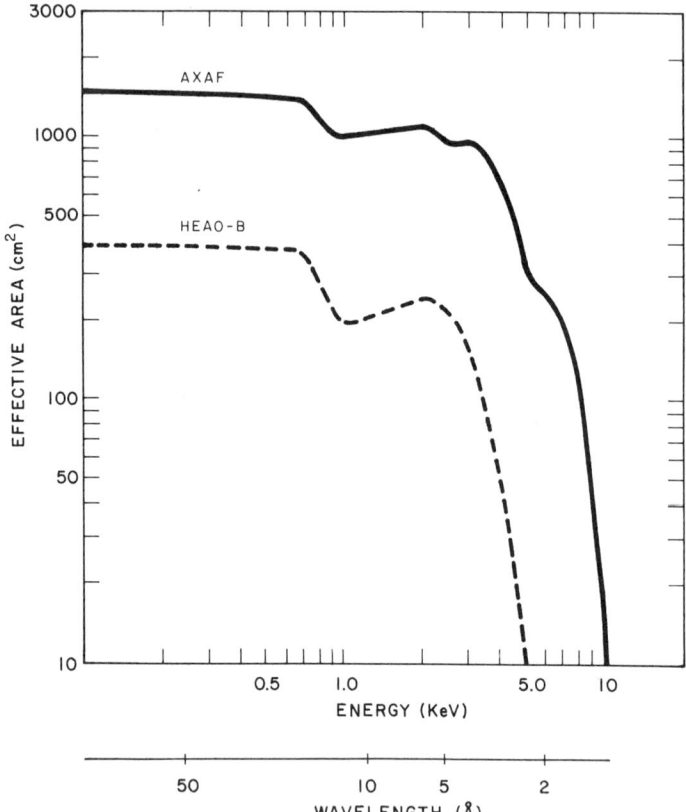

FIGURE 3. A comparison of the AXAF and Einstein Observatory on-axis effective collecting areas as a function of energy (wavelength).

AXAF and HEAO-2. At low energies, AXAF has about 1700 cm^2 or about four times the area of HEAO-2. The AXAF area still exceeds 1000 cm^2 at 3 keV. Probably the most significant difference results from the 10-m focal length of AXAF (versus 3 m for HEAO-2); this greatly increases the area at higher energies, extending AXAF beyond 8 keV with more than 100 cm^2 (whereas the HEAO-2 high energy cutoff was around 4 keV). This AXAF capability opens the opportunities for carrying out plasma diagnostics on iron emission and absorption features in the 6–7 keV range and also extends the energy band available for making broadband, continuum spectral measurements. It also is relevant to note that the ROSAT, scheduled to fly in 1987, will have about 1200 cm^2 at low energies, with a high energy cutoff at about 2 keV.

AXAF will be more sensitive than HEAO-2/Einstein by a factor of 50 to 100. The increased collecting area is one of three major factors that lead this increase in sensitivity. A second factor is the improved detector performance, particularly with higher quantum efficiency and lower background. The third factor is higher angular resolution due to increased requirements on mirror figure and surface smoothness.

FIGURE 4 shows the predicted AXAF performance and the actual HEAO-2 resolution at 2.5 keV by plotting the fraction encircled energy versus the radius in arc seconds. For AXAF, 60% of the signal is to be concentrated in a ½" radius, which is a factor of almost 20 times better than HEAO-2. At lower energies, where scattering is less important and slope tolerances or mirror figure determine performance, the AXAF FWHM is expected to be about seven times better than HEAO-2. The concentration of signal means that for point X-ray sources, the detection-cell size and its associated background can be greatly reduced, thereby increasing sensitivity.

We are presently constructing two pairs of X-ray test mirrors to demonstrate that the desired performances can be achieved. Optical data on the first set of mirrors indicate that this is, in fact, the case and we expect to verify this performance with X-ray test data in the spring of 1985.

Experience obtained in these technology programs will be used to set specifications, determine procedures, and establish realistic cost and schedules for the flight program.

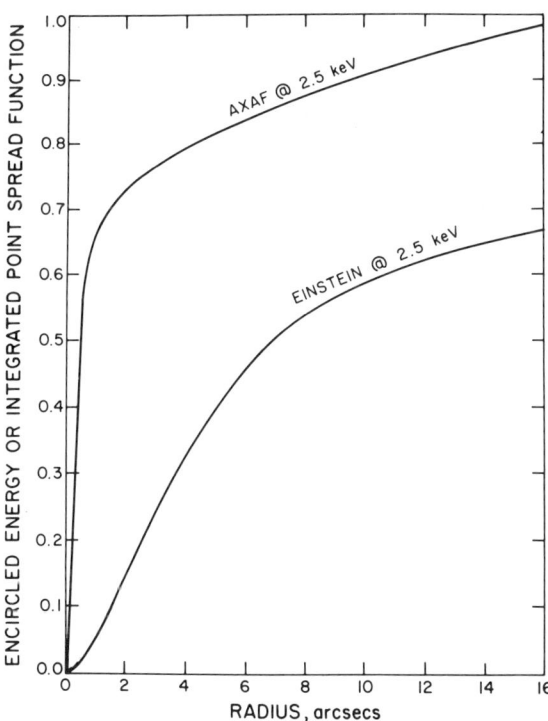

FIGURE 4. The integrated on-axis point spread functions for the AXAF and the Einstein Observatory for 2.5-keV photons.

The approach is intended to minimize risks in state-of-the-art technology areas, while providing assurance that required performance will be obtained.

AXAF SCIENTIFIC CAPABILITIES

The remainder of this paper covers in some detail four specific examples of fundamental scientific problems that AXAF will address. These are: (1) the use of AXAF to make mass determinations in galaxies and in clusters of galaxies; (2) to study galaxy evolution, the interactions of galaxies with their environment, and the properties of galactic halos; (3) to measure the distance scale independent of cosmological models and then to estimate the age of the universe and project its future; and (4) to accumulate complete, X-ray selected samples of stars, clusters of galaxies, active galactic nuclei, and quasars, as well as to search for new classes of objects, such as protogalaxies.

These four topics represent only a small fraction of the many areas of astrophysics for which AXAF will provide essential information. It should also be noted that the four are tilted to extragalactic astronomy with emphasis on imaging and broadband spectroscopic observations, while AXAF will of course also be a very powerful facility for carrying out galactic studies and for making high resolution spectroscopic observations as well.

Mass Determinations in Galaxies and Clusters of Galaxies

To illustrate the use of X-ray observations to determine mass in galaxies and clusters, we use the Einstein data obtained for M87, the giant elliptical galaxy near the center of the Virgo cluster of galaxies. (See references 1 and 2 for a detailed discussion of the M87 X-ray observations and analysis.) FIGURE 5 is a 0.3–4.0 keV X-ray contour map showing the smooth, extended, approximately azimuthally symmetric emission observed for M87. For an assumed distance of 15 Mpc to M87, 1' corresponds to 4.4 kpc. The map shows that the X-ray emission extends over a region of at least 130 kpc in diameter, a size considerably larger than the optically observed galaxy. Data from adjacent X-ray exposures show that the X-ray emission associated with M87 extends beyond a 60' radius or over a region at least ½ Mpc in diameter. The key point is that the X-ray emitting gas traces the underlying gravitational potential arising from the total mass present.

We follow the work of Fabricant, Lecar, and Gorenstein (1980)[1] to derive the equations used to determine mass. As seen in FIGURE 5, the data are roughly symmetric for different directions, thus allowing us to average azimuthally and to simplify the calculations to be a function of a radial coordinate only. (Note that this is not necessary in order to apply the method, in general.) The gas is assumed to be in hydrostatic equilibrium, with the outward directed force of thermal gas pressure gradient at radius, r, balanced by the inward directed gravitational force due to mass interior to r. This assumption is supported by the radial profile of the gas, which is significantly flatter than expected either for gas freely falling in or flowing out of the galaxy. It is also supported by the fact that the gas cooling time is much longer than the free-fall time and by the lack of temperature increase towards the center, which would be expected if the gas were settling or expanding adiabatically. This situation is described by equation (1):

$$\frac{dP_{gas}}{dr} = -\frac{GM(<r)\rho_{gas}}{r^2}, \tag{1}$$

where P_{gas} is the gas pressure, ρ_{gas} is the gas density, G is the gravitational constant, and $M(<r)$ is the total mass interior to the radius, r. We can also relate gas pressure,

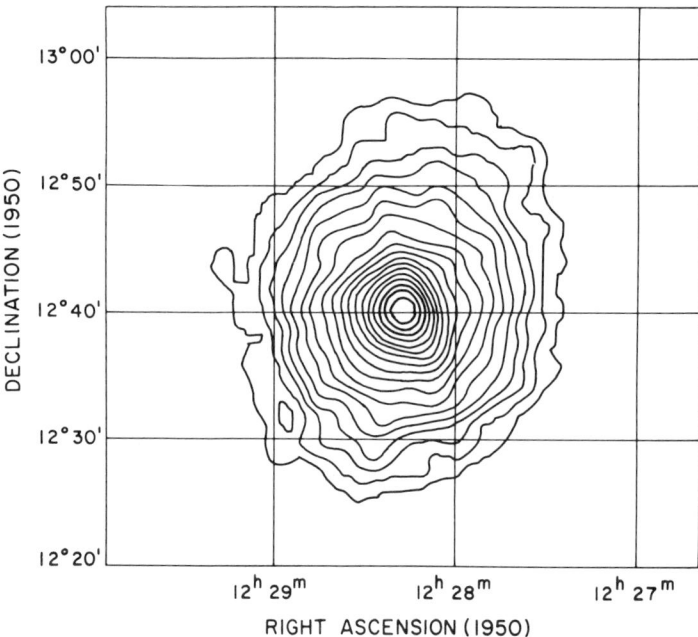

FIGURE 5. A 0.3–4.0 keV X-ray contour map made using data from the field centered on M87. Background has been subtracted and a correction has been made for the vignetting of the telescope optics.

density, and temperature through the ideal gas law:

$$P_{gas} = \frac{\rho_{gas} k T_{gas}}{\mu m_H}, \tag{2}$$

where μ is the mean molecular weight of the gas (0.6 here), m_H is the mass of the hydrogen atom, and k is Boltzmann's constant. Differentiating the second equation and substituting into the first and simplifying, we express $M(<r)$ in terms of the gas temperature and the logarithmic gradients of gas density and gas temperature:

$$M(<r) = -\frac{kT_{gas}}{G\mu m_H}\left[\frac{d\log \rho_{gas}}{d\log r} + \frac{d\log T_{gas}}{d\log r}\right]r. \tag{3}$$

The required quantities on the right-hand side of equation (3) can be determined from the X-ray observations.

A quantitative result obtained from the M87 Einstein X-ray image is presented in FIGURE 6, which shows the surface brightness (summed in azimuth) in erg cm^{-2} s^{-1} arcmin^{-2} versus angular radius in arc min. The data are shown along with representative error bars. The conversion from observed counts to 0.2–4 keV flux makes use of the average temperature measured and is not overly sensitive to the value used. An analytic expression, introduced by Cavaliere and Fusco-Femiano (1976)[3] to describe a hydrostatic, isothermal gas in a spherical gravitational potential, can be conveniently fit to the observations as shown by the solid curve and represented by:

$$S(r) \propto [1 + (r/a)^2]^{-n}. \qquad (4)$$

A χ^2 fit to the data determines $a = 1\rlap{.}'62 \pm 0\rlap{.}'28$ and $n = 0.81 \pm 0.01$ (90% error limits). Note that an analytical fit is not really necessary since a numerical analysis also can be

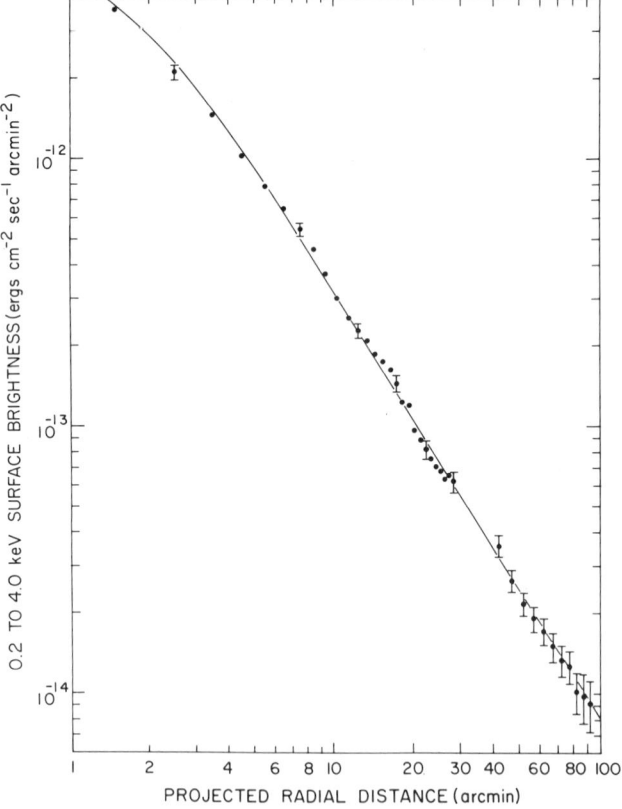

FIGURE 6. The 0.2–4.0 keV surface brightness profile of M87. Background has been subtracted and a correction has been made for the vignetting of the telescope optics.

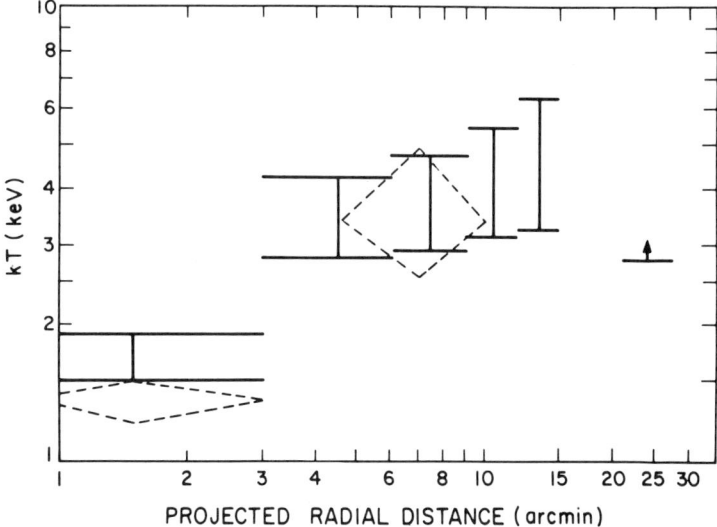

FIGURE 7. The temperatures (and errors) derived by isothermal fits of Raymond and Smith (1977,[5] 1982[6]) thermal spectra to the IPC radially binned spectral data for M87 are shown as solid lines. No correction has been made for projection effects. Data from the solid-state spectrometer[4] are shown for comparison (dashed diamonds).

carried out. The surface brightness profile, $S(r)$, is then deprojected to determine the density profile, $\rho_{gas}(r)$, which when folded through the telescope and detector would produce the observed surface brightness profile. For these data, an isothermal gas profile is sufficiently accurate, but in a more general case the temperature versus radius can affect the deprojection and an iterative process may be needed. The resultant gas density profile is given by

$$\rho_{gas}(r) \propto [1 + (r/a)^2]^{-(n+1/2)/2} \quad (5)$$

and the logarithmic density gradient is computed from this expression as

$$\frac{d \log \rho_{gas}}{d \log r} = \frac{-(n + \frac{1}{2})(r/a)^2}{1 + (r/a)^2}. \quad (6)$$

These expressions are evaluated for M87 using the values of n and a given above.

FIGURE 7 presents the measured temperature data versus the radius observed for M87 with two Einstein instruments. The dashed diamonds give the results for the nonimaging solid-state spectrometer (from reference 4) and the solid lines give the temperature and the 90% confidence limits obtained with the imaging proportional counter.[2] In overlapping regions, the agreement is quite good. Data in the inner few arc minutes show evidence for cooling, but while this is interesting scientifically, it is not very relevant here since most of the mass is found at larger radii. The available temperature data extend out to an almost 30′ radius and suggest that temperatures of ~3–4 keV can fit the data. In any case, these data are used to set limits on temperature gradients,

$$-0.4 \le \frac{d \log T_{gas}}{d \log r} \le 0.3,$$

which can then be used in equation (3) along with the density gradient to determine $M(<r)$.

The result for the gravitational mass (in solar units) interior to the radius, r, is plotted versus r (in arc min) in FIGURE 8. The optical measurements of Sargent et al. (1978)[7] extend to ~1' and find a few $\times 10^{11}$ M_0 in the central part of the galaxy, but the velocity-dispersion data provide little information at larger radii. The X-ray data constrain the total mass to lie between the two solid lines shown in FIGURE 8, which incorporates the uncertainties in the measured gradients and the temperature normalization. At $r = 50'$ or about 220 kpc, the mass interior to r is $(3-5) \times 10^{13}$ M_0 or more than 100 times that seen in the visible galaxy. The X-ray emitting gas, itself, accounts for ~3% of the mass.

Most of the mass is not radiating in the X-ray or the optical bands. Therefore, it can be called dark and its nature is still unknown. An important point to note is that the

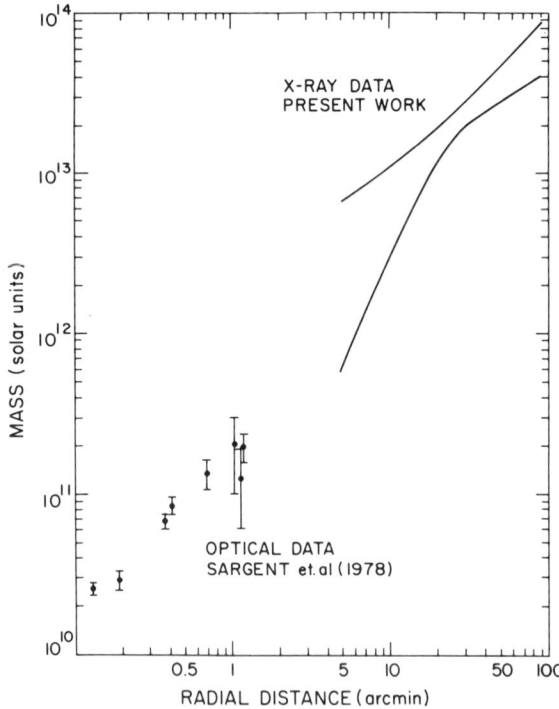

FIGURE 8. The mass of M87 as a function of radius, summarizing both optical and X-ray results. The optical mass measurements extend to ~1' and are derived from velocity dispersion measurements.[7] The allowed mass range derived from the X-ray data is bounded by two solid lines.

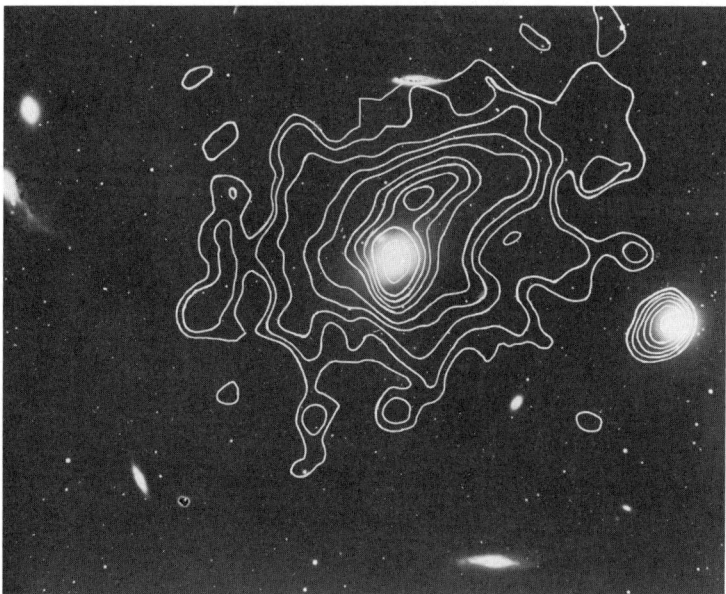

FIGURE 9. IPC X-ray contours superimposed on an optical photograph (from the Kitt Peak 4-m telescope) containing the galaxies M86 and M84 in the Virgo cluster.

X-ray gas traces or illuminates the underlying potential. This analysis could be carried out using Einstein data for M87 because it is a bright and relatively cool X-ray source. However, many clusters have higher gas temperatures (above the high energy cutoff of Einstein) and require the added bandwidth coverage of AXAF. With its increased area and bandwidth, AXAF can greatly improve on measurements of surface brightness and continuum temperature distributions, which will extend these studies to much fainter and more distant objects.

This technique can be applied to galactic halos and to groups and clusters of galaxies to determine distributions as well as total amounts of matter present. This will be useful for determining the distribution of dark matter on different size scales, which can provide critical information for choosing between hot, low-mass particle candidates and cold, massive particle candidates for the composition of the dark matter.

Galaxies

X-ray observations can provide unique insights into galaxies, revealing their evolution, their interactions with their environment, and their underlying mass distributions. FIGURE 9 shows a visible light picture of several bright galaxies (the complex extended white objects) in the central or core region of the Virgo cluster of galaxies. (These galaxies are about 1° away from M87.) We note the rather similar optical appearance of the two bright, early-type or elliptical galaxies, M86 and M84. (M86 appears slightly larger and slightly brighter than M84.)

Superimposed on the optical images of M86 and M84, we show the Einstein X-ray contours obtained by Forman et al. (1979).[8] The X-ray images of M86 and M84 are quite different in appearance. Note the contours show a very extended, asymmetric plume of X-ray emission for M86, which is about six times more luminous in X rays than M84. On the basis of these X-ray data (plus optical radial velocities), we believe that M86 has spent $\gtrsim 10^9$ years away from the core and has accumulated gas via stellar evolution. M86 has recently entered the cluster core and encountered the dense cluster gas already there. Ram pressure forces are stripping the reservoir of gas that M86 had accumulated and held. In contrast, M84 probably spends most or all of its time moving at a relatively high velocity in the dense cluster core and is continuously being ram pressure stripped of its gas. Thus, M84 has no opportunity to build up a large gaseous halo that could be suddenly stripped to create a plume like M86. This is one way in which the X-ray data can be used to trace a galaxy's history and interpret its interaction with its environment.

These gaseous halos were discovered early in the Einstein program and several more were suggested for galaxies in the more distant cluster A1367.[9] However, separation of galaxy and cluster properties required data on field galaxies before we could understand fully the importance of these observations.

The critical data for field galaxies are summarized in a recent paper by Forman, Jones, and Tucker (1985).[10] In FIGURE 10, Einstein X-ray contours are shown superimposed on the optical picture of the elliptical field galaxy NGC1395. The X-ray emission is clearly extended over a radial distance of at least 10', corresponding to 100 kpc. (The instrument response of ~1' can be seen for a point source to the west of NGC1395.) The X-ray emission corresponds to a 0.5–4.5 keV luminosity of about 10^{41} erg s^{-1} (approximately one-half that observed for M86) and the X-ray emitting corona has a temperature of between 0.5 and 0.9 keV. The mass of the X-ray emitting gas is about $2 \times 10^{10} M_0$. As for M87, the X-ray gas can be used to study the underlying potential. The flatness of the gas radial profile again suggests hydrostatic equilibrium is valid and when combined with the assumption of an isothermal temperature distribution, this yields $M_{total} \approx 5 \times 10^{12} M_0$ out to a radius of ~100 kpc. This is from five to ten times the mass accounted for in the visible galaxy. Once again, we find much more unseen or dark mass.

This is a general Einstein result for elliptical field galaxies. They have massive, dark halos that can be traced by X-ray emitting gas. This is very important since the lack of radio emission at large radii in ellipticals prevents HI rotation curves from being used to search for massive halos (which is the method often used for spiral galaxies). Discovery of these X-ray emitting coronae shows that, contrary to previous beliefs, early-type galaxies do not currently have substantial galactic winds. The data suggest that coronal gas can be provided by current stellar evolution (although much more could have been produced by a very early generation of stars and although that gas probably did escape via a galactic wind). The observed coronae can be powered by supernova explosions, although more detailed calculations are needed in this area. Gaseous material lost from early phases of stellar evolution in elliptical field galaxies could provide an intergalactic medium with a density about 10^{-8} cm^{-3}, far short of the amount of matter needed to close the universe, but nonetheless a significant amount of material. Typical ratios of total mass to luminous mass for elliptical field galaxies are comparable to values for clusters of galaxies, implying that cluster dark matter may have originated in halos of member galaxies.

These studies were just begun with Einstein. AXAF can greatly refine them for nearby galaxies by obtaining more precise surface brightness and temperature profile measurements (which are required for accurate measurements of the distributions and total amounts of halo material). AXAF can actually detect coronae such as these out to distances comparable to 3C273 ($z \sim 0.15$), although detailed maps and analyses will probably be limited to objects within a few hundred Mpc. This represents a volume of space ~ 500 times larger than could be studied with Einstein.

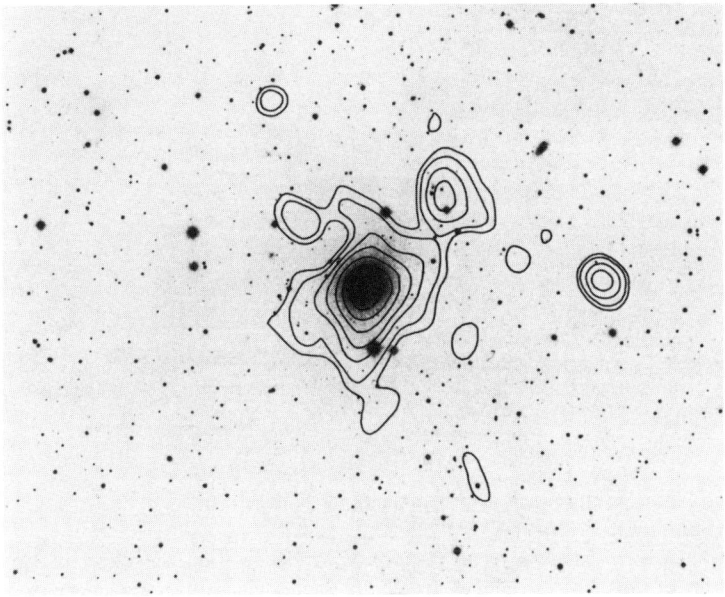

FIGURE 10. Isointensity plots of the X-ray emission are shown for NGC1395. The emission is smoothed with a Gaussian of 45". Contour levels are 2.85E-4, 3.65E-4, 5.41E-4, 6.33E-4, 8.30E-4, 1.28E-3, and 1.78E-3 counts/sec/square arcmin for significance levels, 3σ, 4σ, 5σ, 6σ, 10σ, 14σ, and 20σ, respectively.

Cosmological Measurements

AXAF can be used to make cosmological measurements via the Sunyaev-Zel'dovich effect, combining X-ray and microwave observations of clusters of galaxies to obtain cosmology independent distance determinations. This is schematically indicated in FIGURE 11.

X-ray observations provide measurements of cluster temperature, T_E, X-ray flux, F_x, and core radius, θ_c. Redshifts can be determined from optical or X-ray data. The linear core radius, R_c, can be expressed in terms of the angular radius times the distance. Here D_L is the luminosity distance. Note the $(1 + z)^{-2}$ factor relating the angular diameter distance to the luminosity distance. Clusters emit by thermal bremsstrahlung and therefore luminosity (given by flux times 4π times the distance

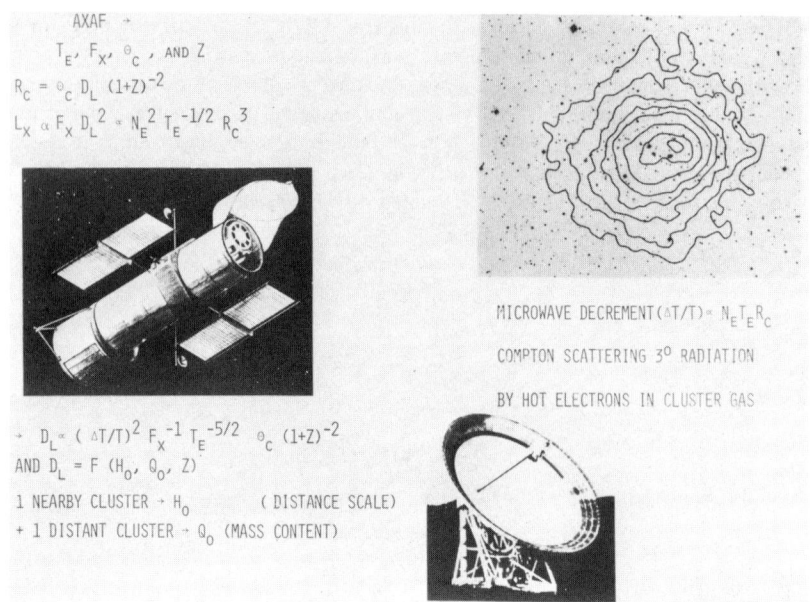

FIGURE 11. Cosmological measurements that can be made with AXAF using the Sunyaev-Zel'dovich effect. Independent distance determinations can be arrived at by combining X-ray and microwave observations of clusters of galaxies.

squared) can be determined as proportional to the electron density squared times the volume times $T_E^{-1/2}$.

Ground-based microwave observations provide data on the effect of Compton scattering of the 3° background radiation by the hot electrons in the cluster gas, with the microwave decrement ($\Delta T/T$) proportional to the product of electron density times electron temperature times radius. These observations then can be combined to determine the distance to the cluster, as is indicated in FIGURE 11. The key element in the use of this method to determine absolute distances is the different functional

TABLE 1. Analysis of Three Clusters Used for Distance Determination[a]

Cluster	z	AXAF Exp Time	1σ Fractional Errors			Fractional Distance Error[b]			
						X-Ray Only	Total		
			$F_{x/j}$	T_E	θ_c		Long λ	2mm	1mm
A2256	.060	10^4	.009	.016	.009	.034	.149	.112	.054
A2218	.174	3×10^4	.014	.022	.012	.047	.144	.111	.061
CL0016 + 16	.541	10^5	.022	.031	.022	.064	.143	.115	.074

[a]The current results[12] at 1.5 cm are $\Delta T = -1.58 \pm 0.26$ mK for 0016 + 16 and $\Delta T = -0.51 \pm 0.14$ mK for A2218. They need an improvement of a factor of four to eight in microwave measurements plus AXAF to obtain the distances and Hubble constant to ≤15%. At $z = 1.0$, $D_L(q_o = 0.5)/D_L(q_o = 0.0) \sim 0.8$, which needs accurate measurements at mmλ.

[b]This assumes microwave beam FWHM = 2.4 and microwave decrement error of 10^{-5} microwave background.

dependence of density times radius in the two separate measurements—X-ray and microwave.

Once we have cosmology independent distance measures, the results can be compared to values obtained for a particular cosmological model, such as a Friedmann Universe characterized by Hubble constant, H_o, and deceleration parameter, q_o (and cosmological constant = 0). The data on one cluster can give H_o and, in principle, for one more distant cluster, q_o as well. This provides key information on the size, age, and ultimate fate of the universe.

What factors are likely to limit the precision of these measurements? Van Speybroeck (1984)[11] has carried out a thorough analysis to assess this question via simulations based on Einstein observations and currently available microwave data. The results are summarized in TABLE 1. Three clusters are analyzed at redshifts of 0.060, 0.174, and 0.541, respectively, with corresponding planned AXAF exposures of 10^4, 3×10^4, and 10^5 seconds. The table lists 1σ measurement errors for the X-ray flux normalization, $F_{x/j}$ (1–2%), temperature, T_E (1.6–3.1%), and core radius, θ_c (1–2.2%). The next column gives the predicted fractional contribution to the distance measurement uncertainty (1σ) from the AXAF X-ray data only (3.4–6.4% for these three clusters). In this analysis, the distance actually scales with $1/T_E^2$ and this factor contributes most to the X-ray uncertainty. (The additional factor of $T_E^{1/2}$ is subsumed in the flux normalization.) With a microwave beam FWHM of 2.4' and a decrement measurement uncertainty of 10^{-5} times the background itself, we can expect to measure distances to 5–15% accuracy of 1σ. The results depend on the frequency for the microwave measurements, with more precise results obtained if the microwave data are at shorter wavelengths, as is shown in the last three columns of TABLE 1.

The best microwave data currently available[12] show that the decrements are real (results significant at several σ level), but measurement uncertainties up to four to eight times smaller still are needed. Private discussions indicate that these are feasible at long wavelengths (cm). Then we can expect to determine distances and H_o to 15% precision. (If high precision microwave measurements can eventually be made at 1 mm, then 5% precision can be obtained for distance determinations.)

At $z = 1$, the difference in distance between $q_o = 0.5$ and $q_o = 0.0$ is about 20%, so either we need data for some clusters at $z \geq 1$ along with mm data or we need to use a number of clusters to reduce the overall statistical uncertainty to obtain useful estimates of q_o. Estimates of q_o are, of course, needed to carry out cosmological calculations of the age of universe [$1/H_o$ versus ($2/3$) ($1/H_o$) for $q_o = 0.0$ versus $q_o = 0.5$] and to predict the ultimate fate of the universe (open for $q_o < 0.5$ versus closed for $q_o > 0.5$).

The simulations show the accuracy to which AXAF can make the needed X-ray observations and they indicate the directions and limits to which the microwave observations should be pursued. These are very important measurements because they allow us to establish the distance scale independently from cosmology and then to determine values for cosmologically interesting parameters.

Deep Surveys

AXAF also can be used to carry out deep X-ray surveys. Some results can be anticipated on the basis of Einstein observations. The result of a half-day effective exposure with Einstein to a previously blank or source-free region of the X-ray sky is

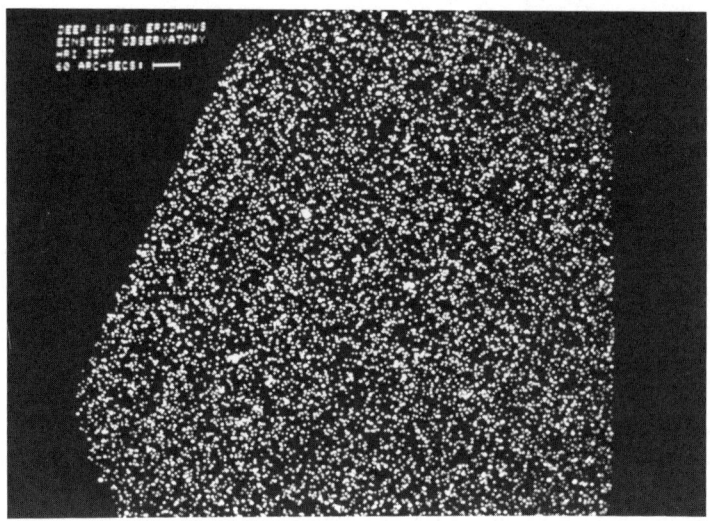

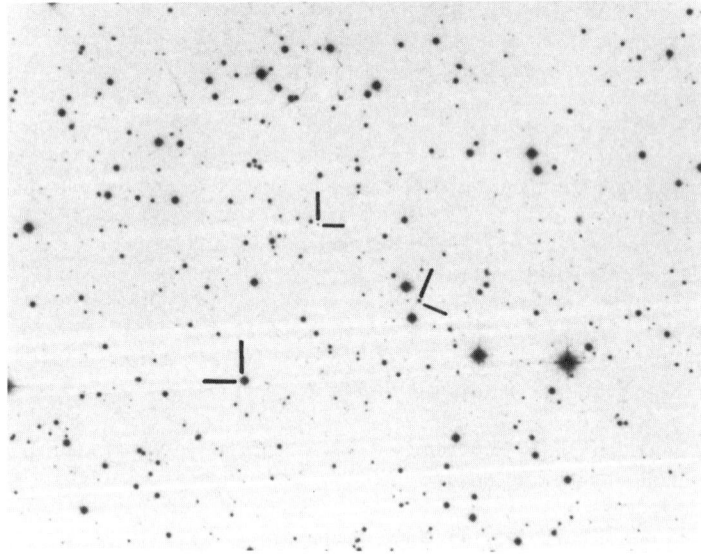

FIGURE 12. Top figure: An HRI image obtained during Einstein deep X-ray survey of field in Eridanus. Three sources—two quasars and one star—are visible in the X-ray data. Bottom: 48" Schmidt plate showing the visible light photograph corresponding to the X-ray exposure of the top figure. The optical counterparts of the three X-ray sources are indicated.

illustrated in the upper half of FIGURE 12, which shows three discrete sources (in a portion of the field of view) superposed on a low level detector background (partly cosmic ray induced, partly internal to the detector, and partly diffuse sky). These three sources had not been previously reported because they are about 100 times fainter than could be detected by instruments available prior to the Einstein observations.

The optical counterparts to these three X-ray sources are indicated in the lower half of FIGURE 12 by the three arrows. Among the hundreds of optical objects in the same field, Einstein picked out a 13th magnitude G-star with an X-ray corona similar to that of our own sun and two faint, previously uncataloged quasars—one an 18th magnitude, $z = 2$ quasar and the other a 20th magnitude, $z = 0.5$ quasar. This is a typical result since Einstein deep surveys are comprised of $\sim 1/3$ galactic stars and $2/3$ extragalactic objects, which are most often active galactic nuclei and quasars. These results indicate the power of X-ray observations for selecting quasars from among the many objects present in a given region of the sky.

With AXAF, we also will carry out planned deep surveys (as well as collect serendipitous sources found in long exposures) to reach fluxes 50–100 times fainter than Einstein. With these, we can obtain complete samples of active galaxies, quasars, clusters of galaxies, and various classes of normal stars, once we have identified the sources with their optical counterparts. For extragalactic objects, we can study luminosity and evolution functions, and we may be able to reach the epoch of cluster formation since Coma-like clusters can be detected to $z \sim 2$ with AXAF. Prospects also exist for discovering new classes of objects such as protogalaxies, which may be luminous X-ray sources powered by high rates of supernova explosions.

AXAF sensitivity should allow us to reach fluxes where 500–1000 sources per square degree can be expected. The all-sky X-ray background may be totally resolved into individual sources or else strong evidence for a truly diffuse component may be established. Einstein data directly show that discrete sources constitute at least 35% of the 2-keV background and suggest that quasars contribute from 50 to 100% of the total background (but optical models and extrapolations are required to obtain this estimate). AXAF also can provide much needed spectra for faint sources in the 2–6 keV band; the few presently available quasar X-ray spectra do not match the observed spectrum of the X-ray background.

SUMMARY

TABLE 2 lists a sampling of fundamental questions that AXAF will address and it indicates very briefly the basis for our expecting AXAF to provide the needed insight and information. Several of these questions have been discussed in some detail in the preceding sections; discussion of others is available in various NASA documents (such as the report of the AXAF Science Working Group: NASA TM-78285).

X-ray observations are particularly sensitive to violent events and allow us to study explosions, extremely high magnetic fields, energetic particles, and hot gases on many different scales. AXAF will be a powerful X-ray observatory that, with the Hubble Space Telescope, the Gamma-Ray Observatory, the Space Infrared Telescope Facility, ground-based radio arrays, and large optical-infrared telescopes, will enable us to penetrate deeply into the universe and study objects over the entire electromagnetic spectrum.

TABLE 2. Fundamental Questions That AXAF Will Address

(1) Are quasars powered by accretion onto supermassive black holes?
Time variability of X-ray emission determines source size and efficiency of energy production for comparisons with predictions from black hole models.

(2) Is a substantial fraction of the matter in the universe in the form of superheated ($\sim 10^9$ K) gas?
The X-ray background that is possibly due to such gas will be resolved into discrete sources if they, rather than hot gas, are responsible.

(3) Are there large amounts of dark matter hidden in galaxies and clusters of galaxies, and is the dark matter distributed according to predictions of grand unified theories of particle physics?
The distribution and temperature of X-ray emitting gas in the halos of galaxies and in the clusters of galaxies directly measure the gravitational potential and hence the total amount of matter (both visible and dark) on scales from kiloparsecs to megaparsecs. Different versions of grand unified theory predict different distributions of dark matter on different scales.

(4) Did galaxies form first and clusters later (as predicted by grand unified theories if most of the matter in the universe is in the form of "cold" particles like axions or photinos) or vice versa (as predicted if the matter is made of "hot" particles like low-mass neutrinos)?
The X-ray emission by intracluster hot gas permits its distribution, density, temperature, and chemical composition to be determined as a function of redshift. Different theories of galaxy formation predict different results.

(5) Do galaxies form at redshifts of 3 to 10?
Detection of the X-ray emission from gas heated by supernova explosions predicted to take place in newly formed galaxies may provide the best way of finding such objects.

(6) Is the lack of exploding galaxies (QSOs) early in the evolution of the universe indicated by optical studies real?
AXAF can detect QSOs out to $z = 5$–10, selecting rare objects for detailed follow-up observations and identifications.

(7) What is the age and ultimate fate of the universe?
Simultaneous X-ray and microwave observations of hot gas in clusters will permit determination of both the Hubble constant, H_o, and the deceleration parameter, q_o, and thus, the age of the universe. Measurements of q_o will provide information on the ultimate fate (open or closed) of the universe.

(8) Is solar activity, important for life on earth, due to a self-excited dynamo in the sun, and if so, how does this dynamo work in detail?
Virtually all later-type (F,G,K, and M) stars exhibit X-ray emission characteristic of activity like the sun's. By studying the dependence of such emission upon the mass, angular momentum, and age of the star, we can develop acceptable theories of the solar dynamo.

(9) Are the masses, radii, and interior structures of neutron stars in conflict with theoretical models based upon nuclear physics and general relativity?
Study of X-ray binaries yields orbital parameters (hence masses) and spin-up rates (which depend on internal structure). Observations of X-ray spectral lines yield gravitational redshifts and hence radii. Recent results for radii seem to be in conflict with theoretical models.

(10) Do low-mass or "brown dwarf" stars contribute a major part of the mass of globular clusters and of galaxies?
Detection of coronal X-rays from such stars may provide a sensitive way for finding them; they are believed to be numerous and may contribute to a substantial fraction of the mass.

(11) What occurs during supernova explosions and what are the properties of the progenitor stars?
The X-ray emission from the different layers of material in young shell-like supernova remnants enables us to deduce the chemical composition of the ejected material as well as the total mass of the pre-supernova star.

(12) Do all supernova explosions leave a neutron-star remnant?
In some supernova remnants, X-rays have been detected from neutron stars—sometimes as pulsed radiation and sometimes not. Absence of such radiation in young remnants could imply the absence of a neutron star, suggesting that a black hole formed instead or that the star was totally disrupted.

ACKNOWLEDGMENTS

The author wishes to acknowledge helpful discussions with his colleagues in preparing this talk, especially M. Weisskopf, W. Forman, C. Jones, G. Field, L. Van Speybroeck, M. Zombeck, and M. Birkinshaw.

The author wishes to thank D. Fabricant and P. Gorenstein for their generous permission to reproduce FIGURES 5–8 and W. Forman, C. Jones, and W. Tucker for FIGURE 10. The *Astrophysical Journal* also kindly consented to the use of these figures.

REFERENCES

1. FABRICANT, D., M. LECAR & P. GORENSTEIN. 1980. Astrophys. J. **241**: 552.
2. FABRICANT, D. & P. GORENSTEIN. 1983. Astrophys. J. **267**: 535.
3. CAVALIERE, A. & R. FUSCO-FEMIANO. 1976. Astron. Astrophys. **49**: 137.
4. LEA, S., R. MUSHOTZKY & S. HOLT. 1982. Astrophys. J. **262**: 24.
5. RAYMOND, J. C. & B. W. SMITH. 1977. Astrophys. J. Suppl. Ser. **35**: 419.
6. RAYMOND, J. C. & B. W. SMITH. 1982. Private communication.
7. SARGENT, W. L. W., P. J. YOUNG, A. BOKSENBERG, K. SHORTRIDGE, C. R. LYNDS & F. D. A. HARTWICK. 1978. Astrophys. J. **221**: 731.
8. FORMAN, W., J. SCHWARZ, C. JONES, W. LILLER & A. C. FABIAN. 1979. Astrophys. J. Lett. **234**: L27.
9. BECHTOLD, J., W. FORMAN, R. GIACCONI, C. JONES, J. SCHWARZ, W. TUCKER & L. VAN SPEYBROECK. 1983. Astrophys. J. **265**: L27.
10. FORMAN, W., C. JONES & W. TUCKER. 1985. Astrophys. J. **293**: 102.
11. VAN SPEYBROECK, L. 1984. SAO Proposal P1394-2-84 to NASA.
12. BIRKINSHAW, M., S. F. GULL & H. HARDEBECK. 1984. Nature **309**: 34.

POSTER PAPERS

Strong Pulsar Waves

E. ASSEO,[a] X. LLOBET,[b] AND R. PELLAT[a]

[a]*Centre de Physique Théorique*
École Polytechnique
91128 Palaiseau, France
[b]*Institute for Fusion Studies*
University of Texas
Austin, Texas 78712

Our interest in the propagation of large amplitude electromagnetic waves is related to the presence of these waves in the pulsar atmosphere beyond the light cylinder distance. The self-consistent plasma wave solution in spherical geometry is a priori more appropriate to describe the pulsar environment than the solutions in plane geometry. Sphericity effects are considered[1] as perturbative relatively to the homogeneous, superluminal, linearly polarized plane wave of an e^+e^- plasma described earlier.[2,3] A WKB-type method is used with two characteristic scale lengths: a short scale, of the order of the wavelength, on which the phase of the wave varies and a long scale, characteristic of the variations of the wave parameters, on which spherical geometry effects do occur. The particle number flux, the energy flux, and the momentum flux, phase averaged over one wave period, are constants of the motion. When the wave propagates in a spherical plasma, its amplitude varies and its phase velocity and wavelength are changed. Consequently, the mean properties of the plasma, which depend on the characteristics of the wave itself, are also modified. Periodic deformations, superimposed on the large amplitude sawtooth-shaped wave solution of the homogeneous case, result from sphericity effects. They are dependent on the gradients of the physical quantities associated with the sawtooth, vary slowly over the long-distance scale, vanish at large distances, and consequently do not alter the characteristic shape of the large amplitude wave. At the same time, the variations of the strength parameter characteristic of the amplitude of the wave constrain its propagation. In the pulsar case, our conclusion is that the effect of the spherical medium does not prevent the large amplitude wave to propagate over significant distances in the pulsar atmosphere while conserving all its characteristics. However, different mechanisms considered elsewhere may damp[4] or even destroy[5] these large amplitude waves.

REFERENCES

1. ASSEO, E., X. LLOBET & R. PELLAT. 1984. Astron. Astrophys. **139**: 417.
2. ASSEO, E., F. C. KENNEL & R. PELLAT. 1975. Astron. Astrophys. **44**: 31.
3. KENNEL, F. C. & R. PELLAT. 1976. J. Plasma Phys. **15**: 335.
4. ASSEO, E., F. C. KENNEL & R. PELLAT. 1978. Astron. Astrophys. **65**: 401.
5. ASSEO, E., X. LLOBET & G. SCHMIDT. 1980. Phys. Rev. **A22**: 1293.

Mutually Interacting Quantum Fields in an Expanding Universe: Decay of a Massive Particle

J. AUDRETSCH AND P. SPANGEHL

Fakultät für Physik
Universität Konstanz
Postfadh 5560
D-7750 Konstanz, Federal Republic of Germany

We give an analysis of quantum field theory of mutually interacting quantum fields in a 3-flat expanding Robertson-Walker universe. The classical curved space-time acts as an external gravitational field, the influence of which is taken into account without any approximation. As a case study, we give a mathematically rigorous calculation of the decay of a massive Φ-particle into two massless Ψ-particles in first order of the $(\lambda/a)\Phi\Psi^2$ mutual interaction (a is the cosmic scale factor). The particles are conformally coupled scalar particles. The interaction picture and the in-out scheme based on an S-matrix approach are used. No renormalization is needed. Because the external field is capable of producing pairs of massive Φ-particles from the vacuum, the specification of the outcome of the mutual interaction can only be based on the registration of massless particles in the out-region. A detailed analysis of the corresponding added-up probability is given. It sheds some light on the imperfection of the usual in-out approach. We use our result to obtain a limit for the local applicability of flat space-time results in a cosmological situation. For details, see *Class. Quantum Grav.* **2**: 733–753 (1985).

Thermodynamics and General Relativity Could Determine the Symmetry of the Universe

SELÇUK Ş. BAYIN
Department of Physics
Canisius College
Buffalo, New York 14208

Behavior of black hole parameters (area, surface gravity, and so on), like certain thermodynamic quantities (entropy, temperature, and so on), motivated Bekenstein[1] to conjecture the existence of black hole thermodynamics. Later, the discovery of black hole radiation by Hawking established the physical link between these parameters and their thermodynamic counterparts.[1] However, despite the success of black hole thermodynamics, the relation between general relativity and thermodynamics remains to be established for more general metrics.[2] In this paper,[3] in order to explore this relation we consider the possibility of the Bianchi symmetry of a Friedmann model changing as the universe evolves.

It is well known that symmetry changes in thermodynamics can be associated with phase transitions.[3] A typical example is the phase transition from α-iron to γ-iron, where the symmetry of the lattice over which the iron atoms are distributed changes from *bcc* to *fcc*.[3] It would be interesting if the gas of galaxies, distributed over curved space, displayed an analogous phase transition when the Bianchi symmetry of the universe changed. Since the Bianchi symmetry refers to the symmetry of the three-dimensional space, Friedmann models are particularly useful in demonstrating the possible existence of such phase transitions.[4] Phase transitions are also due to the collective behavior of gas molecules. Hence, they are governed by the entropy criteria rather than the energy. A well-known example of this fact is the existence of white tin in stable form at 298°K, a form that should be unstable at that temperature according to the energy criteria.[3] In order to demonstrate the existence of similar phase transitions among Friedmann models, we need to identify the corresponding quantities that play the roles of temperature, entropy, and so on.

The suggestive model we use is the one in which the radius of curvature of the three-dimensional space is treated like the inverse of the temperature and where $\rho(P, T)$ plays the role of the Gibbs potential energy density. We show that for the transitions between Bianchi I and V and Bianchi I and IX symmetric Friedmann models, there is only one Gibbs function and the transformation is of second order. For the transformations between Bianchi V and Bianchi IX symmetric models, we have two distinct Gibbs functions and in general this leads us to first order phase transitions. These conclusions are obtained independently of the details of the local equation of state. We also discuss two specific cases to demonstrate some of the properties of the model. One of these properties is that this model gives us a new way of determining the symmetry of the universe. By using a well-known equation of state ($P = \alpha\rho$), we show that with respect to the thermodynamics we have defined, it is advantageous for the universe to be open (Bianchi V symmetric).

The existence of such a direct relation between general relativity and thermodynamics can be understood by remembering the fact that thermodynamic quantities are (in principle) defined in the thermodynamic limit, where the volume and the total number of particles go to infinity, while their ratio remains finite. However, for most practical problems, thermodynamic variables can be considered well defined long before the size and mass of the system approaches infinity. On the other hand, in cosmology, where we are dealing with a gas of galaxies, when the thermodynamic limit is achieved the total mass of the system becomes so large that the underlying geometry of the space-time can no longer be assumed to be flat.[5]

NOTES AND REFERENCES

1. BEKENSTEIN, J. D. 1980. Phys. Today (January).
2. BONNOR, W. B. 1985. Phys. Lett. **112A:** 26.
3. BAYIN, S. Ş. 1986. Astrophys. J. Feb. 15.
4. Also remember that time in Friedmann models is classical (universal).
5. Consider that one liter of air at room temperature contains $\sim 10^{12}$ times more molecules than the number of galaxies in the universe.

The Differential Approach to Spinors and Their Symmetries

I. M. BENN[a] AND R. W. TUCKER[b]

[a]*Department of Natural Philosophy*
University of Glasgow
Glasgow, United Kingdom

[b]*Department of Physics*
University of Lancaster
Lancaster, United Kingdom

A formulation of spinor analysis in space-time is given in terms of smooth sections of a real Clifford bundle. Its relation to the two-component complex calculus for spinor components is elucidated. Treating spinors in terms of inhomogeneous differential forms carrying PIN(3,1) and SPIN(3,1) representations enables the discrete covariances of the Maxwell-Dirac system to be induced naturally from smooth isometries of the space-time metric. Attention is drawn to the distinction between the Dirac and Kähler equations in curved space when expressed in this geometric formulation.

Vacuum Energy in Cosmic Dynamics[a]

H. J. BLOME AND W. PRIESTER

Sonderforschungsbereich Radioastronomie
und
Institut für Astrophysik
und Extraterrestrische Forschung
Auf dem Hügel 71
D-53 Bonn, Federal Republic of Germany

The analysis of the Th/U ratio in meteorites favors values of the cosmic age between $(18–22) \times 10^9$ years. This evidence, together with a Hubble parameter of $H_o > 70$ (km/s · Mpc) = $(14 \times 10^9 \text{ years})^{-1}$, cannot be reconciled in a Friedmann model with $\Lambda = 0$. It requires a cosmological constant on the order of 10^{-56} cm^{-2}, which is equivalent to a vacuum density of $\rho_v = 10^{-29}$ g · cm^{-3}.

Friedmann-Lemaître models ($\Lambda > 0$) with a hot big bang have been calculated. They are based on a present value of the baryonic matter density of $\rho_o = 0.5 \times 10^{-30}$ g · cm^{-3}, as derived from the primordial ^{4}He and ^{2}H abundances.

For a Hubble parameter of $H_o = 75$ (km/s · Mpc), our analysis favors models that can be represented by a model with Euclidean metric (deceleration parameter, $q_o = -0.93$; age, $t_o = 19.7 \times 10^9$ years) and by a closed model with perpetual expansion ($q_o = -1.0$; $t_o = 22 \times 10^9$ years).

The possible behavior of the vacuum density is discussed with the help of Streeruwitz' formulae in the context of the closed model with an additional inflationary phase at very early times.[1,2]

REFERENCES

1. BLOME, H. J. & W. PRIESTER. 1984. Urknall und Evolution des Kosmos: I. Einstein-Friedmann Kosmos und das Neutrino Problem; II. Inflationär Modifizierter Urknall und Eschatologie des Kosmos. Naturwissenschaften (J. Springer. Berlin, Heidelberg, New York) **71**: 456–467; **71**: 515–527.
2. BLOME, H. J. & W. PRIESTER. 1984. Vacuum energy in a Friedmann-Lemaître cosmos. Naturwissenschaften **71**: 528–531.

[a]This paper appeared in full length in: BLOME, H. J. & W. PRIESTER. 1985. Vacuum energy in cosmic dynamics. *Astrophys. Space Sci.* **117**: 327–335.

The Hubble Parameter
An Upper Limit from QSO 0957 + 561 A,B

U. BORGEEST AND S. REFSDAL

Hamburger Sternwarte
2050 Hamburg 80, Federal Republic of Germany

It is shown that the cluster of galaxies that contributes to the light deflection by the double quasar can only reduce the time delay, Δt, between light variations of the images. Assuming a King-type deflecting galaxy, we can give an upper limit for the Hubble parameter of

$$H_o \leq 200 \text{ km s}^{-1} \text{ Mpc}^{-1}/\Delta t \text{ (yrs)}$$

and a best estimate of

$$H_o \approx 120 \text{ km s}^{-1} \text{ Mpc}^{-1}/\Delta t \text{ (yrs)}.$$

R. Florentin Nielsen and K. Augustesen have recently reported (*I.A.U. Circular* no. 3945; *Astron. Astrophys.* **138:** L19) that continued photographic photometry indicates that $\Delta t \approx 1.6$ years. If this is correct, we get

$$H_o < 125 \text{ km s}^{-1} \text{ Mpc}^{-1}$$

and a best estimate of

$$H_o \approx 75 \text{ km s}^{-1} \text{ Mpc}^{-1}.$$

Ultrahigh Energy Gamma Rays and Cosmic Rays from Accreting Degenerate Stars[a]

K. BRECHER[b] AND G. CHANMUGAM[c]

[b]*Department of Astronomy*
Boston University
Boston, Massachusetts 02215

[c]*Department of Physics*
Louisiana State University
Baton Rouge, Louisiana 70803

Ultrahigh-energy (UHE) gamma-ray emission with photon energies up to 2×10^{16} eV has been detected from the galactic X-ray source, Cygnus X-3. This implies acceleration of charged particles to even greater energies. The well-known binary X-ray sources, Her X-1 and Vela X-1, have also been reported to be sources of UHE gamma rays. We present a model for acceleration of particles to high energies in binary systems containing accreting magnetized neutron stars. It is based on a unipolar induction model previously proposed to provide the mechanism for acceleration of particles in active galactic nuclei through accretion onto massive black holes. If correct in the present application, the model implies that many of the currently known pulsating X-ray sources and magnetized cataclysmic variables should also emit UHE gamma rays. This process has a number of very appealing features that make it a strong candidate for the source of UHE galactic cosmic rays as well.

[a]This research was carried out at NASA/GSFC and was supported by a NRC-NASA Senior Research Associateship to K.B. and by NSF grant no. AST-8219598 to G.C.

Physical Determination and Meaning of the Law of Hubble

ALEXANDRU CEAPA

Poste Restante
Bucharest 1, Romania

A physical rederiving of the law of Hubble (LH) is based on the operational approach of the Lorentz transformations (LTs).[1] Just as the origin of the LH in the star motion relative to Earth, as predicted by the Doppler relativistic formula (Drf), requires a starlight traveling along their relative velocity, it follows not only that Drf is inadequately chosen [being involved by LTs equation (1) in reference 1], but also that the time-equation (2) in reference 1 is that which gives LH with the Hubble constant of $H = v/\lambda$. Confronted with the empirical value of $H^{-1} \sim (4.1 \pm 2)\,10^{17}$ s, it predicts a velocity of $v \sim 10^{-24}$ m s^{-1}, by which LH does not support the galaxies receding from Earth, but instead supports a recoil that every source of light must undergo at the emission of a wave train. The recoil and the continuity of the signal imply a broadening vT of a wavelength at each new wavelength emitted by the source. Consequently, the nth measured wavelength in signal underwent n broadenings since its emission and its $\Delta\lambda$ is given by $\Delta\lambda = n\,v\,\lambda/c$, where $n\,\lambda = x$.

This result removes the discrepancy between the ratios of the terrestrial wavelengths identified with those coming from a star and the ratios of the suitable $\Delta\lambda$'s[2] (equal by LH), which cannot be attributed to measuring errors because of the dependence of v on both the number of photons in a plane transverse to the wave vector and their energy. It can be checked with a laser that operates continuously by deflecting its light repeatedly (e.g., between the moon and a satellite).

REFERENCES

1. CEAPA, A. 1983. *Paper in* Tenth International Conference on GRG, vol. 2. B. Bertotti *et al.*, Eds.: 904. Padova, Italy.
2. JAUNCEY, D. L. *et al.* 1984. Astrophys. J. **286**: 498.

Quantized Magnetic Bremsstrahlung and Gamma-Ray Bursts

TAI L. CHOW

Department of Physics
California State University, Stanislaus
Turlock, California 95380

In the last several years, gamma-ray bursts have continued to be an intriguing phenomenon. An overall analysis of the observational data suggests that the bursts are produced in regions with a gravitational potential of $(0.1–0.2)c^2$ and a magnetic field of $\sim 2 \times 10^{12}$ gauss.[1] This leads us to explore quantized magnetic bremsstrahlung in intense magnetic fields from the surface layers of neutron stars as a possible burst mechanism. In intense magnetic fields, when the gyrating radius of the electron is comparable to its de Broglie wavelength, the classical description of electrons is not valid and a quantization process takes place, resulting in discrete energy levels (known as Landau levels) in the motion perpendicular to the magnetic field:

$$E(n, P_z) = \left(n + \frac{1}{4}\right)\frac{e\hbar}{mc}H + \frac{P_z^2}{2m}, \tag{1}$$

where H is the homogeneous magnetic field in the z-direction (homogeneous over the gyrating orbit, which is of the order of the de Broglie wavelength and much less than 10^{-8} cm, and constant in time over the gyrating period, which is much less than 10^{-8} sec). The criteria for application of the quantum description of electron motion is given by[2]

$$e\hbar H/mc > kT \quad \text{or} \quad 10^{-3} H > T. \tag{2}$$

In a nondegenerate medium, the electrons satisfy a Maxwellian-type distribution and the electron population at different energy levels is given by[3]

$$N(n, E_z) = \frac{\exp[\alpha] - \exp[-\alpha]}{(\pi k T E_z)^{1/2}} N \exp\left[\frac{-E(n, E_z)}{kT}\right] \tag{3}$$

and

$$\frac{N(n, E_z)}{N(n-1, E_z)} = \exp[-2\alpha], \quad \alpha = \frac{e\hbar}{mc}\frac{H}{kT}. \tag{4}$$

Thus, observing from equations (2) and (4), the quantization of electron states should take place in the nondegenerate outer layers of neutron stars where the temperature is around 10^{7}°K, the magnetic fields are of the order of 10^{12}G, and most of the electrons are in the very lowest levels. The transitions between the different levels are accompanied by either the emission of absorption of radiation. The characteristic energy of the radiation is dependent on the angle of emission or absorption of photons with respect to the axis of the magnetic field; in a first approximation, it is given by[4]

$$\nu_{n,n-1} = \frac{E(n, E_z) - E(n-1, E_z)}{h} = 2.8 \times 10^5 \, H \, H_z \qquad (5)$$

and the energy emitted in unit time in connection with the transition, $n \to n-1$, is

$$E_{n,n-1} = \hbar \omega_{n,n-1} A(n, n-1), \qquad (6)$$

where $A(n, n-1)$ is the transition probability,

$$A(n, n-1) = \frac{4}{3} \frac{e^2 \omega_{n,n-1}^3}{hc^3} |r_{n,n-1}|^2, \qquad (7)$$

and $r_{n,n-1}$ is the matrix element of the linear oscillator and is equal to $(nh/2m\omega_{n,n-1})^{1/2}$.

The total energy emitted in unit time is, after summation over all energy states, n,

$$E = \sum_n N(n, E_z) I_{n,n-1} = \frac{2}{3} \frac{e^4 k}{m^3 c^5} NH^2 T \text{ erg s/s}, \qquad (8)$$

from which we observe at $H \sim 10^{12} G$ and $T \sim 10^{7} {}^\circ K$ that the total energy released in unit time is approximately

$$E \sim (10^{39} \div 10^{40}) \text{ ergs/s}$$

if N is taken to be $10^{33} \div 10^{34}$.

The observed emission and absorption features in the gamma-ray burst spectra are therefore roughly consistent with the quantized magnetic bremsstrahlung processes in intense magnetic fields of neutron stars. The bursts are probably the results of some sudden process of heating of electrons at the surface of neutron stars by magnetic instabilities, which leads to a balance with radiative losses. Although the energy available in thermal electrons, at $T \sim 10^{7} {}^\circ K$, is much lower than needed for supporting powers of $\sim 10^{40}$ ergs/s, there is, however, a large amount of magnetic energy available. We are now looking for a possible mechanism that transforms magnetic energy into kinetic energies of the electrons.

REFERENCES

1. MAZETS, E. P. et al. 1981. Nature **290:** 378–382.
2. ZIMAN, J. M. 1960. Electrons and Phonons, p. 521. Oxford University Press. London/New York.
3. CHOW, T. L. et al. 1973. J. Phys. **A6:** 1285–1288.
4. CANUTO, V. et al. 1971. Space Sci. Rev. **12:** 3.

Observations and FRW Models

A. A. COLEY[a] AND B. O. J. TUPPER[b]

[a]*Department of Mathematics, Statistics,
and Computing Science
Dalhousie University
Halifax, Nova Scotia, Canada B3H 4H8*

[b]*Department of Mathematics and Statistics
University of New Brunswick
Fredericton, New Brunswick, Canada E3B 5A3*

It is known that different sources may (formally) give rise to the same gravitational field. In particular, it has been shown that the gravitational field (that is, the space-time metric) in an FRW model can be generated as a solution of Einstein's field equations in which the matter distribution is (1) that of a perfect fluid (as in the standard case), (2) that of an imperfect fluid with or without an electromagnetic field, or (3) that of two fluids in which one is a comoving (with the hypersurface orthogonal observer) radiating perfect fluid and the second is either a comoving perfect fluid (again, as in the standard case) or a non-comoving (tilting) imperfect fluid. It is a general feature of the new, nonstandard FRW models that they expand out of a pure radiation state to a final dust state.

The observational consequences of these models are investigated. Observations in the present epoch, such as the mass-luminosity relation, are found to be essentially the same as in the standard models. However, in the new models, the large tilting velocities at the early times after the radiation stage lead to quite distinct differences. Therefore, the behavior of the new models at earlier epochs and, more specifically, close to the singularity, is investigated in detail. In particular, it is believed that such models may be of use in the galaxy formation problem since dissipative processes (such as viscosity) can be treated more appropriately as perturbations in imperfect fluid cosmological models.

Analysis of Weyl-Affine Theories of Gravity in Terms of the Gravitational Frequency Shift Effect

A. A. COLEY[a] AND A. SARMIENTO G.[b]

[a]*Department of Mathematics, Statistics,*
and Computing Science
Dalhousie University
Halifax, Nova Scotia, Canada B3H 4H8

[b]*Instituto de Astronomiá, UNAM*
Apartado Postal 70-264
México 04510, D.F., México

A subclass of nonmetric theories of gravity, called Weyl-affine theories of gravity (WATGs), is analyzed by calculating their predictions for the gravitational frequency shift undergone by a wave signal in a planned solar probe. The analysis is carried out using a formalism in a spherically symmetric and static gravitational field. One of the advantages of the formalism is that any possible "nonmetricity" is contained in an arbitrary function, λ, of the Newtonian gravitational potential, U. The numerical results are calculated for a situation modeling a future experiment in the solar system. In the calculations, the metric components and the function, λ, are expanded up to third order in U. Within the limits of the gravitational redshift experiments performed to date, it is found that WATGs must coincide with their metric counterparts (i.e., λ is unity). It is hoped that the planned solar probe will test the nature of the theories under investigation to a higher degree of accuracy.

Image Separation Statistics for Multiply Imaged Quasars

C. C. DYER

Scarborough College
David Dunlap Observatory
University of Toronto
Canadian Institute for Theoretical Astrophysics
Toronto, Canada

The gravitational lens effect for multiply imaged quasars is considered for various mass distributions in the lensing galaxies. It is shown that the presently observed statistics of maximum separation are not in strong disagreement with predictions based on these realistic lens models after some account is taken of the observational limitations of present searches. It is also shown how these image separations may be used to place certain constraints on various properties of the lensing objects and their distributions in space. This is possible since the shape of the image-separation probability curves is quite dependent on the number of lensing galaxies as a function of epoch and on the intrinsic properties of these galaxies (such as their scale size) as a function of epoch.

Symbolic Tensor Manipulation on Personal Microcomputers

C. C. DYER AND J. F. HARPER

Scarborough College
David Dunlap Observatory
University of Toronto
Canadian Institute of Theoretical Astrophysics
Toronto, Canada

A software package for the formal manipulation of tensors and related objects for general relativity is described. This system is written in muSIMP, a derivative of LISP, from Stoutemeyer and Rich, and runs on a large number of Intel 8088/8086 based personal computers, like the IBM-PC and the DEC RAINBOW, preferably with 256 kbytes of memory. A broad range of functions exists, both for explicit component oriented manipulation and for purely symbolic formal manipulation of abstract tensors. The somewhat small amount of memory available on this class of machines is partially offset by the memory management policies of LISP-based systems and by the short word size of these machines with few wasted bits in pointer structures. For a broad range of problems, experience has shown that this software typically performs at speeds within a factor of three of an unloaded VAX-750 running one of the larger algebra packages for the same task. Given the actual speed obtainable on such a supermini in a normally loaded system, the system described here has many desirable features.

Gamma Rays from a Hot Plasma

Application of the Models to 3C 273 and Geminga

F. GIOVANNELLI,[a] S. KARAKUŁA,[b] AND W. TKACZYK[b]

[a]*Istituto di Astrofisica Spaziale, CNR*
00044 Frascati, Italy

[b]*Institute of Physics*
University of Łódź
PL 90-236 Łódź, Poland

In this paper, we analyze some processes producing high-energy X rays and gamma rays from a relativistic plasma, namely: (i) proton-proton interactions for thermal and nonthermal models; (ii) e-e and e-p bremsstrahlung in the thermal case and e-p bremsstrahlung in the nonthermal case.

The derived models were used in order to fit experimental data from 3C 273 and Geminga. Our thermal model fits quite well with the data of 3C 273, but it fails the fit of Geminga experimental points. On the contrary, data of Geminga are fitted very well by our nonthermal model. This result strongly supports the presence of a neutron star as the gamma-ray source in the error box of Geminga.

On Kaluza-Klein Cosmologies

M. GLEISER

Department of Mathematics
King's College
London, United Kingdom

We have analyzed possible cosmological solutions for the D-dimensional Einstein-Maxwell models of $N = 1$, $D = 11$ and $N = 2$, $D = 10$ supergravity theories with the $(4 + D)$-dimensional space-time having a product structure of $M^4 \times B^D$, where B^D is a D-dimensional compact Einstein space and M^4 is a Minkowsky space. For the Einstein-Maxwell models, we obtained a constant internal space, while for the $N = 1$, $D = 11$ supergravity with the various possible compactified solutions, a slowly expanding internal space is possible for a flat seven sphere. Our results for the $N = 2$, $D = 10$ supergravity with product space, $M^4 \times S^2 \times CP^2$, have led us to examine more carefully the effects of a scalar field in the search for more realistic solutions. A first attempt was made for a six-dimensional Einstein-Yang-Mills-Higgs model with a $SO(3)$ monopole in the internal space. We found that the inclusion of a Higgs field together with a matter contribution admitted an exact solution for the internal radius and consequently for the time variation of the coupling constants of the model. We are now involved in calculating one loop quantum effects at finite temperature for these models and hope to present preliminary results.[1,2]

REFERENCES

1. GLEISER, M., S. RAJPOOT & J. G. TAYLOR. 1984. Phys. Lett. **138B:** 377;. Phys. Rev. **D30:** 756; King's College Report (January 1984); Ann. Phys. **160:** 299.
2. GLEISER, M. & J. G. TAYLOR. 1984. King's College Report (August 1984); 1985. Phys. Rev. **D31:** 1904.

Gas in Cosmic Voids

P. M. GONDHALEKAR[a] AND N. BROSCH[b]

[a]Rutherford Appleton Laboratory
Chilton, DIDCOT
Oxfordshire OX11 OQX, England

[b]Wise Observatory
Tel Aviv University
Tel Aviv, Israel

We have identified several absorption lines seen in UV spectra of nearby quasars as being produced by matter in foreground voids. The lines (Lyman α, CIV, and SiIV) indicate metal enrichment of gas in regions lacking appreciable amounts of luminous material. The simplest scenarios for the formation of large-scale structures in the universe, the pancake theory and the hierarchical clustering, do not explain simultaneously the existence of voids with a typical scale and their internal metal enrichment by material processed in stars. The alternative may be offered by Ostriker's proposal of an explosion-dominated universe.

Considerations concerning the Definition and Distribution of Gravitational Energy

I. GOTTLIEB AND N. IONESCU-PALLAS

Einstein's gravitational field equations may be exactly written in a covariant form in a flat space-time as

$$D_q D_k (H^{kq} H^{rs} - H^{rq} H^{ks}) = \frac{8\pi G_N}{c^2} (\gamma S \rho U^r U^s + V^{rs}), \quad (1)$$

where $S = \sqrt{-g}/\sqrt{-a}$, $dS_R = g_{jk} dx^j dx^k$, $dS_M = a_{jk} dx^j dx^k$, $g_{jk} = a_{jk} + \Psi_{jk}$ (relationship between Riemannian and flat universes), D_k is the covariant derivative in flat space-time, G_N is the Newton's constant, $dS_R > 0$, $dS_M > 0$, and $\gamma^{-1} = \sqrt{1 + \Psi_{st} U^s U^t}$. The expression of the tensor, V^{jk}, is

$$V^{jk} = (c^2/16\pi G_N) \left\{ D_i H^{jk} D_q H^{iq} - D_i H^{ji} D_q H^{kq} + \frac{1}{2} H^{jk} F_{iq} D_s H^{ih} D_h H^{sq} \right.$$
$$- H^{ji} F_{qh} D_s H^{kh} D_i H^{qs} + H^{ki} F_{qh} D_s H^{jh} D_i H^{qs} + F_{is} H^{hs} D_h H^{ji} D_s H^{kq}$$
$$\left. + \frac{1}{8} (2H^{ji} H^{kq} - H^{jk} H^{iq})(2F_{hs} F_{rt} - F_{sr} F_{ht}) D_i H^{ht} D_q H^{rs} \right\}, \quad (2)$$

where $H^{ri} F_{it} = \delta^r_t$. The tensor in the left side of equation (1) is symmetrical and its covariant divergence is identically vanishing. It follows that outside the mass distribution $D_s V^{rs} = 0$. Motion equations may be written in one of the following forms (ensuring the fulfillment of the equivalence principle):

$$D_s(\gamma S \rho U^r U^s + V^{rs}) = 0, \qquad \delta \int \gamma^{-1} dS_M = 0, \quad (3)$$

while the total invariant energy density in space acquires the expression,

$$\rho_E = (c^4/16\pi G_N)(H^{kq} D_q D_k H - D_k H^q_s \cdot D_q H^{ks}), \quad (4)$$

where $H = a_{rt} H^{rt}$ and the covariant field calibration condition of $D_r H^{rt} = 0$ was used. For a single pointlike gravitational source at rest, of mass, M_o, we obtain from equation (4), $E = M_o c^2$. For the case of Cartesian coordinates, agreement is obtained with Landau-Fock formulation of gravitational energy.

Gravitational Collapse and Quantum Gravity

P. HAJICEK

Institute for Theoretical Physics
University of Bern
Sidlerstrasse 5
CH-3012 Bern, Switzerland

We study the simplest model of gravity in which a collapse to a black hole can take place: the so-called Berger-Chitre-Moncrief-Nutku model. The causal structure in this model, therefore, depends strongly on the dynamical processes themselves and cannot be considered as given a priori. There are, however, properties of the causal structure that follow just from the constraints and gauge propagating equations; these are, in a sense, independent from the dynamics. We find, in particular, that the invisibility of apparent horizons from infinity and the non-timelike signature of their world tubes are so characterized. We also show that there are gauges that lead to foliations covering a part of the space-time that is globally hyperbolic and asymptotically Minkowskian, along with not containing any apparent horizon. Altogether, this means that the canonical quantization is applicable to black holes. Moreover, we make plausible that some of these properties survive the quantization because they can be formulated as properties of the configuration space. This seems to contradict some aspects of the currently believed scenario for Hawking effect.

A Compact Object in the Bimetric Theory

AMOS HARPAZ[a] AND NATHAN ROSEN

Department of Physics
Technion–Israel Institute of Technology
Haifa, Israel

The bimetric theory of relativity (BTR) assumes the existence of a background metric, $\gamma_{\mu\nu}$ (corresponding to a space-time of constant curvature), in addition to the ordinary metric, $g_{\mu\nu}$, determined by the matter distribution. With a suitable choice of $\gamma_{\mu\nu}$, the field equations can be solved for $g_{\mu\nu}$ in the case of a static, spherically symmetric body. For a compact object (radius equal to Schwarzschild radius), calculations show that far from the object the results agree with those of the general theory of relativity (GTR). However, close to the radius of the object, the two theories give different results. Inside the object, the results are totally different: In GTR, no equilibrium configuration exists in this region and the matter collapses to a singularity at the center. In BTR, a configuration in hydrostatic equilibrium exists for a collapsed star inside its Schwarzschild radius. Such configurations were constructed for two equations of state and for different masses.

---

[a] Also the University of Haifa, School of Education of the Kibbutz Movement, Oranim.

Microwave Measurement of the Galactic Helium-3 Abundance

G. M. HEILIGMAN[a] AND D. G. YORK[b]

[a]NAS/NRC Resident Research Associate
Jet Propulsion Laboratory
California Institute of Technology
Pasadena, California 91109

[b]University of Chicago
Chicago, Illinois 60637

We report the progress of a program to measure the abundance of helium-3 in the interstellar medium. We are using antennae of NASA's Deep Space Network[c] to detect the $^3\mathrm{He}^+$ $F = 0-1$ hyperfine emission line at $\lambda = 3.46$ cm from Galactic H II regions. Our observations of W3 give a $^3\mathrm{He}/\mathrm{H}$ number ratio of 1×10^{-4}, which is significantly smaller than the ratio measured by Rood, Bania, and Wilson (*Astrophys. J.* **280**: 629). We confirm that W3, despite its large galactocentric distance, has been enriched in $^3\mathrm{He}$.

We failed to detect $^3\mathrm{He}^+$ in the planetary nebula NGC 6853 to a limit of $^3\mathrm{He}/\mathrm{H} < 2 \times 10^{-4}$. This number ratio suggests that PNs are not a major source of Galactic $^3\mathrm{He}$. Present $^3\mathrm{He}$ measurements place only weak constraints on models of early-universe nucleosynthesis.

---

[c]The NASA/JPL Deep Space Network is operated by the Jet Propulsion Laboratory, California Institute of Technology, under contract no. NAS7-100 sponsored by the National Aeronautics and Space Administration.

A First Order Phase Transition from Inflationary to Big Bang Universe

GERALD HORWITZ

Racah Institute of Physics
The Hebrew University of Jerusalem
Jerusalem, Israel

The microcanonical entropy is calculated for a system of massive, conformally coupled, scalar bosons using a conformal gravitational theory. The resulting entropy is seen to indicate a first order phase transition from an inflationary expansion stage (where the amplitude of the scalar boson follows that of the scale function of the universe and the mass of the scalar boson is the source of the cosmological constant) to a big bang stage (where neither of these conditions hold). Such a first order phase transition involves an entropy increase of some thirty orders of magnitude. In our theory, the invariant temperature (proper temperature times scale function) is not zero, nor is it the Hawking temperature, but it is tens of magnitudes smaller than the corresponding temperature of the big bang stage. A specific model for these bosons that provides the phase transition and serves as the source of the cosmological constant is also examined briefly, where the bosons are identified as spontaneously generated primordial black holes as in the cosmological model of Brout, Englert, and Casher. In that case, the decay of the black holes provides a decaying cosmological constant and an explicit mechanism for heating up the universe.

The Virgo Cluster as a High Energy Cosmic Ray Source

S. KARAKUŁA AND W. TKACZYK

Institute of Physics
University of Łódź
PL 90-236 Łódź, Poland

The extragalactic charged particles are reflecting from the Galaxy by its magnetic field. Assuming the magnetic field in the Galaxy to be quasi-longitudinal, we have evaluated the mean transparency of the Galaxy for extragalactic protons defined as a fraction of particles at a given energy from a given direction passing by the galactic plane. The dependence of $\langle \sin b^{II} \rangle$ on the energy observed by Haverah Park was analyzed and the possible contribution to the high energy cosmic ray flux from the Virgo Cluster was evaluated.

SS 433 Revisited

W. KUNDT

Institut für Astrophysik der Universität
Auf dem Hügel 71
D-5300 Bonn, Federal Republic of Germany

The galactic X-ray and radio binary star SS 433 ranges among the most fascinating astrophysical objects of this century mostly because of its sinusoidally moving spectral lines (of H and He) and its relativistic twin-jets [and also because of its obvious association with the (old) supernova remnant, W 50]. Its kinematic properties are believed to be well understood, such as its distance, $d = (5.5 \pm 1)$ kpc, orbital inclination, $i = 79°$, beam precession angle, $\theta = 20°$, ejection velocity, $v = 0.26\,c$, and the geometry of its accretion disk (e.g., Margon, 1984),[1] but a number of details remain unexplained or even inconsistent. Instead, I have convinced myself that a more satisfactory description of the system is obtained for $d = (3 \pm 0.5)$ kpc, $i = 67°$, $\theta = 53°$, $v = 0.9999\,c$, and with no signature of an accretion disk in the light curve so far.[2] In view of the fact that SS 433 is considered a miniature QSO, which should help us understand the more distant extragalactic sources, this ambiguity in the interpretation ought to be sorted out soon.

REFERENCES

1. MARGON, B. 1984. Annu. Rev. Astron. Astrophys. **22:** 507.
2. KUNDT, W. 1981. Vistas Astron. **25:** 153.

Large-Scale Anisotropy of the Cosmic Background Radiation at 3 mm

P. LUBIN[a] AND T. VILLELA[b]

[a]*Lawrence Berkeley Laboratory*
University of California, Berkeley
Berkeley, California 94270

[b]*INPE/CNPq São Jośe dos Campos*
São Jośe, Brazil

The large-scale structure of the cosmic background has been measured at 3 mm wavelength using a balloon-borne liquid-helium cooled radiometer. Flights from both hemispheres have achieved a sky coverage of over 85% with very little galactic contamination. Other than a first order dipole and a possible faint galactic contribution, no other structure is apparent. All quadrupole components are 100 μK or less, with errors of 50–80 μK. A 90% confidence level upper limit of 6×10^{-5} RMS is placed on a quadrupole component. A possible detection of galactic dust has been obtained with a cosec (b) (b-galactic latitude) model giving 40 ± 11 μK at the pole. This galactic component contributes less than 100 μK to the quadrupole component. A map of the sky has been prepared.

The Properties of a Generalized Inflation

F. LUCCHIN[a] AND S. MATARRESE[b]

[a]*Dipartimento di Fisica*
Via Marzola 8
I 35100 Padova, Italy

[b]*I.S.A.S.*
Strada Costiera 11
I 34100 Trieste, Italy

Cosmological problems, like horizon, flatness, and origin of primordial density perturbations, are investigated in the framework of a generalized (nonexponential) inflationary model. In particular, we studied in some detail a toy model in which the scale factor, R, grows like $R \propto t^p$, with $p > 1$ (Power-Law Inflation; see, e.g., Abbott and Wise. 1984. *Nucl. Phys.* **B244:** 541), while considering the dynamics of a scalar field whose potential can give such an expansion. We found the constraints on the model related to the solution of the horizon, flatness, reheating, and perturbation spectrum problems. We also considered the limits on the model coming from the observational data on the microwave background anisotropy.

A Hybrid Model for the Active Source at the Center of Our Galaxy

A Very Massive Star Coupled with a Black Hole

LEONID M. OZERNOY

Department of Theoretical Physics
Lebedev Physical Institute
Moscow 117924, Union of Soviet Socialist Republics

Recent observations by Geballe *et al.* (1984) of broad HeI and HI lines from IRS 16 located at the galactic center indicate strongly an outflow of matter from it. One can show that usual stars fail to provide a wind with the parameters observed. The source of the wind cannot be a supermassive ($M \sim 10^6 \, m_\odot$) black hole (BH) either, since a necessary condition for producing the radiative matter outflow from a BH, $L/L_{\text{Edd}} \gtrsim 1$, yields $M_{\text{BH}} \lesssim 300 \, m_\odot$. Most likely, both the HI + HeI wind and a significant part of the total radiation of the IRS 16 center can be attributed to a very massive star (VMS) of the mass $\sim 300 \, m_\odot$, i.e., to a source like η Carinae. At the same time, both the positrons for the 511-keV annihilation line and the gamma continuum radiation from the galactic center can be most naturally explained by a moderate-mass black hole (MMBH) of the mass $\sim (10-300) m_\odot$. It is argued here that if IRS 16 and the compact radio source Sgr A* are spatially separated from each other, a hybrid symbiotic couple consisting of an outflowing VMS and a MMBH feeding by the VMS wind seems to be able to explain the most principal features of the galactic center's emission. The VMS may result from a recent burst of star formation at the galactic center region, while the MMBH may be a remnant of the more remote past.

An X-Ray Study of M51 (NGC 5194) and Its Companion (NGC 5195)

G. G. C. PALUMBO,[a] G. FABBIANO,[b] C. FRANSSON,[c] AND G. TRINCHIERI[b]

[a]*Istituto TE.S.R.E./CNR*
Bologna, Italy

[b]*Harvard-Smithsonian Center for Astrophysics*
Cambridge, Massachusetts 02138

[c]*Stockholm Observatory*
Stockholm, Sweden

X-ray observations of M51 (NGC 5194) and its companion (NGC 5195) with the HRI aboard the Einstein Observatory led to the detection of both galaxies with (0.2–4.0 keV) X-ray luminosities of $L_x = 3.0 \times 10^{40}$ erg s^{-1} and 2.2×10^{39} erg s^{-1}, respectively. The integrated X-ray luminosities and the X-ray to optical and X-ray to radio flux ratios of these two galaxies are consistent with the values expected for normal spiral and irregular galaxies. Three bright pointlike sources are detected in M51 above the extended unresolved emission of the galactic plane. The radial profile of the X-ray surface brightness of this galaxy follows the optical (blue-light) disk profile rather than the spiral arm light distribution, indicating that most of the X-ray emission is associated with the disk population. The same distribution is followed by the nonthermal radio continuum emission, which is in agreement with previous results that suggest a link between X-ray sources and cosmic ray production and radio emission in spiral galaxies. The nucleus of M51 is a bright extended X-ray source with (0.2–4.0 keV) $L_x = 8.4 \times 10^{39}$ erg s^{-1}. This emission could in part be due to hot gas outflowing from the nucleus, as seen in the X-ray image of NGC 253, or to a starburst region, older than that at the nucleus of M83. We can set a firm limit of $L_x < 1.5 \times 10^{39}$ erg s^{-1} on the observed luminosity of a pointlike nonthermal nuclear source. This upper limit implies the presence of a large amount of extinction in the nucleus. An additional softer continuum component is needed to explain the observed H_2 intensity with photoionization from a nuclear source.

Relativistic Motion in the Quasar 3C147?

E. PREUSS,[a] W. ALEF,[a] N. WHYBORN,[b]
P. N. WILKINSON,[b] AND K. I. KELLERMANN[c]

[a]*Max-Planck-Institut für Radioastronomie*
Bonn, Federal Republic of Germany

[b]*Nuffield Radio Astronomy Laboratory*
Jodrell Bank, United Kingdom

[c]*National Radio Astronomy Observatory*
Green Bank, West Virginia 24944

The results from the recent VLBI observations of 3C147 with milliarcsecond resolution are in contrast with what one expects from a simple relativistic beam model for compact radio sources. Contrary to the predictions based on low frequency variability and X-ray observations, we find no evidence for structural changes over the past five years.

Analogies between Kruskal Space and de Sitter Space

WOLFGANG RINDLER

Physics Department
The University of Texas at Dallas
Richardson, Texas 75080

Kruskal space is the analytic completion of Schwarzschild space and it consists of two outside and two inside Schwarzschild regions. Under suppression of the two angular coordinates, this space is usually diagrammed in terms of the Kruskal coordinates, u, v, much like Minkowski space is in terms of x, y. In particular, radial light paths correspond to $\pm 45°$ lines, the hyperbolas of $u^2 - v^2 = a^2$ represent uniformly accelerated particles (these being at rest in outer Schwarzschild space), and Lorentz transformations in u, v map the space into itself. Hermann Weyl first gave the analytic completion of de Sitter space as a hyper-hyperboloid $u_1^2 + u_2^2 + u_3^2 + u_4^2 - v^2 = a^2$ in five-dimensional Minkowski space, which also contains two outside and two inside de Sitter regions. In a Weyl diagram, u_3 and u_4 are suppressed. There are many analogies: Lorentz transformations in u_i, v map Weyl space into itself, the $\pm 45°$ generators are light paths, timelike plane hyperbolic sections are uniformly accelerated particles, and the horizon structure relative to each free worldline is analogous to the absolute horizon structure in Kruskal space.

Connection between Einstein Equations, Nonlinear Sigma Models, and Self-Dual Yang-Mills Theory

NORMA SANCHEZ AND BERNARD WHITING[a]

ER 176 CNRS
Département d'Astrophysique Fondamentale
Observatoire de Meudon
92195 Meudon Principal Cedex, France

We analyze the connection between nonlinear sigma models, self-dual Yang-Mills theory, and general relativity (self-dual and non-self-dual, with and without killing vectors), both at the level of the equations and at the level of the different type of solutions (solitons and calorons) of these theories. We give a manifestly gauge invariant formulation of the self-dual gravitational field analogous to that given by Yang for the self-dual Yang-Mills field. This formulation connects in a direct and explicit way the self-dual Yang-Mills and the general relativity equations. We give the "R gauge" parametrization of the self-dual gravitational field (which corresponds to modified Yang's-type and Ernst equations) and analyze the correspondence between their different types of solutions. No assumption about the existence of symmetries in the space-time is needed. For the general case (non-self-dual), we show that the Einstein equations contain an $O(2,1)$ nonlinear sigma model. This connection with the sigma model holds irrespective of the presence of symmetries in the space-time. We found a new class of solutions of Einstein equations depending on holomorphic and antiholomorphic functions and we relate some subclasses of these solutions to solutions of simpler nonlinear field equations that are well known in other branches of physics, like sigma models, SineGordon, and Liouville equations. They include gravitational plane wave solutions.

We generalize the approach given previously in the first paragraph to formulate in a consistent way (massive and interacting) Quantum Field Theories in a wide class of accelerated coordinates defined by holomorphic (and/or antiholomorphic) mappings in Euclidean space (imaginary time). We include uniform and nonuniform acceleration, a characterization of global and asymptotic thermal equilibrium situations in terms of the properties of the space-time itself, and a unicity theorem concerning the exponential mapping defining the maximal analytic of a manifold. We analyze the response of different accelerated quantum detector models, compare them to the case when the detectors are inertial in an ordinary Planckian gas at a given temperature, and discuss the anisotropy of the detected response for Rindler observers.

---

[a]B. Whiting only contributed to the second paragraph of this poster paper.

Accretion from a Medium Containing a Density Gradient

NOAM SOKER AND MARIO LIVIO

Department of Physics
Technion - Israel Institute of Technology
Haifa 32000, Israel

The problem of accretion (by a compact object) from a medium containing a density gradient has been studied by using a three-dimensional numerical calculation in the PIC method.

Our main purpose is to determine the amount of angular momentum that can be accreted by the compact object. The calculation does not include pressure effects. (The interaction between the particles conserves angular momentum explicitly.) The density profile is of the form, $\rho = \rho_o (1 - y/H)$.

The program was tested in the symmetric (homogeneous) case and gave velocity profiles consistent with the analytic results of Bondi and Hoyle.

Preliminary results in the case of a density gradient indicate that:
(1) The accretion rate is between 0.8 $\dot{M}_{BH}$ and 1.35 $\dot{M}_{BH}$, where $\dot{M}_{BH}$ is the Bondi-Hoyle (symmetric) result, for a wide range of interactions assumed.
(2) The accreted angular momentum is only about 0.1 L_s, where L_s is the angular momentum of matter entering the Bondi-Hoyle (symmetric) accretion cylinder.

The main reasons for the decrease in the accreted angular momentum are a displacement of the accretion cylinder and angular momentum transport by friction.

These results can have important consequences for both the question of the possibility to form a disk from wind accretion and for spin-up of accreting neutron stars.

Statistic of Voids of Galaxies

ANDRZEJ SOŁTAN

Copernicus Astronomical Center
00-716 Warsaw, Poland

Three-dimensional analysis of the distribution of galaxies in the Center for Astrophysics' galaxy redshift survey is carried out. Voids of various sizes are isolated in the objective, statistical way. Distribution of voids in the real universe is compared with simulations developed by Soneira and Peebles. It is shown that simulations that were devised to match low order correlation functions produce similar voids as observed in the CfA catalog. Space distribution of galaxies around empty regions is analyzed. No significant enhancement of galaxy density around voids is found. Morphological types of galaxies adjacent to voids are not distinct from the whole population. The statistical approach suggests that voids of galaxies are simply underdense regions caused by clustering rather than separate well-defined entities produced by some peculiar process in the early universe.

Unicity of General Relativity in the Field Theoretic Approach

GIANCARLO SPINELLI

Dipartimento di Matematica
Politecnico di Milano
20133 Milano, Italy

It is well known that general relativity can also be obtained by an iterative procedure starting in an ideal pseudo-Euclidean space-time in which the field is described by a tensor potential. To each step of the iterative procedure, one has to introduce the energy-momentum tensor of that order as the source of the field equations. When measured by real rods and clocks, the observable space-time turns out to be Riemannian and general relativity is obtained as the theory to which the method converges. Such a convergence theorem was shown assuming proper energy-momentum tensors to each order. The aim of the present work is to show that, to each order, one can take the most general energy-momentum tensor required by the iterative procedure and that the convergence to the general theory of relativity still holds.

Recent Developments in the Pulsating Universe

FRANK R. TANGHERLINI

Physics Department
College of the Holy Cross
Worcester, Massachusetts 01610

The pulsating cosmological model described previously (11th Texas Symposium) has been studied further with regard to the dimensionality of the space problem.[1] The Friedmann-Robertson-Walker line element and the corresponding Einstein field equations are generalized to n-spatial dimensions (as others have done)[2,3] and it is found that our previous method of avoiding the singularity at the big bang does not yield a singularity-free, stable, closed, pulsating universe for spatial dimensionality of $n > 3$. Since the modified Einstein equations are Newtonian in form, the argument is similar to that given previously by Ehrenfest and as generalized by the author to the Schwarzschild field.[4] The problem of deriving our modified Einstein field equations from first principles remains unsolved. The closed, pulsating universe, however, is very compatible with the present apparent stability of the proton.

NOTES AND REFERENCES

1. TANGHERLINI, F. R. 1986. Nuovo Cimento. In press.
2. GIDDINGS, S., J. ABBOT & K. KUCHAR. 1984. General Relativity and Gravitation **16:** 751.
3. KOIKA, T. & M. YOSHIMURA. 1985. Phys. Lett. **155B:** 137.
4. For a very recent review of the dimensionality of space problem, see: BARROW, J. D. & F. J. TIPLER. 1986. The Anthropic Cosmological Principle. Oxford University Press. London/New York; see also the earlier review by JAMMER, M. 1969. Concepts of Space. Harvard University Press. Cambridge, Massachusetts.

The Evidence for Large Gravitational Redshift in Seyfert Galaxy NGC 4151

W. TKACZYK AND S. KARAKUŁA

Institute of Physics
University of Łódź
PL 90-236 Łódź, Poland

The photon spectra of two Seyfert galaxies, NGC 4151 and MCG 8-11-11, are well established experimentally in the X-ray and gamma range. This caused a large number of theoretical speculations about their origin. The photon spectrum of NGC 4151 in the energy range from soft X ray to $\simeq 1$ MeV is a power-law type with varying in time of the spectral index (-1.6 in HEAO 1/A4 experiment;[1] -1 in MISO observations[2,3]). Recently, the observed cutoff in the soft X-ray range was interpreted as the absorption in the strong magnetic field, but for explaining the spectral break at MeV energies, there have been proposed many other models. (For details, see reference 4.) In this paper, we propose the annihilation of the high temperature positrons with cold electrons as the possible mechanism of photon productions in the Seyfert galaxy NGC 4151. The photon spectrum NGC 4151 with its observed features from the soft X-ray to the gamma-ray range can be very well described by the annihilation of positrons and electrons at temperatures of 3×10^{12} K and 10^8 K, respectively. Moreover, the photon spectra from annihilation of unthermalized plasma with the above parameters should be shifted to the lower energy by a redshift of $z \simeq 100$. In that case, the source of photons should be placed closely to the black hole horizon ($r \simeq 1.0001\ r_g$). The source of high temperature positrons can be decays of secondary π^+'s from pp interactions[5,6] or Penrose pair production processes.[7] The simultaneous observations in the X-ray and gamma ranges can support this scenario.

REFERENCES

1. BAITY, W. A. *et al.* 1984. Astrophys. J. **279:** 555.
2. PEROTTI, F. *et al.* 1981. Astrophys. J. **247:** L63.
3. PEROTTI, F. *et al.* 1983. Adv. Space Res. **3:** 117.
4. BASSANI, L. *et al.* 1984. The Manchester Conference on AGN.
5. GIOVANNELLI, F. *et al.* 1983. Astron. Astrophys. **125:** 121.
6. GIOVANNELLI, F. *et al.* 1983. Astron. Astrophys. **125:** 126.
7. LEITER, D. & M. KAFATOS. 1978. Astrophys. J. **226:** 32.

On Hydrodynamics of Astrophysical Jets

R. TUROLLA,[a] L. NOBILI,[b] AND M. CALVANI[c]

[a]*International School for Advanced Studies*
Strada Costiera, 11
35014 Trieste, Italy

[b]*Department of Physics*
University of Padova
Via Marzolo, 8
35128 Padova, Italy

[c]*Institute of Astronomy*
University of Padova
Vicolo dell'Osservatorio
35100 Padova, Italy

A model for jet acceleration in the funnels of a geometrically thick, radiation-supported disk is constructed. The dynamics of matter lost from the funnels' surface is studied in a general relativistic background and by means of a fully hydrodynamical treatment, including viscosity. Hydrodynamic equations governing the gas flow are derived either for an optically thick or for an optically thin medium. The general behavior of the solutions is discussed with particular emphasis to their existence and uniqueness.

Index of Contributors

Alef, W., 387
Asseo, E., 358
Audretsch, J., 359
Avni, Y., 71–87

Bahcall, N. A., 108–122, 331–337
Bayin, S. Ş., 360–361
Begelman, M. C., 51–70
Benn, I. M., 362
Blome, H. J., 363
Borgeest, U., 364
Brecher, K., 365
Brosch, N., 375

Calvani, M., 395
Ceapa, A., 366
Chanmugam, G., 365
Chow, T. L., 367–368
Christodoulou, D., 147–155
Coley, A. A., 369, 370

de Lapparent, V., 123–135
Dyer, C. C., 371, 372

Eichler, D., 205–214

Fabbiano, G., 386
Fransson, C., 386

Geller, M. J., 123–135
Giovannelli, F., 373
Gleiser, M., 374
Gondhalekar, P. M., 375
Gottlieb, I., 376

Hajicek, P., 377
Harpaz, A., 378
Harper, J. F., 372
Heiligman, G. M., 379
Horwitz, G., 380

Ionescu-Pallas, N., 376
Israel, M. H., 188–204

Karakuła, S., 373, 381, 394
Kellermann, K. I., 387
King, A. R., 320–330
Königl, A., 88–107
Kundt, W., 382
Kurtz, M. J., 123 135

Lingenfelter, R. E., 215–242
Livio, M., *ix,* 390
Llobet, X., 358
Lubin, P., 383
Lucchin, F., 384

Matarrese, S., 384
Mayle, R., 267–293

Nobili, L., 395
Nomoto, K., 294–319

Olive, K. A., 1–25
Ozernoy, L. M., 385

Palumbo, G. G. C., 386
Partridge, R. B., 36–50
Pellat, R., 358
Penrose, R., 136–146
Piran, T., 247–266
Preuss, E., 387
Priester, W., 363

Ramaty, R., 215–242
Refsdal, S., 364
Rindler, W., 388
Rosen, N., 378

Sanchez, N., 389
Sarmiento G., A., 370
Shaviv, G., *ix*
Soifer, B. T., 156–162
Soker, N., 390
Sołtan, A., 391
Spangehl, P., 359
Spinelli, G., 392
Stark, R. F., 247–266

Tanaka, Y., 163–187
Tananbaum, H., 338–357
Tangherlini, F. R., 393
Tkaczyk, W., 373, 381, 394
Trinchieri, G., 386
Tucker, R. W., 362
Tupper, B. O. J., 369
Turolla, R., 395

Vilenkin, A., 26–35
Villela, T., 383

Weaver, T., 267–293
White, S. D. M., 243–246
Whiting, B., 389
Whyborn, N., 387

Wilkinson, P. N., 387
Wilson, J. R., 267–293
Woosley, S. E., 267–293

York, D. G., 379